BIOLOGICAL MEMBRANES

BIOLOGICAL MEMBRANES

Edited by

DENNIS CHAPMAN

Department of Biochemistry and Chemistry,
Royal Free Hospital School of Medicine,
University of London,
England

VOLUME 5

1984

ACADEMIC PRESS
(*Harcourt Brace Jovanovich, Publishers*)

London · Orlando · San Diego · San Francisco · New York
Toronto · Montreal · Sydney · Tokyo

ACADEMIC PRESS INC. (LONDON) LTD.
24/28 Oval Road,
London NW1 7DX

United States Edition published by
ACADEMIC PRESS INC.
Orlando, Florida 32887

British Library Cataloguing in Publication Data

Biological membranes.
   Vol. 5. Trigger processes in cell biology
   1. Cell membranes
   I. Chapman, Dennis, 1927–
   574.87'5   QH601

   ISBN 0–12–168546–2

   LCCCN 67–19850

Typeset by Bath Typesetting Ltd., Bath
and printed in Great Britain by
Thomson Litho Ltd., East Kilbride, Scotland

Contributors

D. CHAPMAN, *Department of Biochemistry and Chemistry, Royal Free Hospital School of Medicine, University of London, Rowland Hill Street, London NW3 2PF, England.*

G. BENGA, *Department of Cell Biology, Medical and Pharmaceutical Institute, 6 Pasteur Street, Cluj-Napoca, Roumania.*

J. H. CROWE and L. M. CROWE, *Department of Zoology, University of California, Davis, California 95616, U.S.A.*

E. SACKMANN, *Physik Department E22, Technical University of Munich, 8046 Garching-Munich, West Germany.*

C. M. KEMP, *Department of Visual Science, Institute of Ophthalmology, University of London, Judd Street, London WC1H 9QS, England.*

YU. A. OVCHINNIKOV, *Shemyakin Institute of Bioorganic Chemistry, U.S.S.R. Academy of Sciences, 117988 Moscow, U.S.S.R.*

N. G. ABDULAEV, *Shemyakin Institute of Bioorganic Chemistry, U.S.S.R. Academy of Sciences, 117988 Moscow, U.S.S.R.*

A. V. KISELEV, *Shemyakin Institute of Bioorganic Chemistry, U.S.S.R. Academy of Sciences, 117988 Moscow, U.S.S.R.*

R. M. STROUD, *Department of Biochemistry and Biophysics, University of California School of Medicine, San Francisco, California 94143, U.S.A.*

W. KREUTZ, *Institut für Biophysik und Strahlenbiologie, Universität Freiburg im Breisgau, Albertstrasse, 23, D-7800 Freiburg, Federal Republic of Germany.*

F. SIEBERT, *Institut für Biophysik und Strahlenbiologie, Universität Freiburg im Breisgau, Albertstrasse, 23, D-7800 Freiburg, Federal Republic of Germany*

K. P. HOFMANN, *Institut für Biophysik und Strahlenbiologie, Universität Freiburg im Breisgau, Albertstrasse, 23, D-7800 Freiburg, Federal Republic of Germany.*

P. B. GARLAND, *Department of Biochemistry, University of Dundee, Dundee DD1 4HN, Scotland.*

B. D. GOMPERTS, *Department of Experimental Pathology, University College School of Medicine, London WC1E 6JJ, England.*

D. T. EDMONDS, *Department of Physics, Clarendon Laboratory, University of Oxford, South Parks Road, Oxford OX1 3PU, England.*

W. M. ARNOLD, *Lehrstuhl für Biotechnologie, Universität Würzburg, Röntgenring 11, D8700 Würzburg, West Germany.*

U. ZIMMERMANN, *Lehrstuhl für Biotechnologie, Universität Würzburg, Röntgenring 11, D8700 Würzburg, West Germany.*

Introduction

Biomembrane research continues to be very active. Naturally the emphasis of the research shifts as the unresolved problems in the field become clear or changes in scientific instrumentation open up new possibilities for exploration. A particular feature of current research is the considerable interest in the molecular structure and function of biomembrane proteins. Trigger processes, the processes of ion transport, are also highly topical, and new techniques such as kinetic infrared spectroscopy and electric fusion have been developed recently. This volume of *Biological Membranes* tries to match these changes of emphasis of current research and at the same time provides a balanced spread of overall activity in this field.

The first chapter, D. Chapman and G. Benga, discusses the concept of biomembrane fluidity and its modulation and the usefulness of the concept. New techniques for studying lipid phase transitions and fluidity such as difference and Fourier transform spectroscopy are described, and a comparison of the modulation of lipid chain dynamics by cholesterol and intrinsic proteins indicated. This technique is clearly an important addition to the existing probe methods in that it is a method which is non-perturbing and operates on a different time-scale than techniques such as NMR or ESR spectroscopy. Methods for modulating biomembrane fluidity by hydrogenation and polymerization are also among the topics described in this chapter.

Everyone concerned with biological structures realizes the importance of water for determining the structure and function of living systems. Yet there are examples of certain micro-organisms and nematodes which can withstand considerable dehydration and can then be cooled to very low temperatures or raised to high temperatures (greater than 100°C) and yet can be subsequently rehydrated and recover all their original activity. This is a process termed *anhydrobiosis*. How is this possible? What protects the biomembranes from damage? Can we learn from these living systems techniques which could be useful in cryobiology? The topic of how biomembranes may be stabilized whilst undergoing extensive dehydration is discussed by John and Lois Crowe in Chapter 2.

The study of the structure and function of biomembranes is a fascinating one because it allows scientists of many different disciplines to contribute. The approach of the physicist to its study with ideas based on concepts from solid state physics or from metallurgy is indicated in Chapter 3 by E. Sackmann. He describes the basics of various trigger processes which occur with

vii

biomembranes. A particularly important trigger process involving bio-membrane structures is that involving visual transduction. How does light activate rhodopsin, and what are the various subsequent processes which occur both at the biophysical and biochemical level? These topics are discussed by C. Kemp in Chapter 4.

A related topic which has been receiving much attention recently involves questions concerning the structure and molecular mechanisms involved with bacteriorhodopsin, the major protein present in the purple membrane. Can the amino acid sequence be arranged satisfactorily to match the structure determined by electron diffraction? What are the consequences of this for the light-activated proton pumping activity? These are some of the questions discussed by Yu. A. Ovchinnikov in Chapter 5.

Related to trigger processes involving light, as with bacteriorhodopsin and rhodopsin, are other important trigger systems involving receptor molecules. Among these receptor molecules the acetylcholine receptor is particularly important. In Chapter 6, R. M. Stroud describes recent attempts, particularly using electron microscope methods, to obtain information at the molecular level of its structure and then to relate this to its important trigger function.

New techniques are being developed to study the kinetic changes involved in biomembrane structure and function. Among these new techniques is kinetic infrared spectroscopy. The development of this technique and its application to trigger processes such as those involved with rhodopsin and bacteriorhodopsin are described by W. Kreutz in Chapter 7.

The possibility of developing optical probes for detecting conformational changes in membrane proteins is discussed briefly by P. B. Garland in Chapter 8.

Ion transport is a key event with biomembranes whether it be ions such as sodium, potassium or calcium. Calcium appears to play a particularly important role in many trigger mechanisms. The role of calcium and cellular activation is discussed in Chapter 9 by B. D. Gomperts.

Despite the major role of ion transport across biomembrane structures, we are still uncertain about the molecular structure and organization of mem-brane ion channels. In Chapter 10, D. T. Edmonds puts forward his own creative and imaginative models involving ordered water to indicate how ion transport may take place and points to future experiments which could be developed for testing such models.

Biomembrane fusion and its control has become increasingly important in recent years, particularly as a result of interest in monoclonal antibody production. The new technique of electric field induced fusion of cell bio-membranes developed by U. Zimmermann is described by him and W. M. Arnold in Chapter 11.

It should be clear from the range of topics discussed in this volume that considerable activity exists in the field of biomembrane research. It is particularly fascinating to see ways in which basic science may immediately provide a springboard into useful technology, e.g. knowledge of biomembrane structures may bring with it an advance in the production of new biocompatible surfaces for prosthetic devices, a knowledge of nematode dehydration-protective mechanisms may result in new cryobiology agents useful for heart and kidney transplants, and fundamental studies of the rotation and fusion of cells in the presence of electric fields may bring new technologies for separating cells for antibody production and new diagnostic methods.

I hope that this volume will be found provocative, stimulating and useful for research in this important field.

D. CHAPMAN *May 1984*

Contents

Chapter 1

Biomembrane Fluidity—Studies of Model and Natural Biomembranes

DENNIS CHAPMAN

Department of Biochemistry and Chemistry
Royal Free Hospital School of Medicine, University of London, England

AND

GHEORGHE BENGA

Department of Cell Biology, Faculty of Medicine
Medical and Pharmaceutical Institute, Cluj-Napoca, Romania

I. The Concept of Biomembrane Fluidity

Chapman and co-workers in 1966 postulated a *fluidity concept* for biomembranes which proposed "the particular distribution of fatty acyl residues which occurs with a particular biomembrane is present so as to provide the appropriate membrane fluidity for a particular environmental temperature to match the required diffusion rate or rate of metabolic processes required for the tissues".

BIOLOGICAL MEMBRANES Vol. 5
ISBN 0 12 168546 2

It was also suggested "there appears to be biosynthetic feed-back mechanisms by which a cell will attempt to retain a constant membrane fluidity", (Chapman *et al.*, 1966). Examples to support this idea were given instancing poikilothermic organisms which alter their membrane fluidity to match different environmental temperatures.

This fluidity concept was based on a range of studies. An important contribution to the concept of biomembrane fluidity was a study of the detailed molecular nature of the major thermotropic phase transition of long-chain amphiphilic molecules. This was delineated by infrared spectroscopy studies, (Chapman, 1958). Studies in the years 1964–1966 of pure phospholipids with techniques such as infrared spectroscopy and nuclear magnetic resonance (NMR) spectroscopy pointed to and emphasized the extent of molecular mobility which can be associated with phospholipid molecules. This is particularly the case when these molecules exceed a certain critical transition temperature (see Byrne and Chapman, 1964; Chapman *et al.*, 1966, 1967). The reduction of line-width observed by proton NMR spectroscopy which occurs at the critical transition temperature with a 1,2-dimyristoyl phosphatidyl ethanolamine pure anhydrous phospholipid is shown in Fig. 1.

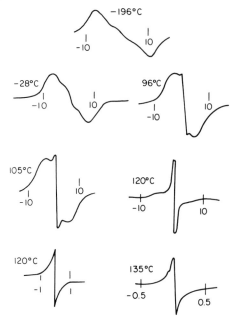

Fig. 1. PMR derivative absorption spectra of 1,2-dimyristoyl-DL-phosphatidylethanolamine at various temperatures. The abscissas are marked in gauss. From Chapman and Salsbury (1966).

These and later studies showed the molecular details underlying the membrane fluidity concept. Above a phase transition temperature, T_c, characteristic for a particular lipid of given chain length:

(i) The lipid chains show flexing and twisting of the methylene CH_2 groups and a marked increase of rotational isomers;

(ii) The oscillations and rotational disorder of the methylene groups are most marked at the methyl end of the lipid chains (Chapman and Salisbury, 1966);

(iii) In addition to the chain motion, other parts of the molecule, e.g. the polar groups of the lecithin molecules, exhibit a marked increase in mobility (Veksli et al., 1969);

(iv) Lipid self-diffusion occurs when sufficient water is present to weaken any ionic linkages between the polar groups (Penkett et al., 1968);

(v) The transition temperature from ordered to fluid "melted" chains is related to the chain length and degree of unsaturation of the lipid (Chapman et al., 1967);

(vi) The fluidity of the biomembrane can be modulated by molecules which penetrate the lipid bilayer such as cholesterol (Chapman and Penkett, 1966; Ladbrooke et al., 1968).

An example of the main lipid phase transition and a pretransition observed using differential scanning calorimetry for dimyristoyl phosphatidyl choline is illustrated in Fig. 2.

Since these early studies, many workers have confirmed, extended and quantified the mobility, order and diffusion of lipid molecules in model and natural biomembranes, and as a result of many more studies the concept of fluidity has become more sophisticated. Various theoretical studies have been developed, e.g. several authors have been concerned with the physical principles which determine lipid chain conformation and packing in bilayers above the gel/liquid–crystal phase transition (Pink and Chapman, 1979).

Once it was appreciated that a particular biomembrane would have its own characteristic fluidity, various methods for providing a measure for this were developed. All these methods aim to evaluate an average molecular motion within a particular membrane domain, since this in turn reflects membrane fluidity. The methods fall in two categories. Some of them, such as calorimetry, X-ray and NMR analyses, detect the molecular parameters of the intrinsic membrane. Other methods rely on the introduction in the membrane of ESR components (spin labelled) or fluorescent labelled molecules.

The concept of biomembrane fluidity is now an integral part of popular models for biomembrane structure and function. Many studies have attempted to relate certain disease conditions with modifications of biomembrane fluidity. A recent summary of abstracts associated with oncology show the continued appreciation of the importance and relevance of biomembrane fluidity (see Cooper: *Membrane Fluidity and Cell Surface Receptor Mobility*, 1981).

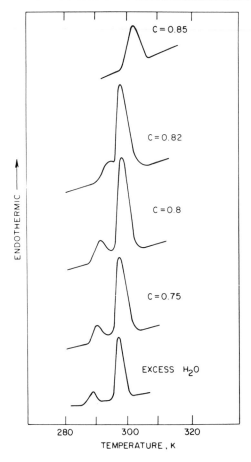

FIG. 2. Differential scanning calorimetry curves for the DMPC–H_2O system as a function of lipid concentration $c = $ wt lipid/(wt lipid + wt water).

II. Non-lamellar Structures

Above the main lipid phase transition temperature in the fluid state, phospholipids can form several different types of phase organization, e.g. some phospholipids give lamellar- and hexagonal-type structures. Studies have shown that the lecithins appear to exhibit a lamellar-type structure over a large range of water concentration. Other phospholipids, such as phosphatidylethanolamines and samples of brain lipid, appear to be able to exist in both the hexagonal and lamellar types of organization, depending upon the concentration in water. Studies by Gulik-Krzywicki et al. (1967) showed that four phases can be distinguished with a mitochondrial lipid extract. The composition of the extract from beef heart mitochondria was 34% lecithin, 29% phosphatidylethanolamine, 10% phosphatidylinositol, 20% cardiolipin, 2% cholesterol, and 5% neutral lipids. Hexagonal phases were observed to occur as well as various lamellar phases. As yet, no certain relationships between these phase properties and their biological function have been established, though there are many speculations about this. Nevertheless, these phase properties must be considered and kept in mind during discussions of lipid and biomembrane behaviour (see Chapter 2).

III. Fluidity and Lipid Dynamics

A number of non-perturbing techniques have been used to determine lipid chain order in lipid bilayer systems.

Some physical techniques for measuring lipid chain order:

Calorimetry On the basis of calorimetric studies, Chapman and co-workers (1969) suggested that the fluid character of the bilayer matrix of biomembranes is different from that of a simple paraffinic melt. (This contrasted with the view based on X-ray diffraction studies that the interior chains were in a *random* chaotic state; see Luzzati, 1968.) The ΔH values associated with the gel to liquid–crystal transition are lower, in fact, than the ΔH values from melting of pure hydrocarbons. The same holds true for the entropy change in the process. The incremental ΔS per CH_2 group is only about 1 e.u. for the gel to liquid–crystal transition of bilayers but is almost twice as large for the melting of simple paraffins (Phillips et al., 1969). These thermodynamic results show that the hydrocarbon chains in the bilayer core are not as disordered as they are in a pure liquid hydrocarbon.

Nuclear magnetic resonance Although the early studies with proton NMR (Chapman and Salsbury, 1966: Veksli et al., 1969) were valuable in emphasizing the mobility of the lipid molecules, another useful innovation was the introduction of non-perturbing deuterium probes for deuterium magnetic resonance studies, by Oldfield et al. (1972a). This has been particularly valuable for studying the details of the lipid dynamics within model and natural biomembranes. It has been exploited by Seelig and co-workers (1974) (Seelig and Seelig, 1980) to examine the *order parameter* of the lipid chains and to show that in the region of constant parameter *gauche* conformations can occur only in complementary pairs leaving the hydrocarbon chains essentially parallel to each other. The deuterium probe results differ from those obtained using spin-labelled molecules (Hubbell and McConnell, 1971), e.g. the spin labels detect a continuous decrease of the order parameter, whereas the deuterium probe shows that the order parameter remains approximately constant for the first nine segments.

The number of order profiles reported in the literature is still small. [²H]-NMR order profiles using selectively deuterated phospholipids have been established now for 1,2-dipalmitoyl-*sn*-glycero-3-phosphocholine (DPPC) (Seelig and Seelig, 1974, 1975), 1,3-dipalmitoyl-*sn*-glycero-3-phosphocholine (Seelig and Seelig, 1980), 1,2-dimyristoyl-*sn*-glycero-3-phosphocholine (Oldfield *et al.*, 1978a,b), 1-palmitoyl-2-oleoyl-*sn*-glycero-3-phosphocholine (POPC) (Seelig and Seelig, 1977; Seelig and Waespe-Sarcevic, 1978) and 1,2-dipalmitoyl-*sn*-glycero-3-phosphoserine (DPPS) (Seelig and Browning, 1978; Browning and Seelig, 1980). In addition, egg-yolk lecithin with perdeuterated palmitic acid acyl chains intercalated physically (Stockton *et al.*, 1976), perdeuterated 1,2-dipalmitoyl-*sn*-glycero-3-phosphocholine (Davis *et al.*, 1979) and a glycolipid (Skarjune and Oldfield, 1979) have been studied. Though limited in number, these order profiles comprise a fairly representative collection of different lipid classes, including saturated and unsaturated fatty acyl chains as well as different polar groups.

A comparison of the rotational correlation times with the deuterium order parameter as a function of the labelled segment has been made. The shapes of the correlation time and order parameter are similar, with correlation times ranging from about 8×10^{-11} for the plateau region to about 3×10^{-11}s for the C-15 methylene segment (Brown *et al.*, 1979).

The deuterium magnetic resonance technique has also been applied to the study of the lipid dynamics of natural biomembranes. This was first carried out by Oldfield *et al.* (1972a) with *Acholeplasma laidlawii* cells and later studied by Stockton *et al.* (1977) with selectively deuterated palmitic acid. Quite recently, similar studies have also been made with *Escherichia coli* membranes. The fatty acyl chains in elaidate 9-10-d₂-enriched *E. coli* inner and outer membranes are less ordered by 10–20% than in the corresponding model systems (Gally *et al.*, 1980).

This characteristic order parameter signature of model membranes is carried over into biological membranes. The agreement between the order profile of *Acholeplasma laidlawii* and those of the pure phospholipid membranes is striking. The incorporation of *cis*-double bond promotes larger changes in the order profile than, for example, the introduction of a net negative charge in the polar head group (DPPS) or the incorporation of proteins into the membrane. The divergence of the POPC order profile between carbon atoms 5 to 9 can be explained by a specific stiffening effect of the *cis*-double bond (Seelig and Seelig, 1977).

Measurements of the deuterium quadrupole splittings, $\Delta\nu_Q$, furnishes information about the time-average *orientation* of the segments involved. In contrast, measurement of the deuterium NMR relaxation times gives information on the *rate* of segmental motion. The correlation between chain order and chain mobility is not well understood as yet, but the distinction between time-averaged structural parameters (such as relaxation times, correlation times and microviscosity) refers to biomembranes in general and is independent of the specific technique employed (see Seelig and Seelig, 1980).

In studies of *deuterium order parameters* with phospholipids of different transition temperatures, a normalisation process was carried out (Seelig and Browning, 1978). They were referred to a reduced temperature $\theta = (T - T_c)/T_c$, where T is the measuring temperature ($^\circ$K) and T_c is gel to fluid transition temperature. The data for a range of lipids (having T_c from -5 to 66°C) were in narrow bands supporting the idea of a *similar physical state existing at a given θ temperature*.

Infrared spectroscopy is also useful for providing information about lipid chain order and dynamics in lipid bilayer systems.

The parameters commonly employed in infrared spectroscopy are the frequency (ν) and the half-bandwidths ($\Delta\nu_{\frac{1}{2}}$) of the individual vibrational modes. These parameters have the advantage of relating to specific phenomena at the molecular level and can be used to discriminate between different motions of the absorbing groups.

The *frequency* of a vibrational absorption is determined by the nature of both the vibrational mode and the vibrating group. The C–H and C–D stretching vibrations in acyl chains are decoupled from other vibrations occurring within the molecule: the bulk of the acyl groups are electronically decoupled from the polar head groups. Consequently, in the all-*trans* conformational state, each of these absorptions is observed at its characteristic frequency, regardless of the terminal functional groups, and is shifted only slightly as a result of interactions with molecules of a different class (e.g. proteins). However, the effects of changes in conformation, specifically the introduction of *gauche* conformers, are large, producing shifts to higher frequencies. Consequently, one can use these shifts to determine the introduction of *gauche* conformers of chain melting, the magnitudes of the shifts being related to the number of *gauche* bonds (see Cameron *et al.*, 1980; Snyder *et al.*, 1982).

In principle, a considerable amount of information regarding molecular motion can be obtained from the absorption band contour by means of moment analysis and the study of correlation functions (Bailey, 1974). This is difficult in the case of systems such as hydrated membranes, because the spectrum of the particular species must be isolated from the curved background of the water absorption. However, the *half-bandwidth* is much less susceptible than the above methods to the introduction of errors due to imperfect background correction. The width of the band contour results from rotational, translational and/or collisional effects. Thus, the half-bandwidth monitors the freedom of motion of the absorbing group, i.e. the amplitudes and rates of motion within its immediate environment. The half-bandwidths will therefore reflect the main thermal transition where a considerable change in this environment occurs. It is also sensitive to other changes which do not introduce *gauche* conformers, such as a decreased freedom of vibrational or torsional motion of the chains.

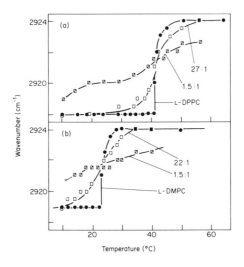

FIG. 3. Temperature dependence of the maximum wave number of the CH_2 asymmetric stretching vibrations (□) in (a) L–DPPC–cholesterol and (b) L–DMPC–cholesterol at the molar ratios indicated. The temperature dependence for the pure lipids (●) is also given (Cortijo *et al.*, 1982).

Infrared spectroscopic studies of the main endothermic phase transition of the phospholipid–water systems results in an abrupt change in the band parameters of both methylene bands (Asher and Levin, 1977; Cameron *et al.*, 1980; Cortijo and Chapman, 1981). The frequency of the band maximum has been used to monitor changes in the lipid conformation. The temperature profiles of the band-maximum frequencies for the methylene asymmetric stretching vibrations of L-DPPC and L-DMPC are shown in Fig. 3. As has been observed, by the use of other different physical techniques such as calorimetry (Alonso *et al.*, 1982) slight impurities in the samples make the main transition less sharp (T_c for pure L-DPPC is 41°C and for L-DMPC is 23°C). Studies of the symmetric stretching vibrations around 2850 cm^{-1} give practically the same results. The minor overall change of the band-maximum frequencies for the symmetric vibration (only 3–4 cm^{-1}) during the lipid chain melting transition make it less suitable for these types of studies due to the greater relative influence of the experimental error in the band position (~ 0.5 cm^{-1})

The results obtained with the pure lipid–water systems are in accord with those of various workers, using a variety of physical techniques including calorimetry and NMR spectroscopy. The way in which the abrupt endo-thermic lipid phase transition (Chapman et al., 1967) is indicated by the shift of the asymmetric methylene band at the appropriate T_c temperature (see Fig. 3) is reassuring for the application of this technique.

Cameron et al. (1980) have also used infrared spectroscopy to study the pretransition observed in calorimetric studies of saturated lecithins. They used Fourier Transform infrared spectroscopy to study the infrared-active acyl chain vibrational modes of fully hydrated multibilayers of 1,2-dipalmit-oyl-sn-glycero-3-phosphocholine (L-DPPC) over the temperature range 0–55°C. Frequencies, bandwidths and other spectral parameters were measured as a function of temperature for the methylene scissoring, rocking and wagging modes as well as for the C–H stretching modes, and they were used to monitor the packing of the acyl chains. Particular emphasis was placed on determining the nature of the pretransition event. They showed that between 36 and 38°C the spectral changes are indicative of a phase change in the acyl chain packing from an orthorhombic to a hexagonal subcell. It was concluded that in the gel phase, at all temperatures below the main transition, the acyl chains are predominantly in all-*trans* conformations and that the temperature-dependent variations of spectral parameters result from changes in interchain interactions.

Mantsch et al. (1981) have shown that phase transitions of phospholipids related to the lamellar $\rightarrow$ hexagonal transitions can be studied using infrared spectroscopy. They studied the egg yolk phosphatidylethanolamine–water system which exhibits this type of phase transition.

Two phase transitions were monitored in this system, a bilayer to bilayer phase transition centred at 12°C (T_m), and a bilayer to nonbilayer phase (H_{II}) transition centred at 28°C (T_h). The spectral changes observed in the acyl chain bands at T_m are indicative of a transition from a conformationally and motionally ordered gel phase to a disordered liquid–crystalline phase containing a high population of *gauche* conformers. There are also consider-able changes in the relative intensities of the carbonyl bands at T_m, while changes in the phosphate bands are minimal. The spectral changes at T_h are indicative of the introduction of additional conformational and motional disorder into the acyl chains. While the transition to the H_{II} phase is promoted by an increase in the degree of unsaturation, it is not observed in sonicated PE. The transition T_m is not affected by the sonication and is less sensitive to the fatty acid composition. On the basis of the spectral changes at T_m and T_h, it was proposed that the bilayer to nonbilayer phase transition results from the fact that the bilayer structure is no longer able to accommodate the simultaneous requirements of a continuous head-group surface and the volume required by the highly disordered acyl chains, the driving force being the introduction of additional conformational disorder into the already disordered liquid–crystalline bilayer phase.

The bands resulting from the phosphate and ethanolamine groups are rather insensitive to both phase changes whereas those resulting from the ester linkages show large changes both at T_m and T_h. The temperature profiles derived from the ester group C=O and C–O absorption bands show the same general form. The temperature profiles derived from this band are given in Fig. 4. The $\Delta A/\Delta T$ plot identifies the two temperature intervals within which there is an increased overall rate of change and which define T_m and T_h. Again, the symmetric shape of T_h contrasts with the ramplike shape of the acyl chain melting transition T_m. Neither of these two transitions is very sharp: T_m occurs over a temperature range of about 10°C; T_h covers a range of about 13°C.

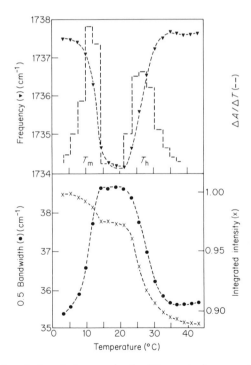

FIG. 4. Temperature-induced changes in the IR spectral parameters of the $C = O$ stretching vibration of egg-yolk phosphatidylethanolamine–water system (Mantsch *et al.*, 1981).

The bandwidth and intensity of the C=O stretching band also exhibit large changes with temperature. As in the case of the frequencies, the bandwidths are nearly identical in the gel-phase bilayer and in the inverted hexagonal phase, while a substantially different value is observed for the bilayer in the liquid-crystalline phase. Both phase changes result in a decrease in integrated intensity (a total decrease of about 12%), the larger decrease being observed at T_h. The complexity of the changes in this band contour results partly from the fact that it is comprised of two components.

Mechanical calorimetry measurements (Evans and Kwok, 1982) have made micromechanical tests of phospholipid vesicles to determine the equilibrium changes in area of a dimyristoyl–phosphatidylcholine (DMPC) bilayer vesicle as produced by variable bilayer tensions at discrete temperatures in the range of 20–30°C. The micromechanical tests involved aspiration of a large ($> 2 \times 10^{-3}$ cm diameter) vesicle with a small suction pipette ($\sim 8 \times 10^{-4}$ cm diameter). Because the internal volume was held fixed by the 100 mM sucrose solution, the tension in the vesicle membrane (created by the pipette suction pressure) caused a displacement of the vesicle projection inside the pipette that was proportional to the increase in total area of the vesicle. From the tension-area isotherms, the elastic compliance (area compressibility) of the membrane was determined throughout the temperature range of the main liquid–crystal–crystalline phase transition of DMPC. This is shown in Fig. 5. Similarly, the area–temperature relations at constant membrane tension were obtained. These data yielded values for the elastic area compressibility in the liquid state (~ 30°C) of 0.007–0.008 cm dyn^{-1}, in the solid state (~ 20°C) of less than 0.001 cm dyn^{-1}, and in the coexistence region (~ 24.2°C) of 0.03–0.05 cm dyn^{-1}. Likewise, the thermal area expansivity (i.e. fractional change in area with change in temperature at constant membrane tension) values ranged from $(4–6) \times 10^{-3}/$°C in the liquid state to 0.1–0.2/°C in the coexistence region to $(5–8) \times 10^{-4}/$°C in the solid state. The only area transition that was seen occurred at about 24°C; the area increase from the solid to liquid state was 12–13%. The convolution approach coupled with the "Clausius-Clapeyron" equation for the membrane surface provided the means to derive the thermal properties of the transition from the elastic compliance vs. temperature data. The results of the analysis gave an expectation value for the transition temperature of 24.2°C, the statistical width of the transition of 0.3°C, and the heat of the transition of about $7 = 0.7$ kcal mol^{-1}.

Lipid diffusion Quantitative studies of lipid diffusion in model bilayers were made by Träuble and Sackmann (1972) and Devaux and McConnel (1972), and values reported were of the order $D = 1.8 \times 10^{-8}$ cm^2 s^{-1}. Triplet probes have also been used (Naqvi *et al.*, 1974) to measure the diffusion coefficient of lipids in the fluid state and a value of 1.6×10^{-8} cm^2 s^{-1} was derived by this method.

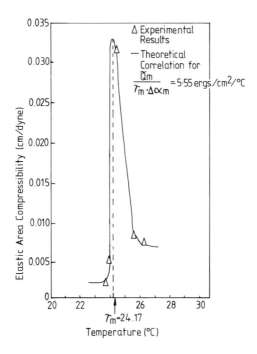

FIG. 5. Elastic area compressibility of DMPC bilayer vesicle versus temperature in the vicinity of phase transition. The data were obtained from the reciprocal tangent to the tension-area dilation isotherms in the limit of zero area dilation. The curve is the theoretical correlation derived from the area convolution equation. This correlation provided the thermoelastic shift (Clausius–Clapeyron effect), 5.5 dyn/(cm °C), the expectation value for the transition temperature approx. 24.2°C, and the statistical width of the transition, approx. 0.3°C. With the increase change in area at the transition (approx. 13 %), the transition temperature (approx. 24.2°C), and the thermoelastic shift, the heat of the transition was calculated to be about 210 ergs/cm² or approximately 7.5 kcal/mol of DMPC (Evans and Kwock 1982).

Similar studies have been carried out using a variety of methods, including NMR spectroscopy (Kuo and Wade, 1979), fluorescence photobleaching, and spin label photochemical methods (Sheats and McConnell, 1978).

A technique of pattern photobleaching has been used to measure the diffusion of fluorescent-labelled phospholipids (Smith and McConnell, 1978). In this technique, the light from an argon-ion laser is passed through a microscope and through a Runci ruling on to a sample under the objective of the microscope. The sample is in the form of phospholipid multilayers. The fluorescent molecules are photobleached in a striped pattern. The time-independent intensity of the stripes is recorded photographically and used to determine the diffusion coefficients below and above the T_c phase transition temperature. Values of 10^{-10} cm^2 s^{-1} were obtained below T_c and $> 10^{-8}$ cm^2 s^{-1} above this temperature.

Studies of the antibodies bound to lipid haptens in model biomembranes have been observed, using the pattern photobleaching method, to diffuse as rapidly as the lipids themselves (Smith *et al.*, 1979) and ranged in value from 10^{-11} to 10^{-8} cm^2 s^{-1}. The lateral diffusion coefficients of fluorescent lipid analogues have been measured as a function of water content. It was shown that an 8-fold decrease in the diffusion coefficient accompanies water removal from a multilamellar system (McCown *et al.*, 1981).

Correlation with monolayer properties Prior to the introduction of the concept of fluidity in 1966, attempts had been made to relate the monolayer properties at the air–water interface with biomembrane structure. However, these attempts were rather unsatisfactory, mainly because no clear view was available as to the molecular details of the monolayer, e.g. what was the molecular organization or dynamics underlying the "expanded state of a monolayer"?

The situation became clarified when Chapman *et al.* (1966) and Phillips and Chapman (1968) pointed out that a correlation exists between the lipid monolayer properties at the air–water interface of lipids and the properties of the lipid bilayers in aqueous dispersions. The "condensed monolayer" corresponds to the crystalline or gel phase and the expanded state to the "fluid" or melted state which occurs above the lipid transition temperature. Similar thermotropic phase changes occur with the monolayers as occurs with lipid bilayers. The isotherms observed at different temperatures with dipalmitoyl phosphatidylcholine are shown in Fig. 6.

All monolayer states are possible with the saturated lecithin and phosphatidylethanolamine homologues (Phillips and Chapman, 1968). It is apparent that if the hydrocarbon chains are sufficiently long, condensed monolayers are formed, whereas with shorter chains liquid-expanded films occur. These two limiting states are sufficiently well defined so that at any particular temperature only one of the homologues studied exhibits the transition state. The data indicate that variations in hydrocarbon chain length which do not give rise to change in monolayer state do not have a significant effect on the π–A curves. Temperature changes can also give rise to the condensed and expanded states for a monolayer of a single homologue. Obviously, a sufficiently low temperature causes the film to become completely condensed, whereas at higher temperatures it is fully expanded. Monolayers in the two limiting states are more or less invariant with temperature, and it is the sensitivity of the phase transition to temperature that leads to the variety of isotherms.

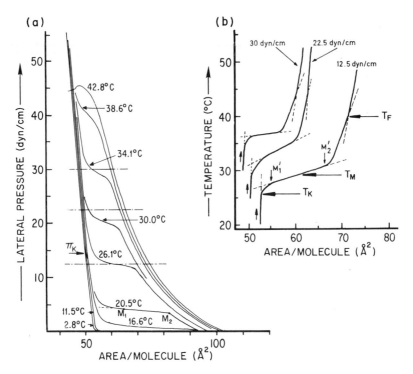

FIG. 6. (a) Continuously recorded isotherms of L-α-dipalmitoyl lecithin (DPPC) on pure water substrate. Note the arrow at Π_K, which indicates a break in the isotherm 2.8°C. (b) Continuously recorded isobars of monomolecular film of L-α-DPPC. The curves were recorded at increasing temperatures. From Albrecht *et al.* (1978), with permission.

The molecules in a completely condensed phosphatidylethanolamine monolayer are much more closely packed than are those in the equivalent lecithin monolayer. This also correlates with the lipid bilayer behaviour. The lecithins have a lower transition temperature in a bilayer structure than does the equivalent chain length phosphatidylethanolamine. This presumably arises from steric factors associated with the large polar groups on the lecithin molecules. Marcelja (1974) has used this correlation between the monolayers and lipid–water dispersions to calculate the lateral pressure in the hydrocarbon chain region. He estimates this to be of the order of 20 dynes cm⁻¹ for each half of the bilayer.

Phase separation Most biomembranes contain a range of lipid classes and a variety of acyl chain lengths and degrees of unsaturation. This led Ladbrooke and Chapman *et al.* (1969) and Phillips *et al.* (1970) to carry out calorimetric studies of mixtures of lipids within model biomembrane structures.

The first studies on phase separation of lipid–water systems were discussed by Ladbrooke and Chapman (1969), who reported studies of binary mixtures of lecithins using calorimetry (Phillips *et al.*, 1970). These authors examined mixtures of distearoyl and dipalmitoyl lecithin (DSL–DPL) and also distearoyl lecithin and dimyristoyl lecithin (DSL–DML). With the DSL–DPL mixtures the phase diagram shows that a continuous series of solid solutions are formed below the T_c line. It was concluded that compound formation does not occur and that with this pair of molecules having only a small difference in chain length co-crystallization occurs.

With the system DSL–DML monotectic behaviour was observed with limited solid solution formation. Here the difference in chain length is already too great for co-crystallization to occur *so that as the system is cooled migration of lecithin molecules occurs within the bilayer to give crystalline regions corresponding to the two compounds* (Ladbrooke and Chapman, 1969).

Examination of a series of fully saturated lecithins with dioleoyl lecithin gave similar results with phase separation of the individual components taking place (Phillips *et al.*, 1970). Later calorimetric studies were reported by Clowes *et al.* (1971) on mixed lecithin–cerebroside systems and on lecithin-phosphatidylethanolamine mixtures (reviewed by Oldfield and Chapman, 1972; Chapman *et al.*, 1974). The lecithin–phosphatidylethanolamine systems of the same chain length give a wide melting range with some separation of the different lipid classes.

The use of spin labels such as TEMPO to examine phase separation of mixed lipid systems was reported by Shimshick and McConnell (1973) on similar lipid mixtures. Other phase separation properties have been observed by Ito and Ohnishi (1974) and have shown that lipid phase separation can occur in phosphatidic acid–lecithin membranes due to the effects of Ca^{2+}. Butler *et al.* (1974) have shown that the probe stearic acid spin label tends to migrate to the more fluid lipid phase in multiphase systems. This confirms the earlier conclusions of Oldfield *et al.* (1972b) and shows that measurements of membrane fluidity in heterogenous systems are not necessarily representative of the entire membrane.

Metal ion and pH effects Metal ion interactions have been known for some years to affect the thermotropic phase transition of soap systems. The thermotropic phase transition of stearic acid occurs at 114°C with the sodium salt and at 170°C with the potassium salt. These phase transitions can be linked to the monolayer characteristics.

Early studies of stearic acid monolayers (Harkins and Anderson, 1937; Shanes and Gershfeld, 1960) showed that interaction with Ca^{2+} ions caused an increase in surface pressure (i.e. condensation) and also decreased the permeability to water. The same effect has been observed with phosphatidylserine monolayers (Rojas and Tobias, 1965), but Na^+ and K^+ addition gave no such condensation. Later, more extensive studies showed that a variety of acidic phospholipid monolayers undergo an increase in surface potential and decrease in surface pressure on addition of Ca^{2+} and other bivalent cations (Papahadjopoulos, 1968). Phosphatidylserine is found to be more selective than phosphatidic acid but for both systems the order of cation effectiveness is:

$$Ca^{2+} > Ba^{2+} > Mg^{2+}$$

The formation of linear polymeric complexes was proposed to account for these findings.

Cationic charge has been observed to be important in some bilayer studies of phosphatidylserine (Ohki, 1969). Black membranes formed in the presence of Ca^{2+} ions are found to be more stable with a higher electrical resistance than those formed in the presence of Na^+ only. The concentrations of cationic species required to produce charge reversal in phosphatidylserine dispersions have been determined, together with association constants for the species formed (Barton, 1968). The results obtained agree well with those obtained previously (Blaustein, 1967) with the exception of uranyl cation UO_2^{2+}. Studies with this cation (Chapman et al., 1974) indicate that this ion causes the thermotropic phase transition temperature of lecithins to increase. Two main phase transitions were observed corresponding to the presence of complexed and uncomplexed lipid. When the titration is complete, only the higher melting transition remains. The studies by Chapman et al. (1974) indicate that the interaction between cations and phosphatidylserine causes greater shifts of transition temperature than is observed with lecithin molecules. All the cations studied were found to shift the phase transition temperature of the phospholipids to higher values.

A comparison of the effects of various ions on phosphatidylcholine bilayer membranes investigated by deuterium and phosphorus magnetic resonance (Akutsu and Seelig, 1981) showed that addition of metal ions led to a structural change at the level of polar groups. The glycerol backbone or the beginning of the fatty acyl chains was not affected. The strength of interaction increased with the charge of the metal ion in the order:

$$Na^+ < Ca^{2+} < La^{3+}$$

However, distinct differences were noted between ions of the same charge and a strongly hydrophobic tetraphenylammonium ion induced a similar change as La^{3+}. Recent studies with liposomes containing anionic phospholipids have indicated that the interaction with Ca^{2+} is exothermic, causes a large increase in the transition temperature, produces a phase separation in vesicles of mixed phospholipid composition, induces the formation of non-bilayer phases and promotes vesicle fusion.

The precise nature of the interaction between ions and phospholipids is still open to doubt. There is some evidence that charge neutralization is the prime interaction of charged phospholipids with divalent cations (Verkleij *et al.*, 1974; Träuble and Eibl, 1974). Divalent cations were found to increase the transition temperature and the monovalent cations to fluidize the bilayer. Some authors believe that the primary effect of the cation on lecithins may be on the aqueous portion of the lipid bilayer (Gottlieb and Eanes, 1972; Ehrstrom *et al.*, 1973; Godin and Ng, 1974). A recent study of an extensive range of salts with lecithin bilayers (Chapman *et al.*, 1977a) indicates that the anion present has a very large effect in determining the state of fluidity of the bilayer, the results obtained being best explained by a thermodynamic treatment based on relative association constants. Verkleij *et al.* (1974) have shown that the thermotropic phase transition of a synthetic phosphatidyl–glycerol is influenced by pH, Ca^{2+}, and a basic protein of myelin. Träuble (1972) has shown that pH can affect lipid transition temperatures, particularly lipids such as the phosphatidylethanolamines and phosphatidic acids. Träuble (1976) has pointed out that surface charges tend to fluidize or expand lipid biomembranes. Fluidization of a biomembrane can be induced either by an increase in pH or by an increase in salt concentration, whilst rigidification can be achieved by a decrease in pH or (when the lipid is fully ionized) by an increase in ionic strength.

Electric field-induced phase change Electric fields change the transition temperature T_t of a phase transition if the dielectric constants (ε_1 and ε_2) in the two phases are different. The shift in transition temperature is given (Sackman, 1979) by:

$$\Delta T = \frac{T_t(\varepsilon_2 - \varepsilon_1)E^2}{\Delta H \, 8\pi\rho} \tag{1}$$

where ΔH is the heat of transition and ρ the density. For the chain-melting transition ($\Delta H = 42$ kJ mol^{-1}, $\rho = 1$ g cm^{-3}, $\varepsilon_2 - \varepsilon_1 \leqslant 10$), one expects $\Delta T \simeq 0.1°C$ at $E = 10^5$ V cm^{-1}. This is a very small effect indeed and has not yet been observed.

pH-Induced lipid phase transitions (Träuble and Eibl, 1974). Dramatic conformational changes may be effected in membranes of charged lipids by variations in the pH of the aqueous phase. Under favourable conditions (low ionic strength) phase changes may also be triggered by variations in the concentration of monovalent ions. The essential features are summarized in Fig. 7. (1) The transition temperature T_t is strongly pH-dependent. For phosphatidic acid T_t varies abruptly between pH 8 and pH 10 and is lower by about 10°C in the twofold charged state (Fig. 7a). (2) At low ionic strength T_t is also a sensitive function of the salt concentration. For DMPE (Fig. 7b)

T_t decreases with increasing ionic strength in the uncharged state (pH 4), whereas the reverse is valid for the fully charged state (pH 8.5).

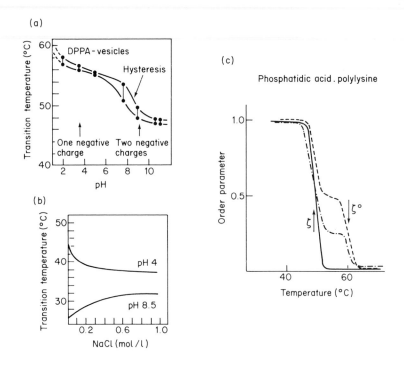

FIG. 7. Charge-induced lipid phase transitions. (a) pH dependence of transition temperature of dipalmitoylphosphatidic acid (DPPA). (b) Influence of ionic strength on transition temperature of dimyristoylphosphatidic acid methyl ester (DMPE). (c) Shift in transition temperature (from T_t to T'_t) of DPA by absorption of polylysine: (—) free lipid, (— —) 50% lysine, (— · — ·) 30% lysine. By permission of Sackmann (1979).

The origin of the pH effects is the sensitive dependence of the chain-melting transition temperature on the lateral packing density. The latter decreases if fixed charges appear at the polar head group. This charge-induced lateral expansion is best seen in monolayer studies. It is a consequence of an additional (expansive) lateral pressure $\Delta\pi_{el}$ from the repulsion of the fixed lipid charges. According to Gouy–Chapman theory, $\Delta\pi_{el}$ is related to the surface potential ψ_0 (Payens, 1955) by the relation

$$\Delta\pi_{el} = 6.1\sqrt{c}[\cosh(e\psi_0/2kT) - 1], \qquad (2)$$

where c is the bulk concentration of ions, and the value 6.1 applies for T

= 300 K and $\varepsilon = 80$. The value of ψ_0 depends on the fraction α of charged lipids, that is, on the degree of dissociation:

$$\psi_0 = (2kT/e) \, \text{arcsinh}(134\alpha/A\sqrt{c}), \tag{3}$$

where A is the area per lipid molecule. The important point is that $\Delta\pi_{el}$ is an additional internal pressure that tries to push the lipid molecules apart. Because the main lipid phase transition leads to an increase in area, $\Delta\pi_{el}$ will decrease the transition temperature. This decrease may be estimated by application of the Clausius–Clapeyron equation:

$$\Delta T = (\Delta a/\Delta s)\Delta\pi_{el} \tag{4}$$

where Δa and Δs are the changes in area and entropy at the phase transition, respectively. These differences were determined by monolayer studies (Albrecht et al., 1978) and by calorimetry (Jacobson and Papahadjopoulos, 1975). The quantities Δa and Δs depend on the lipid charge. For phosphatidic acid, the ratio $\Delta a/\Delta s$ does not depend to any considerable extent on α($\Delta a/\Delta s \simeq 1 \deg \text{cm dyn}^{-1}$). For small values of α (and thus ψ_0) it follows that $\Delta\pi_{el} \simeq 8\alpha$ (in dynes per centimetre). A transition temperature difference of $\Delta T \simeq 8°C$ is expected between the singly and doubly charged lipid. The experimental value is $\Delta T = 10°C$.

A second important point is the dependence of $\Delta\pi_{el}$ on the ionic strength c of the aqueous phase. The reason for this is that the degree of dissociation α depends on ψ_0 according to the equation

$$\alpha/(1 - \alpha) = [\text{H}^+]^{-1}K_0 \exp(e\psi_0/kT), \tag{5}$$

where K_0 is the dissociation constant of isolated lipid molecules. At increasing salt concentrations a Stern layer of positive ions accumulates at the lipid surface. The Stern layer decreases and interferes with the access of protons to the ionizable groups (Vaz et al., 1979). If the lipid is fully ionized ($\alpha = 1$), an increase in c will decrease $\Delta\pi_{el}$. Consequently, the low temperature shift in T is diminished, in agreement with the result shown in Fig. 7b.

IV. Fluidity and its Modulation by Cholesterol

Condensation and fluidity For some years it was known that cholesterol could affect and apparently condense monolayers (at the air–water interface) with certain unsaturated phospholipids. The meaning of this was, however, obscure and controversial, some workers believing that a *cis*-double bond was essential for this and invoking unusual structures and complexes between the lipid and cholesterol. Chapman and co-workers showed, however, that phospholipids containing *trans* double bonds and even saturated phospholipids could exhibit these effects.

Proton NMR studies indicated that cholesterol molecules affected the fluid lipid chains within a model biomembrane system (Chapman and Penkett, 1966).

Later studies using ESR probes (Barratt *et al.*, 1969; Hubbell and McConnell, 1971) and fluorescent probes confirmed the interpretation of these experiments. Studies using deuterium NMR have recently been made with model biomembranes containing various amounts of cholesterol (Rice *et al.*, 1979). Addition of cholesterol at the equimolar level (about 33 wt %) to the sample results in an increase in quadrupole splitting from 3.6 to 7.8 kHz corresponding to an increase in molecular order parameter (from $S_{mol} = 0.18$ to $S_{mol} = 0.41$). Cooling the sample to a temperature some 5°C below that of the gel to liquid–crystal phase transition temperature ($T_c = 23°C$) has little effect on the quadrupole splitting, consistent with previous data (Ladbrooke *et al.*, 1968; Oldfield *et al.*, 1971). Cooling the pure lipid to the same temperature, however, results in hydrocarbon chain crystallization into the rigid crystalline gel phase (Chapman *et al.*, 1967), and a broad, rather featureless spectrum with $\Delta v_Q \sim 14.0$ kHz is observed (Oldfield *et al.*, 1978a). Analysis of this result in terms of a molecular order parameter is not possible since the details of the motion of the rest of the hydrocarbon chain are unclear.

Similar ordering effects of cholesterol are seen when a C6-labelled phospholipid, 1-myristoyl-2-(6,6-dideuteriomyristoyl)-*sn*-glycero-3-phosphocholine (DMPC-6,6-d_2), is used.

Infrared spectroscopic studies have been made of lipid–water systems when cholesterol is included into the bilayer structure. The effect observed by incorporation of cholesterol into the lipid bilayers provides results which are in general accord with those obtained by the use of a range of physical techniques (Ladbrooke *et al.*, 1968; Oldfield and Chapman, 1972), i.e. there is an increase in the number of *gauche* conformers below T_c (less order) and a decrease in this number above T_c (more order). This effect increases with greater amounts of cholesterol in the bilayer. At very high cholesterol concentrations (see Fig. 3) almost no change occurs with temperature in the relative population of *gauche* and *trans* conformers of the lipid chains.

A feature of the inclusion of cholesterol into biomembrane structures is that the presence of large amounts of cholesterol prevents *lipid chain* crystallization and hence removes phase transition characteristics. This was shown by Ladbrooke *et al.* (1968), who described studies on lecithin–cholesterol–water interactions by differential scanning calorimetry (DSC) and X-ray diffraction. The 1,2-dipalmitoyl-L-phosphatidylcholine (DPPC)–cholesterol–water system was studied as a function of both temperature and concentration of components. This particular lecithin was used because it exhibits the thermotropic phase change in the presence of water at a convenient temperature (41°C). The addition of cholesterol to the lecithin in water lowers the transition temperature between the gel and the lamellar fluid cystalline phase, and decreases the heat absorbed at the main transition. No transition is observed with an equimolar ratio of lecithin with cholesterol in water. This ratio corresponds to the maximum amount of cholesterol that can be introduced into the lipid bilayer before cholesterol precipitation occurs.

X-ray evidence (Ladbrooke *et al.*, 1968) indicates that a lamellar arrangement occurs and that at 50% cholesterol an additional long spacing pattern occurs due to the separation of crystalline cholesterol. These results may be interpreted in terms of penetration of the lipid bilayer by cholesterol. In the lamellae of aqueous lecithin, the chains are hexagonally packed and tilted at 58°. It can be envisaged that penetration will be facilitated when the chains are vertical. This causes an increase in the X-ray long-spacing. At concentrations of cholesterol greater than 7.5% the long-spacing decreases. Above this critical concentration, a reduction occurs of the cohesive forces between the chains producing chain fluidization.

Below the lipid T_c transition temperature, calorimetric studies show that the main lipid endotherm is removed with increasing amount of cholesterol. The presence of the cholesterol molecules is to modulate the lipid fluidity above and below the transition temperature of the lipid. The first studies of Ladbrooke *et al.* (1968) suggested that the enthalpy was totally removed at 50 mol %; later studies (Hinz and Sturtevant, 1972) suggested that this occurred at 33 mol %. The latter conclusions led to the concept that cholesterol existed as a 2 : 1 lipid–cholesterol complex.

Very recent studies using sensitive scanning calorimeters confirm (Mabrey *et al.*, 1978) the early conclusion that the enthalpy is in fact removed at 50 mol % of cholesterol to lipid as originally envisaged. Many monolayer studies have been made of phospholipid–cholesterol mixtures. Results of a recent study are shown in Fig. 8.

Few studies have investigated the effects of low concentrations of cholesterol (<10 mol %) in membranes; the majority of studies have examined concentrations between 10–50 mol %. This is despite the fact that several biological membranes, including some types of endoplasmic reticulum, synaptic vesicles and mitochondrial membranes, have cholesterol contents of less than 10 mol % (Ness, 1974, Benga *et al.*, 1978). The separation of lipids into different domains in membranes (Karnovsky *et al.*, 1982) suggests that cholesterol-poor domains could feasibly exist in cholesterol-rich membranes. Consequently, one objective of a recent study using spin labelling ESR (Benga *et al.*, 1983) was to examine in detail the effects of cholesterol at low concentrations. The second objective was to study in parallel with the effects of cholesterol the ordering effects of low and high concentrations of 25-hydroxycholesterol on egg PC fatty acyl chains. The compound 25-

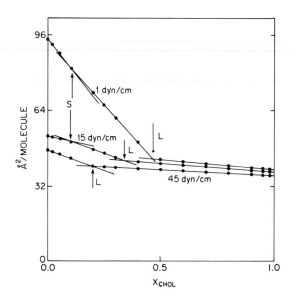

FIG. 8. Area concentration curves of the DPPC-cholesterol system below (1 dyn/cm), just above (15 dyn/cm) and above (45 dyn/cm) the main transition of DPPC, showing the condensing effect of cholesterol on DPPC. From Albrecht *et al.* (1981).

hydroxycholesterol, one of the major autooxidation products of cholesterol (Smith, 1981), is an extremely potent atherogenic factor and low concentrations are toxic to cultured, aortic smooth muscle cells (Peng *et al.*, 1979). Benga *et al.* (1983) have found that very small concentrations of 25-hydroxycholesterol may decrease significantly the fluidity of lipid bilayer, even more than the corresponding concentrations of cholesterol and that above 1 mol % increasing concentrations of 25-hydroxycholesterol do not decrease membrane fluidity. The interrelationship between temperature, sterol concentration, and the order parameter of 5- and 12-doxylstearic acid was examined. With cholesterol, a continuous increase in the order parameter occurred up to a concentration of 50 mol %. With 25-hydroxycholesterol an increase in the order parameter of 5-doxylstearic acid occurred only in the concentration range 0.1 to 5 mol %. With 12-doxylstearic acid as a probe, an ordering effect of 25-hydroxycholesterol was observed at concentrations only up to 1 mol %. The greatest difference between cholesterol and 25-hydroxycholesterol was observed with 16-doxylstearic acid. Whereas increasing concentrations of cholesterol markedly altered spectra, increasing concentrations of 25-hydroxycholesterol produced no significant changes. These results are in agreement with other studies (Butler *et al.*, 1970; Lecuyer and Dervichian, 1969) that have shown that specific structural requirements for cholesterol and its analogues are necessary before significant changes in the degree of order are observed. The sterol requirements are a fairly planar steroid nucleus, a 3 β-hydroxyl group and a hydrocarbon chain at C-17. Minor changes in the steroid nucleus have little effect upon the ability of the sterol to order the bilayers (Semer and Gelerinter, 1979). Other authors have shown that a sterol hydrocarbon chain is necessary to induce appreciable order in the bilayer (Butler *et al.*, 1970). It appears that a hydroxyl group in the 25-position limits the interaction of the steroid nucleus with phospholipid acyl chains.

The relationship between the cholesterol content of a multilamellar lipid–water system and the permeability of the bilayer to water has been studied by Blok *et al.* (1977). These workers studied the osmotic shrinkage of the dipalmitoyl-phosphatidylcholine liposomes in glucose solutions. Below the transition temperature the liposomes are relatively impermeable to water.

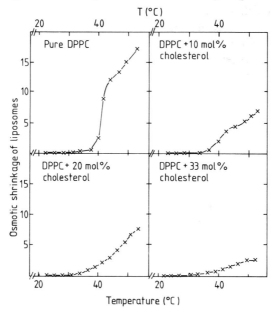

FIG. 9. Cholesterol seals lipid vesicles. Liposomes of pure phospholipids, as well as being leaky to ions and other small molecules, are permeable to H₂O. This is conveniently demonstrated by following their osmotic shrinkage in glucose solutions. DPPC liposomes in the gel state are relatively impermeable to water. However, there is a dramatic increase in permeability over the lipid phase transition. This is due to the formation of interfacial lipid. Above this temperature permeability continues to increase with rising temperature, but at a slower rate. The continued addition of cholesterol leads to a progressive reduction in the permeability of the liposomes to water. Reprinted with permission from Blok, M. C., Van Deenen, L. L. M and De Gier, J. (1977).

Above the phase transition there is a dramatic increase in water permeability. Addition of cholesterol leads to a progressive reduction in water permeability which is associated with the ordering effect that cholesterol has upon the lipid chains. These effects are illustrated in Fig. 9.

Theoretical models Engelman and Rothman (1972) suggested that each cholesterol molecule is separated from a nearest neighbour cholesterol by a single layer of lipid chains which is thereby removed from the cooperative phase transition. They arranged the structure so that ΔH became zero at ≈ 0.33, in agreement with the results of Hinz and Sturtevant (1972). This result is now acknowledged to be incorrect due to instrumental difficulties (see above). Forslind and Kjellander (1975) constructed a structural model of the phosphatidylcholine (PC)–cholesterol–water system for $T > T_c$. They were, however, concerned neither with a phase diagram nor thermodynamic quantities, but did obtain good agreement with measured values of the lamellar repeat distances as a function of c. Martin and Yeagle (1978) extended the model of Engelman and Rothman (1972) to allow for the possibility that each cholesterol could be surrounded by its own "annulus" of lipid chains, and to allow for cholesterol dimer formation with both their annulus model and the chain-sharing model of Engelman and Rothman. They predict that phase boundaries could occur at $c \approx 0.22$, $c \approx 0.31$–0.35 and $c \approx 0.47$. Scott and Cherng (1978) performed a Monte Carlo simulation to study the effect on hydrocarbon chain-order parameters of a rigid cylinder (representing the cholesterol molecule) immersed to various depths in the lipid layer. They considered eight 10-link chains (each link representing a C–C bond) surrounding the cylinder. Depending on the depth of penetration, the order parameters can be increased or decreased. They considered only excluded volume effects (hard core repulsion) and did not consider Van der Waals forces. Pink and Carroll (1978) considered the PC chains or cholesterol molecules to occupy the sites of a triangular lattice, approximated the chain states by two states: an all-*trans* state and a melted state, as well as steric repulsions represented by an effective pressure (Marcelja, 1974). They did not calculate a phase diagram but studied the dependence of ΔH on c, predicting the shape that has now been measured (see above). Subsequently, Pink and Chapman (1979) used the same model to calculate a phase diagram and various thermodynamic quantities. They confirmed the dependence of ΔH on c and calculated that a phase boundary existed at $c \approx 0.2$ for a model of DPPC-cholesterol bilayers, as is now widely accepted. The region $0 \leqslant c \leqslant 0.2$ was found to be generally a single-phase region for $T < T_c$ (see above). They also showed that the average area per chain increased (decreased) as c increased for $T < T_c$ ($T > T_c$) and that the area per chain adjacent to a cholesterol was greater (smaller) than the average for $T < T_c$ ($T > T_c$).

V. Fluidity and Protein Dynamics

The fact that many biological membranes are built upon a fluid bilayer leads to the related phenomenon that in some biomembranes the intrinsic proteins are able to exhibit both rotational and lateral diffusion.

Rotational diffusion Early attempts to study the rotational diffusion of proteins in lipid bilayers were made using the technique of fluorescence depolarization (Tao, 1971; Wahl *et al.*, 1971). However, although this method had previously been used to study the rotational diffusion of protein free in solution (Weber, 1952), it proved to be of limited use for studying the proteins in membranes.

Later work on the protein rhodopsin showed that this molecule did rotate within the membrane, but in a much slower time-domain than the initial studies had investigated. This was made possible by using the properties of the intrinsic chromophore, retinal, present in this protein (Cone, 1972). Following absorption of light, this protein undergoes a series of changes which result in changes in its absorption spectrum. By following these polarised absorbance changes, information about the protein dynamics could be obtained.

About the same time as the investigations in rhodopsin were performed, Naqvi and co-workers (Naqvi *et al.*, 1973) performed similar studies on the protein bacteriorhodopsin. This unusual protein is found as the sole protein component in the patches of purple membrane present in certain Halobacteria. Like rhodopsin, bacteriorhodopsin is capable of absorbing light and consequently undergoing a series of photochemical changes. However, whereas rhodopsin normally bleaches on exposure to light, the photochemistry of bacteriorhodopsin is such that it undergoes a cyclical process resulting in the regeneration of the ground state once more. This cyclical sequence of events occurs over a few milliseconds and therefore can also be utilized to measure slow rotational motion. In contrast to the finding for rhodopsin, however, bacteriorhodopsin was found to be immobile when present in the purple membrane. This is consistent with what is known about the structure of this specialised membrane. The proteins have been shown to be present in the form of a hexagonal lattice with strong protein–protein interaction occurring (Henderson, 1975). Such an ordered structure is unlikely to allow significant mobility of the protein constituents.

The successful studies on rhodopsin and bacteriorhodopsin both relied on the presence of a naturally occurring chromophore. Most proteins, though, do not possess such a suitable chromophore. It was therefore necessary to develop a suitable artificial chromophore which would have had an excited state lifetime of the order of milliseconds. Naqvi *et al.* (1973) suggested that molecules such as eosin, which have a long-lived triplet state, might be used for this purpose. These workers then went on to show that it was indeed possible to measure the rotational mobility of proteins free in solution using these triplet probes.

Since these early studies, other developments have included the synthesis of other triplet probes which can be covalently linked to the protein being examined (Cherry *et al.*, 1976), and the preparation of probes other than eosin derivatives which have higher triplet yields (Moore and Garland, 1979). Recently, methods have been described for detecting the triplet state by means of the phosphorescence emitted as the triplet state decays back to the ground state (Garland and Moore, 1979). Using this method of phosphorescence depolarization, it has become technically possible to measure protein rotation on samples as small as 6.25 nM. These developments in the technique have meant that many different systems have now been studied, including the Band 3 anion transport protein, sarcoplasmic reticulum ATPase, the acetylcholine receptor, the concanavalin A receptor in lymphocytes, cytochrome P450, cytochrome c oxidase and cytochrome b_5 (Cherry *et al.*, 1976; Hoffman *et al.*, 1979; Austin *et al.*, 1979; Junge and Devaux, 1975; Vaz *et al.*, 1979).

There are other methods which have been developed for measuring rotational mobility of proteins, one of which is the saturation-transfer electron-spin-resonance (ESR) technique. Using this method, studies have been performed on rhodopsin (Baroin *et al.*, 1977) which yielded a rotational correlation time of 20 μs at 20°C, which is in reasonable agreement with the earlier studies of Cone (1972). Other workers have used saturation-transfer ESR to measure the rotational motion of the $Ca^{2+}-Mg^{2+}$-ATPase from sarcoplasmic reticulum. However, the results obtained show a sizeable variation, with one report giving the value of t_2 as 60 μs at 4°C (Thomas and Hidalgo, 1978), whereas another report gave the value of t_2 at 2°C at 800 μs (Kirino *et al.*, 1978). However, both groups of workers report a break in the plot of rotational mobility as a function of temperature at around 15°C. Thomas and Hidalgo (1978) interpret this as being due to a lipid effect, whereas Kirino *et al.* (1978) suggest it is due to a protein conformational change.

Saturation-transfer ESR has also been applied to the study of the rotational motion of the acetylcholine receptors for the electronic organ of *Torpedo marmorata* (Rousselet and Devaux, 1977) and also to the study of muscle proteins (Thomas *et al.*, 1975).

Lateral Diffusion Proteins in some biomembranes are also free to diffuse in a lateral sense within the lipid matrix. However, in some biomembranes the cytoskeleton can restrict such diffusion. Several methods exist for measuring lateral diffusion of membrane components, but the method commonly employed today is the technique of fluorescence photobleaching recovery

(FPR). In this method, a pulse of light is used to bleach a suitable chromophore on the cell surface, thus creating an asymmetrical distribution of chromophores. By measuring the time taken for the chromophores to become symmetrically distributed once more, the rate of lateral diffusion can be deduced. Once again, rhodopsin was the first protein to be investigated in this manner (Poo and Cone, 1974; Liebman and Entine, 1974) because of its intrinsic properties. With this particular example, the distribution of rhodopsin molecules was monitored spectrophotometrically. However, in studies on other proteins it is necessary to attach a suitable probe molecule to the protein under study. Typically, fluorescent probes such as rhodamine or fluorescein are used. These are covalently linked either to a lectin or to antibodies; this enables the fluorescent probe to be directed to specific receptors on the cell surface. Once bound to cell, the system is viewed with a fluorescence microscope. An intense flash from a laser is used to bleach the chromophore in a small area of the cell surface. The laser, now much attenuated, is then used to excite any fluorescence from the bleached area. Initially no fluorescence is observed, as all the chromophores have been bleached; however, the fluorescence intensity increases with time as the unbleached chromophores diffuse back into the bleached area from the rest of the cell surface. From the rate of fluorescence recovery, the lateral diffusion coefficients can be calculated.

An experimental difficulty associated with the FPR method is the problem of keeping the cell position constant, especially when trying to measure very low diffusion coefficients. A variation of the technique termed periodic pattern photobleaching (Smith et al., 1979) may be useful in overcoming this difficulty. With this method, a periodic pattern of parallel stripes is bleached into the surface of the membrane, fluorescence photomicrographs of the cell are made at intervals of time after photobleaching. Diffusion of the fluorescent molecules causes a decay in the contrast of the pattern with time. Fourier analysis of the photographs can then be used to quantify the rate of decay of the contrast and hence to yield information about the diffusion coefficient. The values of the lateral diffusion coefficient obtained for the various proteins are 10^{-19} to 10^{-13} cm^2 s^{-1}. It is not certain whether all or only a part of the membrane, for example, the defect structures and regions of mismatch, is available for lateral diffusion.

Bretscher (1980) has commented on the fact that the values obtained for lateral diffusion of phospholipids and for rhodopsin are very different from those determined for a number of proteins in membranes and cell systems. Thus rhodopsin has a diffusion coefficient of 4×10^{-4} cm^2 s^{-1} at 20°C but values from other systems give values 10–10^4-fold lower.

Redistribution of proteins The fluid nature of the lipid matrix allows redistribution of proteins in the plane of the membrane to occur and can lead to patch formation, which is a passive discrete clustering of macro-molecules into two-dimensional aggregates. Thus interaction with multivalent antibodies induces aggregation of the surface antigens into patches. Aggregation effects are also observed when lectins bind to cell surfaces. Reversible patch formation or particle aggregation has been observed under a variety of conditions such as pH change (da Silva, 1972), temperature shift (Speth and Wunderlich, 1973), addition of anaesthetics (Poste et al., 1975), proteolysis and glycerol treatment. Aggregation induced by these various processes is rapid, reversible and inhibited by glutaraldehyde fixation. It is possible that some of the effects are due to altered cytoplasmic structures. For example, the sensitivity of polymorphonuclear leucocytes for agglutination with concanavalin A (Con-A) is thought to be due to the presence of Con-A receptors in patches in the cell membrane. After treatment with colchicine or vinblastine, however, the agglutination is much reduced, suggesting that microtubules maintain some fixed distribution of membrane constituents (Berlin and Ukena, 1972). In cells interacting with each other in tissues, there is an uneven distribution of plasma membrane proteins. Besides being restricted to only one surface of the cell, some of these proteins are apparently confined to certain regions such as at gap junctions and at synapses. Selective and non-random localization of membrane proteins and morphological specialization (in folds and microvilli) would also be expected in membranes of intestinal, kidney, liver and other epithelial cells.

The redistribution of cross-linked membrane components resulting in polar segregation away from the nucleus is termed capping. During cap formation the cross-linked molecules are segregated from other membrane components by an active process probably dependent on microfilaments and microtubules. The phenomenon of capping can be observed if lymphocytes are treated with fluorescent antibodies to certain membrane antigens (Raff, 1976). In the first place these antigens visibly displace in the membrane forming patches, then they aggregate at one pole of the cell, producing a highly fluorescent cap. At this point the membrane may invaginate into vesicles which are internalized into the cytoplasm by the process of pino-citosis. The capping phenomenon can be inhibited by lowering the temperature so that the lipid bilayer solidifies.

VI. Intrinsic Protein–Lipid Interactions

Considerable confusion has developed in recent years with regard to an understanding of intrinsic protein–lipid interactions, and even the recent review literature contains contradictory statements. It appears that part of the confusion is centred on an over-ready acceptance of conclusions based upon evidence derived from one physical technique. Part of the confusion also springs from attempts to link together a number of different concepts which in fact may not be related, e.g. residual lipid, bound lipid, perturbed lipid, and specific lipid.

Chapman and co-workers (1979) have recently reviewed the situation and point to the following three factors which require consideration:

(1) *The time scale appropriate to the particular physical technique used to study the protein–lipid perturbation* This can be appreciated when we realize that, measured on one time scale, say 10^{-8} s using ESR spin-labelled molecules, a molecule may appear to be rigid, whereas on another time scale, say 10^{-5} s using ^{2}H NMR methods, the same molecule can appear to be mobile. When attempts are made to relate some measured perturbation effects with enzymatic effects, yet another time scale must be considered, i.e. the time interval over which the enzymatic conformational effect occurs.

(2) *The concentration of the protein and its arrangement with the lipid bilayer which is being examined* When we look at the plane of the lipid bilayer, it is immediately apparent that as the concentration of the protein in the lipid bilayer increases, the number of multiple contacts of each lipid with proteins also becomes important. At low protein concentrations single contacts of lipid with the protein are the dominant situation.

(3) *Whether in the system considered the lipid is above or below its transition temperature T_c* The lipids below T_c when they crystallize squeeze the proteins out of the crystalline lipid lattice. The shape and size of the protein cause packing faults in the lattice, and patches are formed of high protein to lipid ratio. These regions have been demonstrated by freeze-fracture electronmicroscopy and are distinct from the remaining crystalline lipid regions. It is clearly important when considering methods for measuring the fluidity of natural biomembranes to consider and examine the effects of intrinsic proteins upon lipid fluidity and dynamics. Such considerations then raise questions concerning the use of probes commonly used to measure biomembrane fluidity. Electron spin resonance probes and fluorescent probes (e.g. diphenylhexatriene) have been commonly used. Nuclear magnetic resonance and infrared spectroscopy have also recently been used on biomembrane systems. An important question is what are these probes measuring in protein–lipid bilayer systems.

ESR studies ESR experiments with cytochrome oxidase and sarcoplasmic reticulum ATPase (time scale $\sim 10^{-8}$ s) showed the existence of "immobile components" in the corresponding spectra. It was proposed that the ESR immobile component was an indication of a special lipid shell (called the boundary layer lipid or the annulus lipid) separate and distinct from the bulk lipid. This concept was extended and generalised so that the annulus lipid was said to control enzyme activities of intrinsic proteins to rigorously exclude cholesterol and to be important for anaesthetic activity.

Various workers pointed out a number of observations which cast doubt on this interpretation of the observed ESR "immobile component" and on the concept of special annulus lipid.

(1) It was shown that the observation of an ESR "immobile component" is not restricted to proteins which in principle may require and possess some captive or tightly bound lipids. It is indeed observed in gramicidin A–lipid–water system (Chapman et al., 1977a). The gramicidin A molecule is a relatively simple polypeptide.

(2) Fatty acid spin labels may not give an accurate picture of protein–phospholipid interactions. This is because some intrinsic proteins themselves can bind fatty acids and spin labelled fatty acids, and hence a partition between the protein and the lipid can occur. An immobilization of the fatty acid spin label may only reflect a binding of the labelled molecule to the protein (Benga and Chapman, 1976).

(3) The "immobile" ESR component is observed at high protein (or polypeptide) content in the lipid bilayer. At such high protein contents mobility of the probe molecule will be expected to be considerably inhibited and multiple contacts of lipid with protein will occur. However, with cytochrome oxidase at the lowest lipid content when spin labelled phospholipid (not fatty acids) are used some mobile component in the ESR spectrum is observed, in addition to the immobile component. The difference in behaviour of the spin labelled fatty acids and phospholipids suggests that the part of the residual lipid of the complex, which in some conditions is apparently immobilized, may sometimes exhibit a high degree of mobility (Benga et al., 1979).

(4) Jost et al. (1973, 1977) assumed that the partition of the spin label faithfully reflects the distribution of lipid between the two domains (the boundary lipid and the fluid bilayer). However, no theoretical basis has been presented for this approach. A theory that takes into account not only the size of the domains in which the spin label molecule distributes itself but also the different affinities of the label for the domains has been developed (Benga et al., 1981).

Recent studies by Hoffmann et al. (1981) and Pink et al. (1981) show that the ESR data on various protein–lipid systems can be interpreted in terms of a random arrangement of protein in the lipid bilayer where intrinsic protein contact with protein increases as the protein concentration is increased, i.e. there is no fixed stoichiometric, complex of protein and lipid in systems such as cytochrome oxidase or calcium ATPase dispersed in lecithin water systems.

NMR spectroscopy studies Nuclear magnetic resonance studies using ^{1}H, ^{19}F or ^{2}H nuclei (time scale 1–0.01 ms) of various biomembranes or reconstituted systems *do not show the occurrence of two types of lipid.* This is the case with ^{1}H NMR studies of rhodopsin in disc membranes, ^{19}F NMR studies of *E. coli* membranes, and ^{2}H NMR studies of the cytochrome and sarcoplasmic reticulum systems. *A continuity between the bulk lipid phase and the "boundary layer" lipids and ready diffusion between these lipids takes place.*

The general observations in *all* ^{2}H NMR reconstitution studies reported is that only *one* homogeneous lipid environment is present *above* T_c even when a substantial amount of protein is present. The ^{2}H NMR experiments give no indication for a strong, long-lived interaction between the membrane protein and the lipid. Instead, data can be explained by a relatively rapid exchange between those lipids in contact with the protein and those further away from it. This exchange must be fast on the ^{2}H NMR time scale (exchange rate 10^4 Hz) in order to produce a single-component ^{2}H NMR spectrum, but slow compared to the ESR time scale (exchange rate 10^7 Hz) in order to account for the two-component spin-label spectrum.

The reduction in the deuterium quadrupole splitting has been ascribed to a disordering effect of the protein interface. In most membrane models present in the literature, the membrane proteins are drawn as smooth cylinders or rotational elipsoids. Even if the protein backbone is arranged in α-helical configuration, the protrusion of acid side-chains could lead to an uneven shape of the protein surface.

The observed disordering effect does not necessarily imply an increase in the configurational space available to the fatty acyl chains. In fact, it appears to be more probable that the total number of chain configurations is lowered, while the statistical probability of more distorted chain conformations increases at the same time (Seelig and Seelig, 1980).

From an increase in *spatial disorder* it is not concluded that the membrane is also *more fluid.* Deuterium T_1 relaxation time measurements suggest a decrease in the rate of segment reorientation in the presence of protein. Such deuterium T_1 measurements have been performed with cytochrome c oxidase and reconstituted sarcoplasmic reticulum. The addition of protein decreases the relaxation time in both cases. (Above T_c, the motion still falls into the fast correlation time regime as shown by the longer T_1 relaxation times at higher temperatures.) It has been suggested that shorter T_1 relaxation times are equivalent to an increase in the microviscosity. This conclusion is supported by ^{13}C NMR experiments with reconstituted sarcoplasmic reticulum (Stoffell *et al.*, 1977). The T_1 relaxation rates of ^{13}C-labelled lipids decrease continuously with increasing protein concentration in the membrane.

Fluorescent probe studies   Recent fluorescent probe studies of reconstituted systems are interesting with regard to protein arrangements in the plane of the lipid bilayer (Hoffmann *et al.*, 1981). A study of the polarization of the probe (in reconstituted systems of an intrinsic polypeptide gramicidin A or of various intrinsic proteins as the concentration of the intrinsic molecule increases) shows that the value of the polarization P reaches a limited value. Empirically, each of the curves has been observed to fit a simple exponential equation. The value of the exponent is in each equation related to the number of probe molecules or lipids which surround the intrinsic molecule.

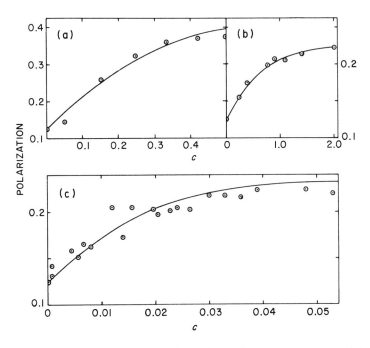

Fig. 10. Polarization of 1,6-diphenyl-1,3,5-hexatriene versus concentration of intrinsic molecules in DMPC at 36°C. The solid lines show the results of calculations as described in the text; the circles show the data. (a) DMPC + cholesterol; (b) DMPC + gramicidin; (c) DMPC + cytochrome oxidase. From Hoffmann *et al.* (1981).

A careful theoretical analysis using probability theory shows that the occurrence of such exponential curves is consistent with the occurrence of a random arrangement of intrinsic molecules. The probe molecule is markedly affected by the presence of the intrinsic molecule: $P_1 = e^{-Mx}$, where M is the number of DPH molecules which can be accommodated around the intrinsic molecule in half of the lipid bilayer and x is related to the concentration of intrinsic molecules in the lipid bilayer matrix. The dependence of $P(M,x)$ upon x reflects the fact that as the concentration of polypeptide or protein increases, the probability of (say) protein–protein contacts increases, and the number of lipid molecules and probe molecules which can contact the protein decreases (Fig. 10).

Spin-label data (Hoffmann *et al.*, 1981) are also observed to fit an exponential equation (Fig. 11) similar to that deduced from the fluorescence data. This could indicate that the number of spin probes which can contact the protein varies with concentration of the protein in the reconstituted system, i.e. since the spin probe is intended to mimic a lipid molecule, the number of lipid molecules which contact the protein varies with protein concentration. Thus the fluorescent probe and ESR probe data are useful for providing an estimate of the number of lipid chains which can contact the protein. The number of lipid chains necessary to surround the hydrophobic core of an intrinsic protein does not necessarily correspond to tightly bound lipid or specific lipid.

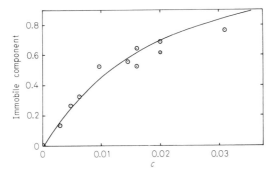

FIG. 11. The immobile component from ESR in DMPC + cytochrome oxidase using a nitroxide probe. Solid line, $1 - P(M,x)$ for $M = 60$. From Hoffmann *et al.* (1981).

Recent studies by Pink *et al.* (1981) consider the various contacts which may take place between intrinsic proteins and lipids as the concentration of proteins increases within the lipid bilayer matrix. These can consist of isolated lipids (not adjacent to any protein-free lipids) making a single contact with the protein (adjacent) and those lipids which are contacted by two or more proteins (trapped lipid) (Fig. 12).

Infrared spectroscopy studies Vibrational spectroscopy (Raman and infrared spectroscopy) is yet another powerful non-perturbing technique operating on a different time scale than ESR or NMR spectroscopy for determining the conformation of lipids and proteins and protein–lipid interactions (Wallach and Winzler, 1974; Wallach *et al.*, 1979). Indeed,

Raman spectroscopy has been used to study the relative population of the *gauche* and *trans* conformers and therefore the order of the lipid chain within lipid bilayer systems (Chapman *et al.*, 1977a; Wallach *et al.*, 1979). However, the high fluorescence levels obtained at high protein concentration of model and natural biomembranes make for difficulties in the general application of this approach. Fourier transform infrared spectroscopy has been used for the study of the influence of cholesterol on lipid structure (Cameron *et al.*, 1980; Umemura *et al.*, 1980) and conventional IR spectroscopy has also been applied to very simple model biomembrane systems (Asher and Levin, 1977; Chapman *et al.*, 1980; Cortijo and Chapman, 1981).

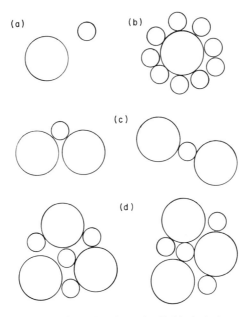

FIG. 12. (a) An isolated protein molecule and a lipid chain in population I. (b) An isolated protein surrounded by nine chains in population II. (c) A pair of protein molecules with one chain "trapped" in population III. (d) A triangle of protein molecules with three chains "trapped" in population III and one in population IV. From Pink *et al.* (1982).

Cortijo *et al.* (1982) recently reported studies of intrinsic protein–lipid interactions as well as membrane protein organization using infrared spectroscopy with an intrinsic polypeptide (gramicidin A) and various intrinsic proteins including Ca–ATPase and bacteriorhodopsin.

The absorption bands arising from water and buffer solutions were eliminated by means of an infrared spectrometer data station. Spectra were examined using H_2O and D_2O aqueous buffer systems. Pure lecithin–water systems, and various model biomembranes containing cholesterol, gramicidin A, bacteriorhodopsin or Ca^{2+}–ATPase were examined. The IR spectra of the reconstituted biomembranes were compared with those of the corresponding natural biomembranes, i.e. the purple membrane of *Halobacterium halobium* and also sarcoplasmic reticulum membranes respectively.

Changes in lipid chain conformation caused by the various intrinsic molecules incorporated within the model lipid bilayer structures were monitored by studying the shifts in frequency (cm^{-1}) of the CH_2 symmetric and asymmetric absorption bands arising from the lipid chains. The effect of gramicidin A and also the intrinsic proteins, as indicated by the shift of band frequencies, are quite different from that of cholesterol at temperatures above the main lipid T_c transition temperature. Cholesterol causes a reduction in *gauche* isomers which increases with concentration of cholesterol within the lipid bilayer.

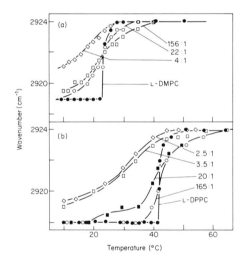

Fig. 13. Difference infrared spectra of (a) L-DPPC and (b) L-DPPC gramicidin A with a 2.5 : 1 (lipid/gramicidin) molar ratio dispersed in the H_2O buffer minus the buffer at 10°C (———) and 51°C (— — —). A flat subroutine was applied to the spectra in order to adjust the absorbances at 2970, 2890 and 2805 cm^{-1} to a parabola. Abex 10 (Cortijo *et al.*, 1982).

The effects observed on the band frequency by incorporation of gramicidin A into the lipid bilayer are more complex than those which occur due to the influence of cholesterol. It can be seen in Fig. 13 that gramicidin A at L/G higher than 10 ($\sim 20\%$ weight) show frequency–temperature profiles similar to those obtained by addition of cholesterol, and it can be deduced that the presence of the intrinsic polypeptide causes an increase in *gauche* isomers below T_c and a decrease in *gauche* isomers above T_c (see ratio of 20 : 1 in Fig. 13). However, when higher polypeptide concentrations are included within the lipid bilayer the band maximum frequencies, above T_c, now increase and can reach the same frequency as that of the pure lipid, i.e. the presence of the polypeptide does not cause a decrease in *gauche* isomers above T_c. The overall effect of cholesterol and low proportion of gramicidin A can be described as effecting the cooperativity of the transition decrease while T_c remains practically constant. However, the samples with high content of gramicidin A appear to show lower T_c than the pure lipid in addition to the observed decrease of phase cooperativity.

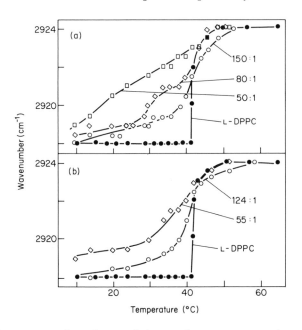

FIG. 14. Temperature dependence of the maximum wave number of the CH_2 asymmetric stretching vibrations in (a) L-DPPC-Ca^{2+}-ATPase and (b) L-DPPC-bacteriorhodopsin at the molar ratios indicated. The temperature dependence for the pure lipids (●) is also given (Cortijo *et al.*, 1982).

The effects observed by incorporation of the intrinsic proteins of bacteriorhodopsin of Ca^{2+}–ATPase into the lipid bilayer (Fig. 14) are more similar to those observed with gramicidin A rather than those observed with cholesterol, i.e. there is little evidence from the observed shifts of band frequency to indicate that a marked increase of *gauche* isomers occurs at temperatures above T_c due to the presence of high concentrations of intrinsic protein.

There is, however, some indication of such an effect when lower concentrations of intrinsic protein are present $\sim 100 : 1$ lipid–protein ratio.

Below the lipid T_c temperature the band frequencies are indicative of an increase in *gauche* isomers i.e. the presence of the intrinsic protein disturbs the *trans* packed crystalline lipid chains. Both proteins smear out the abrupt change of band frequency associated with the lipid phase transition (see Fig. 14). However, there is evidence at a ratio of 80 : 1 lipid to Ca–ATPase that at least two transitions occur: one is near 30°C; the other is centred around 41°C. A similar clear-cut second transition is not so obvious with the BR–lipid systems.

The frequencies of the methylene bands in the purified SR and purple membrane between 10 and 50°C correspond to those given by the lipids of L-DPPC and L-DMPC above their respective T_c temperature.

Acholeplasma laidlawii biomembranes Many studies have been made using a variety of techniques of the micro-organism *Acholeplasma laidlawii*. This micro-organism is capable of accepting particular fatty acids in its growth medium and studies have shown that the membranes of the micro-organism can be enriched up to 95% with a given fatty acid. The micro-organism exhibits a marked phase transition and this has been shown by calorimetric (Steim *et al.*, 1969; Chapman and Urbina, 1971) and X-ray methods (Engelman, 1970). The sharpness of the phase transition is determined by the heterogeneity of the lipid chains present in the membrane. Fourier transform infrared spectroscopy has been applied (Casal *et al.*, 1980) to a study of intact and deproteinated plasma membranes of *Acholeplasma laidlawii* enriched biosynthetically with perdeuteropalmitoyl chains. (The micro-organism readily incorporated deuterium labelled fatty acids into the membranes (Oldfield *et al.*, 1972b; Stockton *et al.*, 1977).)

The temperature-dependent behaviour was monitored via the C–D stretching modes and compared with that observed with a model biomembrane 1,2 di-perdeuteropalmitoyl-*sn*-glycero-3-phosphocholine. A broad transition is observed with the natural biomembrane consisting of two overlapping stages (Fig. 15).

Calorimetry studies When reconstituted protein–lipid systems are cooled to temperatures below the main lipid T_c transition temperature, on addition of protein the transition gradually broadens and the transition enthalpy decreases. At very high protein concentrations the phase transition may not be detectable at all (Curatolo *et al.*, 1977; Chapman *et al.*, 1977b; van Zoelen *et al.*, 1978; Mombers *et al.*, 1979). A broadening of the phase transition has also been confirmed by other methods such as, for example, fluorescence spectroscopy (Gomez-Fernandez *et al.*, 1980; Heyn, 1979).

Some workers have interpreted this observed reduction in enthalpy with protein concentration as a measure of the boundary lipid associated with the

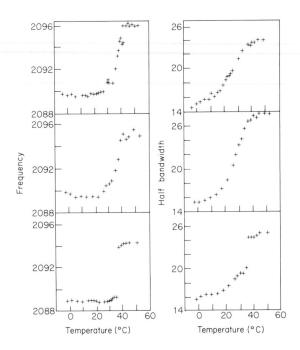

FIG. 15. (a) Temperature dependence of the frequency (in cm⁻¹) of the maximum of the CD_2 symmetric stretching vibration of the perdeuteriopalmitoyl chains in (A) intact plasma membranes, (B) deproteinated plasma membranes, and (C) DPPC-d_{62} model membranes. (b) Temperature dependence of the half-bandwidths (in cm⁻¹) of the symmetric CD_2 stretching vibration of the perdeuteriopalmitoyl chains in (A) intact plasma membranes, (B) deproteinated plasma membranes, and (C) DPPC-d_{62} model membranes (Casal *et al.*, 1980).

protein. This assumes that each individual protein remains separate (as it is above the lipid T_c transition temperature) during the lipid crystallization process. Boggs and Moscarello (1978) have attempted to determine how the amount of boundary lipid depends on fatty acid chain length with lecithin vesicles containing the hydrophobic protein from myelin proteolipid. These authors used scanning calorimetry and determined the enthalpy of the transition as a function of protein : lipid molar ratio. They suggest that 21–25 molecules correspond to boundary lipid for fatty acid chain lengths 14–18 carbons. The boundary lipid was, however, only 16 molecules per molecule of lipophilin for lipids containing fatty acids of chain length 12, or for one with a *trans* double bond. (Their calculations suggest that 70 phospholipids would surround the circumference of the lipophilin protein if both chains touch the protein.)

However, when the temperature of a reconstituted system is lowered below the lipid T_c transition temperature, the lipid chains crystallize, proteins are squeezed out, and patches are formed of high protein–lipid content. As yet more protein is included to the lipid system, the aggregated patch increases in size at the expense of the remaining crystalline region. This is why lipid–water systems show a broadening of the main melting transition and a lowering of the enthalpy as protein is incorporated. The decreased enthalpy value may correspond to an *average perturbed lipid.*

Where the intrinsic protein remains *isolated*, it is possible that the enthalpy reduction could provide information about the number of lipid molecules which *surround* the intrinsic hydrophobic segment of the protein; on the other hand, more than one layer might be perturbed.

In some cases, the proteins in the patches do not remain isolated. This appears to be the case with reconstituted purple membrane protein where a structure is formed similar to that which occurs in the membrane itself. In other cases, the high protein–lipid patches are associated with hexagonally packed lipid which can be detected by X-ray diffraction (Hoffmann et al., 1980). A marked increase of protein rotation can occur below the lipid T_c transition temperature. This is because a melting of the protein–lipid patches can be determined by the melting of the lipids in these patches. This can mean that protein rotation increases dramatically (Hoffmann et al., 1980), as does enzymatic activity sometimes at about 10°C below the transition temperature T_c of the pure lipid, e.g., at 30°C with ATPase incorporated with dipalmitoyl-lecithin–water systems ($T_c = 41$°C). A rapid increase of rotation of the ATPase protein at 30°C was observed using a triplet probe.

Non-lammellar structures Recent interest has been aroused in the role of those lipids which in water can form various types of non-lamellar structures such as the hexagonal phase. Many biological membranes contain these lipids (e.g. the phosphatidylethanolamines), and a question of some importance is which factors determine how these lipids contribute to form a stable lamellar structure in the natural biomembrane. Related to this question is that of the contribution made by the proteins to the stability of the lamellar structure.

Most biological membranes contain lipids that are able to adopt non-bilayer phases (Cullis and de Kruijff, 1979; de Kruijff et al., 1980) of which the hexagonal H_{II} phase is the predominant one. It has been suggested that a temporary occurrence of non-bilayer structures in biological membranes may serve to explain various functional abilities of those membranes such as fusion and transport processes, and it has been shown that temperature, lipid composition, pH and divalent cations may be important parameters. It has been suggested (Van Echteld, 1981) that membrane proteins may also influence the structural organization of the lipids. This is demonstrated by the apparent bilayer stabilizing role of the intrinsic membrane protein rhodopsin (De Grip et al., 1979). Using model membranes, it was observed that the extrinsic membrane protein cytochrome c specifically induced non-bilayer structures in cardiolipin-containing membranes (De Kruijff and Cullis, 1980b), whereas another basic polypeptide, poly-L-lysine, inhibits the formation of such structures. However, little is known about the influence of reconstituted intrinsic membrane proteins on lipid polymorphism. It has been found that glycophorin from human erythrocytes exerts a strong bilayer stabilizing influence when reconstituted with dioleoyl-phosphatidylethanolamine (Taraschi et al., 1980). Recent studies have been with gramicidin, the hydrophobic pentadecapeptide, which is a model for the hydrophobic segment of intrinsic membrane proteins and mimic signal peptides. It has been shown that the NH_2-terminal to NH_2-terminal π_6 (L,D) helical dimer model (Urry, 1971) for the organization of the aqueous pore forming gramicidin is the major conformation occurring in phosphatidylcholine bilayers (Weinstein et al., 1980).

When gramicidin is incorporated in dispersions of unsaturated phosphatidylethanolamines, it substantially decreases the bilayer to hexagonal H_{II} phase transition temperature (Van Echteld et al., 1981). Apparently gramicidin A possesses a strong hexagonal H_{II} phase-promoting ability. This observation is reinforced by the property of gramicidin to induce a hexagonal H_{II} phase organization even in dioleoylphosphatidylcholine, which is known to exclusively adopt a bilayer organization itself. In former studies on gramicidin with dimyristoylphosphatidylcholine and dipalmitoylphosphatidylcholine (1 : 10 molar ratio), only indications for bilayer organization have been observed (Chapman et al., 1977). This suggests that fatty acid chain lengths with corresponding hydrophobic volumes and mismatch with the length of the gramicidin dimer may be important parameters for the hexagonal H_{II} phase inducing property of gramicidin.

Van Echteld et al. (1981) have indeed shown that gramicidin A is able to induce a hexagonal H_{II} phase organization in diacylphosphatidylcholines in a fatty acid chain length-dependent way.

The chain length-dependent formation of the hexagonal H_{II} phase by gramicidin is not restricted to unsaturated diacylphosphatidylcholines. ^{31}P NMR spectra of dispersions of gramicidin with various liquid crystalline saturated diacylphosphatidylcholines in a 1 : 10 molar ratio are presented in Fig. 16. Again, the onset of the induction of the hexagonal H_{II} phase by gramicidin coincides with a fatty acid chain length of 18 carbon atoms. For the shorter chain lengths a complete lamellar phospholipid organization is found.

The onset of the fatty acid chain length-dependent formation of the hexagonal H_{II} phase by gramicidin has been shown to coincide with a length exceeding 16 carbon atoms long fatty acids. The bilayer thickness of liquid

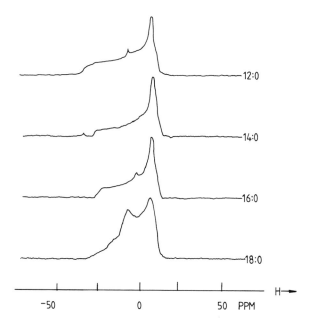

FIG. 16. Proton noise decoupled 81.0 MHz ^{31}P NMR spectra of aqueous dispersions of mixtures of gramicidin with various saturated liquid crystalline diacylphosphatidylcholines in a 1 : 10 molar ratio. Fatty acid composition as indicated in the figure. Recording temperatures: 12 : 0 at 30°C, 14 : 0 at 40°C, 16 : 0 at 50°C and 18 : 0 at 60°C (Van Echteld, 1982).

crystalline 16 : 0, 16 : 0-phosphatidylcholine may be estimated to be 4.2–4.3 nm (42–43 Å). Assuming the thickness of the polar part to be about the same as in the gel state, approximately 0.6 nm (6 Å), a thickness of the hydrophobic part results of 3.0–3.1 nm (30–31 Å). When incorporated in a lipid environment in peptide : lipid molar ratios < 1 : 5, gramicidin adopts a dimeric conformation which is different from the one in organic solvents. The length of the gramicidin channel in a lipid environment in the presence of monovalent cations seems larger than in the corresponding crystal structure (2.6 nm; 26 Å), but shorter than the crystal structure without monovalent cations. The apparently ion-dependent length of lipid associated gramicidin (Wallace et al., 1981) of approximately 3.0 nm (30 Å) remarkably concurs with the hydrophobic thickness.

When the hydrophobic part of the phospholipid molecules exceeds this length, apparently an unstable situation arises which results in the formation of a hexagonal H_{II} phase for part of the molecules. The longer the fatty acid chains, the more lipid molecules enter the hexagonal H_{II} phase. From the concentration-dependence of the gramicidin-induced hexagonal phase a gramicidin enrichment in this phase may be inferred.

Theoretical models   There have been a number of theoretical studies to estimate the extent to which an intrinsic protein may affect the surrounding lipid in a lipid bilayer structure. Marcelja (1976) has published a microscopic mean field model of order in lipid bilayers based on chain conformation. Schroeder (1977) described a method of incorporating lipid–protein inter-actions into a pre-existing mean field treatment of lipid bilayers. The protein acted formally as an external field on the lipids. In both the above studies, the attractive lipid-mediated interaction between two identical proteins was demonstrated.

Jahnig (1977) also has developed a microscopic mean field model based on chain conformation. Owicki *et al.* (1978) studied the order perturbation as a function of temperature and lateral pressure using Landau–de Gennes theory and a variational procedure. They conclude that for a given lateral pressure there is a temperature dependence and that the greater the amount of boundary layer is present at the lipid T_c transition temperature.

Pink and Chapman (1979) have used a lattice model and examined the lipid system where the proteins interact only via Van der Waals interactions and also systems where the proteins have bound or attached lipids on their circumference. These calculations have been used to examine the melting temperatures of eutectic protein–lipid patches.

VII. The Restriction of Movement and Fluidity of Biomembranes

There are a number of ways in which selective restriction of membrane component dynamics can be accomplished. There are various approaches:

(1) *Photo-oxidation of proteins*   The triplet probes commonly used to measure the rotational diffusion of membrane proteins are very potent agents for causing photo-oxidation of biological materials. Under normal circum-stances, these undesirable side effects can be avoided by keeping labelled samples in the dark. In addition, because the samples for flash photolysis are always kept under anaerobic conditions during the course of the measure-ments, no damage to the system under study is normally observed. We have recently attempted (Restall *et al.*, 1981) to make use of this property of photo-oxidation to study the role of the thiol groups in the Mg^{2+}–Ca^{2+}–ATPase from sarcoplasmic reticulum. Our recent studies have shown that after labelling the ATPase protein and illuminating the sample under anaero-bic conditions, a steady decrease in the ATPase activity is observed. Associ-ated with this loss of activity we have established, both from freeze-fracture electronmicroscopy and from measurements of the protein dynamics, that

aggregation of the proteins is occurring. Although the effects of oxidation on lipid composition and fluidity have been extensively studied, the effects of oxidation on membrane proteins have received little attention. These studies so far have all utilised the photo-oxidative properties of a triplet probe covalently attached to the protein. For future work other oxidizing agents may be used which may give greater selectivity in what is affected. In addition, it is of interest to examine the state of protein oxidation in certain pathogenic conditions, notably the protoporphyrias which are sometimes associated with photosensitivity and also as a function of tissue ageing.

(2) *Catalytic hydrogenation* New approaches aimed to markedly affect biomembrane fluidity have been introduced; the first of these is catalytic hydrogenation.

Homogeneous catalysts containing complexes of the transition metal, rhodium I have been used to hydrogenate *cis*-double bonds in the fatty acyl chains of phospholipids in cell and artificial membranes (Chapman and Quinn, 1976). This leads to decreases in bilayer fluidity which have been demonstrated to inhibit the function of transport proteins, enzymes and agglutination reactions occurring in the membranes. The early studies were performed with a catalyst which had to be dissolved in dimethylsulphoxide (DMSO) or tetrahydrofuran in order to introduce them into the system. This suffered from the dual disadvantages that the solvents themselves perturbed the membranes and also that once the catalyst was incorporated into the bilayer it was extremely difficult to remove. More recently, however, a water-soluble homogeneous catalyst has been devised which works with great success. This method now shows great promise as a tool for manipulating membrane fluidity provided that suitable conditions are available both to make the membrane susceptible to penetration, and hence action of the catalyst, and for removal of the catalyst after hydrogenation. Unfortunately, cholesterol severely reduces the amount of catalyst able to partition into the membrane. This, together with the ability of cholesterol to complex unsaturated phospholipids, reduces the degree of hydrogenation that can be achieved. It has the advantage, however, that it does not result in changes in the ratios of phospholipid species or in cholesterol content of the membranes and so is potentially an excellent method to study the fluidity effects on integral proteins in biological membranes.

(3) *Polymerization* It has recently been shown that polymerization and cross-linking of phospholipids can be accomplished using diacetylene groups in the acyl chains. Polymerization in liposomes and Langmuir-Blodgett films on various substrates has been accomplished (Johnston *et al.*, 1980; Albrecht *et al.*, 1982). The polymerization is triggered by UV radiation. The polymers are coloured because the polymerization process causes a conjugated linkage to be formed. The appearance of a new absorption band can

be seen when layers of phospholipid on quartz are irradiated for increasing lengths of time with UV radiation (see Fig. 17). Extensive polymerization has also been produced in certain natural biomembranes, e.g. *Acholeplasma laidlawii*, merely upon irradiation with UV radiation (Leaver *et al.*, 1983). In this case the diacetylenic fatty acid is incorporated in large amounts (90 %) biosynthetically into the biomembrane prior to irradiation.

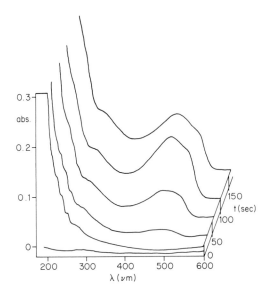

FIG. 17. Visible spectra of 86 layers (43 layers on each side of a quartz slide) of diacetylenic phosphatidylcholine at various irradiation times. After Albrecht *et al.* (1982).

(4) *Lipid peroxidation* Several biochemical changes concomitant with, or due to, lipid peroxidation have been investigated (Plaa and Witschi, 1976; Chance *et al.*, 1979). The mechanism of microsomal lipid peroxidation has been extensively studied (Svingen *et al.*, 1979). The first step is the abstraction of a hydrogen atom from an allylic methylene group, resulting in a lipid radical. Subsequent chain reactions result in a breakdown of polyunsaturated fatty acids and the production of stable end products (ethane, propane, pentane), and also the production of malondialdehyde. Spectral studies of fluorescent probes (Dobretsov *et al.*, 1977) indicated that lipid peroxidation decreases the fluidity of microsomal membranes, whereas Eichenberger *et al.* (1982), using steady state fluorescence anisotropy of diphenylhexatriene, showed that lipid peroxidation in rat liver microsomes increases the order of the microsomal phospholipid acyl chains. This increase in order might be due to the formation of covalent bonds between adjacent lipid radicals.

The treatment of erythrocyte membranes with phenylhydrazine also causes the peroxidation of unsaturated fatty acids in endogenous membrane phospholipids and this results in a decreased bulk lipid fluidity compared with normal untreated membranes (Rice-Evans and Hochstein, 1981). It was proposed that changes in the physical state of plasma membranes subsequent to the peroxidation of membrane lipids may contribute to a loss of erythrocyte deformability and to enhanced sequestration during normal and accelerated ageing.

(5) *Anaesthetics* A property that local anaesthetics have is the ability to perturb bilayer fluidity. The neutral, local anaesthetics are particularly useful, as their neutrality precludes any electrostatic interactions with either the phospholipids or the proteins in the membranes. Examples of the compounds which act to increase bilayer fluidity are benzyl alcohol, short-chain ($< C10$) *n*-alkanols and carbocaine. These molecules are all soluble to various extents in water and are thus readily added to test systems, where they partition into the membrane bilayer. Washing the treated membranes with buffer readily removes the anaesthetic, allowing the reversibility of any effects to be demonstrated.

Treatment of synthetic phospholipid vesicles with such compounds causes a characteristic depression of the lipid phase transition by virtue of the fact that they act as an impurity in the bilayer achieving a "classical depression of the freezing point". In biological membranes they depress the temperature at which lipid phase separations occur and increase the fluidity of the bilayer. Benzyl alcohol (40 mM), for example, increases the fluidity of rat liver plasma membranes by an amount equivalent to a 6°C rise in temperature. Such an increase in bilayer fluidity caused by benzyl alcohol has very different effects on various integral proteins associated with this membrane.

Charged anaesthetics possess the ability to perturb one or other half of the bilayer selectively since in the majority of cases, the acidic phospholipids predominate in the inner facing half of the lipid bilayer. Thus cationic anaesthetics preferentially act on one half of the bilayer, whereas the anionic anaesthetics will preferentially act on the other half. This asymmetric perturbation of the bilayer can lead to changes both in the shape of cells and also in the direction in which cilia beat in the *Paramecium*. In the latter case it is the interaction of cationic anaesthetics with the inner surface of the plasma membrane that opens a voltage-sensitive Ca^{2+}-gate. This leads to Ca^{2+} influx and ciliary reversal. However, anionic anaesthetics are ineffective. Ciliary reversal cannot be effected in the *Paracemium Pawn* mutant because this organism lacks a functional voltage-sensitive Ca^{2+}-gate oriented within its membrane. The ability to fluidize preferentially one or other half of the bilayer yields a tool for selectively modulating the activity of particular proteins. An interesting example of this is adenylate cyclase, which is inserted asymmetrically in the plasma membrane with its functioning, globular portion localized at the cytosol side, where the negatively charged phospholipids predominate. Under these conditions cationic anaesthetics activate the enzyme by increasing the fluidity, whereas anionic ones have no effect.

VIII. Alterations of Membrane Fluidity in Certain Selected Systems

There are many reports of changes in membrane fluidity induced by physical and chemical factors, exogenous agents or experimental manipulations of membrane lipid composition. Changes in membrane fluidity occurring in a number of diseases have also been described.

Hydrostatic pressure was found to affect membrane lipid fluidity as indicated by the linear increase in the T_c temperature of phospholipids with increasing hydrostatic pressure (Centerick et al., 1978). Moreover, the well-established lipid fluidizing effect of general anaesthetics (Seeman, 1972) can be reversed by pressure, presumably via a similar mechanism (Johnson and Miller, 1970; Trudell et al., 1973).

Osmotic pressure causes an increase of membrane fluidity in liposomes and plant protoplasts (Borochov and Borochov, 1979). On the contrary, fluidity decreases with membrane potential in a nonlinear manner, irrespective of the potential direction (Corda et al., 1982).

Interferon influences biomembrane fluidity. Pretreatment of human foreskin cells with homologous interferon significantly inhibits cell fusion induced by *Sendai* virus, and this refractory state is accompanied by a decrease in cell plasma membrane fluidity (Chatterjee et al., 1982). On the contrary, a significant increase of the plasma membrane fluidity was observed in mouse-fibroblasts upon treatment with mouse β interferon. The effect is dose dependent and is an early one (maximum for a 30 min treatment), and may be directly related to the initiation of antiviral activity (Kuhry et al., 1983).

The membrane fluidity of synaptosomal membranes was decreased by *in vitro* incubation with either cholesterol hemisuccinate or stearic acid, resulting in an up to 5-fold increase in the specific binding of serotonin. Fluidization of membrane lipids by treatment with lecithin or linoleic acid caused a small but significant decrease in serotonin binding. These changes were considered compatible with the concept of vertical displacement of membrane proteins influencing the accessibility of serotonin binding sites (Heron et al., 1980).

The normal alcohols have commonly been used as membrane perturbing agents. Like most other anaesthetics they expand the membrane surface area and cause a fluidization or disordering of the membrane lipids. These effects have been shown in various membranes with the use of spin labels (Lenaz et al., 1976) or fluorescent probes (Zavoico and Kutchai, 1980). In artificial membranes a broadening and a lowering of the phase transition was induced by normal alcohols (Eliasz et al., 1976).

The effects of chronic ethanol consumption on the physical properties of liver mitochondrial membranes have been investigated by spin labelling ESR (Waring et al., 1982). The increased rigidity of membranes was considered to be an adaptive adjustment of the phospholipid composition of mithochondrial membranes.

A number of studies have been dedicated to changes in membrane fluidity in various physiological circumstances or to differences among species.

Spin label studies of lipid fluidity and organization in the mitochondria of the brown adipose tissue from warm and cold-adapted rats (nonhibernators) and hamsters (hibernators) indicated that the degree of order is slightly lower, and the fluidity greater in the lipids of rats relative to those of hamsters (Cannon et al., 1975). There was apparently no connection between the fluidity of lipids in the mitochondria of brown adipose tissue and the ability of hamsters to lower their body temperature on cold-adaptation.

Ageing of rats was found to be associated with a progressive increase in the molar ratio of cholesterol/phospholipid in hepatic mitochondria. A highly significant ageing-dependent increase in the lipid structural order parameter was found in the hepatic mitochondria isolated from the aged animals which correlated well the lipid alterations (Vorbeck et al., 1982).

In contrast to many studies of rat liver mitochondria investigations on subcellular fractions from human liver are scarce. Benga et al. (1978) found a peculiar composition of human liver mitochondria: their membranes contain twice as much lipid as those from the rat and a significantly higher fraction of hydrophobic amino acids in proteins (Benga and Ferdinand, 1977). Interesting differences in essential fatty acid composition of total lipids and phospholipids between rat and human liver mitochondria were found: human liver mitochondria contain more linoleic acid and less arachidonic acid than those of the rat; such a pattern of distribution of fatty acids in liver mitochondria has not been reported for any other species. When spin labelled lipids were used to probe the fluidity in human and rat liver mitochondrial membranes, a higher fluidity in human membranes was found.

The lipid apparent microviscosity of the rat liver microsomal membrane on the first day after birth was found to be half of that observed on the last day of life (Kapitulnik et al., 1979). This remarkable perinatal fluidization of the membrane resulted from a marked increase in the molar ratio of phospholipids to cholesterol. A similar increase in membrane fluidity was also reported for the plasma membrane of chick heart between the last day of embryonic life and adulthood (Kutchai et al., 1976). It was suggested that the changes in membrane fluidity may contribute to developmental changes in the uptake of sugars, amino acids, and urea by heart cells.

Detailed lipid analyses of human and rat liver microsomes revealed interesting differences (Benga et al., 1983b). Human liver microsomes contain twice as much lipid as those from the rat. This increased lipid content is not associated with an increase in content of a particular lipid class; human and rat liver microsomes differ especially in the essential fatty acid composition of total lipids and phospholipids: human liver microsomes contain more linoleic acid and less arachidonic acid than those of the rat. Such a pattern of distribution of fatty acids is similar to that previously reported for human liver mitochondria and has not been reported for other species. Although the unsaturation of lipids is lower in human than in rat liver microsomes, spin label studies revealed a higher fluidity in human membranes. It is suggested that this might arise from a lesser immobilization of lipids by proteins in human liver subcellular membranes.

Manipulations of lipid composition in a variety of membranes can be induced in different ways. Variations in environmental temperature produce changes in lipid composition and fluidity of cell membranes. Marr and Ingraham (1962) have shown that the relative amounts of fatty acids found in phospholipids of membranes can be altered by growing bacteria at different temperatures, and this alteration of the fatty acids leads to modifications in the microviscosity of the membranes (Miller and Koshland, 1977). The huge literature concerning regulation of membrane fluidity in prokaryotes has recently been reviewed (Melchior, 1982). Therefore we will only mention here the mechanisms for temperature modulation of membrane fatty acid composition. The composition is responsive to temperature in such a way that the membrane is totally or almost totally fluid at the temperature of growth. This necessity requires a control mechanism that senses temperature and the physical state of the membrane and directs the incorporation of proportionally more unsaturated or other low-melted fatty acids into membrane lipids as the temperature decreases. The suggestion that membrane fluidity is maintained within narrow limits (Chapman et al., 1966) has been supported by data regarding the control of activity of membrane-bound fatty acid desaturases. It has become clear that control can take place at several, possibly interrelated levels. In some cases enzyme synthesis is affected by temperature. Fatty acid desaturase is not synthesized in B. megaterium (Fulco, 1970) at 35°C but is strongly induced at 20°C. Temperature may have a direct effect on the protein itself, which, once synthesized at low temperature, undergoes rapid irreversible inactivation at higher temperatures. In addition to effects of temperature on the biosynthesis of unsaturated fatty acids, another level of control appears to be at the site of phospholipid syntheses in the membrane. Temperature-dependent selection of saturated and unsaturated fatty acid CoA by membrane-bound acyl-transferase, which catalyses the esterification of glycerophosphate has been demonstrated by Sinensky (1971).

It should be emphasized that a feature of the temperature-dependent selection process at the membrane level is that the fatty acids appear to be selected on the basis of melting point, a thermodynamic property that only indirectly reflects molecular structure. Although unsaturation is the usual route to low melting point, the same goal is attained in many bacteria by employing branched chains in Gram-positive bacteria or by incorporation of shorter chain fatty acids into bacterial membrane lipids (Melchior, 1982).

The protozoan *Tetrahymena pyriformis* has been profitably exploited for studying the mechanism of cellular acclimation to low environmental temperatures. It is believed that a vital response to chilling in *Tetrahymena*, as in many other organisms, is the alteration of membrane lipid composition so as to overcome the rigidifying effect that low temperature invariably has on membrane lipids (Thompson, 1980). In recent studies Dickens and Thompson (1982) showed that in *Tetrahymena pyriformis* cells chilled from 39 to 15°C, fatty acids of the microsomal membrane phospholipids increased significantly in unsaturation over a 15 h acclimation period. However, during the first hour of exposure to 15°C the extent of fatty acyl group rearrangement by deacylation–reacylation overshadowed acyl chain desaturation as a means of altering lipid structure. In the light of these findings, selective phospholipid deacylation–reacylation is indicated as a mechanism which may be of pivotal importance in achieving rapid homeoviscous adaptation.

Supplementation of culture medium with specific fatty acids is another way of manipulating the fatty acid composition of cell membranes. One animal cell system that has received considerable attention has been LM cells. The fluidity of mitochondrial, microsomal, and plasma membranes isolated from LM cells grown in tissue culture was determined with a fluorescent probe (Gilmore et al., 1979). When fatty acid composition was altered by adding linoleate to the growth medium, a decrease in the rotational relaxation time of DPH in all three membrane fractions occurred. However, approximately the same ratios of the rotational relaxation times of DPH were maintained between the different membranes. This suggests that this parameter is closely regulated.

Nutritional methods may be used to alter lipid composition of the tissues of rats and to study the consequences upon membrane fluidity and activity of various enzymes. Decreased sterol biosynthesis *in vivo* was elicited in rat intestinal microvillus membranes by feeding sodium taurocholate or by fasting the rats, whereas increased synthesis has been induced by biliary ligation or feeding cholestyramine, a bile salt binding resin.

Cholesterol-rich erythrocytes may be prepared by incubation with cholesterol-rich phospholipid dispersions. The resulting cells have several abnormal properties including decreased membrane fluidity (Cooper et al., 1978). In contrast, an increase in phosphatidylcholine/sphingomyelin ratio is associated with increased erythrocyte membrane fluidity (Cooper et al., 1977).

Alterations of membrane fluidity of human erythrocytes have been reported in a variety of diseases. In patients with liver disease, abnormalities in the composition of plasma lipoproteins are associated with corresponding changes in the erythrocyte membrane lipid composition. The membranes are enriched in cholesterol and phosphatidylcholine and both the cholesterol phospholipid and phosphatidylcholine/sphingomyelin molar ratios are raised. Membrane fluidity is significantly decreased in such patients. The fluidity of lipid from the membranes of patient erythrocyte was also decreased, suggesting that decreased membrane fluidity was a consequence of altered lipid composition rather than protein abnormalities (Owen et al., 1982).

Several recent developments have stimulated membrane studies in genetic neuropsychiatric diseases. One development has been the emerging concept that some neuropsychiatric diseases may be associated with membrane disorders. Second, the awareness that the genetic defect may be present in all cells has promoted the investigation of easily accessible peripheral tissues as model systems. Recent reviews of studies of membranes of human fibroblasts, lymphocytes and erythrocytes in genetic neuropsychiatric diseases have been published (Pettergrew et al., 1981; Butterfield, 1981).

IX. Conclusions

The concept of membrane fluidity has been and continues to be a useful one with its emphasis on the dynamic character of biomembrane structure. This is quite distinct from the previous static structure proposed and emphasized on the basis of electron microscope studies. It has revolutionized our thinking and understanding of many biomembrane functions and may prove valuable for understanding certain disease conditions. Some of these implications are discussed in other chapters in this book.

X. Acknowledgements

We wish to thank the Wellcome Trust, the Medical Research Council (UK) and the Academy of Medical Sciences (Romania) for financial support and to thank many research colleagues for helpful discussions and collaboration on our studies of biomembrane structure and function.

References

Akutsu, H. and Seelig, J. (1981). *Biochemistry* **20**, 7366–7373.
Albrecht, O., Gruler, H. and Sackmann, E. (1978). *J. Phys (Paris)*. **39**, 301–313.
Albrecht, O., Gruler, H. and Sackmann, E. (1981). *J. Colloid Interface Sci.* **79**, 319.
Albrecht, O., Johnston, D. S., Villaverde, C. and Chapman, D. (1982). *Biochim. Biophys. Acta* **687**, 165–169.
Alonso, A., Restall, C. J., Clark, A. D., Turner, M., Gomez-Fernandez, J. C. and Chapman, D. (1982). *Biochim. Biophys. Acta* **689**, 283.

Asher, I. M. and Levin, I. W. (1977). *Biochim. Biophys. Acta* **468**, 63.
Austin, R. H., Chan, S. S. and Jovin, T. M. (1979). *Proc. Natl Acad. Sci. USA* **76**, 5650–5654.
Bailey, R. T. (1974). *Molec. Spectroscopy* **2**, 273.
Baroin, A., Thomas, D. D., Osborne, B. and Devaux, P. (1977) *Biochem. Biophys. Res. Commun.* **78**, 442–447.
Barton, P. G. (1968). *J. Biol. Chem.* **243**, 3884–3890.
Barratt, M. D., Green, D. K. and Chapman, D. (1969). *Chem. Phys. Lipids* **3**, 140–144.
Benga, Gh. and Chapman, D. (1976). *Rev. Roum. Biochim.* **13**, 251–261.
Benga, Gh. and Ferdinand, W. (1977) *Int. J. Biochem.* **8**, 17–20.
Benga, Gh., Hodarnau, A., Bohm, B., Borza, V., Tilinca, R., Dancea, S. and Petrescu, I. (1978) *Europ. J. Biochem.* **84**, 625–633.
Benga, Gh., Hodarnau, A., Tilinca, R., Portni, D., Dancea, S., Pop, V. and Wrigglesworth, J. (1979). *J. Cell Sci.* **35**, 417–429.
Benga, Gh., Hodarnau, A., Ionescu, M., Pop, V. I., Frangopol, P. T., Strujan, V., Holmes, R. and Kummerow, F. A. (1983). *Ann. NY Acad. Sci.* **414**, 140–152.
Benga Gh., Pop, V. I., Ionescu, M., Hodarnau, A., Tilinca, R. and Frangopol, P. T. (1983b). *Biochim. Biophys. Acta* **750**, 194–199.
Benga, Gh., Porumb, T. and Wrigglesworth, J. M. (1981) *J. Bioenergetics Biomembranes* **13**, 269–284.
Berlin, R. D. and Ukena, T. E., (1972) *Nature (London)* **238**, 120–122.
Blaustein, M. P. (1967) *Biochim. Biophys. Acta* **135**, 653–668.
Blok, M. C., Van Deenen, L. L. M. and de Gier, J. (1977). *Biochim. Biophys. Acta* **464**, 509.
Boggs, J. M. and Moscarello, M. A. (1978). *Biochemistry* **17**, 5374–5379.
Borochov, A. and Borochov, H. (1979). *Biochim. Biophys. Acta* **550**, 546–549.
Bretscher, M. S. (1980). *TIBS* October, pp. vi and vii.
Brown, M. F., Seelig, J. and Haberlen, U. (1979) *J. Chem. Phys.* **70**, 5045–5053.
Browning, J. L. and Seelig, J. (1980) *Biochemistry* **19**, 1262–1270.
Butler, K. W., Smith, I. C. P. and Schneider, H. (1970). *Biochim. Biophys. Acta* **219**, 514–517.
Butler, K. W., Tattrie, N. H. and Smith, I. C. P. (1974) *Biochim. Biophys. Acta* **363**, 351–360.
Butterfield, D. A. (1981). *J. Neurol. Sci.* **52**, 61–67.
Byrne, P. and Chapman, D., (1964). *Nature (London)* **202**, 987–988.
Cameron, D. G., Casal, H. L. and Mantsch, H. H. (1980). *Biochemistry* **19**, 3665
Cannon, B., Polnaszek, K. W., Eriksson, L. E. G. and Smith, I. C. P. (1975). *Arch. Biochem. Biophys.* **167**, 505–518.
Casal, H. L., Smith, I. C. P., Cameron, D. G. and Mantsch, H. H. (1979). *Biochim. Biophys. Acta* **550**, 145.
Casal, H. L., Cameron, D. G., Smith I. C. P. and Mantsch, H. H., (1980). *Biochemistry* **19**, 444.
Centerick, F., Peeters, J., Heremans, K., De Smedt, H. and Olbrechts, H., (1978). *Europ. J. Biochem.* **87**, 401–407.
Chance, B., Sies, H. and Boveris, A. (1979). *Physiol. Rev.* **59**, 527–605.
Chapman, D. (1958). *J. Chem. Soc.* **152**, 784–789.
Chapman, D. and Penkett, S. A., (1966). *Nature (London)* **211**, 1304–1305.
Chapman, D. and Quinn, P. J., (1976) *Proc. Natl Acad. Sci. USA* **73**, 3971–3975.
Chapman, D. and Salsbury, N. K., (1966). *Trans. Farad. Soc.* **62**, 2607–2621.
Chapman, D., Byrne, P. and Shipley, G. G., (1966). *Proc. Roy. Soc. Ser. A* **290**, 115–142.

Chapman, D., Williams, R. M. and Ladbrooke, B. D. (1967) *Chem. Phys. Lipids* **1**, 445–475.

Chapman, D., Owen, N. F., Phillips, M. C. and Walker, D. A. (1969). *Biochim. Biophys. Acta*, **183**, 458–465.

Chapman, D. and Urbina, J. (1971). *FEBS Lett.* **12**, 169.

Chapman, D., Keough, K. and Urbina, J. (1974). *J. Biol. Chem.* **249**, 2512–2521.

Chapman, D., Cornell, B. A., Eliasz, A. W. and Perry, A. (1977a). *J. Mol. Biol.* **113**, 517–538.

Chapman, D., Kingston, B., Peel, W. E. and Lilley, T. H. (1977b). *Biochim. Biophys. Acta* **464**, 260–275.

Chapman, D., Gomez-Fernandez, J. C. and Goni, F. M., (1979). *FEBS Lett.* **98**, 211–223.

Chapman, D., Gomez-Fernandez, J. C., Goni, F. M. and Barnard, M. (1980). *J. Biochem. Biophys. Methods* **2**, 315–323.

Chatterjee, S., Cheung, H. C. and Hunter, E. (1982). *Proc. Natl Acad. Sci. USA* **79**, 835–839.

Cherry, R. J., Cogoli, A., Oppliger, M., Schneider, G. and Parish, G. R. (1976). *Nature (London)* **263**, 389–393.

Clowes, A., Cherry, R. J. and Chapman, D. (1971). *Biochim. Biophys. Acta* **249**, 301–307.

Cone, R. A. (1972) *Nature (London)* **236**, 39–43.

Cooper, R. (1981). "Membrane Fluidity and Cell Surface Receptor Mobility." U.S. Dept. of Health.

Cooper, R. A., Leslie, M. H., Fischkoff, S. and Shinitzky, M., (1978) *Biochemistry* **17**, 327–331.

Cooper, R. A., Durocher, R. J. and Leslie, M. H., (1977). *J. Clin. Invest.* **60**, 115–121.

Corda, D., Pasternak, C. and Shinitzky, M., (1982). *J. Membrane Biology* **65**, 235–242.

Cortijo, M., Alonso, A., Gomez-Fernandez, J. C. and Chapman, D. (1982). *J. Mol. Biol.* **157**, 597.

Cortijo, M. and Chapman, D. (1981). *FEBS Lett.* **131**, 245.

Cullis, P. R. and De Kruijff, B. (1979). *Biochim. Biophys. Acta* **551**, 399–420.

Curatolo, W., Sakura, J. D., Small, D. M. and Shipley, G. G. (1977). *Biochemistry* **16**, 2312–2319.

Davis, J. H., Nichol, C. P., Weeks, G. and Bloom, M. (1979). *Biochemistry* **18**, 2103–2112.

De Grip, W. J. and Bovee-Geurts, P. H. M. (1979). *Chem. Phys. Lipids* **23**, 321–335.

De Kruijff, B., Cullis, P. R. and Verkleij, A. J. (1980). *Trends Biochem. Sci.* **5**, 79–83.

De Kruijff, B. and Cullis, P. R. (1980a). *Biochim. Biophys. Acta* **602**, 477–490.

De Kruijff, B. and Cullis, P. R. (1980b). *Biochim. Biophys. Acta* **601**, 235–240.

Devaux, P. F. and McConnell, H. M. (1972). *J. Am. Chem. Soc.* **94**, 4475–4481.

Dickens, B. F. and Thompson, G. A. (1982). *Biochemistry* **21**, 3604–3611.

Dobretson, G. E., Borschevskaya, T. A., Petrov, V. A. and Vladimirov, Y. A. (1977). *FEBS Lett.* **84**, 125–128.

Ehrstrom, M., Eriksonn, L. E. G., Israelachvili, J. and Ehrenberg, A. (1973). *Biochem. Biophys. Res. Commun.* **55**, 396–402.

Eichenberger, K., Bohni, P., Winterhalter, K. H., Kawato, S. and Richter, C. (1982). *FEBS Lett.* **142**, 59–62.

Eliasz, A. W., Chapman, D. and Ewing, D. F., (1976). *Biochim. Biophys. Acta* **448**, 220–233.

Engleman, D. M. (1970). *J. Mol. Biol.* **47**, 115.

Engleman, D. M. and Rothman, J. E., (1972). *J. Biol. Chem.* **247**, 3694–3697.

Evans, E. and Kwok, R. (1982). *Biochemistry* **21**, 4874.

Forslind, E. and Kjellander, R. (1975). *J. Theror. Biol.* **51**, 97–109.

Fulco, A. J. (1973). *Annu. Rev. Biochem.* **43**, 215–241.

Gally, H. U., Pluschke, G., Overath, P. and Seelig, J. (1980). *Biochemistry* **19**, 1638–1643.

Garland, P. B. and Moore, C. H. (1979). *Biochem. J.* **183**, 561–572.

Gilmore, R., Cohn, N. and Glaser, M. (1979). *Biochemistry* **18**, 1042–1049.

Godin, D. V. and Ng, T. W. (1974). *Mol. Pharmacol.* **9**, 802–819.

Gomez-Fernandez, J. C., Goni, F. M., Bach, D., Restall, C. and Chapman, D. (1980). *Biochim. Biophys. Acta* **598**, 502–516.

Gottlieb, M. H. and Eanes, E. D. (1972). *Biophys. J.* **12**, 1533–1548.

Gulik-Krzywicki, T., Rivas, E. and Luzzati, V. (1967). *J. Mol. Biol.* **27**, 303–332.

Harkins, W. D. and Anderson, T. F. (1937). *J. Am. Chem. Soc.* **59**, 2189–2197

Henderson, R. (1975). *J. Mol. Biol.* **93**, 123–138.

Heron, D. S., Shinitzky, M., Hershkovitz, M. and Samuel, D. (1980). *Proc. Natl Acad. Sci. USA* **77**, 7463–7467.

Heyn, M. P. (1979). *FEBS Lett.* **108**, 359–364.

Hinz, H. J. and Sturtevant, J. M. (1972). *J. Biol. Chem.* **247**, 3697–3700.

Hoffmann, W., Sarzala, M. G. and Chapman, D. (1979). *Proc. Natl Acad. Sci. USA* **76**, 3860–3864.

Hoffmann, W., Sarzala, M. G., Gomez-Fernandez, J. C., Goni, F. M., Restall, C. J., Chapman, D., Heppeler, G. and Kreutz, W. (1980). *J. Mol. Biol.* **141**, 119–132.

Hoffman, W., Pink, D. A., Restall, C. and Chapman, D. (1981) *Europ. J. Biochem.* **114**, 585–589.

Hubbell, W. L. and McConnell, H. M. (1971). *J. Am. Chem. Soc.* **93**, 314–326.

Ito, T. and Ohnishi, S. (1974). *Biochim. Biophys. Acta* **352**, 29–37.

Jacobson, K. and Papahadjopoulos, D. (1975) *Biochemistry* **14**, 152.

Jahnig, F. (1977). Dissertation, Max-Planck-Institut für Biophysikalische Chemie, Gottingen-Nikolausberg, FRG.

Jahnig, F. (1981). *Mol. Cryst. Liq. Cryst.* **63**, 157–170.

Johnson, S. M. and Miller, K. W. (1970). *Nature* **228**, 75–76.

Johnston, D. S., Sanghera, S., Pons, M. and Chapman, D. (1980). *Biochim. Biophys. Acta* **602**, 57–69.

Jost, P., Griffith, O. H., Capaldi, R. A. and Vanderkooi, G. (1973). *Proc. Natl Acad. Sci. USA* **70**, 480–484.

Jost, P., Nadakavukaren, K. K. and Griffith, O. H. (1977). *Biochemistry* **16**, 3110–3114.

Junge, W. and Devault, D. (1975). *Biochim. Biophys. Acta* **408**, 200–214.

Kapitulnik, J., Tsherdshedsky, M. and Barenholz, Y. (1979). *Science* **206**, 843–844.

Karnovsky, M. J., Kleinfeld, A. M., Hoover, R. L. and Klavsner, R. D. (1982). *J. Cell. Biol.* **94**, 1–6.

Kirino, Y., Ohkuma, T. and Shimizu, H. (1978). *J. Biochem.* **84**, 111–115.

Kuhry, J. G., Poindron, Ph. and Laustriat, G. (1983). *Biochem. Biophys. Res. Commun.* **110**, 88–95.

Kuo, A. L. and Wade, C. G. (1979). *Biochemistry* **18**, 2300–2308.

Kutchai, H., Barenholz, Y., Ross, T. F. and Wermer, D. E. (1976). *Biochim. Biophys. Acta* **436**, 101.

Ladbrooke, B. D. and Chapman, D. (1969). *Chem. Phys. Lipids* **3**, 304–367.

Ladbrooke, B. D., Williams, R. M. and Chapman, D. (1968). *Biochim. Biophys. Acta* **150**, 333–340.

Leaver, J., Alonso, A., Durrani, A. and Chapman, D. (1983). *Biochim. Biophys. Acta* **727**, 327–332.

Lecuyer, H. and Dervichian, D. G. (1969). *J. Mol. Biol.* **45**, 39–57.

Lenaz, G., Bertoli, E., Curatola, G., Mazzanti, L. and Bigi, A. (1976). *Arch. Biochem. Biophys.* **172**, 278–288.

Liebman, P. A. and Entine, G. (1974). *Science* **185**, 457–469.

Luzzati, V. (1968). *In* "Biological Membranes" (ed. D. Chapman) p. 71. Academic Press, London, Orlando and New York.

Mabrey, S., Mateo, P. L. and Sturtevant, J. M. (1978). *Biochemistry* **17**, 2464–2468.

McCown, T. M., Evans, E., Diehl, S. and Craig Wiles, H. (1981). *Biochemistry* **20**, 3134.

Mantsch, H. H., Martin, A. and Cameron, D. G. (1981). *Biochemistry* **20**, 3138.

Marcelja, S. (1974). *Biochim. Biophys. Acta* **367**, 165–176.

Marcelja, S. (1976). *Biochim. Biophys. Acta* **455**, 1–7.

Marr, A. G. and Ingraham, S. L. (1962). *J. Bacteriol.* **84**, 1260–1267.

Martin, R. B. and Yeagle, P. L. (1978). *Lipids* **13**, 594–597.

Melchior, D. L. (1982). *In* "Current Topics in Membranes and Transport", (eds F. Bronner and A. Kleinzeller), pp. 263–316. Academic Press, New York and London.

Miller, J. B. and Koshland, D. E. (1977). *J. Mol. Biol.* **111**, 183.

Mombers, C., Verkleij, A. J., de Gier, J. and Van Deenen, L. L. M. (1979). *Biochim. Biophys. Acta* **551**, 271–281.

Moore, C. H. and Garland, P. B. (1979). *Biochem. Soc. Trans.* **7**, 945–946.

Naqvi, R. K., Gonzalez-Rodriguez, J., Cherry, R. J. and Chapman, D. (1973). *Nature (London)* **245**, 249–251.

Naqvi, R. K., Behr, J. P. and Chapman, D. (1974). *Chem. Phys. Letts* **26**, 440–444.

Ness, W. R. (1974). *Lipids* **9**, 596–612.

Ohki, S. (1979). *Biophys. J.* **9**, 1195–1205.

Oldfield, E. and Chapman, D. (1972). *FEBS Lett.* **23**, (3), 285–297.

Oldfield, E., Chapman, D. and Derbyshire, W. (1971). *FEBS Lett.* **16**, 102.

Oldfield, E., Keough, K. and Chapman, D. (1972a). *FEBS Lett.* **20**, 344–346.

Oldfield, E., Chapman, D. and Derbyshire, W. (1972b). *Chem. Phys. Lipids* **9**, 69.

Oldfield, E., Gilmore, R., Glaser, M., Gutowsky, H. S., Hshung, J. C., Kang, S. U., King, T. E., Meadows, M. and Rice, D. (1978a). *Proc. Natl Acad. Sci. USA* **75** 4657–4660.

Oldfield, E., Meadows, M., Rice, D. and Jacobs, R. (1978b). *Biochemistry* **17**, 2727–2740.

Owen, J. S., Bruckdorfer, K. R., Day, R. and McIntyre, N. (1982). *J. Lipid Research* **23**, 124–132.

Owicki, J. C., Springgate, M. W. and McConnell, H. M. (1978). *Proc. Natl Acad: Sci. USA* **75**, 1616–1619.

Papahadjopoulos, D. (1968). *Biochim. Biophys. Acta* **311**, 330–348.

Payens, T. A. (1955). *Philips Res. Rep.* **10**, 425.

Peng, S. K., Tham, P., Taylor, C. B. and Mikkelson, B. (1979). *Am. J. Clin. Nutr.* **32**, 1033–1042.

Penkett, S. A., Flook, A. G. and Chapman, D. (1968). *Chem. Phys. Lipids* **2**, 273–290.

Pettergrew, J. W., Nichols, J. S. and Stewart, R. M. (1981). *In* "Genetic Research Strategy for Psychobiology and Psychiatry" (eds E. S. Gershon, S. Matthysse, X. O. Breakefield, R. D. Ciaranello), pp. 171–186. The Boxwood Press, New York.

Phillips, M. C. and Chapman, D. (1968). *Biochim. Biophys. Acta* **163**, 301–313.

Phillips, M. C., Williams, R. M. and Chapman, D. (1969). *Chem. Phys. Lipids* **3**, 234–244.

Phillips, M. C., Ladbrooke, B. D. and Chapman, D. (1970). *Biochim. Biophys. Acta* **193**, 35–44.

Pink, D. A., Georgallas, A. and Chapman, D. (1981). *Biochemistry* **20**, 7152.

Pink, D. A. and Carroll, C. E. (1978). *Phys. Letts* **66A**, 157–160.

Pink, D. A. and Chapman, D. (1979). *Proc. Natl Acad. Sci. USA* **76**, 1542–1546.

Plaa, G. L. and Witschi, H. (1976). *Ann. Rev. Pharmacol.* **16**, 125–141.

Poo, M. M. and Cone, R. A. (1974). *Nature (London)* **247**, 438–441.

Poste, G. D., Papahadjopoulos, D., Jacobson, K. and Vail, W. J. (1975). *Nature (London)* **253**, 552–554.

Raff, M. C. (1976). *Sci. Amer.* **234**, 30.

Restall, C., Arrondo, J. L. R., Elliot, D. A., Jaskowska, A., Weber, W. V. and Chapman, D. (1981). *Biochim. Biophys. Acta* **670**, 433–440.

Rice, D. M., Meadows, M. D., Scheiman, A. O., Goni, F. M., Gomez-Fernandez, J. C., Moscarello, M. A., Chapman, D. and Oldfield, E. (1979). *Biochemistry* **18**, 5893–5903.

Rice-Evans, C. and Hochstein, P., (1981). *Biochem. Biophys. Res. Commun.* **100**, 1537–1542.

Rojas, E. and Tobias, J. M. (1965). *Biochim. Biophys. Acta* **94**, 394–404.

Rousselet, A. and Devaux, P. F. (1977). *Biochem. Biophys. Res. Commun.* **78**, 448–454.

Sackman, E. (1979). "Light Induced Charge Separation in Biology and Chemistry" (eds. E. Gerischer and J. J. Katz). Verlag Chemie, Weinheim.

Schroeder, H. (1977). *J. Chem. Phys.* **67**, 1617–1619.

Scott, H. L. and Cherng, S. L. (1978). *Biochim. Biophys. Acta* **510**, 209–215.

Seelig, A. and Seelig, J. (1974). *Biochemistry* **13**, 4839–4845.

Seelig, A. and Seelig, J. (1975). *Biochim. Biophys. Acta* **406**, 1–5.

Seelig, A. and Seelig, J. (1977). *Biochemistry* **16**, 45–50.

Seelig, J. and Browning, J. L. (1978). *FEBS Lett.* **92**, 41–44.

Seelig, J. and Seelig, A. (1980). *Q. Rev. Biophys.* **13**, 19–61.

Seelig, J. and Waespe-Sarcevic, N. (1978). *Biochemistry* **17**, 3310–3315.

Seeman, P. (1972). *Pharmacol. Rev.* **24**, 583–655.

Semer, R. and Gelerinter, E. (1979). *Chem. Phys. Lipids* **23**, 201–211.

Shanes, A. M. and Gershfeld, N. L. (1960). *J. Gen. Physiol.* **44**, 345–363.

Sheats, J. and McConnell, H. M. (1978). *Proc. Natl Acad. Sci. USA* **75**, 4661.

Shimshick, E. J. and McConnell, H. M. (1973). *Biochemistry* **12**, 2351–2360.

da Silva, P. R. (1972). *J. Cell Biol.* **53**, 777–787.

Sinensky, M. (1971). *J. Bacteriol.* **106**, 449–455.

Skarjune, R. and Oldfield, E. (1979). *Biochim. Biophys. Acta* **556**, 208–218.

Smith, B. A. and McConnell, H. M. (1978). *Proc. Natl Acad. Sci. USA* **75**, 2759–2763.

Smith, B. A., Clark, W. R. and McConnell, H. M. (1979). *Proc. Natl Acad. Sci. USA* **76**, 5641–5644.

Smith, L. M., Smith, B. A. and McConnell, H. M., (1979). *Biochemistry* **18**, 2256–2259.

Smith, L. I. (1981. "Cholesterol Autoxidation." Plenum Press, New York.

Snyder, R. G., Strauss, H. L. and Elliger, C. A. (1982). *J. Phys. Chem.* **86**, 5145.

Speth, V. and Wunderlich, F. (1973). *Biochim. Biophys. Acta* **291**, 621–628.

Steim, J. M., Tourtellote, M. E., Reinert, J. C., McElhaney, R. N. and Radar, R. L. (1969). *Proc. Natl Acad. Sci. USA* **63**, 104–109.

Stockton, G. W., Johnson, K. G., Butler, K. W., Tulloch, A. P., Boulanger, Y., Smith, I. C. P., Davis, J. H. and Bloom, M. (1977). *Nature* **269**, 267.

Stockton, G. W., Polnaszek, C. F., Tulloch, A. P., Hasan, F. and Smith, I. C. P. (1976). *Biochemistry* **15**, 954–966.

Stockton, G. W., Johnson, K. G., Butler, K., Tulloch, A. P., Boulanger, Y., Smith, I. C. P., Davis, J. H. and Bloom, M. (1977). *Nature* **269**, 268.

Stoffel, W., Zierenberg, O. and Scheefers, H. (1977). *Hoppe-Seyler's Z. Physiol. Chem.* **358**, 865–882.

Svingen, B. A., Buege, J., O'Neal, F. O. and Aust, S. D. (1979). *J. Biol. Chem.* **254**, 5892–5899.

Tao, T. (1971). *Biochem. J.* **122**, 54.

Taraschi, T., De Kruijff, B., Verkleij, A. J. and van Echteld, C. J. A. (1982). *Biochim. Biophys. Acta* **685**, 153–161.

Thomas, D. D. and Hidalgo, C. (1978). *Proc. Natl Acad. Sci. USA* **75**, 5488–5492.

Thomas, D. D., Seidel, J. C., Hyde, J. S. and Gergely, J. (1975). *Proc. Natl Acad. Sci. USA* **72**, 1729–1733.

Thompson, G. A., Jr. (1980). "The Regulation of Membrane Lipid Metabolism". CRC Press, Boca Raton, FL.

Träuble, H. (1972). *In* "Biomembranes" (F. Kreutzer and J. F. G. Slegers, eds) Vol. 3, 197ff. Plenum Press, New York.

Träuble, H. (1976). *In* "Structure of Biological Membranes" (eds S. Abrahamsson and I. Pascher), Nobel Foundation Symposium **34**, 509–550. Plenum Press, New York.

Träuble, H. and Eibl, H. (1974). *Proc. Natl Acad. Sci. USA* **71**, 214–219.

Träuble, H. and Sackmann, E. (1972). *J. Am. Chem. Soc.* **94**, 4499–4510.

Trudell, J. R., Hubbell, W. L. and Cohen, E. N. (1973). *Biochim. Biophys. Acta* **291**, 328–334.

Umemura, J., Cameron, D. G. and Mantsch, H. H. (1980). *J. Phys. Chem.* **84**, 2272.

Urry, D. (1971). *Proc. Natl Acad. Sci. USA* **68**, 672–676.

Van Echteld, C. J. A., Van Stigt, R., De Kruijff, B., Leunissen-Bijielt, J., Verkleij, A. J. and de Grier, J. (1981). *Biochim. Biophys. Acta* **648**, 287–291.

Van Echteld, C. J. A. (1982). Ph.D. Thesis entitled "Modulation of Membrane Lipid Polymorphism.

Van Zoelen, E. J. J., Van Dijck, P. W. M., De Kruijff, B., Verkleij, A. J. and Van Deenen, L. L. M. (1978). *Biochim. Biophys. Acta* **514**, 9–24.

Vaz, W. L. C., Austin, R. H. and Vogel, H. (1979). *Biophys. J.* **26**, 415–426.

Veksli, Z., Salsbury, N. J. and Chapman, D. (1969). *Biochim. Biophys. Acta* **183**, 434–446.

Verkleij, A. J., De Kruijff, B., Ververgaert, P. H. J. Th., Tocanne, J. F. and Van Deenen, L. L. M. (1974). *Biochim. Biophys. Acta* **339**, 432–437.

Vorbeck, M. L., Martin, A. P., Long, J. W., Smith, J. M. and Orr, R. R. (1982). *Arch. Biochem. Biophys.* **217**, 351–361.

Wahl, P., Kasai, M. and Changeux, J. P. (1971). *Europ. J. Biochem.* **18**, 332–341.

Wallace, B. A., Veatai, W. R. and Blout, E. R. (1981). *Biochemistry* **20**, 5754–5760.

Wallach, D. F. H., Verma, S. P. and Fookson, J. (1979). *Biochim. Biophys. Acta* **559**, 153.

Wallach, D. F. H. and Winzler, R. J. (1974). "Evolving Strategies and Tactics in Membrane Research". Berlin-Heidelberg-N.Y. Springer.

Waring, A. J., Rottenberg, H., Ohnishi, T. and Rubin, E. (1982). *Arch. Biochem. Biophys.* **216**, 51–61.

Weber, G. (1952). *Biochem. J.* **51**, 145–155.

Weinstein, S., Wallace, B. A., Blour, E. R., Morrow, J. S. and Veatch, W. (1980). *Proc. Natl Acad. Sci. USA* **76**, 4230–4234.

Zavoico, G. B. and Kutchai, H. (1980). *Biochim. Biophys. Acta* **600**, 263–269.

Chapter 2

Effects of Dehydration on Membranes and Membrane Stabilization at Low Water Activities

JOHN H. CROWE AND LOIS M. CROWE

*Department of Zoology, University of California,
Davis, California 95616, U.S.A.*

BIOLOGICAL MEMBRANES Vol. 5
ISBN 0 12 168546 2

I. Introduction

Water is indispensable for maintenance of structural and functional integrity of biological membranes and macromolecules (Tanford, 1978; Franks, 1982). Removal of the water in the vicinity of a membrane, for example, irreversibly violates its structural (Crowe, L., and Crowe, 1982) and functional (Crowe, J., et al., 1983a,b) integrity. The present paper will review some aspects of what we know about how dehydration effects this damage to biological membranes. In addition, we will provide a review of some of our own work and that of others which has shown that membranes can, under certain conditions, be stabilized in the absence of water in a state which permits return to normal structure and function when the membrane is returned to water.

The impetus for the work we will describe comes from our investigations over the past two decades on organisms such as seeds of plants, spores, cysts of certain crustaceans, and some species of soil-dwelling animals which all possess the remarkable ability to survive total dehydration. These organisms, said to be in a state of "anhydrobiosis" ("life without water"), may persist in the dehydrated state for years. When free water again becomes available to them, they rapidly swell and resume active metabolism, often within minutes. We have suggested in our studies on these organisms that elucidation of the mechanisms whereby cellular components of these organisms such as membranes escape irreversible damage from dehydration will be important in determining the role of water in maintaining structural integrity of membranes in fully hydrated cells. We will summarize our progress and that of others in this regard in this review.

To understand the consequences of dehydration for mixtures of phospholipids or for a biological membrane beyond a phenomenological level, we must first understand the interactions of water with phospholipids and membranes, so we begin this review with a discussion of hydration of important membrane phospholipids.

II. Hydration of Phospholipids

A major factor influencing the associations of amphiphiles in aqueous solution is the low solubility of the hydrocarbon chains in water (Tanford, 1980). However, despite the dominating effect of the hydrocarbon chains on the lipid–lipid associations, the hydration state of the polar head group may also influence such associations. In this and the following sections we will describe the hydration state of the polar head group and the consequences of removal of that water. We will not attempt to provide a comprehensive review of this field; not only is such a review beyond the scope of this paper, but also excellent reviews of this subject are readily available (Hauser, 1975a,b; Hauser et al., 1981; Worcester, 1976).

A. HYDRATION SHELLS

The usual approach in attempts at evaluating the physical status of the water that associates with phospholipids is to hydrate dry phospholipids and to study the consequences for both the phospholipid and added water, using the several physical techniques that can be applied for this purpose.

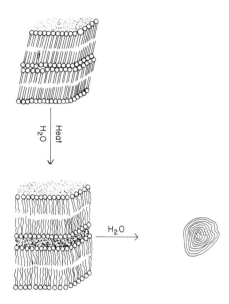

FIG. 1. Hydration of dry PC (a), which exists in a gel state, results in swelling of the bilayers (b). Addition of more water leads to formations of multilamellar liposomes, which may be broken up into small, unilamellar liposomes by several methods.

The results of such studies have shown that when water is added to dry phospholipids like phosphatidylcholine, the adjacent bilayers swell, and water is incorporated between the bilayers (Fig. 1). The amount of swelling depends upon the nature of the polar head group. In addition to inserting between bilayers, a significant fraction of the added water binds to the phospholipid. It is this fraction of the water, variously known as "bound" (Hauser, 1975a,b), "abnormal" (Garlid, 1978), or "vicinal" (Drost-Hansen, 1982), with which we are concerned here. The extent to which water penetrates the bilayer and binds to phospholipids apparently not only depends on the nature of the polar group (cf. Hauser, 1975b; Hauser et al., 1981) but may also be affected by the nature of the hydrocarbon chains (Stumpel et al., 1983), the composition of lipid mixtures (Simon et al., 1981; Ter-Minassian-Sarasa and Madelmont, 1983), and on the chemical composition of the surrounding bulk phase aqueous medium (Chen et al., 1981; Papahadjopoulos

TABLE I

Hydration of various phosphatidylcholines

Bound water		Trapped water		Total hydration		Physical technique	Reference
(moles H_2O)/(mole lipid)	(gH_2O)/(g lipid)	(moles H_2O)/(mole lipid)	(gH_2O)/(g lipid)	(moles H_2O)/(mole lipid)	(g H_2O)/(g lipid)		
6	0.14	12	0.30	18	0.44	H_2O vapour adsorption	Elworthy (1961)
13.6	0.33					H_2O distribution in two-phase system	Henrickson (1970)
12–16	0.29–0.39	11	0.26	25	0.61	Hydrodynamic measurement	Hauser (1975a,b)
11–15	0.25–0.37					Differential scanning calorimetry	Chapman et al. (1967)
12	0.29	16–22	0.38–0.50	30	0.73	X-ray diffraction	Small (1967)
10–12	0.24–0.29					^{1}H-NMR	Veksli et al. (1972)
12	0.29	11	0.26	23	0.55	^{2}H-NMR	Finer and Darke (1974)
11	0.25	10	0.25	21	0.50	Diffusion of ^{3}HHO	Rigaud et al. (1972)
10–12	0.25–0.32			21		Differential thermal analysis	Crowe and Crowe (1982)
3.4						Differential scanning calorimetry	Ter-Minassian-Saraga and Madelmont (1982)
				24		X-ray diffraction	Rand (1981)

et al., 1977). Most of these effects will be considered in the following paragraphs, but we begin with what is known of the binding of water to the only phospholipid that has been most extensively characterized in this regard—phosphatidylcholine.

Phosphatidylcholine (PC) binds significant amounts of water, as has been demonstrated by numerous workers, using various techniques (summarized in Table I). These techniques differ greatly in level of organization at which the measurements were made; they were made with planar bilayers in some cases (water vapour adsorption, X-ray diffraction, differential scanning calorimetry, deuterium-NMR), with vesicles in others (hydrodynamic studies), and with inverted micelles in still others (two phase measurements). Despite the differences in these techniques, with the single exception of the adsorption studies, which have long been known to give lower values than the other physical techniques listed (cf. Labuza, 1975), there is reasonable agreement that about 0.25–0.30 g water is bound by 1.0 g of phospatidylcholine (PC), which amounts to about 10–12 water molecules/PC. This amount of water apparently constitutes the main hydration shell of the few phospholipids that have been studied, and is the limiting water content above and below which important transitions occur in phospholipid associations. We also point out that the data from D-NMR studies are consistent in this regard with those provided by other physical techniques, suggesting that D_2O associations with the phospholipid may be similar to those shown by H_2O. In addition, recent measurements of dielectric properties of bilayers in the presence of D_2O and H_2O did not indicate any significant differences between the two hydrating media (Coster *et al.*, 1982). This observation is of some interest since D-NMR and neutron diffraction using D_2O as a marker for water have been instrumental in determining the nature of the association between the phospholipid and water, as illustrated in the following paragraph.

TABLE II

Hydration shells of various phospholipids (adapted from Finer and Darke (1974) and Hauser *et al.* (1981)).

Hydration shell	Moles H_2O/Mol Phospholipid		
	PC	PE	PS
(1) inner shell	1–2	0	1–2
(2) main shell	11	11	10
(3) weakly bound H_2O	0	0	12
(4) trapped H_2O	11	0	120
(5) hydration at which free H_2O observed	> 23	> 12	> 140

Finer and his colleagues (Finer, 1973; Finer and Darke, 1974; Finer *et al.*, 1972) have made studies of the hydration of three phospholipids with D-NMR, and have concluded from measurements of the quadrupole splitting that five kinds of water are found in association with large multilamellar vesicles (Table II). The evidence for the magnitude of the inner hydration shell, about which there seems to be good agreement, comes from several sources: Jendrasiak and Hasty (1974) showed from their studies of water vapour adsorption to dry egg PC that this inner shell consists of about 2.5 water molecules/phospholipid. Similarly, Misiorowski and Wells (1973) reported that when egg PC was dissolved in diethylether/methanol containing small amounts of H_2O, the PC bound about 1 molecule of the H_2O/molecule of phospholipid. Chapman *et al.* (1967) found that egg PC is highly hygroscopic and hydrated to a level of about 1 water molecule/PC even over P_2O_5. Thus, the existence of an innermost hydration shell consisting of one or two water molecules/PC which can be removed only by extreme treatment seems reasonable. There is good agreement, as noted above, between the many physical techniques with which the main hydration shell has been measured (Table I; Fig. 2).

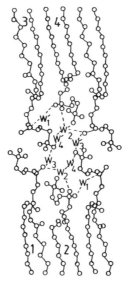

FIG. 2. Conformation of crystalline PC and the first four water molecules associated with the head group. W_1-W_4 = water molecules 1–4, respectively. (Redrawn from Hauser *et al.*, 1981.)

The extent or even the existence of additional hydration shells varies with the nature of the polar group. For example, phosphatidylethanolamine (PE), unlike PC, binds hardly any water at all beyond the main hydration shell, and swells little in excess water, an explanation for which will be provided in the section on sites of hydration. Phosphatidylserine (PS), by contrast, influences the physical properties of enormous amounts of water compared with either PC or PE (Table II), a phenomenon that appears to be related to the charge on PS. When X-ray spacing of the repeat units in dry PS was recorded with increasing hydration, a rapid increase in spacing was seen at water contents exceeding about 0.3 g H_2O/g PS (Atkinson *et al.*, 1974). On the other hand, PS has been shown not to take up water at low pH, when the carboxyl group is protonated. Thus, the uptake of the large amount of water seen in PS is probably related to electrostatic repulsion by neighbouring PS molecules, leading to a loose packing and increased trapping of water.

B. SITES OF HYDRATION IN PHOSPHOLIPIDS

1. *The Main Hydration Shell*

Similar amounts of water in the main hydration shell are seen in the few phospholipids that have been studied, which suggests that there is probably a common structure in all these phospholipids which is responsible for the water binding. The most likely candidate is the phosphodiester group, and there is good evidence that it is the site around which the main hydration shell is centred. The evidence in support of the phosphate group as the main hydration site lies in four categories:

(1) *Deuterium-NMR* Finer *et al.* (1972) have shown that deuteron exchange with the NH group of PE yields a doublet splitting in D-NMR spectra that is distinctly different from that of the D_2O in the main hydration shell. The existence of this unique splitting due to deuteration of the NH_3 (which does not appear in spectra of phospholipids undergoing hydration) may be taken as evidence that the NH_3 group is not the centre of hydration.

(2) *Water replacement* It has been shown by several workers that divalent cations bind to phosphodiester groups in phospholipids (Hauser and Phillips, 1973). Some of the same workers have shown that binding of these cations displaces water from the phospholipid, and the suggestion has been made, based largely on binding kinetics, that divalent cations and water compete for the same binding sites. Some of the most convincing evidence in this regard has been obtained by Misiorowski and Wells (1973), who studied binding of water to PC in anhydrous methanol. They found that PC bound 1 Ca, Ms, or Ce/molecule of PC. The cation binding appeared to be competitively inhibited by water and was completely inhibited when 2 molecules of H_2O were bound/PC. More recently, Papahadjopoulos *et al.* (1977) have shown that the

X-ray diffraction patterns for crystalline PS and PS in bulk solution with Ca added are nearly indistinguishable. They concluded from these findings that addition of Ca dehydrates this phospholipid, probably by binding to the phosphate group, thus displacing water.

(3) *Neutron diffraction* Locations of water molecules hydrating the head groups of various PCs have been studied, using H_2O–D_2O exchange. With all these PCs at different temperatures and water contents the main site of hydration was shown to be associated with the polar head group (Worcester, 1976).

(4) *X-ray diffraction* Hauser *et al.* (1981) have done elegant X-ray diffraction studies on single crystals of several phospholipids, with the most detailed results to date. They have shown that when DMPC is crystallized from anhydrous solvents, the crystal contains a unit cell of four PCs, with two water molecules/PC. Four of these water molecules (W_1, W_1', W_2, and W_2') link the phosphate groups in ribbons, providing stability parallel to the plane of the bilayer (Fig. 3). The remaining four water molecules (W_3, W_3', W_4, and W_4') form hydrogen bonds with phosphate oxygen. At the same time, W_3 and W_4 form hydrogen bonds with the corresponding water molecules across the bilayer interface (Fig. 3). It has not yet been possible to study the disposition of additional added water, but presumably the remaining water in the main hydration shell (6–8 more water molecules) are hydrogen bonded to these first four water molecules. In any case, there seems to be general agreement that the main hydration shell is centred around the phosphate head group.

In contrast with PC, the head groups of adjacent PEs are not hydrogen bonded to each other by water. Instead, the head group of PE shows compact molecular packing both in the hydrated and dehydrated states due to direct intermolecular hydrogen bonding. Instead of a water link between the PE head groups, the ammonium groups link together the unesterified phosphate oxygens by short bonds, resulting in extremely close packing of the head groups, which occupy only about $0.39 \, \text{nm}^2$ ($39 \, \text{Å}^2$) compared with nearly $0.50 \, \text{nm}^2$ ($50 \, \text{Å}^2$) for PC. Similar intermolecular bonding in PC is apparently inhibited by the presence of the bulky choline.

The X-ray diffraction data of Hauser *et al.* (1981) indicate that the head groups of the single crystals of the PCs and PEs which they have studied are oriented more or less parallel to the plane of the bilayer, although the angle of inclination varies with the head group. The consensus emerging from studies using several other physical techniques strongly and perhaps somewhat unexpectedly suggests that this orientation does not change appreciably either with state of hydration or transition from the liquid-crystalline to gel state (Worcester, 1976; Buldt *et al.*, 1979; Buldt and Wohlsemuth, 1981).

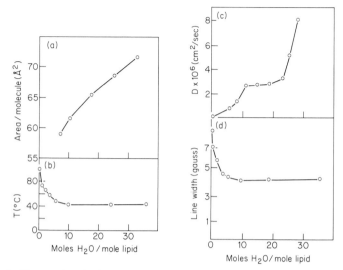

FIG. 3. Changes in physical parameters of PC with hydration. (a) Lateral expansion of PC bilayers (from Small, 1967). (b) Decrease in transition temperature with increase in water content (from Chapman et al., 1967). (c) Diffusion coefficient of water (from Rigaud et al., 1972). (d) Changes in line width of deuterium NMR spectra (from Veksli et al., 1969).

2. Outer Hydration Shells

The physical status of the other fractions of water may be of less importance in our description of the effects of dehydration on phospholipid associations, but this subject is of such interest that we wish to comment further. The status of the trapped water is controversial; for example, free water has variously been estimated to be present when the number of water molecules exceeds 10–12 or 23/PC. The former values are derived from physical measurements (as presented in Table I), which suggest that water added at hydration values exceeding 10–12/PC has most of the physical properties of bulk water. The limiting value of 23 water molecules/PC for appearance of free water is based on the D-NMR data of Finer and colleagues reviewed earlier, X-ray diffraction data (reviewed in Rand, 1981), and calorimetric data of Ter-Minassian-Saraga and Madelmont (1981, 1982). Finer and his colleagues found that at hydration levels greater than 23 water molecules/PC the D-NMR spectra gave a narrow line imposed on doublets that was not seen at lower water contents. The narrow, isotropic line (assumed to be due to the presence of free water undergoing rapid, isotropic motion) increased in intensity and became increasingly narrow with rising water content. The extent and location of this unusual water is as yet unclear. One possibility is that water near the polar head groups but not part of the main hydration shell is restricted in its motion. Evidence for this sort of long range ordering of water has been described based on studies of the viscosity of fluids contained in aqueous compartments in membrane vesicles. For example, Berg *et al.* (1979) used electron spin resonance to measure motion of the water soluble spin probe TEMPAMINE in the aqueous compartment of spinach thylakoids. Along with a set of carefully conducted control experiments, they reported that the mobility of the probe was significantly restricted in the intravesicular water, which they calculated to have a viscosity approximately 10 times that of water. More recently, Clement and Gould (1980) carried out similar experiments, using unilamellar phospholipid liposomes in which the water-soluble fluorescent probe pyranine was trapped in the aqueous compartment. Using fluorescence anisotropy, they found that rotational relaxation times for the probe in the liposomes to be about three times those for the probe in bulk solution. They concluded that water inside the liposomes is probably ordered to some extent. Although the physical properties of phospholipids that lead to such ordering of water are not understood, there have been numerous suggestions that ordered water in association with membranes may play important roles in metabolic processes. Anomalies in observations of the effects of internalized buffers on kinetics of proton transport across membranes of chloroplasts (Ort *et al.*, 1976) have lead some workers to suggest that this ordered water may play a mechanistic role in transport of protons (Ort *et al.*, 1976; Mitchell, 1979) and other cations (Wiggins and Bowmaker, 1982; Negendank, 1982; Edmonds, 1983). The potential importance of this water in metabolic regulation seems to have escaped the attention of most biochemists, but it has led others (Clegg, 1981, 1982) to postulate the existence of a structured water layer near membranes and other macromolecular structures in cells that could play a key role in organization of water-soluble enzymes into multi-enzyme complexes.

3. *Penetration of Phospholipid Bilayers by Water*

There is also no general agreement as yet on the depth to which water penetrates the hydrocarbon chains (Wennerstrom and Lindman, 1979). The kinds of measurements that have been done include:

(1) *Electron spin resonance* Griffith *et al.* (1974) determined the polarity profile across liver microsomal membranes and inferred from the profile that water penetrates about one-third the distance from each surface into the bilayer. However, the influence of the bulky nitroxide radical (used by Griffith *et al.* in these studies) on phospholipid packing density is suspect (Cadenhead and Muller-Landau, 1974), so these results must be viewed with caution.

(2) *Dielectric studies*   The dielectric thickness for several planar bilayers has been determined from capacitance measurements (White, 1977; Ebihara *et al.*, 1979), with the conclusion that water probably penetrates about 2.5 nm (25 Å) from each surface, or to about the level of the first methylene of the hydrocarbon chains.

(3) *Neutron diffraction*   Using this technique, water penetration into partially hydrated samples of dipalmitoyl phosphatidylcholine : cholesterol (2 : 1) has been measured (Zaccai *et al.*, 1979), and these workers reported that water penetrates to the level of the PC carbonyl group. Recently, Simon *et al.* (1982) have reported that addition of cholesterol to PE bilayers decreases the depth to which water penetrates the bilayers, a finding that is of particular interest in view of the well-known influences of cholesterol on thermotropic phase transitions in phospholipid bilayers (cf. Chapman, 1968; Chapman and Plenkett, 1966; Darke *et al.*, 1972).

III. Hydration-dependent Phase Transitions in Complex Mixtures of Phospholipids

The preceding discussion of the hydration state of phospholipids in bilayers suggests that dehydration should profoundly affect the nature of lipid–lipid interactions. Because the head group of PC is hydrogen bonded to the adjacent head group by water, removal of that water would be expected to decrease the intermolecular distance between the polar groups, leading to increased van der Waals interactions in the hydrocarbon chains. Such increased interactions are characteristic of certain types of cooperative phase transitions. We will not review the enormous amount of literature that deals with phase transitions: such reviews are readily available (e.g. Lee, 1977a,b; Pink, 1981; Quinn and Chapman, 1980). Instead we will comment here only on those findings that are of interest in explaining the hydration-dependent phase transitions.

A. LIQUID CRYSTALLINE TO GEL TRANSITIONS

Probably the most studied of all lipid phase transitions is the gel to liquid crystalline transition first suggested for phospholipids by Chapman and his colleagues (reviewed by Quinn and Chapman, 1980). This sort of transition involves ordered packing of the hydrocarbon chains, which are predominantly in the *trans* configuration and packed in a hexagonal array (but see Stumpel *et al.*, 1983 for an alternative interpretation). Upon heating, a gradual increase in mobility of the hydrocarbon chains is evident (as seen with infrared spectroscopy, X-ray diffraction, and NMR) until at the phase transition temperature the chains show a sudden increase in mobility with further increases in temperature. The molecular events responsible for this effect have been extensively studied and continue to constitute the subject of numerous publications. Of particular interest here is the finding that hydration state can also effect a transition of lipid bilayers from the liquid crystalline to gel states. Chapman *et al.* (1967) and Kodama *et al.* (1982) showed that when phase transition temperatures were recorded for PC containing variable amounts of water the transition temperature fell to a limiting value as water content rose to 0.3 g H_2O/g dry lipid. At higher water contents there was no further decrease in the transition temperature (Fig. 2). Thus, with dehydration PC exists in a gel phase at temperatures that would permit in the hydrated lipid to exist in a liquid crystalline phase. Martonosi (1974) has presented similar data for sarcoplasmic reticulum membranes.

B. BILAYER TO NON-BILAYER TRANSITIONS

A number of non-bilayer phases have been demonstrated in phospholipids, beginning with the classic work of Luzzati and his colleagues, who described these phases using X-ray diffraction (Luzzati, 1968; for a more recent perspective on their work in this field see Luzzati, 1982). One of the most interesting of these phases in the present context is the hexagonal II (H_{II}) phase, in which the phospholipids form tubes arranged in a hexagonal packing order, with the polar groups oriented into an aqueous core. Most phospholipids will assume a hexagonal II conformation under specialized conditions (such as extreme dehydration and high temperatures in the case of PC; Luzzati, 1968). In some phospholipids H_{II} phase has been thought to be restricted to such circumstances. However, there is recent evidence (Van Echteld *et al.*, 1982) which demonstrates the presence of H_{II} phase in PC in the presence of gramicidin. The mechanism of induction of hexagonal phase in the presence of the protein is not understood. In addition, there is other evidence (Ulmius *et al.*, 1982) which suggests that addition of cholate to lecithin membranes perturbs the chain order, inducing a hexagonal I structure (tubes of phospholipid arranged in a hexagonal array, with the polar head groups oriented outward, into the aqueous phase). Thus, a body of evidence is being developed which indicates that at least some phospholipids that were formerly thought to exist only in the lamellar phase may under certain circumstances assume a hexagonal structure. A few phospholipids (most notably phosphatidylethanolamine, cardiolipin, and monoglucosyl diglyceride) prefer the H_{II} phase at physiologically relevant temperatures (Williams and Chapman, 1970) both in excess water (Cullis and De Kruijff, 1978, 1979; Rand and SenGupta, 1972b; Wieslander *et al.*, 1981) and when dehydrated (Williams and Chapman, 1970). Ca is required for formation of H_{II} phase by cardiolipin for reasons that will be evident from the following discussion.

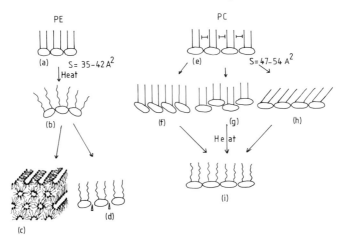

FIG. 4. Hydration and thermal behaviour of PE and PC. Hauser *et al.* (1981) have shown that the head group of PE occupies a comparatively small area (S on the diagram) which can accomodate hexagonal packing of the hydrocarbon chains in the gel state. When PE is heated above the hydrocarbon chain melting temperature, the hydrocarbon chains occupy a greater area, leading to a curved structure (b), which could result in formation of hexagonal II structures (c). Hauser *et al.* (1981) also suggest that PE could be maintained in a bilayer in the liquid-crystalline state by insertion of a polar molecule between the head groups (d). The larger area occupied by the head group of PC (a result of the presence of the bulky choline and intermolecular hydrogen bonding via water) leads to unoccupied space between the hydrocarbon chains in the gel state (e), which could be accommodated by shift in the head group associations and angle of the hydrocarbon chains with respect to the plane of the bilayer (f–h). In the liquid-crystalline state the special head group associations are not required due to the increase in area occupied by the hydrocarbon chains (i). (Modified from Hauser *et al.*, 1981.)

Hauser *et al.* (1981) have provided a persuasive analysis of the tendency of PE to form H_{II} type structures that depends on measurements of conformation of the head group. His analysis is based largely on the theoretical work of Israelachvili *et al.* (1976). As we previously described, the head group of PE shows compact molecular packing both in the hydrated and dehydrated states due to intermolecular hydrogen bonding. Incidentally, this close packing in PE will almost certainly increase Van der Waals interactions between the hydrocarbon chains and results in the elevated thermal transition temperatures that have been observed for saturated PEs (Williams and Chapman, 1970; Harlos, 1978; Van Dijck, *et al.*, 1976). In such saturated

PEs and in less saturated ones (which show chain melting temperatures at more physiologically relevant temperatures), at the chain melting temperature the close packing of the head groups appears to be preserved, the result being that the volume increase accompanying increased fluidity would be restricted to the hydrocarbon chains. Thus, since the mobility of the head group is restricted by hydrogen bonding to the neighbouring head group, if the volume occupied by the hydrocarbon chains increases, formation of structures of high curvature is inevitable. Such structures might associate into H_{II} tubes (Fig. 4).

C. HEXAGONAL II STRUCTURES IN COMPLEX MIXTURES OF PHOSPHOLIPIDS AND IN BIOLOGICAL MEMBRANES

1. *Evidence for the Existence of Non-bilayer Structures in Membranes*

Hexagonal II phase phospholipids have been detected by X-ray diffraction (Luzzati and Husson, 1962; Rand and SenGupta, 1972a), freeze fracture (Deamer et al., 1970; Van Venetie and Verkleij, 1981; Vail and Stollery, 1979; Crowe, L. and Crowe, 1982; Crowe, J. and Crowe, 1982; Hui et al., 1981), conventional transmission electronmicroscopy (Stoeckenius, 1962; Tinker and Pinteric, 1971), and ^{31}P-nuclear magnetic resonance (Cullis and De Kruijff, 1979; Cullis and Hope, 1980; Seelig, 1978). Some importance has been ascribed to such studies in recent years since the controversial (cf Bearer et al., 1982; Wilschut et al., 1982) suggestion that H_{II} or inverted micellar structures could reflect the bilayer instability that should be involved in biologically important processes such as membrane fusion (Cullis and Hope, 1978; Verkleij et al., 1979; Van Venetie and Verkleij, 1981), trans-bilayer movement of phospholipids (Cullis and De Kruijff, 1979) and in modulation of membrane-bound enzyme activity (Hui et al., 1981; Van Venetie and Verkleij, 1982). In view of the potential importance of H_{II} and reversed micellar structures in membranes, we wish to comment briefly on the methodology being used to study these structures. Cullis and De Kruijff (1978, 1979) showed that it is possible to use ^{31}P-NMR to detect H_{II} or micellar structures in membranes, using phosphatidylethanolamine as a model lipid. As noted above, PE exists in H_{II} phase at temperatures above the hydrocarbon chain melting temperature and assumes a lamellar phase when cooled below the hydrocarbon chain melting temperature. When Cullis and De Kruijff recorded ^{31}P-NMR spectra of PEs at different temperatures (and thus in H_{II} or lamellar phase), they obtained the results similar to those shown in Fig. 5. They reported a distinctive reversal in the chemical shift anisotropy with the transition, which suggests that H_{II} phase phospholipids can easily be recognized by this technique. Subsequently, Seelig (1978) provided a theoretical explanation for the reversal of the chemical shift anisotropy. In view of these results, it is tempting to use ^{31}P-NMR as a major tool for detecting H_{II} lipids in membranes. However, we believe that some caution is necessary, for the following reasons. (1) Cullis and Hope (1978) showed that when red blood cell membranes are forced to undergo fusion non-bilayer structures were detected both with freeze fracture and with ^{31}P-NMR. However, a puzzling feature of this finding is that even though most of the membrane phospholipids were in bilayers during the fusion, the ^{31}P-NMR spectra suggested they were predominantly in H_{II} phase. (2) Recently, Noggle *et al.* (1982) showed that phosphatidylglycerol forms bilayers according to several other criteria, while the NMR spectra were characteristic of H_{II} phase. (3) Hui *et al.* (1981) have shown that ^{31}P-NMR spectra can be obtained from lipid phases that are not true H_{II} structures. Thus, we believe that at least for the present the ^{31}P-NMR data must be viewed with some caution and that additional techniques should be used to detect non-bilayer structures in the membrane whenever possible. Similar precautions have been noted by others (Hui *et al.*, 1981).

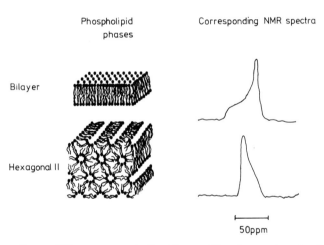

Phospholipid phases   Corresponding NMR spectra

Bilayer

Hexagonal II

50ppm

FIG. 5. Lamellar and H_{II} structures in liquid crystalline (bottom) and gel phase PE (top) and corresponding ^{31}P-NMR spectra. Adapted from Cullis and De Kruijff (1978). See text for further comment.

2. *A Biological Role for Non-bilayer Structures in Membranes?*

The only lipid structure that is compatible with a nonleaky membrane is the bilayer. Nevertheless, most biological membranes contain significant quantities of phospholipids like phosphatidylethanolamine, which prefers to be in the H_{II} phase in aqueous solution at temperatures above the hydrocarbon chain melting temperature. A biological role for lipids like phosphatidylethanolamine in membranes has thus been sought, as indicated above, with results that are as yet suggestive but not completely satisfying. For example, PE is rapidly translocated across the membrane of the bacterium *Bacillus megaterium*, independent of lipid synthesis and protein and sources of energy, and with a rapidity that would probably exclude a "flip-flop" mechanism (Langley and Kennedy, 1979). The suggestion has been made that transbilayer movement of PE is effected by lateral diffusion through "hairpin" structures in the vicinity of hydrophilic membrane channels. Since the hairpin bend would have a small radius of curvature, it could be stabilized by lipids like PE that form highly curved aggregates (Langley and Kennedy, 1979). The recent finding that addition of gramicidin to phospholipids that do not normally form H_{II} structures induces them to do so seems consistent with this notion. Even if a biological role for lipids like PE exists in membranes, however, it is clear that a balance between those lipids that tend to form bilayer structures and those that tend to form non-bilayer structures must be maintained if the membrane is to have a bilayer structure. This appears to be the case, based on the few studies that have been done on phase behaviour of complex mixtures of phospholipids. Cullis and DeKruijff (1978) have shown that mixtures of phosphatidylcholine and phosphatidylethanolamine exist in a predominantly bilayer phase. However, when the ratio of PC : PE was reduced below a critical level, H_{II} aggregates were formed. More recently, Hui *et al.* (1981) have shown in a careful study using X-ray diffraction, freeze fracture, and ^{31}P-NMR that at 30°C when mole ratios of PC : PE were decreased from 30 to less than 10 hexagonal phase lipids were detected. At intermediate ratios, however, a transition zone was found in which micellar structures with isotropic NMR spectra existed. At even lower mole ratios of PC : PE micellar structures were found that showed NMR spectra characteristic of H_{II} phospholipids. Interestingly enough, Kagawa *et al.* (1973) had previously found that when a mitochondrial ATPase was reconstituted in various mixtures of PC and PE, low activity was found in either pure PC or pure PE, in other words in pure bilayer or pure hexagonal phase lipids. However, when the ATPase was reconstituted in mixtures of PC and PE, activity was found to increase with increasing PE content, up to about 85% PE, above which activity declined. This maximal activity coincides with the mixture of PC and PE at which the transition zone of micellar structures exists. This intriguing finding raises the possibility that this transition phase modulates the activity of the enzyme in some way. In related studies, Wieslander *et al.* (1980) have presented convincing theoretical arguments derived from the work of Israelachvili *et al.* (1976) that lipids that normally form hexagonal phase may be important in modulating membrane fluidity. In support of this argument, they have grown *Acholeplasma laidlawii* at 37°C for 12 hours, followed by a shift to 17°C for an additional 12 hours. During the 12 hours at 17°C the cells were seen to synthesize excess monoglucosyldiglyceride, a phospholipid which prefers H_{II} phase to bilayer.

D. HYDRATION-DEPENDENT PHASE BEHAVIOUR OF COMPLEX MIXTURES OF PHOSPHOLIPIDS AND BIOLOGICAL MEMBRANES

As yet we still know very little about hydration dependent phase behaviour of complex mixtures of phospholipids, although existence of both lamellar and hexagonal II structures in dehydrated mixtures of phospholipids from biological membranes has been confirmed by both X-ray diffraction (Luzzati and Husson, 1962) and freeze fracture (Deamer et al., 1970). However, some suggestions of the consequences of dehydration may be taken from the literature (admittedly still sparse) on thermal effects on such mixtures and membranes. Good evidence exists that at temperatures below the liquid crystalline to gel transition temperature mixtures of dipalmitoyl phosphatidyl-choline (DPPC) and dipalmitoyl phosphatidylethanolamine (DPPE) are immiscible and may show separation of these lipids in the plane of the membrane (Lee, 1977a; Fockson and Wallach, 1978; Luna and McConnell, 1978; Arnold et al., 1981). Evidence from freeze fracture is consistent with this proposition; it is well known that if samples are not cooled at a sufficiently rapid rate that lateral phase separation of intramembrane particles occurs (e.g. Hoffman et al., 1980). This separation of intramembrane particles into discrete patches in the plane of the membrane has been attributed to phase separation of phospholipids as they form their respective gel phase crystals, from which the intramembrane particles are excluded (Verkleij and Verver-gaert, 1975; Van Heerikhuizen et al., 1975).

Thus, we can make predictions concerning the hydration dependent phase behaviour of PC/PE mixtures and of biological membranes: As dehydration progresses, PC should enter the gel phase, with a concurrent lateral phase separation of other lipids and proteins. The PE would be expected to enter the hexagonal II phase upon separation from the PC. The PE hexagonal II phase crystals would also be expected to leave the plane of the membrane as dehydration progresses. The available evidence from X-ray diffraction is consistent with this hypothesis; the hydration dependent phase diagram for a complex mixture of lipids (from human brain) presented by Luzzati and Husson (1962) shows that in excess water only bilayer structures were detec-ted. As the water content was reduced, the interlamellar spacing fell, until in the region of 0.25 g H_2O/g lipid a discontinuity in the decrease in spacing was seen, concurrent with the first appearance of H_{II} lipids. As water content was decreased further, the lamellar spacing fell and the diameter of the H_{II} tubes fell.

In summary, we suspect that dehydration of a membrane should lead to phase separations of lipid classes, resulting in formation of complex crystals such as H_{II}. In addition, we would expect to find aggregation of membrane proteins. Both processes are likely to be damaging to membrane integrity. In the next section we investigate that hypothesis.

IV. Hydration-dependent Phase Transitions in a Biological Membrane

We have chosen as a model membrane for our studies on hydration dependent phase transitions Ca-transporting microsomes isolated from muscle. Much of what we report here has previously been published in various forms (Crowe, J. and Crowe, 1982; Crowe, J. *et al.*, 1982, 1983a,b; Crowe, L. and Crowe, 1982; Crowe, L. *et al.*, 1983). Related results on other membranes have been reported by Gruner *et al.* (1982). We chose these particular membranes because they are relatively simple membranes, in which most of the

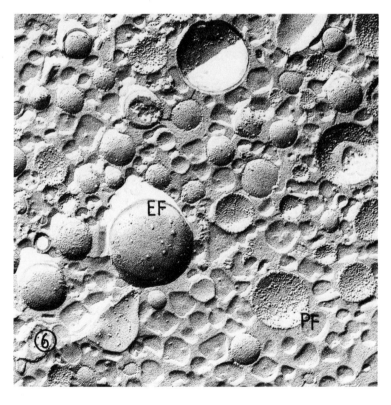

FIG. 6. Freeze fracture of sarcoplasmic reticulum (SR) vesicles, showing intramembrane particles on the P (PF) face. From Crowe *et al.* (1983).

membrane protein is the Ca-ATPase. During freeze fracture the intra-membrane particles, which are identified with this ATPase, distribute in a characteristic pattern, with most of the intramembrane particles associated with the protoplasmic face (PF face, Fig. 6). Thus, these microsomes afford two markers for changes in membrane integrity which may be studied as a function of hydration state: distribution of the intramembrane particles and ability of the microsomes to accumulate Ca.

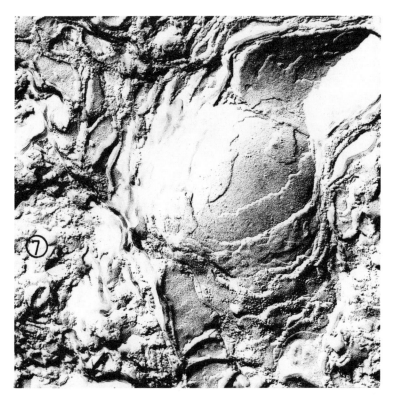

FIG. 7. Partially dehydrated SR, illustrating lateral phase separation of lipids and proteins and fusion between vesicles, forming multilamellar structures.

When the membranes are partially dehydrated before being frozen and fractured, images like that in Fig. 7 are obtained. It is clear from these images that as the membranes are dehydrated, fusion between vesicles occurs, leading to formation of multilamellar structures (Fig. 7). Accompanying these processes there is apparently a lateral phase separation of intramembrane particles from the lipid, leading to the development of lipid domains devoid of particles and clumps of particles that may or may not reside in the plane

FIG. 8 Cholesterol crystals formed during dehydration of SR.

FIGS 9–10. Longitudinal (Fig. 9) and cross (Fig. 10) fractures through hexagonal phase rods of phospholipid. From Crowe *et al.* (1982).

of the membrane (although we suspect the former in view of the fact that the Ca-ATPase is an extremely hydrophobic protein). As the dehydration progresses, crystals are formed, with several characteristic types: (1) Lamellar phases are easily identified, even in the driest preparations. From what is known about phase behaviour of pure phospholipids, we suspect that these lamellar phases are in a gel state. (2) Cholesterol crystals, identified based on their characteristic morphology (Fig. 8). Crystallization of cholesterol accompanying dehydration of a membrane has previously been noted by Ladbrooke *et al.* (1968a,b). (3) Hexagonal phase, probably hexagonal II phase. These crystals are seen in both longitudinal (Fig. 9) and cross (Fig. 10) fracture. According to Luzzati (1968), the centre to centre spacing of hexagonal II rods in pure phospholipid preparations is about 4.5 nm (45 Å). The rods in these preparations measure about 43 Å centre to centre, which supports the suggestion that they are in hexagonal II phase. (4) Complex crystals, devoid of intramembrane particles (Figs 11–12). We know nothing as yet of the composition of these crystals or of their structure, although there is some suggestion from the freeze fracture images that the crystals may be composed of rods.

The crystallization of membrane lipids during dehydration may be followed by thermal analysis, using a simple experimental device which we designed (Fig. 13). It is possible to measure simultaneously with this device weight changes (which may be translated into water content) and thermal activity. During dehydration the membranes are seen to be cooled by evaporative water loss, until they are dried to about 0.3 g H_2O/g dry weight. Beginning at about this water content, which corresponds to the main hydration shell of the phospholipids that have been studied, a series of exotherms is seen (Fig. 14), suggesting that removal of additional water results in crystallization of the membrane components. When the membranes are rehydrated using the same device, they are seen to be warmed by adsorption of water vapour and in the region of 0.3 g H_2O/g dry weight a series of endotherms is seen (Fig. 14). We suspect that these endotherms represent a decrystallization process as the main hydration shell is formed about polar head groups in the phospholipids.

^{31}P-NMR can also be useful in describing phase transitions in these membranes during dehydration, with the cautionary provisos mentioned earlier. The freshly isolated muscle microsomes are seen with NMR to possess a chemical anisotropy characteristic of lamellar phase phospholipids (Fig. 15). The dry membranes show a broad, featureless signal, as one might suspect since the phosphate-containing head group probably has restricted mobility at this low hydration state (Fig. 15). When a small amount of water was adsorbed to the sample from the vapour phase, the partially rehydrated membranes were seen to have a reversed anisotropy and a chemical shift

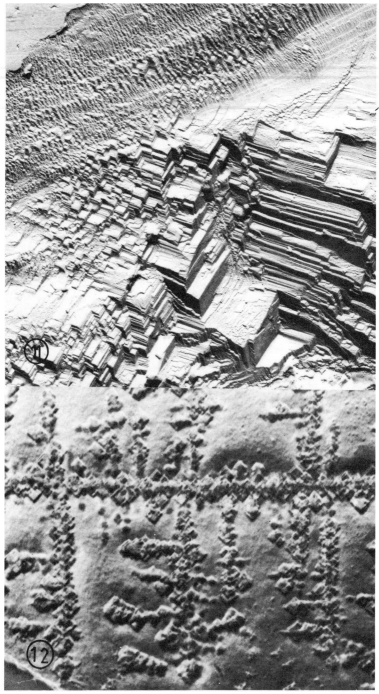

FIGS 11–12. Freeze fracture (Fig. 11) and light micrograph with Normarski optics (Fig. 12) of complex lipid crystals formed during dehydration of SR.

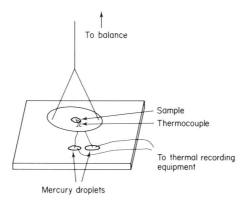

FIG. 13. Diagram of thermal analysis equipment used to measure thermal activity in SR during dehydration. The mercury droplets beneath the balance pan provide a flexible electrical connection, permitting continuous recording of thermal activity and weight of the sample (which can be expressed as water content).

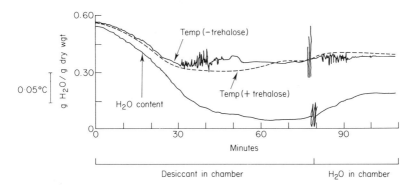

FIG. 14. Thermal activity and water content of SR vesicles during dehydration. Note the thermal activity in the region of 0.2–0.3 g H_2O/g membrane both during dehydration and rehydration. From Crowe and Crowe (1982); Crowe et al. (1983).

similar to that characteristic of hexagonal II phase (Fig. 15). When the temperature was decreased from 21°C to —20°C, the spectrum showed a gradual reversal of this anisotropy and chemical shift to yield a spectrum similar to that of the fully hydrated membranes (Fig. 16). Since PE is known to undergo a phase transition from hexagonal II phase to lamellar phase with decreased temperature, this finding is consistent with our original suggestion

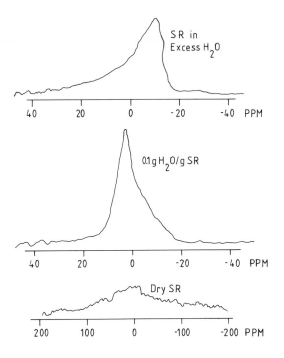

FIG. 15. ^{31}P-NMR spectra of dry SR (bottom), partially rehydrated SR (middle), and SR in excess water (top). The partially rehydrated membranes show a spectrum characteristic of H_{II} phospholipids and those in excess water show a spectrum characteristic of lamellar phase. From Crowe *et al.* (1983).

that the hexagonal phase lipids that are likely to be found in dehydrated membranes or dehydrated complex mixtures of lipids represent PE that has phase separated from bilayer preferring phospholipids like PC during the dehydration process. A puzzling feature of this finding is that the freeze fracture indicates that much of the phospholipid in these membranes is in a lamellar, gel phase. Nevertheless, the NMR spectrum suggests that all the phospholipids are in hexagonal II phase. Several explanations are possible: (1) For reasons that are not understood, head group mobility may be restricted in the dry lamellar phase phospholipids, leading to line broadening for that part of the spectrum which is hidden underneath the predominant hexagonal II phase spectrum. However, we have conducted scans for several hundred p.p.m. both upfield and downfield of the H_{II} signal and have seen no evidence of the hypothesized broad signal. (2) The hexagonal phase rods could be preferentially hydrated when water is added from the vapour phase, leading to increased mobility of the polar head groups of these phospholipids

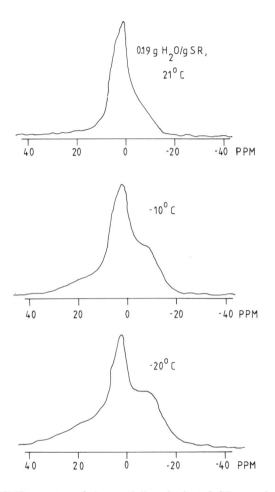

FIG. 16. ^{31}P-NMR spectra of the partially rehydrated SR membranes (Fig. 15, middle) exposed to low temperatures. We interpret the changes in these spectra to represent a transition from H_{II} at the higher temperature to lamellar phase at the lower. The final spectrum in the series appears to consist of a mixture of H_{II} and lamellar phases.

first, and hence a hexagonal II phase spectrum. However, if the hexagonal phase lipids seen in these preparations are indeed PE, as we suggested above, this possibility does not seem likely; PE does not bind water strongly, and swelling of multilamellar PC is known to occur preferentially to hydration of PE (cf. Hauser, 1975a). (3) The head group of lamellar phase phospholipids without the main hydration shell may undergo conformational changes which yield an NMR spectrum similar to the hexagonal II phase spectrum.

If that is the case, the thermal effect on the spectrum (Fig. 16) is difficult to explain. (4) It is possible that most of the phospholipids in the dry preparations are in hexagonal phase or a micellar structure that is thought to be intermediate between the lamellar phase and hexagonal II phase (Hui *et al.*, 1981; Van Venetie and Verkleij, 1981). At least some micellar structures

FIG. 17. Freeze fracture of rehydrated SR. Note the large vesicles, formed due to fusion during dehydration, and redistribution of intramembrane particles on the two faces of the membrane.

have been shown to possess an NMR spectrum similar to that of hexagonal II phase (Hui *et al.*, 1981). There is also some evidence that this micellar phase reverts to lamellar phase with a decrease in temperature (Hui *et al.*, 1981). Thus, if all the phospholipids in these dry preparations are in either hexagonal II or micellar phases, the NMR signal could have the shape and chemical shift we have seen. The difficulty with this explanation is that micelles are not seen with the freeze fracture. However, this problem may easily be explained if the transition of the micellar phase lipids to lamellar phase is faster than the freezing rate during the preparation for freeze fracture. If that is the case, the micellar phase lipids would appear in freeze fracture as lamellar phase. Because reorganization of hexagonal phase II lipids to lamellar phase would require more extensive reorganization, these lipids would be expected to appear in freeze fracture as hexagonal II phase. We are currently conducting experiments designed to distinguish between these possibilities.

Upon rehydration, the dry preparations of membranes are seen with the light microscope to form the familiar myelin figures described by others during rehydration of dry phospholipids. When such preparations are freeze fractured, they are seen to consist of large vesicles, with intramembrane particles found in aggregates on both faces of the membrane (Fig. 17). These vesicles show no ability to accumulate Ca. They often possess a high ATPase activity, but Ca uptake is uncoupled from the ATP utilization.

V. Survival of Dehydration by Intact Cells and Organisms: Anhydrobiosis

A. MEMBRANES IN ANHYDROBIOTIC ORGANISMS

As we noted in the introduction, although water is thought to be required for maintenance of membrane structure and, as we have demonstrated here, dehydration leads to irreversible damage to membranes, many organisms are capable of surviving complete dehydration. The dry organisms, said to be in a state of anhydrobiosis (Keilin, 1959), include living systems like seeds, bacterial spores, mosses, cysts of certain crustaceans, and certain soil-dwelling microorganisms (some species of rotifers, tardigrades, and nematodes; see Bewley, 1979; Crowe and Clegg, 1973, 1978; Mazur, 1980; Womersley, 1981 for reviews.). In these concluding sections we intend to examine the potential mechanisms by which they escape irreversible damage to their membranes.

The few attempts to examine the phase properties of membranes in cells and tissues that survive dehydration that have been done are suggestive.

The techniques that have been applied fall into the following categories.

(1) *Permeability properties*   Simon and his colleagues (reviewed in Simon, 1978) discovered several years ago that when dry seeds were plunged into water their internal constituents leaked into the surrounding medium. They also found that the leakage did not continue unabated, but ceased when the water content rose above about 0.3 g H_2O/g dry weight (g/g). They also found that when the seeds were rehydrated over water vapour before being placed in water, if the water content was greater than 0.3 g/g before the seeds were placed in water the rate of leakage was greatly reduced. Since this water content represents the water content at which major phospholipid phase transitions occur, Simon suggested that membrane phospholipids in the dry seeds might exist in a non-bilayer structure (he suggested hexagonal II phase, although a micellar phase now seems more likely), reverting to bilayers upon rehydration. During the interval required for reversion of the lipids to the bilayer structure, the cells would, Simon reasoned, leak. We have produced similar data for other dry organisms (Crowe *et al.*, 1979), with some important differences; we studied leakage rates from soil-dwelling nematodes that may be completely dehydrated under specialized conditions. These animals must be dried slowly initially, during which time they convert up to 20% of their dry weight to the carbohydrate, trehalose. The special properties of this molecule that make it important in this process will be discussed below. If the animals are dried too rapidly, they are killed by the dehydration. When we studied rates of leakage of cellular constituents from these animals during rehydration, we found that those dried rapidly leaked at a much greater rate than those dried slowly. In both groups of organisms, however, the rate of leakage declined to insignificant levels when they were hydrated to >0.3 g/g. We concluded from these data that if the cause of leakage is indeed the existence of non-bilayer structures in the cell membranes of the dry organisms, as Simon suggested, such non-bilayer structures are more common in the dry, dead organisms than the dry, viable ones.

(2) *Freeze fracture*   Several workers have searched with freeze fracture for non-bilayer structures in membranes of dry organisms, with equivocal results. There is some evidence that such structures exist in dry cells (but not in membranes; cf. Tovio-Kinnucan and Stushnoff, 1981), but the available evidence can at best be judged as inconclusive.

(3) *X-ray diffraction*   McKersie and Stinson (1980) have studied phase behaviour of phospholipids isolated from seeds. They found that at all water contents the lipids in their extracts were in lamellar phase.

This finding is particularly interesting because they reported that their extracts contained 25–30% PE, a phospholipid known to enter H_{II} readily, as we have seen. It seems likely that phase separation of phospholipids, known to occur in bulk lipids (as we pointed out above), would result in at least some H_{II} during dehydration. The fact that McKersie and Stinson found no evidence of H_{II} in these preparations suggests that lipid composition may affect the phase behaviour. Data of this sort on complex mixtures of phospholipids are clearly needed.

(4) *NMR* Recent studies by Priestly and De Kruijff (1982) on phase behaviour of membrane phospholipids in dry pollen showed that in the dry state the NMR signal was broadened, similarly to the signal for dry SR shown in Fig. 15, bottom. When water was added, the signal progressively narrowed to an isotropic shape and then clearly assumed the shape characteristic of the bilayer. With the precautions pointed out in the section on in mind, these data suggest that H_{II} structures probably do not exist in these dry cells, but the data are consistent with micellar structures.

B. INHIBITION OF PHASE TRANSITIONS IN ANHYDROBIOTIC ORGANISMS

In our opinion, the bulk of the existing evidence suggests that membranes in cells of organisms that survive dehydration (with the possible exception of seeds) are most likely to be found in bilayers when the organisms are in the dry state. But we have shown, as summarized in the previous section, that dehydration induces phase transitions in a biological membrane. Thus, we have sought an explanation for what would seem to be anomalous phase behaviour of the membranes of anhydrobiotic organisms. One potential explanation was pointed out above: phospholipid composition may alter phase behaviour of complex mixtures, a possibility for which the available data are so scanty that comment much further than what has already been said is not possible. Another possibility is that the membranes are stabilized in bilayers by the presence of materials extrinsic to the membrane. A likely candidate is the carbohydrate, trehalose, a compound that is found at high concentrations (as much as 20% of the dry weight) in a wide variety of organisms that survive dehydration. These include: dry gastrulae of the brine shrimp, *Artemia* (Clegg, 1965); dry larvae or adults of certain nematodes (Madin and Crowe, 1975; Womersley and Smith, 1982); dry, active baker's yeast (Payen, 1949); spores of various fungi (Sussman and Lingappa, 1959); and the resurrection plant *Selaginella lepidophylla* (Harding, 1923). Survival in the dry state is correlated with presence of this compound, which is synthesized during induction of anhydrobiosis and degraded during resumption of active metabolism. It is a non-reducing disaccharide of glucose (the *only* non-reducing disaccharide of glucose, in fact), with an α-1,1 linkage between the glucose monomers, an aspect of the chemistry of this molecule that may play an important role in its stabilizing properties on membranes (Fig. 18).

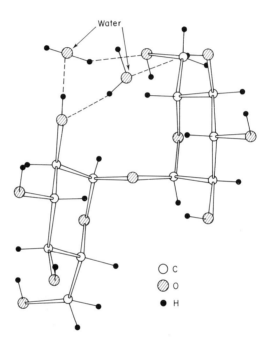

FIG. 18. Diagram of the structure of trehalose. Redrawn from Taga *et al.* (1972)

We have made some progress towards testing the hypothesis that trehalose may stabilize membranes in the dry state, using the SR system. When these membranes are dried in trehalose at concentrations comparable to those found in anhydrobiotic organisms, the vesicles are seen with freeze fracture to exist in the trehalose matrix (the physical state of which is not yet understood) as collapsed cups (Fig. 19). Upon rehydration, these dry membranes yield vesicles that are morphologically indistinguishable from the freshly prepared vesicles. By contrast, when the membranes are dried at lower trehalose concentrations, extensive fusion, phase transitions, and lateral phase separation of intramembrane particles from phospholipid domains were detected.

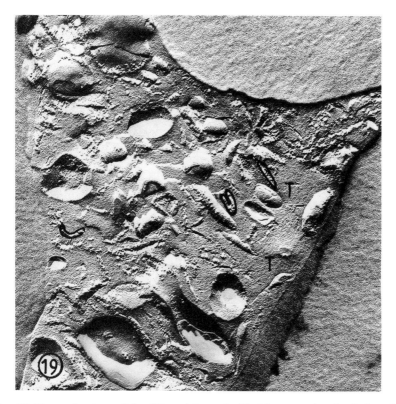

FIG. 19. Freeze fracture of dry SR vesicles embedded in a matrix of trehalose (T). We interpret these images to represent fracture planes through collapsed cups. See Crowe *et al.* (1983b) for a more complete explanation.

Upon rehydration, membranes that had been dried at trehalose concentrations comparable to or greater than those found in anhydrobiotic organisms were seen to possess ATPase activities, Ca transport, and coupling between Ca transport and ATP utilization similar to those of the freshly prepared vesicles (Fig. 20). The trehalose concentration required to stabilize the membrane in the dry state would provide about one trehalose/two phospholipids, an observation that is likely to be important when we consider the mechanism of the stabilization. When thermal analysis of the membranes during drying was conducted, as described above, the thermal activity evident in the membranes dried and rehydrated in the absence of trehalose (Fig. 14) was absent (Fig. 14). We conclude from these studies that the presence of trehalose inhibits hydration dependent phase transitions in these membranes. It is also evident from these data that trehalose inhibits fusion between the vesicles during dehydration. We have previously provided an explanation for this effect (Crowe, J. and Crowe, 1982) which can be summarized as follows: During dehydration the vesicles must become flattened into discs. At the edges of the discs mechanical stresses are likely to exist due to the decreased radius of curvature. In addition, assuming equimolar quantities of phospholipids in both halves of the bilayer, at the edges of the disc the packing density of the polar head groups must be less than packing density towards the centre of the disc, thus exposing the hydrocarbon chains to water. This is an unfavourable thermodynamic state (Tanford, 1980), which it can be shown, taken with the mechanical stress at the edge of the disc, might lead to a rupture of the disc into two bilayer sheets. If such a rupture were to take place, it would inevitably be followed by fusion between neighbouring sheets. Trehalose could inhibit the fusion by preventing physical contact between the vesicles. A similar explanation for the effect of size on fusion between vesicles has been presented by Nir *et al.* (1982).

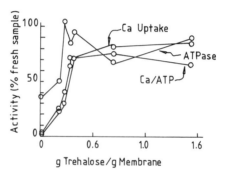

FIG. 20. ATPase activity, Ca transport and coupling between Ca uptake and ATP utilization in hydrated SR that had previously been freeze dried in various concentrations of trehalose. From Crowe *et al.* (1983).

Before proceeding to a discussion of the potential mechanism by which trehalose inhibits phase transitions, we wish to establish that this is a nearly unique property of this molecule. We have tested the stabilizing effects of a wide variety of other carbohydrates on SR in the dry state and have established that trehalose is far more effective in this regard than any of the other sugars tested (Fig. 21). Sugars other than trehalose fall into the following categories:

(1) *Other non-reducing sugars* Sucrose, for example, is effective at preserving membrane integrity in the dry state, as others have previously noted (Sreter *et al.*, 1970), but at least three times as much sucrose is required to achieve the same degree of stabilization as that achieved in the presence of trehalose (Fig. 21).

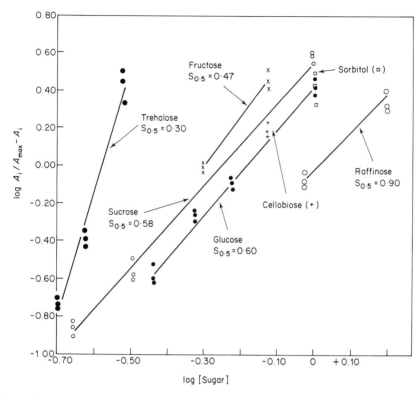

FIG. 21. Ca uptake by rehydrated SR previously freeze dried in the presence of various carbohydrates. These data are a linear transformation, where A_i = activity in rehydrated SR previously dried at (sugar)$_i$ and A_{max} maximal activity in rehydrated SR. From Crowe *et al.*, (1983).

(2) *Reducing sugars* Some reducing sugars such as glucose are initially effective at stabilizing the dry SR, but with the passage of time ATPase activity and Ca transport decline. The coupling ratio between Ca transported and ATP utilized remains constant during this loss of ATPase activity. We interpret this effect to be due to a browning reaction between the reducing sugar and the ATPase. This well known reaction results in denaturation of the protein (cf. Loomis *et al.*, 1979). Thus, we suspect that loss of ATPase activity and Ca transport is due to browning of the Ca-ATPases in the membrane. The high coupling ratio reflects the fact that some ATPases have not yet undergone a browning reaction and continue to transport Ca.

(3) *Polyalcohols* Molecules such as glycerol are well known to have fusogenic properties that might be expected to have deleterious effects on membrane integrity. We have shown this to be the case for all the

polyalcohols we have studied; polyalcohols are the least effective of all the carbohydrates we have tested at stabilizing membrane structure at low water activity.

VI. Interactions of Carbohydrates with Molecules of Biological Interest

A. INTERACTIONS WITH WATER

All but one of the sugars and polyalcohols that have stabilizing effects on SR membranes at low water activities have been shown to interact strongly with water. The single exception is trehalose, which has not been studied in this regard. It has been postulated that these carbohydrates participate in the water lattice in bulk water through hydrogen bonding, and in fact there is some thermodynamic evidence that in at least a few carbohydrates (glucose and sucrose, for example) hydrogen bonding between the sugars and water is stronger than hydrogen bonding between water molecules themselves (Taylor and Rowlinson, 1955). This ability of carbohydrates to interact strongly with water may be related to the spacing of –OH groups on the hydrocarbon backbone of the sugar; Kabayama and Patterson (1958) showed that equatorial –OH groups on pyranose and cyclitol rings could fit precisely into the water lattice in bulk water. They also suggested that since axial –OH groups do not possess the same spacing, the fit into the water lattice should not be as favourable as that of equatorial –OH groups. This appears to be the case; thermodynamic studies have shown that at concentrations up to about 0.3 M, mannitol has a structure-making effect on water, while sorbitol has a structure-breaking effect; these two stereoisomers differ in the position of a single –OH group (Stern and O'Connor, 1972), mannitol possessing –OH groups with the appropriate spacing to fit into the water lattice, and sorbitol not. Along the same lines, Franks and his colleagues (Franks, 1975, Franks *et al.*, 1972; Tait *et al.*, 1972) have shown that hydration of a number of sugars is favoured by the presence of equatorial –OH groups. Such sugars, in turn, have a structure-making effect on water.

B. INTERACTIONS WITH PROTEINS

The notion that sugars and sugar alcohols may increase structuring of water by entering into the water structure is attractive in the present context, since it also suggests that these carbohydrates could substitute for water, particularly that small fraction which makes up the main hydration shell. The evidence for this hypothesis is fragmentary but nevertheless suggestive. Much of the following evidence deals not with membranes but with proteins.

However, we feel that the thermodynamic arguments also apply to membranes, even though the experimental evidence for membranes is not available in the same form. Also, some of the interactions described below are between proteins and glycerol or membranes and glycerol. We are aware that these observations may not be extrapolated to sugar alcohols or even to other polyalcohols, but in the absence of other data this information nevertheless indicates patterns for future research.

A considerable fraction of the surface residues of proteins are hydrophobic, in the sense that they have no atoms available for formation of hydrogen bonds (Chothia, 1974, 1975). These groups are thought to be surrounded by cagelike structures of water molecules hydrogen bonded to each other. This solvent ordering by hydrophobic residues (which results in a decrease in entropy), it has been argued, is the thermodynamic factor which determines the three-dimensional structure of the protein (Tanford, 1978). The polar or ionic groups bind water to them through dipole–dipole or ion–dipole interactions. A number of sugars and sugar alcohols are known to have effects on proteins that may be due to disruption of this protein–water interaction: for example, these compounds increase the thermal stability of proteins, i.e. increase the transition temperature for the protein. Timasheff and his colleagues have shown that the thermal transition for several proteins is elevated in the presence of glycerol. He has ascribed this effect to the fact that glycerol is known to repel hydrophobic substances effectively. In fact, Gekko and Timasheff (1981) have shown that glycerol is apparently excluded from water structures formed around hydrophobic groups. Even more to the point for our present discussion, Tanford (1980) has shown that the free energy for the transfer of a hydrocarbon from a non-polar solvent to glycerol is nearly as large as the free energy for the transfer to water. Exclusion of glycerol from the water structures around hydrophobic groups is an unfavourable thermodynamic state that could be relieved, Gekko and Timasheff (1981) argue, by removal of hydrophobic residues from contact with the solvent, i.e. by folding into the interior of the protein. Since such folding of all hydrophobic residues into the interior of the protein would be hindered by its three-dimensional structure, Gekko and Timasheff suggest that the water and glycerol molecules redistribute in the vicinity of the protein. The arguments they present are thermodynamic ones, based on thermodynamic data, so no statement can be made about the actual distribution of glycerol and water about the protein. However, they suggest that the net chemical potential of glycerol and water about the protein could be retained in equilibrium with the bulk solution if glycerol enters into the water structure about polar groups, achieving a balance between the repulsion from nonpolar regions, attraction to polar regions, and attraction to water. Thus, the formation of water–glycerol structures around polar groups of the protein seems reasonable. Along similar lines, Clegg *et al.* (1982) have shown that when relaxation times for water protons in water–glycerol–protein mixtures were recorded, a decline in mobility of the water protons was seen with a decrease in water content of the system. At the lowest water contents, however, they recorded an increase in water proton mobility. No such increase in water proton mobility was seen in proteins during dehydration in the absence of glycerol. Clegg *et al.* (1982) interpret this effect to be due to hydrogen bonding of glycerol to polar residues in the protein at low water activities, displacing the water. In summary, although the available data are sparse, it seems reasonable to suggest that glycerol may form hydrogen bonds with polar residues in proteins. Similar stabilizing effects of sugar alcohols are known (Back *et al.*, 1979), and the same arguments may thus apply.

C. INTERACTIONS WITH MEMBRANES

Comparable analyses do not exist for membranes, but it does not seem unlikely that an analogous effect of alcohols and sugar alcohols may occur on membranes. What we are suggesting is that a molecule like glycerol or a sugar alcohol might enter into the water structure centred around the phosphate group in a phospholipid. The increase in water structuring could lead to stabilization of the bilayer structure during dehydration. At the lowest water contents, mobility of the polar head group might be restricted by hydrogen bonding of –OH groups in the alcohol directly to head groups of adjacent phospholipids or to any water remaining in the main hydration shell. Support for this hypothesis may be had from only a few sources, but they are suggestive ones. Cadenhead and Demchak (1969) showed, using Langmuir troughs, that DMPC spread on an aqueous surface showed a lateral expansion of the film when glycerol was placed in the subphase. This expansion under the influence of glycerol elevated the maximal packing density from about 0.40 nm² (40 Å²)/molecule to about 0.46 nm² (46 Å²) (our measurements, based on the data in their paper). They ascribed this effect to insertion of glycerol molecules between the phospholipid headgroups. This seems to be a reasonable interpretation, because most membranes show high permeability to glycerol (Brown *et al.*, 1982). In addition, more recent evidence based on X-ray diffraction (McDaniel *et al.*, 1983), monolayer studies (Crowe, J. *et al.*, 1984a) and differential scanning calorimetry (McDaniel *et al.*, 1983; Crowe *et al.*, 1984b) is consistent with the suggestion that glycerol inserts between polar head groups. Leaving the mechanism of the effect for the moment, the lateral expansion of the membrane would be expected to increase the fluidity of the hydrocarbon chains, and there is good evidence in the literature that this is the case. Eliasz *et al.* (1976)

showed that *n*-alcohols containing fewer than 10 carbons all lowered the transition temperature of phospholipid bilayers (i.e. increased the fluidity), while longer-chained *n*-alcohols had the opposite effect. Similar measurements with a series of phenylalcohols indicated that increasing the number of methylene groups between the phenyl group and the terminal –OH decreased the transition temperature (Fig. 22), a finding which seems consistent with the possibility that the alcohol may insert between the polar head groups.

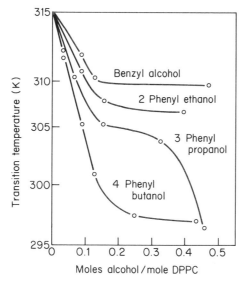

FIG. 22. Effects of various phenyl alcohols on the transition temperature of DPPC. With an increase in the number of methylene groups between the phenyl group and terminal –OH, there is an increase in the effect on transition temperature. Redrawn from Eliasz *et al.* (1976).

Similar data have been published by Hui and Barton (1973) and Ebihara *et al.* (1979), with similar conclusions. To our knowledge, the only comparable data for interactions between carbohydrates and membranes may be found in the recent work of Chen *et al.* (1981), who have published differential scanning calorimetry data on DPPC in the presence of various carbohydrates. They showed, for example, that when the galactose concentration around the lamellar phase lipids was increased, the enthalpy of the gel to liquid crystalline phase transition was apparently decreased. They provided an explanation of this effect based on the proposition that the carbohydrate increases hydrophobic interactions between membrane phospholipids, thereby excluding water from the region of the hydrocarbon chains. They proposed that a significant proportion of the enthalpy change observed at the main transition temperature was due to the melting of this water. Therefore, they reasoned, its exclusion from around the hydrocarbon chains would decrease the enthalpy of the phase transition, without a change in the phase transition temperature. We suspect that an alternative explanation might be more appropriate. Basing a model on penetration of the bilayer is suspect *a priori*, since the available evidence indicates that significant amounts of water are not likely to be found below the level of the first two or three methylene groups of the hydrocarbon chains (see Section III for further discussion).

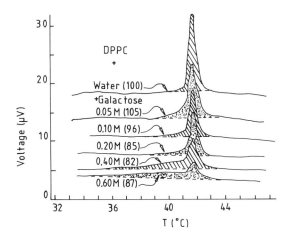

Fig. 23. Differential scanning calorimetry scans of DPPC vesicles prepared in the presence of various concentrations of galactose. Note that the main transition temperature appears not to change and that the enthalpy of the transition is apparently decreased under influence of the carbohydrate. However, broadening of the enthalpy in the low temperature direction suggests a decrease in the transition temperature under influence of the carbohydrate. In order to test this hypothesis, we measured the relative areas under each endotherm, illustrated by the numbers on the curves. It is clear from these measurements that the total enthalpy of the transition is actually changed only by a small amount under influence of the carbohydrate. We interpret this result to mean that fluidity of the DPPC bilayers is increased under the influence of the carbohydrate. Original data from Chen *et al.* (1981). The dashed lines represent base lines we used in measurements of the areas under each endotherm.

In addition, Professor Dennis Chapman pointed out to us that in every case where there is a decrease in the enthalpy of the main transition under the influence of the carbohydrate shown in the data of Chen *et al.* (1981), there is a long tail on the traces on the low temperature side (Fig. 23). If that enthalpy is taken into account, it may be seen that the total enthalpy of the phase transition is changed by at most 20%, although the phase transition temperature is broadened in the low temperature direction (Fig. 23). In fact, we suspect that the enthalpy change may be even less in view of the vagaries involved in drawing the baseline accurately with the broad endotherms seen in Fig. 23. Thus, this decrease in phase transition temperature under influence of the carbohydrate could be explained by a lateral expansion of the membrane and concomitant increase in fluidity. The increase in fluidity could, in turn, be ascribed to interaction of the carbohydrate with the head groups, as previous workers (Cadenhead and Demchak, 1969; Ebihara et al., 1979) have suggested for other alcohols. Recent studies in our laboratory suggest that this is the case.

D. MECHANISM OF STABILIZATION OF MEMBRANES BY TREHALOSE

It appears to us that a case can be made, despite the relative paucity of data, that carbohydrates may stabilize membranes not only due to their effect on increasing ordering of solvent water but also due to hydrogen bonding, which could result in restricted mobility of the polar head group. The location of the carbohydrate near the head group is of course still unknown, but the possibility that it is inserted between head groups does not necessarily preclude hydrogen bonding to water around the head group or directly to the head group. Indeed, from what we know about the conformational state of the head group both at low water activities (Hauser et al., 1981) and bulk solution (Seelig, 1978), insertion between head groups and simultaneous hydrogen bonding to the head group or to water in the main hydration shell is feasible. Experimental data that can test this model have not yet been obtained, nor has rigorous modelling been attempted. However, it is interesting to note that the concentration of trehalose required to stabilize the dry sarcoplasmic reticulum membranes would provide one trehalose molecule/two phospholipids. Our most recent studies with infrared spectroscopy suggest the existence of interactions between –OH groups in trehalose and the $P=0$ of the polar head group in dry phospholipid–trehalose preparations (Crowe, J. et al., 1984b).

Another remaining problem is an explanation of why some carbohydrates, particularly trehalose, are more effective than others at stabilizing membranes at low water activities. Using the rationale that comparisons of the conformation of the various carbohydrates we have used in these studies might lend some clues concerning the mechanism of their stabilizing effects, we have made a study of their conformations, based on X-ray diffraction studies taken from the literature. The conformation of the sugars and sugar alcohols near the head groups is, of course, unknown, but it is more likely to be closer to the crystalline state than the conformation in bulk solution.

TABLE III

Structural characteristics of various carbohydrates

Carbohydrate	-OH groups available for intermolecular hydrogen bonding/nm^3	Equatorial -OH groups available for intermolecular hydrogen bonding/nm^3	Equatorial -OH/total -OH	Reference
Trehalose dihydrate[a]	19	19	1	Brown et al. (1972); Taga et al. (1972)
α-Lactose monohydrate[a]	20	10	0.5	Beevers and Hansen (1971); Fries et al. (1971)
β-Maltose monohydrate[a]	18	15	0.83	Quigley et al. (1970); Gress and Jeffrey (1977)
Cellobiose[b]	19	19	1	Jacobson et al. (1961); Brown (1966), Chu and Jeffrey (1968)
Sucrose[b]	14	11	0.8	Beevers et al. (1952); Brown and Levy (1963)
Glucose[b]	26	20	0.77	McDonald and Beevers (1952); Brown and Levy (1965)
Fructose[b]	27	5	0.2	Kanters et al. (1977); Takagi and Jeffrey (1977)
Sorbitol[b]	30	n.a.		Park et al. (1971)
Raffinose	17	17	1	Berman (1970)
Myo-inositol	30	25	0.83	Rabinowitz and Kraut (1964)
Glycerol	25	n.a.		Calculated from density and molecular weight

We paid special attention to the number and position of –OH groups available for hydrogen bonding, their density in the crystalline structure, as well as the number of axial and equatorial –OH groups. A summary of the data and a list of the sources is shown in Table III, with the carbohydrates listed in relative order of effectiveness at stabilizing the membranes. There appears to be no clear relationship between any of the parameters of the crystal structure and effectiveness of the carbohydrate, but the data do suggest that the number of equatorial –OH groups available for hydrogen bonding has little to do with the relative effectiveness of these compounds. For example, trehalose and cellobiose have the same number of –OH groups available for intermolecular bonding, the same number of –OH groups/unit volume, and all the –OH groups are equatorial, yet trehalose is twice as effective as cellobiose in stabilizing the membrane. Sucrose, glucose, and fructose have 11, 5 and 20 equatorial –OH groups/unit volume, respectively, but all have nearly the same effectiveness. Myo-inositol, which has five of its six –OH groups in an equatorial position (more equatorial –OH groups/unit volume than trehalose) and which forms strong hydrogen bonds with water is one of the least effective carbohydrates tested.

VII. Conclusions

The available evidence indicates that removal of the water in the main hydration shell centred around the polar head groups of at least some phospholipids leads to lateral phase separations of phospholipid classes and membrane proteins. These phase separations result in crystallization of steroids such as cholesterol and phase transitions of some of the lipids such as phosphatidylethanolamine to complex crystalline phases. These upheavals in the organization of membrane lipids and proteins accompanying dehydration irreversibly alter both the structural and functional integrity of the membrane.

Despite the existence of these deleterious effects of dehydration on membrane structure, cells and organisms are commonly capable of surviving complete dehydration in nature. Many of these cells and organisms contain significant quantities of low molecular weight carbohydrates, particularly trehalose. We have shown that membrane vesicles dried in the presence of trehalose retain both their structural and functional integrity. When the carbohydrate is added to the membrane at concentrations comparable to those found in organisms that survive dehydration, the protective effect was found to be unique to trehalose. While the available evidence seems consistent with our hypothesis that this effect is due to replacement of the main hydration shell of the polar head group of phospholipids by –OH groups of trehalose, we are just beginning to test that hypothesis directly.

It is also probable that the lipid and protein composition may affect membrane stability at low water activities. With this thought in mind, it seems worth while pointing out that the results discussed here have been obtained with membranes isolated from organisms that normally do not survive dehydration. Nevertheless, as we have shown, the structural and functional integrity of these membranes is maintained intact when they are dried in the presence of trehalose, a compound found at high concentrations in cells that will survive dehydration. This is particularly interesting because one might think, at first glance, that the ability to survive dehydration— whether exhibited by an isolated membrane, an intact cell, or a whole organism—would involve a multitude of adaptations. It follows that one might expect cells that normally survive dehydration, such as those of seeds, would be equipped through evolutionary history with so specialized a chemical composition and structure that it would be impossible to dehydrate without killing it a cell whose evolutionary history has not generated a similar set of adaptations. We suspect that is not the case; we believe that relatively few modifications may be required to render an intact cell capable of surviving dehydration, and that ultimately it will be possible to do so.

VIII. Acknowledgements

We gratefully acknowledge support by the National Science Foundation through grant 80-04720 and National Sea Grant for grants RA/41 and RA/43. We also acknowledge the helpful discussions we have had with Professors Pieter Cullis, David W. Deamer and Ralph Amey about the physical properties of membranes, the assistance of Dr G. B. Matson and Dr W. B. Busa with the NMR studies, and the able technical assistance of L. C. Cooper, S. A. Jackson, R. Harris, and C. Hall. We are also grateful to our students and post-doctoral associates past and present who have participated in this work. Finally, we are particularly grateful to Professor Dennis Chapman for his many helpful discussions and for his cordial welcome to his laboratory, where this review was written when we were on sabbatical leave in 1983.

References

Arnold, K., Losche, A. and Gawrisch, K. (1981). *Biochim. Biophys. Acta* **645**, 143.
Atkinson, D., Hauser, H., Shipley, G. G. and Stubbs, J. M. (1974). *Biochim. Biophys. Acta* **339**, 10.
Back, J. F., Oakenfull, D. and Smith, M. B. (1979). *Biochemistry* **18**, 5191.
Bearer, E. L., Duzgunes, N., Friend, D. S. and Papahadjopoulos, D. (1982). *Biochim. Biophys. Acta* **693**, 93.
Beevers, C. A. and Hansen, H. (1971). *Acta Cryst.* **B27**, 1323.
Beevers, C. A., McDonald, T. R. R., Robertson, J. H. and Stern, F. (1952). *Acta Cryst.* **5**, 689.

Berg, S. P., Lusczakowski, D. M. and Morse, P. D. II. (1979). *Arch. Biochem. Biophys.* **194**, 138.

Berman, H. M. (1970). *Acta Cryst.* **B26**, 290.

Bewley, D. (1979). *Plant Physiol.* **30**, 195.

Brown, E. F., Sussman, I., Avron, M. and Degani, H. (1982) *Biochim. Biophys. Acta* **690**, 165.

Brown, C. J. (1966). *J. Chem. Soc. A.* **1966**, 927.

Brown, G. M. and Levy, H. A. (1963). *Science* **141**, 921.

Brown, G. M. and Levy, H. A. (1965). *Science* **147**, 1038.

Brown, G. M., Rohrer, D. C., Berking, B., Beevers, C. A., Gould, R. O. and Simpson, R. (1972). *Acta Cryst.* **B28**, 3145.

Buldt, G. and Wohlgemuth, R. (1981). *J. Membrane Biol.* **58**, 81.

Buldt, G., Gally, H. U., Seelig, J. and Zaccai, G. (1979). *J. Mol. Biol.* **134**, 673.

Cadenhead, D. A. and Demchak, R. J. (1969). *Biochim. Biophys. Acta* **176**, 849.

Cadenhead, D. A. and Muller-Landau, J. (1974). *J. Colloid Interface Sci.* **49**, 131.

Chapman, D. (1968). *In* "Biological Membranes". (ed. D. Chapman). Academic Press, London.

Chapman, D. and Plenkett, S. A. (1966). *Nature* **211**, 1304.

Chapman, D., Williams, R. M. and Ladbrooke, B. D. (1967). *Chem. Phys. Lipids* **1**, 445.

Chen, C.-H., Berns, D. S. and Berns, A. S. (1981). *Biophys. J.* **36**, 359.

Chothia, C. H. (1974). *Nature* **248**, 338.

Chothia, C. H. (1975). *Nature* **254**, 304.

Chu, S. S. C. and Jeffrey, G. A. (1968). *Acta Cryst.* **B24**, 830.

Clegg, J. S. (1965). *Comp. Biochem. Physiol.* **14**, 135.

Clegg, J. S. (1981). *J. Exp. Zool.* **215**, 303.

Clegg, J. S. (1982). *In* "Biophysics of Water". (eds F. Franks and S. Mathias). John Wiley, New York.

Clegg, J. S., Seitz, P., Seitz, W., and Hazlewood, C. F. (1982). *Cryobiology* **19**, 306.

Clement, N. R. and Gould, J. M. (1980). *Arch. Biochem. Biophys.* **202**, 650.

Coster, H. G. L., Laver, D. R. and Schoenborn, B. P. (1982). *Biochim. Biophys. Acta* **686**, 141.

Crowe, J. H. and Clegg, J. S. (1973). "Anhydrobiosis". Dowden, Hutchinson and Ross. Stroudsburg, Pa.

Crowe, J. H. and Clegg, J. S. (1978). "Dry Biological Systems". Academic Press, London.

Crowe, J. H. and Crowe, L. M. (1982). *Cryobiology* **19**, 317.

Crowe, J. H., Crowe, L. M. and Deamer, D. W. (1982). *In* "Biophysics of Water". (eds F. Franks and S. Mathias). John Wiley, New York.

Crowe, J. H., Crowe, L. M. and Jackson, S. A. (1983a). *Arch. Biochem. Biophys.* **226**, 477.

Crowe, J. H., Crowe, L. M. and Mouradian, R. (1983b). *Cryobiology* **20**, 346.

Crowe, J. H., O'Dell, S. J. and Armstrong, D. A. (1979). *J. Exp. Zool.* **207**, 431.

Crowe, J. H., Whittam, M. A., Chapman, D. and Crowe, L. M. (1984a). *Biochim. Biophys. Acta* **769**, 151.

Crowe, J. H., Crowe, L. M. and Chapman, D. (1984b). *Science* **223**, 701.

Crowe, L. M. and Crowe, J. H. (1982). *Arch. Biochem. Biophys.* **217**, 582.

Crowe, L. M., Mouradian, R., Crowe, J. H., Jackson, S. A., and Womersley, C. (1984). *Biochim. Biophys. Acta* **769**, 141.

Cullis, P. R. and De Kruijff, B. (1978). *Biochim. Biophys. Acta* **513**, 31.

Cullis, P. R. and De Kruijff, B. (1979). *Biochim. Biophys. Acta* **559**, 399.

Cullis, P. R. and Hope, M. J. (1978). *Nature* **271**, 672.

Cullis, P. R. and Hope, M. J. (1980). *Biochim. Biophys. Acta* **597**, 533.

Darke, A., Finer, E. G., Flook, A. G., and Phillips, M. C. (1972). *J. Mol. Biol.* **63**, 265.

Deamer, D. W., Leonard, R., Tardieu, A. and Branton, D. (1970). *Biochim. Biophys. Acta* **219**, 47.

Drost-Hansen, W. (1982). *In* "Biophysics of Water". (eds F. Franks and S. Mathias) John Wiley, New York.

Ebihara, L., Hall, J. E., MacDonald, R. C., McIntosh, T. J. and Simon, S. A. (1979). *Biophys. J.* **28**, 185.

Edmonds, D. (1983). *In* "Biological Membranes". (ed. D. Chapman). Vol. 4. Academic Press, London.

Eliasz, N. W., Chapman, D. and Ewins, D. F. (1976). *Biochim. Biophys. Acta* **448**, 220.

Elworthy, P. H. (1961). *J. Chem. Soc.* 5385.

Elworthy, P. H. and McIntosh, D. S. (1964). *J. Phys. Chem.* **68**, 3448.

Finer, E. G. (1973). *J. Chem. Soc., Faraday Trans.* II **69**, 1590.

Finer, E. G. and Darke, A. (1974). *Chem. Phys. Lipids* **12**, 1.

Finer, E. G., Flook, A. G., and Hauser, H. (1972). *Biochim. Biophys. Acta* **260**, 49.

Fockson, J. E. and Wallach, D. F. H. (1978). *Arch. Biochem. Biophys.* **189**, 195.

Franks, F. (1975). *In* "Water Relations of Foods". (ed. R. B. Duckworth). Academic Press, London.

Franks, F. and Mathias, S. (eds) (1982). "Biophysics of Water", 400 pp. John Wiley, New York.

Franks, F., Ravenhill, J. R. and Reid, D. S. (1972). *J. Solution Chem.* **1**, 3–16.

Fries, D. C., Rao, S. T. and Sandarlingham, M. (1971). *Acta Cryst.* **B27**, 334.

Garlid, K. (1978). *In* "Dry Biological Systems". (eds J. H. Crowe and J. S. Clegg) Academic Press, London.

Gekko, K. and Timasheff, S. N. (1981). *Biochemistry* **20**, 4667.

Gress, M. E. and Jeffrey, G. A. (1977). *Acta Cryst.* **B33**, 2490.

Griffith, O. H., Dehlinger, P. J. and Van, S. P. (1974). *J. Memb. Biol.* **15**, 159.

Gruner, S. M., Barry, D. T. and Reynolds, G. T. (1982). *Biochim. Biophys. Acta* **690**, 187.

Harding, L. (1923). *Sugar* **25**, 475.

Harlos, K. (1978). *Biochim. Biophys. Acta* **511**, 348.

Hauser, H. (1975a). *In* "Water—A Comprehensive Treatise". (ed. F. Franks). Plenum Press, New York and London. Vol. 4.

Hauser, H. (1975b) *In* "Water Relations of Foods". (ed. R. B. Duckworth). Academic Press, London.

Hauser, H., Pascher, I., Pearson, R. H. and Sundell, S (1981). *Biochim. Biophys. Acta* **650**, 21.

Hauser, H. and Phillips, M. C. (1973). *J. Biol. Chem.* **248**, 8585.

Henrikson, P. K. (1970). *Biochim. Biophys. Acta* **203**, 228.

Hoffman, W., Sarzala, G. M., Gomez-Fernandez, J. C., Goni, F. M., Restall, C. J. and Chapman, D. (1980). *J. Mol. Biol.* **141**, 119.

Hui, F. K. and Barton, P. G. (1973). *Biochim. Biophys. Acta* **296**, 510.

Hui, S., Stewart, T. P., Yeagle, P. L., and Albert, A. D. (1981). *Arch. Biochem. Biophys.* **207**, 227.

Israelachvili, J. N., Mitchell, D. J. and Ninham, B. W. (1976). *J. Chem. Soc. Faraday Trans.* **272**, 1525.

Jacobson, R. A., Wunderlich, J. A. and Lipscomb, W. N. (1961). *Acta Cryst.* **14**, 598.

Jendrasiak, G. L. and Hasty, J. H. (1974). *Biochim. Biophys. Acta* **337**, 79.

Kabayama, M. A. and Patterson, D. (1958). *Can. J. Chem.* **35**, 563.

Kagawa, Y., Kandrach, A., and Racker, E. (1973). *J. Biol. Chem.* **248**, 676.

Kanters, J. A., Roelofsen, G., Albas, B. P. and Meinders, I. (1977). *Acta Cryst.* **B33**, 665.

Keilin, D. (1959). *Proc. Roy. Soc. London* **B150**, 149.

Kodama, M., Kuwabara, M. and Seki, S. (1982). *Biochim. Biophys. Acta* **689**, 567.

Labuza, T. P. (1975). *In* "Water Relations of Foods." (ed. R. B. Duckworth). Academic Press, London.

Ladbrooke, B. D., Jenkinson, T. J., Kamat, V. B. and Chapman, D. (1968a). *Biochim. Biophys. Acta* **164**, 101.

Ladbrooke, B. D., Williams, P. M. and Chapman, D. (1968b). *Biochim. Biophys. Acta* **150**, 333.

Langley, K. E. and Kennedy, E. P. (1979). *Proc. Natl Acad. Sci., USA* **76**, 6245.

Lee, A. G. (1977a). *Biochim. Biophys. Acta* **472**, 237.

Lee, A. G. (1977b). *Biochim. Biophys. Acta* **472**, 285–344.

Loomis, S. H., O'Dell, S. J. and Crowe, J. H. (1979). *J. Exp. Zool.* **208**, 355.

Luna, E. J. and McConnell, H. M. (1978). *Biochim. Biophys. Acta* **509**, 412.

Luzzati, V. (1968). *In* "Biological Membranes". (ed. D. Chapman). Academic Press, London

Luzzati, V. (1982). *J. Physiol. (Paris)* **77**, 1025.

Luzzati, V. and Husson, F. (1962). *J. Cell Biol.* **12**, 207.

Madin, K. A. C. and Crowe, J. H. (1975). *J. Exp. Zool.* **193**, 335.

Martonosi, M. A. (1974). *FEBS Lett.* **47**, 327.

Mazur, P. (1980). *Orig. Life* **10**, 137.

McDaniel, R. V., McIntosh, T. J. and Simon, S. A. (1983). *Biochim. Biophys. Acta* **731**, 97.

McDonald, T. R. R. and Beevers, C. A. (1952). *Acta Cryst.* **5**, 654.

McIntosh, T. J., Simon, S. A. and MacDonald, R. C. (1980). *Biochim. Biophys. Acta* **597**, 445.

McKersie, B. D. and Stinson, R. H. (1980). *Plant Physiol.* **55**, 316.

Misiorowski, R. L. and Wells, M. A. (1973). *Biochemistry* **12**, 967.

Mitchell, P. (1979). *In* "Membrane Bioenergetics". (eds C. P. Lee, C. Schatz and L. Ernster). Addison-Wesley. Reading, Mass. p. 361.

Negendank, W. (1982). *Biochim. Biophys. Acta* **694**, 123.

Nir, S., Wilschut, J. and Bantz, J. (1982). *Biochim. Biophys. Acta* **688**, 275.

Noggle, J. H., Maracek, J. F., Mandal, S. B., Van Venetie, R., Rogers, J., Jain, M. K. and Ramirez, F. (1982). *Biochim. Biophys. Acta* **691**, 240.

Ort, D. R., Dilly, R. A. and Good, N. E. (1976). *Biochim. Biophys. Acta* **449**, 108.

Papahadjopoulos, D., Vail, W. J., Newton, C., Nir, S., Jacobson, K., Poste, G. and Lazo, R. (1977). *Biochim. Biophys. Acta* **465**, 579.

Park, Y. J., Jeffrey, G. A. and Hamilton, W. C. (1971). *Acta Cryst.* **B27**, 2393.

Payen, R. (1949). *Can. J. Res.* **27**, 749.

Pink, D. A. (1981). *In* "Biological Membranes". (ed. D. Chapman). Academic Press, London. p. 131.

Priestly, D. A. and De Kruijff, B. (1982). *Plant Physiol.* **70**, 1975.
Quigley, G. J., Sarko, A., and Marchessault, R. H. (1970). *J. Am. Chem. Soc.* **92**, 5834.
Quinn, P. J. and Chapman, D. (1980). *CRC Crit. Rev. Biochem.* **8**, 1.
Rabinowitz, I. N. and Kraut, J. (1964). *Acta Cryst.* **17**, 159.
Rand, R. P. (1981). *Ann. Rev. Biophys. Bioeng.* **10**, 277.
Rand, R. P. and SenGupta, S. (1972a). *Biochim. Biophys. Acta* **255**, 484.
Rand, F. P. and SenGupta, S. (1972b). *Biochemistry* **11**, 945–949.
Rand, R. P. and SenGupta, S. (1972). *Biochim. Biophys. Acta* **255**, 484.
Rigaud, J. L., Gary-Bobo, C. M. and Lange, Y. (1972). *Biochim. Biophys. Acta* **266**, 72.
Seelig, J. (1978). *Biochim. Biophys. Acta* **515**, 105.
Simon, E. W. (1978). In "Dry Biological Systems". (eds J. H. Crowe and J. S. Clegg). Academic Press, New York and London.
Simon, S. A., McIntosh, T. J. and Latorre (1982). *Science* **216**, 65–67.
Small, D. M. (1967). *J. Lipid. Res.* **8**, 551.
Sreter, F., Ikemoto, N. and Gergeley, J. (1970). *Biochim. Biophys. Acta* **203**, 335.
Stern, J. H. and O'Connor, M. E. (1972). *J. Phys. Chem.* **21**, 3077.
Stoeckenius, W. (1962). *J. Cell Biol.* **12**, 221–229.
Stumpel, J., Eibl, H. and Nicksch, A. (1983). *Biochim. Biophys. Acta* **727**, 245.
Sussman, A. S. and Lingappa, B. T. (1959). *Nature* **130**, 1543.
Taga, T., Senma, M. and Osaki, K. (1972). *Acta Cryst.* **B28**, 3258.
Tait, M. J., Suggett, A., Franks, F., Ablett, A. and Quickenden, P. A. (1972). *J. Solution Chem.* **1**, 131.
Takagi, S. and Jeffrey, G. A. (1977). *Acta Cryst.* **B33**, 3510.
Tanford, C. (1978). *Science* **200**, 1012.
Tanford, C. (1980). "The Hydrophobic Effect." John Wiley and Sons, New York (2nd edition).
Taylor, J. B. and Rowlinson, J. S. (1955). *Trans. Faraday Soc.* 1183.
Ter-Minassian-Saraga, L. and Madelmont, G. (1981). *J. Colloid Interface Sci.* **81**, 369.
Ter-Minassian-Saraga, L. and Madelmont, G. (1982). In "Biophysics of Water". (eds F. Franks and S. Mathias). John Wiley and Sons. New York. p. 127.
Ter-Minassian-Saraga, L. and Madelmont, G. (1983). *Biochim. Biophys. Acta* **728**, 394.
Tinker, D. O. and Pinteric, L. (1971). *Biochim. Biophys. Acta* **183**, 304.
Tovio-Kinnucan, M. A. and Stushnoff, C. (1981). *Cryobiol.* **18**, 72.
Ulmius, J., Lindblom, G., Wennerstrom, H., Johansson, L.B.-A., Fontell, K., Soderman, O. and Arvidson, G. (1982). *Biochem.* **21**, 1553.
Vail, W. J. and Stollery, J. G. (1979). *Biochim. Biophys. Acta* **551**, 71.
Van Dijck, P. W. M., De Kruijff, B., van Deenen, L. L. M., de Grier. I., and Demel, R. A. (1976). *Biochim. Biophys. Acta* **455**, 576.
Van Echteld, C. J. A., De Kruijff, B., Verkleij, A. J., Leunissen, J., and De Gier, J. (1982). *Biochim. Biophys. Acta* **692**, 126.
Van Heerikhuizen, H., Kwak, E. Van Brussen, E. F. J., and Witholt, B. (1975). *Biochim. Biophys. Acta* **413**, 177.
Van Venetie, R. and Verkleij, A. J. (1981). *Biochim. Biophys. Acta* **645**, 262.
Van Venetie, R. and Verkleij, A. J. (1982). *Biochim. Biophys. Acta* **692**, 397.
Verkleij, A. J. and Ververgaert, P. H. J. T. (1975). *Ann. Rev. Phys. Chem.* **26**, 101.

Veksli, Z., Salsbury, N. K. and Chapman, D. (1969). *Biochim. Biophys. Acta* **183**, 434–446.

Veksli, Z., Salsbury, N. K. and Chapman, D. (1972). *Biochim. Biophys. Acta* **183**, 434, 196, 1055.

Wennerstrom, H. and Lindman, B. (1979). *J. Phys. Chem.* **83**, 2931.

White, S. H. (1977). *Ann. N. Y. Acad. Sci.* **303**, 243.

Wieslander, A., Christiansson, A., Rilfors, L. and Lindblom, G. (1980). *Biochemistry* **19**, 3650.

Wieslander, A., Rilfors, L., Johansson, L. B-A. and Lindblom, G. (1981). *Biochemistry* **20**, 730.

Wiggins, P. M. and Bowmaker, G. A. (1982). *In* "Biophysics of Water". (eds F. Franks and S. Mathias). John Wiley and Sons, New York. p. 218.

Williams, R. M. and Chapman, D. (1970). *Prog. Chem. Fats Other Lipids* **11**, 1.

Wilschut, J., Holsappel, M. and Jansen, R. (1982). *Biochim. Biophys. Acta* **690**, 297.

Womersley, C. (1981). *Comp. Biochem. Physiol.* **70B**, 669.

Womersley, C. and Smith, L. (1982). *Comp. Biochem. Physiol.* **70B**, 579.

Worcester, D. L. (1976). *In* "Biological Membranes". (eds. D. Chapman, and D. F. Wallach). Academic Press, London. Vol. 3, p. 1.

Zaccai, G., Buldt, G., Seelig, A. and Seelig, J. (1979). *J. Mol. Biol.* **134**, 693.

Chapter 3

Physical Basis of Trigger Processes and Membrane Structures

ERICH SACKMANN

*Physik Department E22, Technical University of Munich,
8046 Garching-Munich, West Germany*

BIOLOGICAL MEMBRANES Vol. 5
ISBN 0 12 168546 2

I. Introduction

Biomembranes are multicomponent systems of lipids and proteins which are organized heterogeneously in such a way that well-organized functional units are formed. Examples are (1) the electron transfer chains of the inner mitochondria membrane or of the photosynthetic system in the thylakoid membrane, (2) the hormone transducing system comprising receptor, G-protein and adenylate cyclase. The essential membrane processes are mediated by proteins. However, it appears that the lipids play an essential role for the formation of the microscopic regions of well-defined composition and for the modulation of the enzymatic activities.

The optimal performance of the functional units depends critically on the molecular architecture of their local environment. Thus there exists a close relationship between the biochemical function and the physical structure of membranes. This relationship is essential for triggering of membrane processes by local structural changes which may be initiated by local phase transitions of the lipid moiety. It is essential that the conformational changes can be induced chemically, that is, by the binding of ions, charged proteins or by changes in the local pH or ionic strength. Membranes are two-dimensional systems, that is, they are strongly coupled to the aqueous environment. The chemically induced transitions can be triggered by the adsorption of molecular species from the bulk phase via electrostatic or dispersion forces. Due to the coupling between the opposing monolayers, a conformational change on one side of the membrane may lead to a complementary change at the opposite leaflet.

In the following article the physical principles of membrane conformational changes are summarized and some trigger mechanisms are suggested.

II. Physical Properties of Lipid Bilayers

From the point of view of structural changes of membranes it is helpful to divide the phospholipids into two classes: the charged lipids, such as phosphatidylserine, phosphatidylglycerol, phosphatidylinositol, cardiolipin, and the lipids with zwitterionic head groups, such as phosphatidylcholine, phosphatidylethanolamine and sphingomyelin. In membranes containing the first class of lipids, conformational changes may be triggered by the adsorption of bivalent ions such as Ca^{2+} or Mg^{2+} or of extrinsic macromolecules with excess positive charges such as cytochrome c or polylysine (Träuble and Eibl, 1974; Ohnishi and Ito, 1974; Galla and Sackmann, 1975; Birrell and Griffith, 1976). These molecules make up a substantial part of biological membrane lipids ($\sim 15\%$ in nerve cells and $\sim 17\%$ in erythrocyte plasma membranes), and the triggering of membrane processes may well be achieved via charge-induced lipid conformational changes.

A. LIPID ORGANIZATION AND STRUCTURAL PHASE TRANSITIONS OF BILAYERS

Phospholipids have a strong tendency to form bilayers in water. However, the so-called hexagonal phase where the lipids form elongated micelles, is also a stable structure in water (Luzatti et al., 1968; Tardieu et al., 1972). For lecithins it is only observed at high temperatures. Pure ethanolamines form hexagonal phases also at low temperatures. Thus 1,2-dioleylphosphatidylethanolamine undergoes a bilayer to hexagonal transition at 0°C (Verkleij et al., 1980). Most interestingly, the bilayer-to-hexagonal transition can be induced by the binding of divalent cations (Ca^{2+}, Mn^{2+}) to acidic phospholipids such as phosphatidylserine or cardiolipin (Harlos and Eibl, 1980). As a transient state of the bilayer-to-hexagonal transition the formation of inverted micelles intercalated between the two opposing monolayers of the bimolecular leaflets (see Fig. 3c) has been postulated by Verkleij et al. (1980).

As mentioned in the introduction, there is growing evidence that the vast variety of different lipids in biological membranes plays an active role for the association of enzymes into functional complexes and that changes in the composition and structure of the lipid environment of these functional units may modulate their activity. It is very probable that the ability to undergo structural phase transitions is essential for this functional role of the lipid molecules. This was first recognized by Chapman (Chapman, 1971).

The bilayer-to-hexagonal conformational change is such a phase transition. Another type is the transition from the fluid to the crystalline (gel) state of bilayers at which the hydrocarbon chains change from a disordered conformation with a high content of rotational isomers to a fully extended state (Fig. 1).

A common feature of both types of lipid phase transitions for charged lipids is that they can be induced either thermically by changes in temperature as well as isothermically by a variation in the pH and the ionic strength, or by the addition of bivalent ions or charged proteins. There are two remarkable differences:

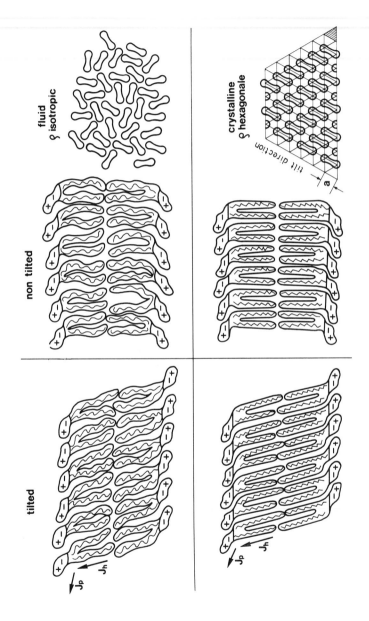

FIG. 1. Schematic representation of symmetry of lipid bilayer phases. The symmetry is determined (1) by the lateral arrangement of the acyl chains which is characterized by the lateral density ρ and (2) by the chain-orientation with respect to the membrane normal.

(1) The chain melting transition temperature of the crystalline-to-fluid transformation of bilayers increases with increasing chain length, while that of the bilayer-to-hexagonal transition decreases with the chain length.

(2) The heat of transition of the chain melting phase change is by an order of magnitude larger than that of the bilayer-to-hexagonal transition, which is remarkably small ($\Delta H \approx 1$ Kcal/mol), and does not depend on the chain length (Harlos and Eibl, 1980).

The bilayer phases are distinguished by their differences in the symmetry of the molecular organization which is characterized (1) by the collective orientation of the chains with respect to the membrane normal and (2) by the distribution of the chains in the plane of the membrane (Fig. 1). A collective tilt of the chain results if the areas occupied in equilibrium by the polar head groups (H) and by the two chains (C) are different (Ladbrooke and Chapman, 1969; Larsson, 1977). The tilted crystalline phase may lead to a corrugated superstructure of the bilayer which is denoted as the ripple phase (Larsson, 1977; Gebhardt *et al.*, 1977; Doniach, 1979; Sackmann and Rüppel, 1980). A tilted configuration may also be induced by rigid boundaries such as non-cylindrical integral protein molecules.

B. MEMBRANE ELASTICITY

Membranes exhibit two types of elasticity which may be exploited by nature for triggering: (1) the lateral elasticity and (2) curvature elasticity. A change ΔA in the area A (e.g. induced by a phase change of membrane bound molecules) leads to a tangential stress in the plane of the membrane of

$$\Delta \pi = \varkappa^{-1} \frac{\Delta A}{A} \tag{1}$$

where $\varkappa$ is the lateral compressibility (in cm/dynes) which may be obtained from film balance measurements (Albrecht *et al.*, 1978). It should be emphasized that a lateral dilatation (or compression) may be relaxed rapidly by the transfer of molecules between the monolayer and the aqueous phase. Thus the lateral elasticity of membranes may be a transient phenomenon.

The mechanical behaviour of membranes and isolated lipid bilayers is dominated by elastic forces associated with changes in the spontaneous curvature of the bimolecular leaflets (Helfrich, 1973; Petrov *et al.*, 1979). The spontaneous curvature of biological membranes is a consequence of the asymmetric distribution of both the lipids and the proteins between the inner and the outer monolayer. These are amphipathic molecules and in equilibrium the areas occupied by the polar head (H) and the hydrophobic core (C) in the plane of the membrane are expected to be different. Molecules with $H > C$ will tend to accumulate in the outer and molecules with $H < C$ in the inner monolayer. From the point of view of triggering processes it is most interesting that the ratio H/C and thus the spontaneous curvature may be changed (1) by the adsorption of macromolecules (e.g. those of the cytoplasmic skeleton) and (2) in the presence of charged lipids (or proteins) by changes in the local pH, the ionic strength or by the adsorption of bivalent ions (Petrov et al., 1979). Deviations from the spontaneous curvature c_0 are associated with an elastic energy per unit area which may be expressed as (Helfrich, 1973; Petrov et al., 1979):

$$G_{el} = \tfrac{1}{2}K(c_1 - c_2 + c_0)^2 + K'c_1c_2 \qquad (2)$$

where c_1^{-1} and c_2^{-1} are the principal radii of curvature of the membrane and where K and K' are the so-called curvature elastic constants. For fluid lipid layers their values differ by about a factor of two $K \sim 2 \times 10^{-12}$ erg K' $\sim 10^{-12}$ erg. The membrane elasticity is dominated by the first contribution. However, for special contours such as myelin-like protrusions of cells (Deuling and Helfrich, 1977) it is $c_1 + c_2 + c_0 = 0$ and the second term comes into play. This contribution may play a very important role for the formation of pores (Petrov et al., 1980), for membrane fusion (Harbich et al., 1978) and for the formation of passages between the extracellular space and partially internalized vesicles such as the coated pits (Pastan and Willingham, 1981).

C. LATERAL PHASE SEPARATION IN LIPID MIXTURES

Membranes composed of mixtures of lipids which (1) differ in molecular structure or (2) form phases of different symmetry, such as crystalline and fluid or tilted and non-tilted ones, tend to exhibit lateral phase separation (Shimshick and McConnell, 1973; Sackmann, 1978). The behaviour is characterized by temperature-composition phase diagrams. A typical example is given in Fig. 2 for two lecithin differing in chain length by two CH_2-groups.

Lateral immiscibility is also possible if both phases are in a fluid state. Evidence for this has been observed in mixtures of lecithin with phosphatidic acid (Galla and Sackmann, 1975) as well as with a phosphatidylethanolamine (Shimshick and McConnell, 1973). An interesting example is the mixture of Fig. 2. Direct evidence for a heterogeneous lateral distribution of the two lecithins in the fluid state was obtained by small angle neutron scattering (Knoll et al., 1981). However, in this case the segregation has been interpreted as critical behaviour caused by a critical point hidden within the coexistence region. ($T_s < T < T_1$ in Fig. 2). Consequently the lipid heterogeneity is rather due to extended concentration fluctuations which form and dissolve rapidly. It is rather probable that this type of segregation is a quite common phenomenon for mixtures of lipids with strongly different chain lengths. If this is true, lateral lipid segregation would occur quite commonly in biological membranes at physiological temperatures.

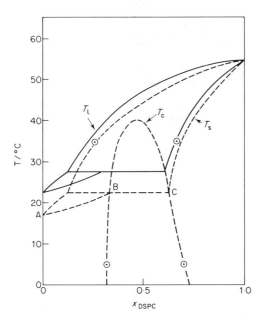

FIG. 2. Phase diagram of mixture of dimyristoylphosphatidylcholine (DMPC) and distearoylphosphatidylcholine (DSPC). Below the liquidus line (temperature T_e) the mixture decomposes into a fluid phase enriched in DMPC and a crystalline phase rich in DSPC. Below the solidus line (T_s) decomposition into two crystalline phases occurs. Most interestingly decomposition occurs also in the fluid state ($T > T_1$). This has been attributed to critical demixing phenomena (according to Knoll et al., 1981).

Lateral phase separation may be induced chemically by the adsorption of external charges (e.g. H^+ and other counterions or basic macromolecules) at the lipid/water interface (Ohnishi and Ito, 1974; Galla and Sackmann, 1975; Sackmann, 1978; Lee, 1977; Berclaz and McConnell, 1981). Actually, the phase separation is caused by the shift in the transition temperature of the charged component which will be discussed below. Clearly, the charge-induced lateral phase separation is most interesting from the point of view of triggering phenomena at the membrane.

D. LATERAL PHASE SEPARATION MAY LEAD TO A METASTABLE DOMAIN-LIKE LIPID ORGANIZATION

Lateral phase separation may lead to a domain-like organization of the components (Sackmann, 1978; Sackmann and Rüppel, 1980). In combination with the possibility of a specific lipid/protein interaction this effect could play an important role for the self organization of functional units in biological membranes. Typical examples observed in model membranes are shown in Fig. 3; (1) for the phase separation in a lecithin–phosphatidylethanolamine mixtures and (2) for the segregation caused by the binding of cytochrome c to phosphatidylcholine/phosphatidylglycerol lamellae. Stable

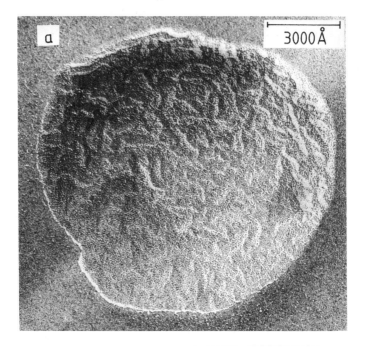

FIG. 3. (a) Freeze fracture electron micrograph of domain structure of vesicle of dimyristoylphosphatidylglycerol induced by the binding of cytochrome c. The dam-like protrusions are formed by the cytochrome c-bound lipid. (b) Domain structure of bilayers of dimyristoylphosphatidylcholine/dimyristoylphosphatidylethanolamine mixture. The regions of the wavy structure are formed by a fluid lecithin-enriched phase; the smooth areas correspond to the solid ethanolamine-enriched phase. (c) Models of possible microstructure of domain. *Left*: lamellar structure of lipid domain embedded in continuous way into bilayer. This structure is observed in the case of Fig. 3b. *Right*: micellar structure of domains. This type of structure has been postulated recently for the case of charge induced phase separation (Verkleij *et al.*, 1980).

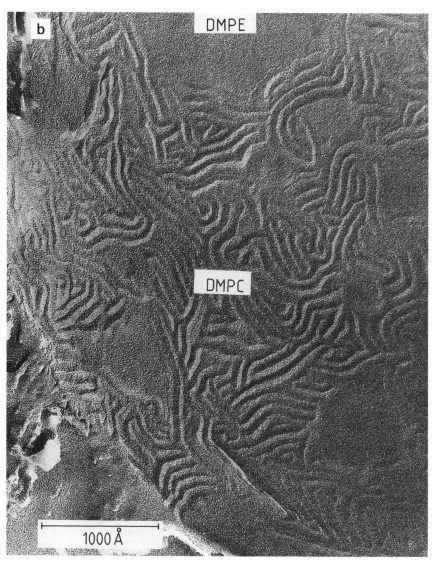

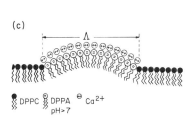

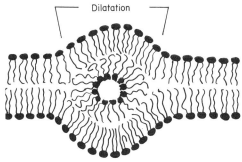

(c)

DPPC
DPPA
pH>7
Ca²⁺

domains with diameters of some 10 nm (100 Å) are formed. The domain structure is an outstanding property of the two-dimensionality of the bilayers (Gebhardt *et al.*, 1977). Two possible structures of the domains have to be considered which are depicted schematically in Fig. 3:

(1) In general, lateral phase segregation will initially lead to a two-dimensional precipitation with a lamellar structure. The local spontaneous curvature of the corresponding bilayer region will in general be different from that of the bulk lipid. A typical example is the lecithine/ethanolamine mixture shown in Fig. 3 (Sackmann and Rüppel, 1980). The limitation of the growth of the precipitation has been explained in terms of the elastic distortion due to the change in lipid orientation at the boundary between the domain and the bulk lipid (Fig. 3b). This distortion prevents the growth of the domains (Gebhardt *et al.*, 1977). Indeed, due to its different curvature the domain is expected to be split off the vesicle if it grows beyond a critical size. This type of instability may be a physical basis for phagocytosis.

(2) A different type of metastable domain structure would result if the precipitated lipid forms a hexagonal phase. In this case spherical or elongated inverted micelles may form if the original lamellar domain reaches a certain size. The micelles may be intercalated between the two opposing monolayers (Fig. 3c). The domain's size is determined by the diameter of the micelles. This type of structure has been postulated for the case of the charge (e.g. Ca^{2+}) induced phase separation. Cullis and De Kruijff, 1979; Verkleij *et al.*, 1980). However, for this system the lamellar type of domain structure seems to form preferentially at least in single-walled vesicles (Sackmann, 1978; Krbecek *et al.*, 1979).

E. DOMAIN FORMATION IS A FAST MEMBRANE PROCESS

The kinetic of the reorganization of a given lateral molecular distribution is determined by the lateral diffusion coefficient, D. The average time T_d needed for the formation of a domain of N molecules may be estimated according to (Gebhardt *et al.*, 1977)

$$T_d = \frac{A}{4\pi D} \frac{N}{x} \tag{3}$$

where A is the area per molecule and where x is the molar fraction of the domain forming lipid. For typical values ($D = 10^{-7}$ cm^2/s; $A = 0.75$ nm^2 (75 Å^2); $x = 0.5$) one obtains $T_d \approx 10^{-6}$ s. From the kinetic point of view domain formation could well play a role for the triggering of fast membrane processes. The gating process which initiates the pore opening is by a factor of 10 slower than the above value of T_d. Note that the time scale of domain formation would be by about a factor of 100 slower if proteins are involved ($D \approx 10^{-9} \, cm^2/s$).

The molecular events of the modulation of membrane processes by anaesthetic drugs is still unknown. The basic question is whether there is a direct interaction with membrane proteins leading to conformational changes of the latter or whether the effect is rather an indirect one which is based on a conformational change of the lipid. A number of model membrane studies demonstrated that local anaesthetics alter the fluidity of lipid bilayers (Trudell, 1977; Papahadjopoulos et al., 1975). An interesting aspect is the reversal of lateral phase separation effects by local anaesthetics. Thus Galla and Trudell (1980) demonstrated that domains of phosphatidic and bound to polymyxin are dissolved to a large extent by clinical concentrations of methoxyfluorane. These findings provide evidence for a previously proposed model of anaesthesia (Trudell, 1977).

III. Lipid–Protein Interaction

Because bilayers are two-dimensional systems, proteins can couple to membranes via hydrophobic, hydrophilic, and electrostatic forces. The first mechanism prevails in the case of typical integral membrane proteins penetrating the lipid bilayer (such as glycophorin or rhodopsin). Electrostatic forces may, however, lead to such a strong binding of the protein that it becomes an integral part of the membrane. A prominent example is the coupling of cytochrome c to membranes containing cardiolipin (Brown and Wüthrich, 1977) or phosphatidylglycerol (Birrell and Griffith, 1976; Kotulla, 1981).

Lipids and proteins form a cooperative system: the state of the lipids controls the configuration and the aggregation of membrane bound enzymes, while the incorporation of the proteins induces conformational changes of the lipid. Several types of structural changes are possible (Fig. 4):

(1) The configuration of the hydrocarbon chains of the lipids in the immediate neighbourhood of the proteins—the so called boundary lipid—is changed (Jost et al., 1973; Stier and Sackmann, 1973).

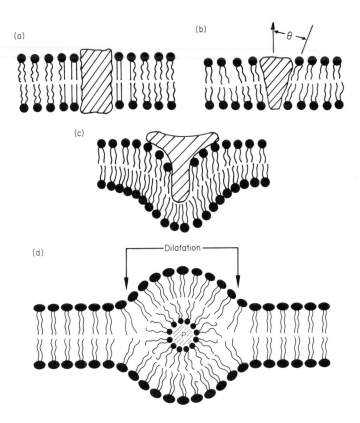

FIG. 4. Possible structural changes of lipid bilayers by the incorporation of proteins. (a) Perturbation of chain configuration of lipid surrounding the protein (bounday lipid). Change in order parameter δs may be positive or negative. (b) (c) Elastic distortion of lipid bilayer. The change in the average lipid orientation (b) or in the local curvature (c) extends over distances of the order of 100 Å. (c) Protein in lipid micelle intercalated between opposing monolayers. Note the dilatation regions which may act as attractive traps for other proteins or substrates.

According to Raman spectroscopy (Lippert and Peticolas, 1972) and calorimetry (Curatolo *et al.*, 1979; Chapman *et al.*, 1979) studies the number of *gauche* configurations is decreased in the fluid and increased in the crystalline state. This change may be described in terms of a change, δS, in the average order parameter, S, of the hydrophobic region (Marčelja, 1976). It is important to realize that the time a lipid molecule spends in the boundary region is rather short, that is, of the order of 10^{-6} s. The change in order parameter as measured by deuterium NMR (Seelig, 1977; Chapman *et al.*, 1979) or spin label techniques (Marsh *et al.*, 1976; Knowles *et al.*, 1981) is a dynamic average over the two states of the lipid in the bulk and the boundary region. Note that this type of disturbance of the lipid orientation is a short range effect. The change in order parameter δS decays exponentially with the distance, r, from the protein surface: $S \propto \exp(-r/G)$ where the range G is of the order of the average lipid distance (~ 0.5 nm; 5 Å) (Marčelja, 1976; Schröder, 1977).

(2) The incorporation of integral proteins may cause an elastic deformation of the bilayer such as a change in the average lipid orientation with respect to the membrane plane (Fig. 4b) or a change in the local curvature (as indicated in Fig. 4c). The elastic disturbance of the lipid bilayer is a long-range effect, that is, the change in order parameter decays slowly with a characteristic range of the order of 10 nm (100 Å). (Gruler and Sackmann, 1977; Prost et al., 1982). This type of membrane disturbance seems to hold for cytochrome b in lecithin model membranes (Füldner, 1981). A more short range elastic deformation is a change in the average length of the lipid molecules (Owicki and McConnell, 1979).

(3) More recently, evidence has been provided that membrane bound proteins may be surrounded by an inverted lipid micelle intercalated between the two opposing monolayers (Stier et al., 1978), as indicated in Fig. 4d. This type of membrane disturbance would also be characterized by a long range of the order 10 nm (100 Å). A possible example is the cytochrome P450 in liver microsomal membranes.

B. THE POSSIBILITY OF SELECTIVE LIPID–PROTEIN INTERACTION

A selective lipid–protein interaction mechanism would be essential for the formation of functional entities composed of lipid/protein aggregates of well-defined composition. Electrostatic forces as well as steric effects could play a role. Examples for the first type are the selective binding of basic macromolecules such as polylysine or cytochrome c to bilayers containing charged lipids. The second type of mechanism could be effective if proteins with a certain length of the hydrophobic core were penetrating the bilayer completely (Owicki and McConnell, 1979; Sackmann, 1978). Clearly lipids with chain length adapted to the length of the core of the protein would accumulate preferentially in its neighbourhood (Sackmann, 1978). Examples have not been found yet. However, in view of the large differences in chain length of membrane lipids (ranging from 16 to 24 C-atoms in the erythrocyte membrane) this mechanism is very probable. In a careful spin label study Knowles et al. (1981) showed that cytochrome oxidase has a distinct preference for cardiolipin and a much weaker one for phosphatidic acid, showing that the specificity is determined largely by the hydrophobic part.

Quite recently a strong hydrophilic interaction of the polysaccharide containing head group of glycophorin with the choline head groups of lecithin was demonstrated (Rüppel *et al.*, 1982) which can be attributed to the specific interaction of sialic acid with the lipid. The interesting aspect is that one protein may couple to about 100 lipid molecules.

C. INDIRECT, LIPID-MEDIATED PROTEIN–PROTEIN INTERACTION MECHANISM

The formation of defined enzyme complexes (such as the hormone receptor/ G-protein/adenylate cyclase complex) requires specific and probably long range protein/protein interactions. A possible mechanism may originate in the disturbances of the lipid matrix by the proteins as described above.

(1) The perturbation of the lipid chains in the halo surrounding the proteins was characterized by a change δS in the order parameter. If two proteins (i and j) approach to such a distance that the halos overlap a mutal force becomes effective. It is characterized by an interaction energy, U_{ij}, which decreases exponentially with the mutual distance, r_{ij} (Marčelja, 1976; Schröder, 1977):

$$U_{ij} \propto \delta S_i \delta S_j \exp \{-2(r_{ij} - R_0)/\lambda\} \tag{4}$$

where R_0 is the radius of the proteins. Obviously an attractive force results if both proteins induce an increase ($\delta S > 0$) or decrease ($\delta S < 0$) in S. A repulsion is expected if the perturbations are opposite. The situation is reminiscent of the electrostatic interaction. The interaction distance is of the order of $\lambda \sim 1$ nm (10 Å).

(2) On the basis of the same arguments, a long-range interaction is expected for the situation of an elastic distortion of the lipid bilayer according to Fig. 4b. The interaction force decays roughly inversely proportional with the mutual distance and may be effective over distances of the order of 10 nm (100 Å) (Gruler and Sackmann, 1977; Prost *et al.*, 1980).

(3) Similarly an elastic long-range force would also be expected between two proteins within inverted lipid micelles according to the model of Fig. 4d. The force is mediated by the dilatation of the bilayer at the equator of the micelle.

IV. Electrically Induced Conformational Changes in Membranes

For the triggering of membrane processes charge and electric field induced conformational changes could play an important role. Some pertinent electrical properties of membranes are summarized below (see also Sackmann, 1979 and Petrov *et al.*, 1979).

A. CHARGE-INDUCED LIPID PHASE TRANSITIONS

As mentioned above, phase transitions of bilayers of charged lipids may be triggered electrically: (1) by variations of the pH or the ionic strength of the aqueous phase or (2) by the adsorption of charged macromolecules or bivalent ions such as Ca^{2+}. Some typical examples are shown in Fig. 5. The charge effects may be traced back to the dependence of the transition temperature on the lipid packing density. The latter depends on the density of the surface charges at the lipid/water interface which leads to an expansive lateral pressure $\Delta\pi_{el}$. According to the Gouy–Chapman theory, $\Delta\pi_{el}$ is related to the electrical surface potential as follows (Payens, 1955):

$$\Delta\pi_{el} = 6.1\sqrt{c}(\cosh\frac{e\psi_0}{kT} - 1) \tag{5}$$

The numerical factor 6.1 holds for $T = 300$ K and a dielectric constant of $\varepsilon = 80$ while the concentration c of the free ions in the aqueous phase is given in mol/l; e is the elementary charge. Following the Clausius–Clapeyron equation, the change in chain melting temperature, ΔT_m, is then related to $\Delta\pi_{el}$

$$\Delta T_m = \frac{\Delta A}{\Delta S}\Delta\pi_{el} \tag{6}$$

where ΔA and ΔS are the changes in the molecular area and the entropy, respectively, at the transition. The essential point is that the surface potential is a function of the degree of dissociation, α, of the lipid (Payens, 1955; Jähnig, 1976):

$$\psi_0 = \frac{2kT}{e}\text{arcsinh}\left\{\frac{134\alpha}{A\sqrt{c}}\right\} \tag{7}$$

where A is the area per lipid molecule.

The charge effects may be understood on the basis of this equation:

(1) The variation of T_m with pH is a consequence of the pH-dependence of α. The theory of chemical equilibrium yields

$$\frac{\alpha}{1-\alpha} = K_0[H_s^+]^{-1} \tag{8}$$

K_0 is the dissociation constant of the lipids and $[H_s^+]$ the local proton concentration at the surface of the membrane. Considering Eqs (8) and (7), ψ_0 and therefore $\Delta\pi_{el}$ are controlled by the pH of the aqueous phase.

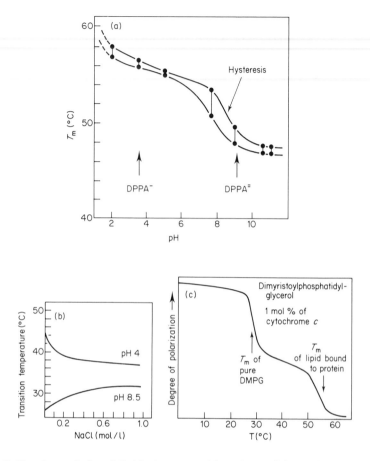

FIG. 5. To charge induced lipid phase transitions in model membranes. (a) pH dependence of chain melting transition temperature, T_m, of dipalmitoylphosphatidic acid. Note the strong hysteresis in the pH range where the lipid dissociates. (b) Influence of ionic strength on transition temperature for dimyristoylphosphatidic acid methylester. (According to experimental results of Träuble and Eibl (1974), and Galla and Sackmann (1975); reproduced from Sackmann (1979)). (c) Change in transition temperature of dimyristoylphosphatidylglycerol (DMPG) by the adsorption of cytochrome c.

(2) The ionic strength effect on T_m is restricted to low ion concentrations c (Fig. 5b) where the strongly adsorbed Stern-layer is built up at the interface. The direction of the change in T_m depends upon whether the lipids are dissociated partially or completely. In the first case the salt effect is due to the fact that the dissociation constant depends on the surface potential ψ_0 (Träuble and Eibl, 1974).

$$K = K_0 \exp\left\{\frac{-|e\psi_0|}{kT}\right\} \tag{9}$$

The adsorption of positive ions leads to a decrease in ψ_0 and diminishes the proton concentration at the lipid/water interface. Eq (5) then predicts a decrease in T_m. This situation holds for the pH 4 curve of Fig. 5b. If the lipid is completely dissociated the addition of ions to the aqueous phase first leads to the build-up of the Stern layer. The corresponding decrease in the electrostatic repulsion of the lipid head group diminishes the contribution to the electrostatic pressure which shifts T_m to higher temperatures.

The high-temperature shift caused by the addition of polylysine or cytochrome c to negatively charged lipid layers (Fig. 5c) is also due to the electrostatic shielding of the lipid head groups by the basic side-groups of the macromolecules. A very interesting observation is that lysine monomers do not influence T_m appreciably, i.e. the binding of the proteins to the lipid bilayer surface is a macromolecular effect.

B. HYSTERESIS (MEMORY EFFECT) OF CHARGE-INDUCED CONFORMATIONAL CHANGES

The pH-induced phase transition exhibits a pronounced hysteresis in the pH region of the dissociation's equilibrium where differently charged lipid species coexist (Fig. 5a; Träuble and Eibl, 1975). An interesting hysteresis effect typical for the charge-induced conformational changes in membranes is related to the domain structure formation (Fig. 6; Sackmann, 1979). It is a consequence of the formation of domains of solid phosphatidic acid. These are formed when the pH is increased until the equilibrium

$$\mathrm{DPPA^- \rightleftarrows DPPA^{2-} + H^+} \tag{10}$$

is shifted to the left. In the centre of equilibrium, domains of solid DPPA$^-$ coexist with fluid regions of the DPPA^{2-}–lecithin mixture. The hysteresis is a consequence of different K-values in the two phases. K depends on the membrane surface potential ψ_0 according to Eq (9). Because of Eq (7) the latter increases with increasing lateral packing density, A^{-1}. Eq (10) then predicts that the dissociation equilibrium is shifted to the left side (that is towards the DPPA$^-$–species) in the solid state and towards the right side (the DPPA^{2-} species) in the fluid state. The hysteresis may also be explained in terms of an allosteric model (cf. Sackmann, 1979).

According to the above considerations hysteresis effects are expected in all cases where conformational transitions are coupled to changes in the charge of the molecular species. Well known examples are the conformational

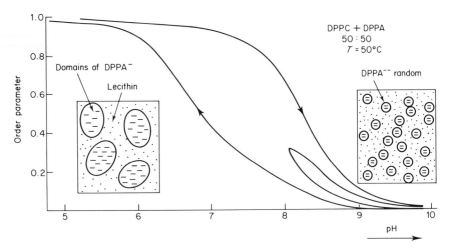

FIG. 6. Hysteresis of pH-induced lateral phase separation in 1 : 1 mixture of dipalmitoyllecithin (DPPC) and dipalmitoylphosphatidic acid (DPPA). At high pH, DPPA is doubly negative; it is in a fluid state and thus forms a homogeneous mixture with DPPC. At low pH the charged lipid has one negative charge; it is in the crystalline state and the lateral phase separation leads to the formation of domains. The temperature was adjusted in such a way that the lipid phase transition occurs at pH 7.5. The ordinate gives the fraction of lipid condensed into domains.

changes of polyelectrolytic biopolymers which were extensively discussed by Katchalsky *et al.* (cf. Neumann, 1973). It is well known that hysteresis effects may be exploited for information storage. In membranes they provide the possibility of switching between two states of functional units, e.g. between a conductive and non-conductive state of an ion channel or between two states of activity of enzyme complexes. Although definite examples are not yet known, it is a most intriguing task to search for them. In particular the above hysteresis effects could play an important role for triggering processes since they open the possibility of initiating drastic changes in membrane configuration by small variations in the local ion concentration.

C. MEMBRANE POTENTIAL AND INSIDE–OUTSIDE DISTRIBUTION OF LIPIDS

As was first noted by McLaughlin and Haray (1974) the voltage drop, V_m, across the bilayer provides an important driving force for the inside–outside distribution of the charged lipids. These lipids determine the charge density σ (charge per unit area) of the inner and outer membrane surface. According to Boltzmann's law one expects

$$\sigma_o/\sigma_i = \exp\{-eV_m/kT\} \tag{11}$$

where the index i and o stand for the charge density in the inner and outer lipid–water interface, respectively. For normal flip-flop times of some 5 hours the lipid distribution could not follow changes in membrane potential. However, if hydrophilic pores could form, the inside–outside lipid distribution could change within a time scale of 10^{-6} s. This is the time a lipid would need to cross the pore by lateral diffusion. Indeed, rather fast flip-flop times are observed for methylated lipids (Hirata and Axelrod, 1980).

D. SPONTANEOUS AND CURVATURE-INDUCED ELECTRIC POLARIZATION

Symmetric lipid bilayers may exhibit an intrinsic electric polarization of two types: (1) spontaneous ferroelectric; (Sackmann, 1979) and (2) curvature-induced polarization (Petrov et al., 1979; Sackmann, 1979). The first effect becomes important if the lipid molecules are cooperatively tilted with respect to the membrane normal $\vec{n}$. The electric dipoles in the polar head group and in the glycerol backbone are fixed with respect to the long molecular axis. Therefore, a residual average dipole moment parallel to this axis arises even if the molecule exhibits a fast rotational tumbling motion. For typical tilt angles of $10°$ to $20°$ a collective polarization in a direction parallel to the plane of the membrane arises which may amount to 1 Debye per molecule. Evidence for such a spontaneous polarization of cooperatively tilted groups comes from dielectric relaxation experiments (Kaatze et al., 1979). In particular such a spontaneous polarization could arise at the boundary of proteins if the lipid molecules are tilted due to a noncylindrical shape of the macromolecule (Fig. 4). It could contribute substantially to lipid/protein interaction (Sackmann, 1979).

Of more general interest is the curvature-induced membrane polarization illustrated in Fig. 7. Its origin is the residual average dipole moment in a direction parallel to the membrane normal $\vec{n}$. In a curved membrane the polarization per unit area is different in the outer $(\vec{P}_o)$ and in the inner $(\vec{P}_i)$ monolayer. It is obvious from Fig. 7 that the curvature-induced polarization is proportional to the local curvature $c : P = e \cdot c$. The so called flexoelectric coefficient e was calculated by Petrov et al. (Petrov and Pavloff, 1979), who distinguished two important situations: (1) free flip-flop and (2) blocked flip-flop of lipid molecules between the inner and the outer monolayer. In the first case the lipid molecules may exchange rapidly enough to compensate for the differences in the lipid packing between the compressed inner and the expanded outer layer. In the second case the change in lipid packing has to be compensated by changes in the head group conformation of the inner and the outer lipids.

The order of magnitude of P is the same for both situations; however, the sign of e may be opposite (Petrov and Pavloff, 1979). The curvature-induced polarization may contribute considerably to the membrane potential. For a curvature of $c = 10^{-3}$ Å^{-1} one estimates a membrane potential of $U = 10$ mV.

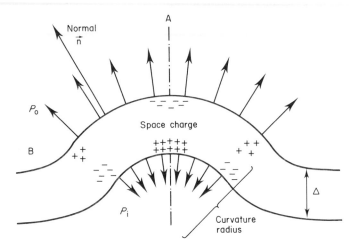

FIG. 7. Polarization and space charge formation in curved bilayers. $\vec{P_o}$ and $\vec{P_i}$ are the average dipole moments per unit area in the outer and the inner monolayer, respectively. (+) and (−) indicate positive and negative excess space charges.

The curvature-induced polarization may lead to a substantial space charge accumulating at the crest, *A*, and at the boundary, *B*, of protrusions (as indicated in Fig. 1) which may amount to 0.1 electron charges per lipid molecule. This space charge could play an important role for the binding of extrinsic proteins to the membrane surface.

V. The Role of Defects in Membrane Structure Discontinuities in Lipid Order

A property which is often underestimated in discussing membrane properties in terms of the fluid-mosaic model (Singer and Nicholson, 1972) is the liquid crystalline nature of lipid layers brought about by the strong tendency of the lipid molecules to orient mutually parallel. The most conspicuous verification of this orientational order is the finding that the average orientation of the hydrocarbon chains is nearly constant upon moving from the polar head group region towards the centre of the bilayer (Seelig, 1977). Any deformation from the equilibrium orientation by an external stimulus leads to elastic restoring forces. However, local metastable deviations from the ideal crystalline structure (called defects in the following) may form e.g. by thermodynamic fluctuations or by the incorporation of large solute molecules into the bilayer. These defects are usually neglected although they are of uttermost importance for the molecular transport within membranes and could play a dominant role for triggering processes. Therefore important types and properties of defects are summarized below.

(1) *Thermally agitated chain defects* (e.g. *gtg*-kinks) create free volume within the hydrophobic region which is essential for the uptake of small solute molecules. These chain defects diffuse very fast (with a diffusion coefficient of $D = 10^{-5}$ cm^2/s) along the molecular axes. They are essential as carriers for both the transport across (Träuble, 1971) and for the lateral transport within (Galla *et al.*, 1979) the membrane. Indeed it was shown recently that the lateral diffusion of small molecules as well as proteins may be explained in a quantitative way on the basis of a free volume model based on these ideas (Galla *et al.*, 1979).

(2) *Grain boundaries* The density of structural defects is especially high in the region of the chain melting transition. This leads to a drastic increase in the permeability of the bilayer for organic molecules (Marsh *et al.*, 1976) as well as ions in this region (Nagle and Scott, 1978). It is by about a factor of 3 higher in the centre of the transition than in the fluid state. This phenomenon can be explained in terms of the coexistence of fluid and solid phase leading to the formation of localized dilatation regions at the boundaries of the two phases. These may act as hydrophobic pores for a facilitated passive transport across the bilayer. The pores may be described as aggregates of defects. The drastic increase in the permeability was also explained in terms of a critical divergence of the lateral bilayer compressibility leading to large density fluctuations (Nagle and Scott, 1978; Jähnig, 1982).

(3) *Hydrophilic pores* are another type of energetically possible defect which could serve for the transmembrane transport. These defects have been postulated on theoretical grounds by several groups (Petrov *et al.*, 1980). Pore formation is indeed observed if erythrocytes (Kinosito and Tsong, 1978) or large vesicles of pure lipid (Harbich and Helfrich, 1979) are exposed to electric fields or strong osmotic pressures (Kinosita and Tsong, 1978; Taupin *et al.*, 1975). The formation of a pore is associated with a high elastic energy due to the sharp change in lipid orientation at the transition between the two monolayers where the lipid is orientated as in micelles. This has been described in terms of an edge energy (γ) which is the energy per length measured along the edge of the pore (Petrov *et al.*, 1980). This energy may be strongly reduced (1) by electric fields, (2) by osmotic tensions, or (3) chemically, that is by the incorporation of impurities with head groups which are large compared to the chain cross-section. The electric-field-induced pore formation has been explained in terms of the gain in energy due to the alignment of the head group dipoles in the field. Stable pores are formed above a critical voltage of about 1 volt where the membrane becomes completely unstable ($=$ dielectric

breakthrough). Pore formation via osmotic tension is the basis of cell lysis. A chemically induced mechanism is the formation of large pores in erythrocytes by the action of lysolecithins (Petrov et al., 1980). The area required by the head group of the lysolecithin (H) is large compared to that (C) of the chain. This molecule will therefore accumulate in the pore and will reduce the edge energy considerably.

A type of defect quite similar to hydrophilic pores are the so-called pits and peaks observed occasionally in bilayers (Harbich et al., 1978) and multi-layers (Kléman et al., 1977). Their existence was predicted by Petrov (Petrov et al., 1980) on the basis of the curvature elasticity of the lipid layers. The possible role of these defects for membrane fusion is discussed below.

(4) *Disclination defects* Discontinuous changes in lipid orientation are a type of defect which could play an important role for triggering processes. Such defects are expected to arise: (1) if proteins with non-cylindrical shape are incorporated (Fig. 8); (2) at the boundary of domains of different curvature (Fig. 3); (3) at the interface between regions of fluid and solid or tilted and non-tilted phases (Sackmann and Rüppel, 1980). The discontinuities in orientation at the boundary of domains of different spontaneous curvature have been made responsible for the stabilization of the domain structure.

(5) *Defects at intercalated micelles* The intercalation of elongated or spherical inverted micelles between opposing monolayers would also lead to a highly defective structure in the regions of the equatorial plane (Fig. 3).

A. DEFECTS AS ATTRACTIVE CENTRES*

At the defects, pockets of low density are formed in the hydrophobic part of the membrane. This is associated with a substantial elastic distortion energy $W(r)$. The energy can be reduced by the incorporation of small (hydrophobic) molecules. For that reason the defects form attractive centres for such molecules. Consider the disclination defect caused by non-cylindrical integral proteins (Fig. 8) as an example. According to the elastic theory of liquid crystals (de Gennes, 1974), the energy to remove a molecule from a protein (radius R) by a distance r is roughly given by

$$W(r) = \pi K \sin^2 \theta \ln (r/R) \qquad (12)$$

near the protein while $W(r) \propto r^{-1}$ at large distances (Prost et al., 1980). Similar arguments hold for other types of defects.

*There is a remarkable analogy between membranes (which are essentially two-dimensional lipid/protein alloys) and metal alloys. The physical properties (e.g. the hardness) of the latter are dominated by their microstructure which is largely determined by the stabilization of defects (dislocations) due to the incorporation of foreign molecules. Like metallurgists, nature seems to have selected the building units of membranes to adjust their physical properties to the biological requirements.

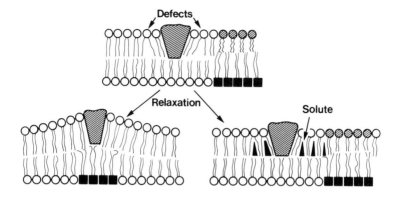

FIG. 8. Schematic representation of two possible mechanisms for relaxation of elastic strain introduced by a discontinuous change in lipid-orientation (= orientational defect) which may be caused by an integral protein of non-cylindrical shape: *Right path*: incorporation of cloud of hydrophobic solute molecules. *Left path*: escape into third dimension by spontaneous local curvature. This curvature could comprise both bilayers or could be restricted to one monolayer. The latter is possible if the void between the monolayers is filled, e.g. by long chain lipids or by hydrophobic molecules.

The above considerations show that defects may be stabilized by the incorporation of solute molecules.

An example is the stabilization of pores by lysolecithin. In general a whole cloud of solute molecules will accumulate in the neighbourhood of defects. The average concentration $c(r)$ of solute at a distance r from the defect centre is given by Boltzmann's law. $c(r) \propto \exp\{-W(r)/kT\}$. This effect could cause the accumulation of local anaesthetics or substrates in the neighbourhood of enzyme complexes. It could then lead to an amplification of the effect of the anaesthetics or an acceleration of enzyme processes.

The elastic distortion will cause long-range attractive forces (F) on the solute molecules: One then expects a directed flow of solute molecules to freshly formed defects with an average flow velocity of $\bar{v} = DF/kT$.

B. DEFECT FORMATION AND TRANSMEMBRANE COUPLING

Figure 8 illustrates that the elastic strain of orientational defects could also be relaxed by a change in local curvature. A closer inspection of freeze fracture pictures of biological membranes indeed provides good evidence that the neighbourhood of some proteins exhibit a pronounced change in curvature (Krbecek *et al.*, 1979). As indicated in Fig. 8, the change in curvature may be restricted to one monolayer if a redistribution of the lipids occurs in such a way that the void forming between the opposing monolayers is filled out, e.g. by long chain lipids or another protein. Clearly, this complementary reorganization in the opposing monolayer would provide a mechanism of transmembrane coupling which is often postulated in discussions of biological membrane processes. (Sheetz and Singer, 1974.)

VI. Membrane Processes may be Triggered by Local Fluctuations in Lipid Density or Fluidity

There are many reports that membrane fluidity is an important physical parameter which determines the efficiency of enzymatic membrane processes. Fluidity is normally characterized by lateral or rotational diffusion coefficients. Recently it was shown that the molecular transport in membranes is determined by the free volume available within the lipid bilayer (Galla *et al.*, 1979). Consequently fluidity and lipid packing density are closely related. Many fluidity effects may thus actually be caused by density changes.

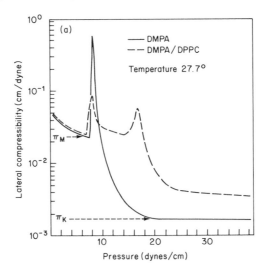

FIG. 9a. Lateral compressibility $\varkappa$ of pure dipalmitoylphosphatidic acid (DMPA) and of a 9:1 mixture of DMPC and DMPA as a function of lateral pressure. $\varkappa$ diverges at the chain melting transition of pure DMPA. The same happens at the solidus and liquidus line of the mixture where it starts to decompose into fluid and solid phases.

Accordingly, the activity of membrane bound enzymes or other membrane processes may be triggered by transient changes in the lipid density. Local density changes may be effectively initiated by lateral phase separation. This is demonstrated in Fig. 9. It shows that the lateral compressibility of a mono-layer may increase by an order of magnitude in the region of a phase boundary (solidus and liquidus line) where the mixture starts to decompose into a solid and a fluid phase. As is well known from the thermodynamics of fluctuations, this divergence in the compressibility will cause large density fluctuations.

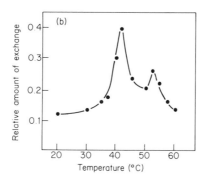

FIG. 9b. Temperature dependence of rate of lipid exchange between vesicles of dipalmitoylphosphatidylcholine (DPPC) and distearoylphosphatidylcholine (DSPC) by exchange protein. The divergence of the exchange rates at the transition temperature of the lipids are clearly due to the density fluctuations at the interfaces of fluid and solid lipid domains coexistent at the phase transitions.

As mentioned above these fluctuations may cause a drastic increase in the permeability of membranes for ions and small molecules. The facilitated ion transport suggests that the pore-like openings in the bilayer which are created by the density fluctuations are not purely hydrophobic. They could also form pathways for a translocation of lipid molecules. The fast transfer of methylated phosphatidylethanolamine from the inner to the outer mono-layer reported by Hirata and Axelrod (1980) could be explained in this way. This is suggested by the finding that the methylation leads to a remarkable increase in membrane fluidity.

The activation of a membrane bound enzyme (e.g. by substrate binding) may lead to an expansion or compression of its hydrophobic core or could shift the molecule into or out of the bilayer. The volume changes may be quite large. Thus the gating of the opening of sodium channels is associated with a relative volume change of about 0.1% (Conti et al., 1980). The activated state is expected to increase if its lipid environment is simultaneously dilated. The latter could be achieved by a local (e.g. charge induced) phase transition or phase separation process. Assume that the change in area associated with the enzyme activation is ΔA. One then expects a relative change in the Michaelis-Menten constant of (Low and Somero, 1975).

$$K = K_0 \exp \{\Delta\pi\Delta A/kT\}$$

$\Delta\pi$ is the change in lateral pressure of the membrane due to a density change of the lipid environment. An interesting aspect of such a concerted process is that the activity of two adjacent integral enzymes may be coupled: a local lateral pressure of the membrane caused by the conformational change of one integral protein could influence the activity of an adjacent enzyme in a cooperative or anti-cooperative way.

An important type of density-dependent enzymatic reaction is the decomposition of phospholipids by phospholipases which was extensively studied by Van Deenen's group. Using the monolayer technique it was found that the activity of phospholipase C has pronounced maxima at definite lateral packing densities. Even more interesting is the observation that a large increase in the phospholipase activity may be triggered by lateral phase separation. Thus Op Den Kamp *et al.* (1977) demonstrated that the lipid hydrolysis in vesicles of binary lipid mixtures is accelerated by a factor of two at the phase boundaries where the mixture starts to separate into fluid and solid domains. The interface between the domains is characterized by a high defect density or low lipid packing density and the rate of hydrolysis is expected to be very high at these zones.

Another example of the modulation of enzyme activity by density fluctuations is shown in Fig. 9b. The rate of lipid exchange between vesicles of pure dipalmitoylphosphatidylcholine (DPPC) and distearoylphosphatidylcholine (DSPC) exhibits peaks at the transition temperature of the lipids. The peaks are clearly due to the density fluctuations occurring at the boundaries between fluid and solid domains coexisting within the transition temperature region. The rather large width of the peaks may be due to the broadness of the phase transition of sonicated vesicles.

It is conceivable that density-controlled enzymatic reactions form the basis of the important role of phospholipid methylation in the transduction of signals through membranes (Hirata and Axelrod, 1980). The cascade of biochemical events evoked in local domains of membranes after the binding of hormones or neurotransmitters are closely related to physical changes of the membrane which are induced by the methylation of phosphatidylethanolamine (Hirata and Axelrod, 1980). One change is an increase in membrane fluidity which could, however, be the consequence of a decrease in lipid chain packing density. This change in the microstructure (1) leads to an activation of the phospholipase A_2 and (2) to an increase in the number of receptors which are able to couple to the hormones (Hirata and Axelrod, 1980). The first effect is another example of the density dependence of the phospholipase activity.

A. LOCAL PHASE TRANSITION AND PHASE SEPARATION PROCESSES MAY INDUCE VARIATIONS OF ENZYME ACTIVITY

The structural phase transition of the lipid bilayers is a central point of model membrane research. An intriguing question therefore is whether phase transitions play also a role for the control of membrane processes. A number of observations point in this direction. As shown several years ago the lipid phase transition has a dramatic effect on the viability of *E. coli* bacteria (Sackmann *et al.*, 1973 for reference) or of mycoplasma (*Acholeplasma laidlawii*; McElhaney and Souzu, 1976). In these studies the lipid moiety of the membrane consisted of a nearly pure lipid fraction with a rather sharp transition. Evidence for more localized lipid transitions is provided by the well known breaks in the Arrhenius plots of the enzymatic activity of integral proteins. A number of typical examples are given in Fig. 10. The temperature at which the apparent activation energy exhibits a break depends on the lipid composition. (Thilo *et al.*, 1977; Bakardjieva *et al.*, 1979a). Thus the break of the adenylate cyclase activity is shifted from 30°C to 22°C if dimyristoyllecithin (DMPC) is incorporated into the plasma membrane of the Chang liver cell (Bakardjieva *et al.*, 1979b). This shows that the break is due to a lipid conformational change. Two types of conformational changes may be responsible for the breaks: (1) local melting of lipid and (2) lateral phase separation into domains of fluid and rigidified lipid.

The first possibility seems to hold for the activity of cytochorme P450 (Fig. 10). A break is only observed for lipophilic (spin-labelled) substrate which approaches the enzyme via the membrane while the reduction of water-soluble molecules exhibits the low value of the activation energy at all temperatures. Moreover, the reduction rate of the lipophilic substrate increases abruptly at the temperature of the break. These observations have been explained in terms of a domain model by postulating that the cytochrome P450 and the cytochrome P450 reductase are incorporated into a lipid domain which melts at the temperature of the break (Stier and Sackmann, 1973). The cluster could also exist as a hexagonal type of structure (Stier *et al.*, 1978). Local melting has also been observed in erythrocyte membranes which leads to a remarkable change in the long range lateral mobility at 18°C (Kapitza and Sackmann, 1980).

It appears that the lateral phase separation is the more common cause for the break in the Arrhenius plot. Thus Thilo *et al.* (1977) showed that the sugar transport rate exhibits two breaks (Fig. 10). One possible explanation is that the lipid phase separation leads to the formation of two modifications of the transport protein: an active form contained in fluid membrane regions and an inactive form contained in the rigidified domains. The same type of

mechanism seems to hold for the breaks of the oxidation activity in mitochondria membranes of chilling sensitive plants (Raison *et al.*, 1977).

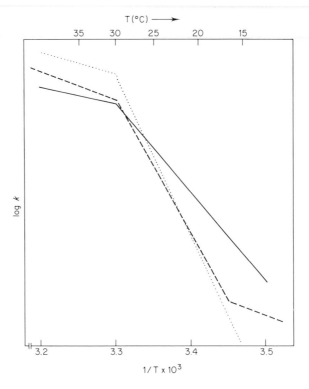

FIG. 10. Typical temperature dependencies (Arrhenius plots) of enzymatic activity observed for membranes which exhibit lipid phase transitions. ....: Rate of reduction of spin-labelled fatty acid by cytochrome P 450 reductase of liver microsomes (after Stier and Sackmann, 1973). ———: Temperature dependence of adenylate cyclase activity in Chang liver cells (after Bakardjiera *et al.*, 1979). (- - - - -: Temperature dependence of rate of active sugar transport in *Escherichia coli* bacteria effected by sugar transport protein (after Thilo *et al.*, 1977).

Lipid methylation together with lateral phase separation provides also a possibility to control the number of hormone receptors which can bind to ligands. Henis *et al.* (1982) have provided evidence that up to 85% of the β-adrenergic receptor on Chang liver cells is organized in domains and is thus immobilized while 15% is mobile. The addition of an agonist (isoproterenol) leads to the release of 70–80% of the receptors into a mobile homogeneously distributed fraction after an incubation time of 30 minutes. Secondly Hirata and Axelrod (1980) showed that the stimulation with agonist leads to a rapid increase in the methylation of phosphtidylethanolamine and the transfer of the methylated lipid from the cytoplasmic side of the membrane to the outer surface. Thus it is indeed probable that the lipid methylation initiates the receptor release.

A possible mechanism could be the following: The receptor molecules are immobilized in patches where they are inaccessible by hormones or agonists. Upon binding of hormones to the few accessible receptors methylation sets in. This could lead to a local increase in lipid-to-protein ratio to such an extent that the patches of receptors dissolve (by reversal of phase separation) which become simultaneously accessible to hormones and mobile. Such a mechanism would provide a positive feedback which would be associated with a sharp amplification of the hormone action.

A further attractive aspect of such a mechanism is that it implies the possibility for a reversal of the amplification process. According to the above-mentioned correspondence between phospholipase activity and lateral packing density of the membrane the increase in phosphatidylcholine concentration is expected to cause an increase in the decomposition of this lipid. This would thus lead to a reversal of the positive feedback mechanism.

The most interesting aspect of the lipid turnover processes is that they provide evidence that the lipid composition of membranes may play an important role for the regulation of cellular events. The basis of this control mechanism is the variation of the membrane microstructures by the lipid composition rather than a direct involvement of the lipids in biochemical processes.

VII. Triggering of Membrane Instabilities and Transmembrane Transport by Vesicles

Macromolecular hormones or toxins are transported across the cell plasma membranes by *endocytosis*. This process of internalization involves several steps: (1) the binding of the ligands to membrane bound receptors, (2) the condensation of the receptor–ligand complexes into domains by lateral diffusion, (3) the invagination of the plasma membrane, and (4) the cleavage of the invagination. In many cases the receptor–ligand complexes condense into specialized regions, known as coated pits. These are preformed invaginations which seem to be stabilized by a macromolecular coat on the cytoplasmic side of the membrane which is made from the protein clanthrin (Roth and Porter, 1964; Pastan and Willingham, 1981). However, the internalization via non-preformed invaginations has been observed as well (Montesano et al., 1982). A physically closely related type of transmembrane transport by vesicles is *exocytosis*, by which cells can segregate pulses of small messenger molecules such as acetylcholine or cAMP (Maeda and

Gerisch, 1977). Both processes involve the same sequence of steps, but in a reversed order. The decisive event is the fusion in the case of the exocytosis and the vesicle cleavage in the case of the endocytosis. One common feature of both steps is a membrane instability at the moment when the vesicle and the plasma membrane separate or merge. The two types of transport processes involve a structural change of a large part of the cell envelope. It appears that two triggering mechanisms play a dominant role: (1) lateral phase separation leading to domain (or patch) formation and (2) changes in membrane curvature. This point will be discussed below.

A. THE ROLE OF PHASE SEPARATION AND CURVATURE

The initial process of the endocytosis of ligand–receptor complexes is their condensation in the strongly curved coated pits where they seem to arrive by random walk. The direction of curvature of the coated pits is opposite to that of the bulk of the plasma membrane, and the trapping of the complexes in the pits suggests that they stabilize this reverse curvature. According to the molecular model of spontaneous curvature (Petrov et al., 1979) this could for instance be achieved by an appropriate ratio of the area occupied by the hydrophobic core (H) and the polar head (C) of the receptor–ligand complex. An important aspect of the endocytosis is that the vesicles (known as receptosomes; Pastan and Willingham, 1981) pinched off the pits are uncoated. This strongly suggests that the cleavage is initiated by lateral phase separation leading to the formation of domains which are free of clathrin but rich in receptor–ligand complexes. Since the latter may strongly favour a high negative curvature the domain could exhibit a much higher curvature than the coated regions. If their size reaches a certain limit the vesicles will be pinched off. This is just the reversal of the domain stabilization discussed above (Gebhardt et al., 1978).

Although there is no definite proof yet it is generally assumed that endocytosis is not driven by surface forces alone but that it also requires metabolic energy (Silverstein et al., 1977). The idea is that endocytosis may be induced by a reorganization of the network of actin filaments associated with the cytoplasmic side of the plasma membrane. Since the cytoskeleton is most probably strongly coupled to membrane bound proteins (Sheetz and Casaly, 1980) this would lead to a complementary reorganization of the membrane components. Such an initiating process would be consistent with the assumption that the initial steps of endocytosis are phase separation followed by a change in spontaneous curvature. Large closed shells of lipid–protein bilayers are rather unstable and tend to decompose into smaller vesicles with diameters of some 100 nm (1000 Å). It is thus very probable that an important function of the cytoskeleton is the stabilization of the rather large cell envelope. The membrane stability would then depend on the mesh size of the filamentous network (Fig. 11a). If a large zone of the bilayer becomes decoupled from the network, it would form an invagination or an extension, depending on its spontaneous curvature. Such a process could be initiated (1) by lateral phase separation (2) by a decomposition of the network by an aggregation of the filaments.

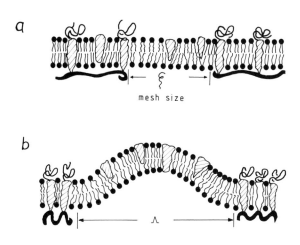

FIG. 11. (a) Usual depiction of cytoskeleton coupled to the membrane via integral proteins (cf. Sheets and Casaly, 1980). The tightness of the two-dimensional network is characterized by the mesh size. (b) Phase separation (or a contraction of the network) leads to the formation of a zone which is decoupled from the cytoplasmic network. This domain will thus exhibit its natural spontaneous curvature. Instability is expected above a critical size of the zone. (c) *Lower picture*: Freeze fracture micrograph of inner monolayer of erythrocyte plasma membrane viewed from the cytoplasmic side. Note the filaments (arrows) to which particles are occasionally bound. These are attributed to the spectrin network. For comparison the fracture face between the opposing monolayer of the same cell is shown in the upper picture. No filaments are observed (Rüppel, D., 1981).

Very strong evidence for such a destabilization mechanism of plasma membranes was provided by experiments of Elgsaeter *et al.* (1976). These authors showed that small vesicles are split off the erythrocyte plasma membrane if the cytoskeleton formed by the spectrin network is aggregated (e.g. by a pH change). In the undisturbed cells, however, the cytoskeleton forms a fine two-dimensional network with an average mesh size of some 100 nm (1000 Å). The network seems to be visible in the freeze fracture micrograph of Fig. 11c.

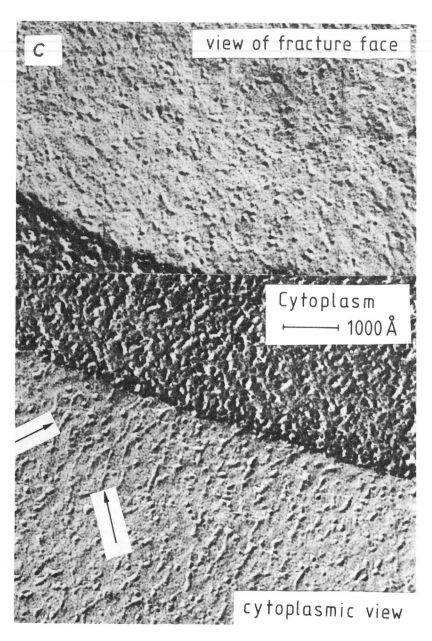

FIG. 11c

Lateral domain formation and membrane curvature seem to play also a dominant role for fusion. Electronmicroscopy studies reveal that before fusion the membranes have to come into close contact over a region of some hundred Å diameter. The zones of contact are free of particles and are surrounded by an annulus containing a high particle density (Satir *et al.*, 1973). It is not known whether the lateral redistribution of the membrane components is achieved by a chemically induced lateral phase separation within the lipid bilayer or by a contraction of the cytoskeleton. The first could be induced by Ca^{2+}. Poste and Papahadjopoulos (Papahadjopoulos *et al.*, 1977) have indeed shown that Ca^{2+} triggers the fusion of vesicles containing charged lipids such as phosphatidylserine. Further evidence for a Ca^{2+}-induced fusion mechanism comes from the finding that the influx of Ca^{2+} into the presynaptic membrane precedes the fusion of the synaptic vesicles with the presynaptic membrane. An important question, however, is whether lateral phase separation would be fast enough to allow for the annulus formation. For typical protein diffusion coefficients of $D \sim 10^{-9}$ cm²/s a fusion zone of $x = 50$ nm (500 Å) diameter could form within a time

$$T = \frac{x^2}{2D} \approx 10^{-1} \text{ s}$$

This is slow compared to the actual fusion time of the synaptic vesicles which is of the order of 10^{-3} s (Heuser *et al.*, 1979). This suggests that preformed fusion sites exist in this case so that no large scale lateral reorganization is needed and that the role of Ca^{2+} is rather to trigger the merging of the membranes. Another important change of membrane topology is a pronounced variation in local bilayer curvature at the zone of membrane contact. Both membranes bend into the same direction, i.e. they become convex towards the site to which the content of the fusing vesicle is emptied (Satir *et al.*, 1973). This local curvature is important for two reasons: first it defines the direction of transport in the case of exocytosis; secondly, as discussed below, it may be an essential step for the process of fusion itself.

B. TRIGGERING OF THE FUSION PROCESS AND THE ROLE OF PORE FORMATION

Fusion is associated with a temporary breakdown of the bilayers. Two different mechanisms are conceivable: (1) a fusion of the two bilayers and (2) a consecutive fusion of the monolayers. The first mechanism is depicted in Fig. 12. It is closely related to the pore formation. A detailed theory of this process is given by Petrov *et al.* (1980) and by Pastushenko *et al.* (1979). These authors showed that the rather high energy associated with the formation of a sharp edge may be reduced drastically in several ways: (1) by an osmotic stress, (2) by the incorporation of amphipathics with large head groups (e.g. lysolecithin, gangliosides), (3) by transmembrane electric fields and (4) by a saddle splay deformation leading to the formation of pits terminating in pores.

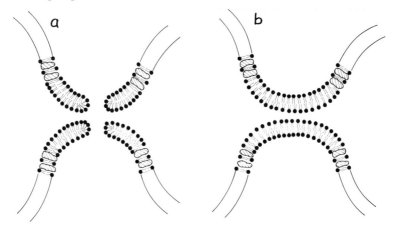

FIG. 12. Possible mechanism of simultaneous fusion of bilayers based on pore formation. (a) Simultaneous formation of pores on the two membranes. (b) Relaxation of the strained state by fusion of bilayers.

The edge energy associated with the formation of a pore of radius r is

$$E_p = 2\pi r\gamma$$

It depends on the difference of the areas occupied by the head (H) and the chains (C) of the lipid molecules in equilibrium: $\gamma \propto (H - 2C)^2$ (Petrov *et al.*, 1980). This is the basis for the reduction of γ by the incorporation of amphipathics with large head groups. A fusion via pore formation could thus be triggered chemically by the action of phospholipase A or by changes in the surface charge density which modifies the H/C ratio.

The pore formation is facilitated by a lateral tension of the bilayer. This is due to the fact that the energy needed to increase the pore radius may be recovered from the gain in energy due to the decrease in the area of the (closed) bilayer. For that reason metastable pores are formed only above a critical diameter. For a lateral tension σ the energy associated with the formation of a pore of the critical radius is (Petrov *et al.*, 1980):

$$E^* = \pi\gamma^2/\sigma$$

According to Taupin *et al* (1975) $\gamma = 0.65 \times 10^{-6}$ erg/cm. For a tension of $\sigma = 10$ dynes/cm E^* is reduced to 10^{-13} erg, which is of the same order of magnitude as the thermal energy ($T = 0.4 \times 10^{-13}$ erg at 300 K).

An osmotic pressure difference could arise rather rapidly in small vesicles fusing with the plasma membrane. This driving force could for instance be responsible for the fusion of mucocysts of *Tetrahymenapyriformis* with the plasma membrane. According to Satir *et al.* (1973), these sacs undergo a transformation from an elongated to a spherical shape just before fusion which may be due to osmotic swelling. A lateral tension leading to pore formation could also be caused by a Ca^{2+}-induced lateral segregation of charged lipids. In this context the annulus of particles surrounding the zones of fusion may play an important role. They are expected to provide a rigid boundary which could confine the membrane tension to the site of fusion.

Membrane fusion by the pore mechanism was observed between vesicles of egg-lecithin by Harbich *et al.* (1978). Evidence for a finite lifetime of pore-like openings in biological membranes comes from scanning electron-microscopy of studies of phagocytosis. Frequently small circular openings are observed at the site where phagocytic vacuoles are pinched off (Satir *et al.*, 1973).

A model of monolayer fusion was first proposed by Satir *et al.* (1973). More recently it was postulated by Cullis and Kruijff (1979) and by Verkleij *et al.* (1980) that this fusion mechanism is brought about by a local bilayer-to-micelle transition. The formation of a micellar structure within bilayers could indeed lead to the membrane destabilization required for the fusion process. This interpretation has been questioned by recent electronmicroscopy work of Hui *et al.* (1981), who provided strong evidence that membrane fusion is rather initiated by the formation of pit type defects with pore-like openings which were predicted by Petrov *et al.* (Petrov *et al.*, 1980; see also Section VI and Fig. 13b).

VIII. Electrical Triggering of Ion Channels

According to Armstrong and Bezanilla (1974), the opening of ion channels is initiated by a capacitive gating current. It is directed outwards and has a very short rise time of $T_r \approx 10^{-5}$ s. The experiment of Armstrong *et al.* (1974) provides evidence that Ca^{2+} is not involved but that the gating is associated with a displacement (e.g. dipole orientation) current in the membrane interior. Several models were proposed (Goldman, 1964; McLaughlin and Haray, 1974; Schwarz and Seelig, 1968). More recently (Sackmann, 1979) the gating was discussed in terms of the liquid crystal properties of the membranes. The elasticity of the lipids allows for a drastic conformational change at relatively weak electric fields if the deformation involves a tilt of the lipids. The model is depicted in Fig. 13. The ion channel is considered to consist of a lipid–protein aggregate while the lipid is assumed to be in a tilted configuration at low electric fields and in a non-tilted one at high fields.

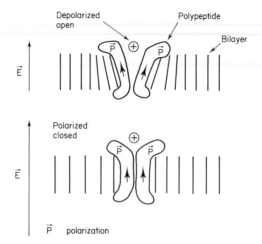

FIG. 13. Electroelastic model of gating of ion-channels. (a) Relaxed state of depolarized state of membrane with easy access of ions to pore opening. (b) Closed
state caused by field-induced tilt deformation.

The aggregate is considered as a continuum which exhibits a spontaneous
excess polarization P. P could be located (but not necessarily) in the hydrophobic core of the proteins. The opening of the pore is associated with a
relaxation of the lipids from an orientation parallel to the membrane normal
n to a tilted one. In order to close the channel, the electric field has to balance
the elastic torque causing the tilt. This implies the existence of a threshold
electric field (Sackmann, 1979)

$$E = \frac{K}{P\lambda^2}$$

where K_{11} is the curvature elastic constant and λ the average lipid distance.
For an electric polarization of $P = 1$ Debye 10 nm^{-2}, $K_{11} = 10^{-12}$ erg and
$\lambda = 1$ nm, one obtains $E_c = 10^3$V cm^{-1}. The threshold depolarization of
nerve membranes is of the order of $E_{th} \approx 10^4$V cm^{-1}, which is considerably
larger than the above value of E_c.

References

Albrecht, O., Gruler, H. and Sackmann, E. (1978). *J. Physique* **39**, 301–313.
Armstrong, C. M., and Bezanilla, F. (1974). *J. Gen. Physiol.* **63**, 533–552
Bakardjieva, A., Galla, H. J., Helmreich, E. J. M. and Levitzki, A. (1979a). *Horm.
Cell. Regul.* **3**, 11–27.
Bakardjieva, A., Galla, H. J. and Helmreich, E. J. M. (1979b). *Biochemistry* **18**,
3016–3023.

Berclaz, T and McConnell, H. M. (1981). *Biochemistry* **20**, 6635–6640
Birrell, G. B. and Griffith, O. H. (1976). *Biochemistry* **15**, 2925–2929.
Brown, L. R. and Wüthrich, K. (1977). *Biochim. Biophys. Acta* **468**, 389–410
Chapman, D. (1971). *Faraday Symposium No. 5.*
Chapman, D., Gomez-Fernandez, J. C. and Goni, F. M. (1979). *FEBS Lett.* **98**, 211–223.
Conti, F., Fioravanti, R., Segal, J. R. and Stühmer, W. (1980). "Developments in Biophysical Research" (A. Borsellino, P. Omodeo, R. Strom, A. Vecli, and E. Wanke, eds). Plenum, New York.
Cullis, P. R., and De Kruijff, B., (1979). *Biochim. Biophys. Acta* **559**, 399–420.
Curatolo, W., Sakura, J. D., Small, D. M., and Shipley, G. G. (1979). *Biochemistry* **16**, 2313–2319.
de Gennes, P. (1974). "The Physics of Liquid Crystals." Clarendon Press, Oxford.
Deuling, H. J. and Helfrich, W. (1977). *Blood Cells* **3**, 713–720.
Doniach, S. (1979). *Chem. Phys.* **70**, 4587–4597.
Elgsaeter, A., Shotton, D. M. and Branton, D. (1976). *Biochim. Biophys. Acta* **426**, 101–122.
Füldner, H. H. (1981). Doctoral thesis, University of Ülm, 1981.
Galla, H. J. and Sackmann, E. (1975). *Biochim. Biophys. Acta* **401**, 509–529.
Galla, H. J., Hartmann, W., Theilen, U. and Sackmann E. (1979). *J. Membr. Biol.* **48**, 215–236.
Galla, H. J. and Trudell, J. R. (1980). *Molec. Pharmacol.* **19**, 432–437.
Gebhardt, C., Gruler, H., and Sackmann, E. (1977). *Z. Naturforsch.* **32c**, 581–596.
Goldman, D. E. (1964). *Biophys. J.* **4**, 167–188.
Gruler, H. and Sackmann, E. (1977). *Croat. Chem. Acta* **49**, 379–388.
Harbich, W., Servuss, R. M. and Helfrich, W. (1978). *Z. Naturforsch.* **33a**, 1013–1017.
Harbich, W. and Helfrich, W. (1979). *Z. Naturforsch.* **34a**, 1063–1065.
Harlos, H. and Eibl, H. (1980). *Chem. Phys. Lipids* **26**, 406–429.
Helfrich, W. (1973). *Z. Naturforsch.* **28c**, 693–703.
Henis, Y. I., Hekman, M., Elson, E. L. and Helmreich, E. J. M. (1982). *Proc. Natl Acad. Sci. USA.* **79**, 2907–2911.
Heuser, J. E., Reese, T. S., Dennis, M. J., Jan, Y., Yan. L., and Evans L. (1979). *J. Cell Biol.* **81**, 275–300.
Hirata, F. and Axelrod, J. (1980). *Science* **209**, 1082–1090.
Hui, S. W., Stewart T. P., Boni, L. T. and Yeagle, P. L. (1981). *Science* **212**, 921–923.
Jähnig, F. (1976). *Biophys. Chem.* **4**, 309–320.
Jähnig, F. (1982). *Biophys. J.* **36**, 329–345.
Jost, P. C., Griffith, O. H., Capaldi, R. A. and Vanderkooi, G. (1973). *Proc. Natl Acad. Sci. USA.* **70**, 480–484.
Kaatze, U., Henze, R. and Pottel, R. (1979). *Chem. Phys. Lipids* **25**, 149–160.
Kapitza, H. G. and Sackmann, E. (1980). *Biochim. Biophys. Acta* **595**, 56–64.
Kinosity, K. and Tsong, T. Y. (1978). *Nature* **272**, 258–261.
Knoll, W., Ibel, K. and Sackmann, E. (1981). *Biochemistry* **20**, 6379–6383.
Knowles, P. F., Watts, A., and Marsh, D. (1981). *Biochemistry* **20**, 5888–5894.
Kléman, M., Williams, C. E., Costello, M. J. and Gulik-Krzywicki, T. (1977). *Phil. Mag.* **35**, 33–50.
Kotulla, R. (1981). Diplomarbeit Ulm.
Krbecek, R., Gebhardt, C., Gruler, H., and Sackmann, E. (1979). *Biochim. Biophys. Acta.* **554**, 1–22.
Ladbrooke, B. D. and Chapman, D. (1969). *Chem. Phys. Lipids* **3**, 304–367.

Larsson, K. (1977). *Chem. Phys. Lipids* **20**, 225–229.
Lee, A. G. (1977). *Biochem. Biophys. Acta* **472**, 285–300.
Lippert, J. L. and Peticolas, W. L. (1972). *Biochim. Biophys. Acta* **282**, 8–17.
Low, P. S. and Somero, G. N. (1975). *Proc. Natl Acad. Sci., USA* **72**, 3014–3018.
Luzatti, V., Gulik-Krzywicki, T. and Tardieu, A. (1968). *Nature* **218**, 1031–1034.
Maeda, Y. and Gerisch, G. (1977). *Exp. Cell Res.* **110**, 119–126.
Marcelja, S. (1976). *Biochim. Biophys. Acta* **455**, 1–7.
Marsh, D., Watts, A. and Knowles, P. F. (1976). *Biochemistry* **15**, 3570–3578.
McElhaney, R. N. and Souza, K. A. (1976). *Biochim. Biophys. Acta* **443**, 348–359.
McLaughlin, S. and Haray, H. (1974). *Biophys. J.* **14**, 200–208.
Montesano, R., Roth, J., Robert, A., and Orci, L. (1982). *Nature* **296**, 651–653.
Nagle, J. F. and Scott, H. L. (1978). *Biochim. Biophys. Acta* **513**, 236–243.
Neumann, E., (1973), Angew. Chem. **85**, 430–450.
Ohnishi, S. and Ito, T. (1974). *Biochemistry* **13**, 881–887
Op Den Kamp, J. A. F., Bevers, E. M. and Zwaal, F. R. A. (1977). *In* "Structural and Kinetic Approach to Plasma Membrane Function" (C. Nicoulau and A. Paraf, eds) Springer-Verlag, Berlin-Heidelberg-New York.
Owicki, J. C. and McConnell, H. M. (1979). *Proc. Natl Acad. Sci. USA.* **76**, 4750–4954.
Papahadjopoulos, D., Jacobson, K., Poste, G. and Shepherd, G. (1975). *Biochim. Biophys. Acta* **394**, 504–519.
Papahadjopoulos, D., Vail, W. J., Newton, C., Nir, S., Jacobson, K., Poste, G. and Lazo, R. (1977). *Biochim. Biophys. Acta* **465**, 579–598.
Pastan, I. H. and Willingham, M. C. (1981). *Science* **214**, 504–509.
Pastushenko, V. F., Arakelyan, V. B. and Chizmadzhev, Y. A. (1979). *Bioelectrochem. Bioenergetics* **6**, 89–95.
Payens, T. A. (1955). *Phil. Res. Rep.* **10**, 425–481.
Petrov, A. G., Mitov, M. D. and Derzhanski, A. I. (1980). Advances in Liquid Crystal, Research and Applications (L. Bata, ed.) pp. 695–738. Pergamon Press, Akadémisi Kiado, Budapest.
Petrov, A. G. and Pavloff, Y. V. (1979). *J. Physique* **40**, C3-455–547.
Petrov, A. G., Seleznev, S. A. and Derzhanski, A. (1979). *Acta Physica Polonica* **A55**, 385–405.
Prost, J., Marcerou, J. P. and Gruler, H. (1982). Personal communication.
Raison, J. K., Chapman E. A. and White, P. Y. (1977). *Plant Physiol.* **59**, 623–627.
Roth, T. F. and Porter, K. (1964). *J. Cell. Biol.* **20**, 313–332.
Rüppel, D. (1982). Thesis, Ülm, 1982.
Rüppel, D., Kapitza, H. G., Galla, H. J., Sixl, F. and Sackmann, E. (1982). *Biochim. Biophys. Acta* (in press).
Sackmann, E., Träuble, H., Galla, H. J. and Overath, P. (1973). *Biochemistry* **12**, 5360–5369.
Sackmann, E. (1978) *Ber. Bunsenges. Phys. Chem.* **82**, 891–909.
Sackmann, E. and Rüppel, D. (1980). *In* "Liquid Crystals of One and Two-Dimensional Order", (W. Helfrich and Heppke, eds). Springer-Verlag, Berlin-Heidelberg-New York.
Sackmann, E. (1979) "Light-Induced Charge Separation in Biology and Chemistry" (H. Gerischer and J. J. Katz, eds), Life Science Research Report 12. Verlag Chemie, Weinheim.
Satir, B., Schooley, C. and Satir, P. (1973). **56**, 153–176.
Schröder, A. (1977). *Chem. Phys.* **67**, 1617–1619.

Schwarz, G. and Seelig, J. (1968). *Biopolymers* **6**, 1263–1277.

Seelig, J. (1977). *Quart. Rev. Biophys.* **103**, 353–400.

Sheetz, M. P. and Singer, S. J. (1974). *Proc. Natl Acad. Sci. USA.* **71**, 4457–4461.

Sheetz, M. P. and Casaly, J. (1980). *J. Biol. Chem.* **225**. 9955–9960.

Shimshick, E. J. and McConnell, H. M. (1973). *Biochemistry* **12**, 2351–2360.

Silverstein, S. C., Steinman, R. M. and Cohn, Z. A. (1977). *Ann. Rev. Biochem.* **46**, 669–722.

Singer, S. J. and Nicholson, G. L. (1972). *Science* **175**, 720–731.

Stier, A. and Sackmann, E. (1973). *Biochim. Biophys. Acta* **311**, 400–408.

Stier, A., Finch, S. A. E. and Bösterling, B. (1978). *FEBS Lett.* **91**, 109–112.

Taupin, Ch., Dvolaitzky, M. and Sauterey, C. (1975). *Biochemistry* **14**, 4771–4775.

Thilo, L., Träuble, H. and Overath, P. (1977). *Biochemistry* **16**, 1283–1290.

Träuble, H. (1971). *J. Membrane Biol.* **4**, 193–208.

Träuble, H. and Eibl, H. (1974). *Proc. Natl Acad. Sci. USA.* **71**, 214–219.

Träuble, H. and Eibl, H. (1975). "Functional Linkage in Biomolecular Systems" (Schmitt, F. O. and Schneider, D. M. eds) pp. 59–101. Raven Press, New York.

Trudell, J. R. (1977). *Anesthesiology* **46**, 5–10.

Verkleij, A. J., Van Echteld, C. J. A., Gerritsen, W. J., Cullis, P. R. and De Kruijff (1980). *Biochim. Biophys. Acta* **600**, 620–624.

Chapter 4

Rhodopsin and its Role in Visual Transduction

Department of Visual Science, Institute of Ophthalmology,
University of London, Judd St., London WC1H 9QS, England

I. Introduction

A. PROPERTIES OF THE VISUAL SYSTEM

The salient properties of the vertebrate visual system are its extreme sensitivity, the breadth of its operating range, which spans light conditions varying by more than a millionfold, its ability to monitor fairly rapidly changing events (up to about 50 per second), its high spatial acuity, and its extremely fine discrimination of colour variations. Many of these properties

BIOLOGICAL MEMBRANES Vol. 5
ISBN 0 12 168546 2

are known to result from processing of the incident light by the retina, which has cells of specific type organized into five fairly clearly defined layers. The inner layers of the retina—from the horizontal cells to the ganglion cells— are concerned with the wealth of information processing which is carried out on the visual signal before it is transmitted via the optic nerve to the brain. Light incident on the retina passes through all of these cell layers before it reaches the receptors, the cells responsible for absorbing quanta of visible light and producing the initial visual signal. The receptors in most vertebrate retinas may be divided on the basis of their shapes into two types; the cones and the rods. The cones are responsible in man for high acuity vision and the perception of colour, and they require fairly high levels of light in order to operate. The rod system functions at lower light levels, but has no ability to differentiate between colours and gives relatively poor spatial discrimination. In addition to their basic role of transducing light, both rods and cones have the ability to adapt in order to change their operating range, that is, to alter their sensitivities according to the light conditions. This adaptation makes a major contribution to the enormous range over which the visual system functions.

The present chapter will be concerned with the rod receptors and will emphasize the properties of rhodopsin (the pigment which absorbs and is photolysed by the incident light), the reactions which occur as a result of photoactivation, and their possible significance in the generation of the physiological response.

B. THE ROD RECEPTOR CELL

The morphology of the receptor is highly specialized (Fig. 1), and the distal region of the cell can be considered as two basic structures connected by a narrow cilium. Of these, the outer segment (OS) is responsible for the light capture. The inner segment (IS), which subserves the energy and biosynthetic requirements of the cell, is connected to the cell nucleus and thence to the synapse, from which the signal is transmitted to secondary neurones.

The OS of the rod is a cylindrical structure, containing a large number of membranes arranged in an essentially parallel stack, in such a way as to provide the maximum surface area to light incident along the long axis of the cell (Fig. 1). Embedded within the OS membranes are the rhodopsin molecules, which harvest the light and trigger the chemical events which culminate in changes in the synaptic activity of the cell. The membrane stack at the base of the OS is formed by invagination of the plasma membrane. Not far from the base of the OS, however, the membranes become pinched off and form closed, flattened vesicles, or discs, separated from and surrounded by the outer cell membrane. Each OS contains up to 2000 discs. Like the discs, the plasma membrane contains rhodopsin (Basinger *et al.*, 1976; Papermaster *et al.*, 1978; Kamps *et al.*, 1982). The existence of the discontinuity between the discs, which contain about 99 % of the visual pigment (Kamps *et al.*, 1982), and the outer plasma membrane is evident anatomically (Fig. 2), but it has been suggested that direct communication may occur between the intradiscal and extracellular space (Schnetkamp, 1980; Kaupp and Schnetkamp, 1982). In cones the OS consists essentially of a single continuously folded membrane, so there is only a single intracellular compartment, the spaces between the lamellae being extracellular, unlike the intradiscal spaces of the rod.

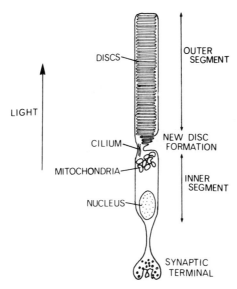

FIG. 1. Diagrammatic representation of a rod. The cilium interconnecting the inner and outer segment incorporates a number of axonemes which extend into the outer segment. The dimensions of the rod OS vary widely with species: in the frog the diameter is approx 5 μm and the length approx. 50 μm; in mammals the respective figures are approx. 1.5 μm and 20–35 μm.

Renewal of rod OS continues throughout life, with the formation of new OS membranes occurring at the junction with the IS (Young and Droz, 1968), and the loss of "old" membranes at the apical tips. These latter, which are shed in the form of small packages of membranes, are engulfed and digested by the adjacent cell layer, the retinal pigment epithelium (Young and Bok, 1969). The OS membranes and the rhodopsin they contain thus progress from the base to the tip of the OS, the total transit time being between 9 and 40 days, depending on species (Young, 1967, 1971).

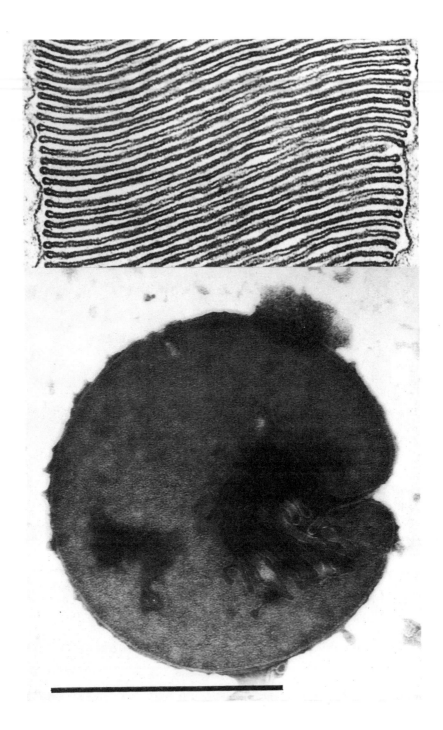

C. THE SIGNALS FROM PHOTORECEPTORS

In order to assess the mechanism by which rhodopsin may act to trigger the photoresponse, it is necessary to describe in simplified outline some of the electrophysiological characteristics of the rod and the signal which it produces.

In the dark, the membrane of the photoreceptor OS is relatively leaky, and there is a constant flow of sodium ions into the cell (Yoshikami and Hagins, 1973). Measurements of the resting potential across the OS plasma membrane show that the interior of the cell is about -30 mV relative to the extracellular medium. The absorption of light by the cell causes the permeability of the membrane to decrease, and as a result the membrane hyperpolarizes, by up to 50 mV (Toyoda et al., 1969). Many studies have shown that the hyperpolarizing response, while arising primarily from the permeability changes in the OS plasma membrane, also reflects secondary conductance changes (Schwartz, 1973) and electrical coupling between adjacent receptors (Schwartz, 1975). However, the decrease in the Na dark current can be monitored directly from individual rods, by sucking them into closely fitting pipettes (Yau et al., 1977). Figure 3 illustrates a series of the responses obtained from a dark-adapted toad rod. There is a clear latent period between the stimulus and the onset of the responses, which themselves last for several seconds. For dim stimuli the amplitude of the response is directly proportional to the light intensity, as would be expected if the resistance of the membrane were being increased in a linear fashion. At higher levels, when the permeability has been almost completely suppressed, the responses no longer increase in amplitude, but instead become more prolonged. The relationship between the intensity of the stimulating light (I) and the amplitude of the photoresponse (V) is described fairly accurately by the hyperbolic function $V/V_{max} = I/(I + s)$, where V_{max} is the maximal, saturated response amplitude obtainable and s is a constant, the value of which is equal to the intensity of a half saturating flash (Baylor and Fuortes, 1970). It has been shown that for the dark-adapted rods of rat, toad and monkey, s has a value of about 30 quanta absorbed per rod (Penn and Hagins, 1972; Baylor et al., 1979a; Nunn and Baylor, 1982). Absorption of a single quantum by one of the total population of some 10^7 (rat) to 10^9 (toad)

FIG. 2. Electron micrographs of rod OS. *Upper*: a cross section of a small portion of a human rod, showing the anatomical separation of the disc and plasma membranes. *Lower*: plan view of an isolated rat disc, showing the single deep incisure and fimbriae characteristic of this species. The discs of many species, particularly those with large diameter OS (e.g. frog) show several deep incisures around their rims. (Scale bar $= 1$ μm.) (Photographs by courtesy of Dr J. Marshall.)

rhodopsin molecules in the rod is sufficient to elicit a response (Baylor et al., 1979b).

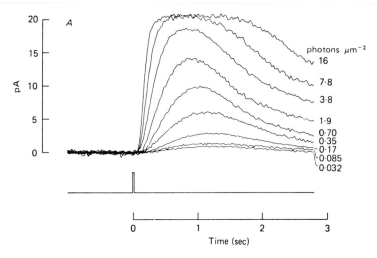

Fig. 3. A series of superimposed rod responses to light flashes of increasing intensity. A toad rod was sucked into a tightly fitting pipette, and the membrane current monitored. Light causes a transient outflow of current, as a result of a *decrease* in the inward flow of current which occurs in the dark. The stimuli were 20 ms green (500 nm) flashes delivered at the time indicated by the pulse in the lower trace, and their intensities are shown alongside the corresponding response. From the dimensions of the rod and its rhodopsin content, it is estimated that the stimulus intensity for a semi-saturating response corresponds to roughly 35 rhodopsin isomerizations per rod. Reproduced from Baylor et al. (1979b).

When the rod is exposed to more prolonged illumination with lights of insufficient intensity to cause significant changes in rhodopsin levels, the membrane potential adjusts to a new steady level more negative than the dark resting voltage, but can still be modulated by superimposed light flashes of increased intensity (Kleinschmidt and Dowling, 1975; Fain, 1976; Bastian and Fain, 1979). This adaptation to background light, which expands the operating range of the rod, differs from the extreme loss of sensitivity which is observed when significant amounts of rhodopsin are bleached (Ernst and Kemp, 1972; Brin and Ripps, 1977). The latter phenomenon, called pigment or bleaching adaptation, is much greater than can be accounted for simply by the reduced probability of quantum capture resulting from the reduced levels of rhodopsin present (Ernst and Kemp, 1972; Grabowski and Pak, 1975). Recovery from this type of adaptation is slow and is incomplete in the absence of pigment regeneration (Ernst and Kemp, 1972; Pepperberg *et al.*, 1976).

The characteristics of the response of the rod have led to the notion that the absorption of light by the visual pigment in the disc membrane leads to a change in concentration of some small diffusible transmitter inside the OS which acts upon the permeability of the Na channels in the plasma membrane. In considering the role of the visual pigment in triggering this secondary change, two basic models have been explored.

In the first of these rhodopsin is considered to act essentially as a gating protein, releasing large numbers of some small transmitter substance from the disc into the intracellular space. It has been proposed that the transmitter is Ca (Yoshikami and Hagins, 1971; Hagins, 1972). Such a hypothesis would explain how activation of a single rhodopsin molecule located in a membrane relatively far away from the cell membrane can cause a relatively large change in the latter's permeability. Many studies have therefore been carried out to determine whether the conductance of OS membranes is light dependent and whether the intracellular concentration of Ca (or some other species) changes by a fairly substantial amount in light-stimulated rods.

An alternative approach to the trigger mechanism of receptor transduction originates from the observation that in the dark the rod photoreceptor contains a relatively high concentration of $3',5'$-cGMP, and that light stimulation causes this to drop (see Section II.A.3, below). The cyclic nucleotides are known to perform a variety of important biological functions involving the activation of protein kinases and the regulation of Ca concentration (Greengard, 1978). It has therefore been suggested that rather than acting as a gating protein, the role of rhodopsin may be that of an activator, setting in train a sequence of enzyme catalysed reactions which in turn culminate in the reduction of the plasma membrane permeability. As in the pore hypothesis, the necessary amplification can be readily accounted for by the postulated enzyme action: with this hypothesis the major problem is that of reaction rates, which must be fast enough to account for the rapidity with which the cell responds.

The contrast between the two basic hypotheses may be highlighted in those properties which it requires of the rhodopsin molecule. Thus the pore model strongly implies that the pigment protein completely traverses the disc membrane, so that it can form a channel through which the particles may flow. The activator model, on the other hand, implies that the molecule undergoes some major conformational change during its photolysis, and thus triggers the enzymic activity, either in another protein bound to the membrane, or in the intracellular medium.

II. Rhodopsin and Photoreceptor Membranes

A. THE RHODOPSIN MOLECULE

Rhodopsin *in situ* may be considered to be a multicomponent entity: a polypeptide apoprotein, opsin, to which a chromophore, retinal, is covalently bound. The protein also contains covalently bound oligosaccharide chains, and its properties are dependent on the non-covalently bound lipids of the disc membrane. Thus, many of the studies which have been carried out on detergent solubilized extracts of the pigment, which have involved stripping the molecule of varying amounts of associated lipid, have yielded results whose variations almost certainly reflect the degree to which the pigment has been removed from its natural environment. In many cases the changes are major (e.g. reduced thermal stability and the loss of regenerability (De Grip, 1982); changes in the absorption properties (Heller, 1968a) and the kinetics of photolysis (Ostroy, 1977); availability of sulphydryl groups (De Grip *et al.*, 1973); loss of optical activity (Shichi, 1971)), and the data obtained in these circumstances must be interpreted with caution in the context of the *in situ* function of rhodopsin. That the partial denaturation of the rhodopsin results from delipidation has been demonstrated by the recovery of function which accompanies either the addition of lipid (Daemen, 1973) or the incorporation of the molecule into appropriate lipid membrane vesicles (O'Brien, 1982a).

1. *Rhodopsin chemistry*

Rhodopsin has a molecular weight of between 35 000 and 40 000 daltons, as has been shown in many studies (Daemen *et al.*, 1972; Robinson *et al.*, 1972; Papermaster and Dreyer, 1974; Frank and Rodbard, 1975; Pober and Stryer, 1975; Hargrave and Fong, 1977). The amino acid composition has also been determined in several laboratories (Heller, 1968a; Zorn and Futterman, 1971; Plantner and Kean, 1976; Sale *et al.*, 1977; Rafferty *et al.*, 1980) and is found to total up to about 350 amino acids. The proportion of hydrophobic amino acids is rather larger than is typically found for proteins (Ebrey and Honig, 1975), though this would perhaps be expected in view of its membrane-bound character. Estimates of the number of sulphydryl groups present have been very varied, ranging from six to ten. This variability is likely to result from the variety of methods by which the determinations were made, since the number of available sulphydryls is known to depend on both the photolytic state of the molecule (Heller, 1968b; DeGrip *et al.*, 1973; Chen and Hubbell, 1978) and its environment (De Grip *et al.*, 1973). In bovine rhodopsin, the most fully studied, the number is almost certainly ten, six of which yield free sulphydryl groups, the others forming two disulphide bridges (DeGrip *et al.*, 1973; Chen and Hubbell, 1978). In rhodopsin in its native environment, one of the sulphydryls is particularly reactive, undergoing modification with iodoacetemidosalicylate (Wu and Stryer, 1972; Sale *et al.*, 1977), while two sulphydryls react with *N*-ethyl maleimide (De Grip *et al.*, 1973) and with 4,4'-dithiopyridine (Chen and Hubbell, 1978).

Rhodopsin consists of a single polypeptide chain, the full amino acid sequence of which has recently been proposed (Ovchinnikov *et al.*, 1982). Other groups have also established, with good agreement, sections of the polypeptide. The N-terminal has been shown to be acetylated (Tsunasawa *et al.*, 1980) and the sequence to position 39 has been established (Hargrave *et al.*, 1980). The peptide chain comprising the first 108 amino acids from the C-terminal has also been sequenced (Hargrave *et al.*, 1980; Pellicone *et al.*, 1980; Findlay *et al.*, 1981; Hargrave *et al.*, 1982).

As indicated earlier, rhodopsin is known to be a glycoprotein (Heller, 1968a). There are two major carbohydrate chains (Hargrave, 1977), each of which is a hexasaccharide consisting of three mannose and three *N*-acetyl glucosamine residues (Heller and Lawrence, 1970; Fukuda *et al.*, 1979). The oligosaccharides are attached to the protein close to the N-terminal, almost certainly at asparagines 2 and 15 (Hargrave, 1977; Fukuda *et al.*, 1979; Liang *et al.*, 1979).

2. The Chromophore and its Linkage to the Protein

It has been known for many years that the chromophoric properties of the visual pigments depend upon the presence of retinal. In rhodopsin, the retinal is in the 11-*cis* isomeric form (Fig. 4). The absorption maximum of 11-*cis* retinal lies at about 380 nm, and there is therefore a major bathochromic shift involved in the formation of rhodopsin, which in most species absorbs maximally at about 500 nm (Fig. 5). (It should be noted that in some species of amphibia and freshwater fishes there is a second class of rod pigments, the porphyropsins, which absorb maximally at slightly longer wavelengths: these are based on the chromophore 3-dehydroretinal (Bridges, 1972) and will not be considered further here.) It is now firmly established that the linkage of the retinal to the protein is by means of Schiff base coupling to the ε-amino group of lysine 53' of the polypeptide (Bownds, 1967; Pellicone *et al.*, 1980; Findlay *et al.*, 1981; Mullen and Akhtar, 1981). There has been much debate as to whether the linkage is protonated, unprotonated, or hydrogen bonded to donor groups on the protein. The notion that the linkage is protonated (Pitt *et al.*, 1955) was initially based on the spectra of model Schiff bases, which, as a result of π-electron delocalisation, show large shifts compared with their unprotonated counterparts, absorbing maximally at about 440 nm. Analyses of resonance Raman spectra appear to confirm that this

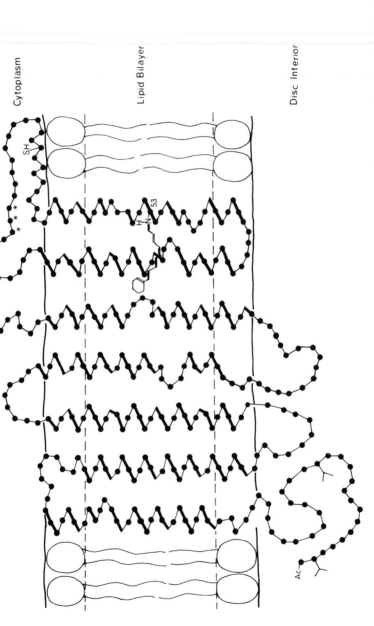

FIG. 4. A model for the rhodopsin molecule in the disc membrane, derived from that proposed by Ovchinnikov *et al.* (1982). The orientation and sequence of the trans-membrane loop adjacent to the C-terminal is based on the structure proposed by Hargrave *et al.* (1982). Also shown (*) are the 7 phosphorylation sites close to the C-terminal, the site of attachment of the chromophore (lysine 53) and two oligosaccharide attachment sites close to the N-terminal.

is the case (Oseroff and Callender, 1974; Mathies *et al.*, 1976; 1977), although some evidence to the contrary has been obtained from studies using infrared (Siebert and Mantele, 1980) and ¹³C-NMR (Abrahamson *et al.*, 1977) spectroscopy. While the balance of evidence appears to favour the protonated linkage, this alone is clearly insufficient to account fully for the observed bathochromic shift which accompanies the formation of the pigment (Ebrey and Honig, 1975). However, it has been shown theoretically that the red shift can be modulated by the distribution of charges on the regions of the protein in the vicinity of the chromophore (Honig *et al.*, 1976), probably located close to the 11,12 double bond (Honig *et al.*, 1979).

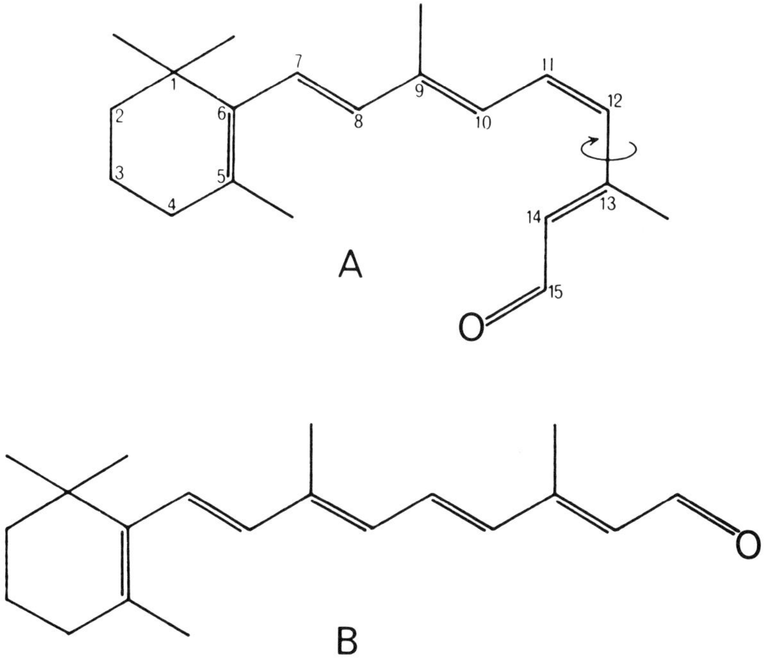

FIG. 5. Isomers of retinal, the chromophore of rhodopsin. (A) 11 *cis*-retinal. The 12 s-*cis* conformation shown is that which is found in crystalline retinal (Gilardi *et al.*, 1971); there is evidence that in Schiff bases the conformation is 12-s-*trans* (Shriver *et al.*, 1979). (B) All *trans* retinal, the isomer released following rhodopsin photolysis.

3. *The Chromophoric Binding Site*

Several lines of evidence have shown that the site to which the 11-*cis* retinal is bound in rhodopsin is both buried within a hydrophobic region of the protein (presumably as a result of conformational changes which accompany the formation of the holoprotein), and dimensionally fairly constrained. Thus, the site is protected from attack by small hydrophilic reagents such as hydroxylamine and sodium borohydride (Bownds and Wald, 1965) but available to the lipophilic cyanoborohydride (Fager et al., 1978), and is left intact when the protein is subjected to limited proteolysis (Saari, 1974; Trayhurn et al., 1974).

The restrictiveness of the site is demonstrated by the limitations which have been found on the ability to form analogue pigments with chromophores other than 11-cis retinal. While it is possible to form artificial pigments with the 9,13-dicis (Crouch et al., 1975), 7-cis, 9-cis, 7,9-dicis, 7,13-dicis and 7,9,13-tricis isomers (De Grip et al., 1974), 13-cis retinal fails to act as a substrate (Crouch et al., 1975). It has therefore been proposed that there is a critical longitudinal chain-length, above which reaction does not occur (Matsumoto and Yoshizawa, 1978).

The requirements of the binding site have also been probed by studying the inhibitory effects of molecules analogous to fragments of the retinal molecule. Regeneration of rhodopsin has been shown to be competitively inhibited by β-ionone, suggesting the existence of a β-ionine ring binding site (Matsumoto and Yoshizawa, 1975). The inhibitory effect is increased for β-ionone ring compounds with attached side-chains of length up to, but not greater than, the native chromophore (Daemen, 1978; Towner et al., 1981). The specificity for the β-ionine site is not severe enough to prevent synthesis of pigment from compounds containing the 3-diazo-aceoxy derivative (Sen et al., 1982) or an aromatic ring (Matsumoto et al., 1980) in its stead. Although it has been shown that pigment generation can occur when even more sterically different substitutions are made for the β-ionine ring, the bathochromic shift is greatly diminished (Blatchly et al., 1980).

The steric specificity of the binding site is also indicated by the optical activity which is displayed by the a and b absorbance bands of the visual pigment (Crescitelli et al., 1966). Because of steric hindrance, the 11-cis retinal molecule would be expected to exist in two twisted enantiomeric forms. In free solution, however, the molecule does not show optical activity since the two isomers are in rapid thermal equilibrium. That rhodopsin shows strong optical activity suggests that one optically active form is preferentially bound within the protein (Ebrey and Honig, 1975). Experiments with analogues which have stable optically active isomers confirm that one enantiomer is preferentially bound (Ebrey et al., 1975).

4. The Location of Rhodopsin in the Membrane

On the basis of its molecular weight and the assumption of a density similar to that for most proteins, the rhodopsin molecule would have a diameter of about 4 nm (40 Å) if it were spherical in shape, which would be insufficient for it to span the membrane (Wu and Stryer, 1972). However, several lines of evidence show that the molecule not only protrudes into the cytoplasm but also into the intradiscal space. The protrusion into the cytoplasmic space has been demonstrated by the susceptibility of the molecule to proteolysis (Trayhurn et al., 1974; Saari, 1974), by its ability to bind antibodies (Blasie et al., 1969), and by X-ray analysis, which show that only some 60% of the molecule is buried within the membrane (Saibil et al., 1976).

Proteolytic digestion of the molecule in the intact disc membranes, using a variety of enzymes, including trypsin, thermolysin, subtilisin, chymotrypsin, or papain, has shown that it can readily be cleaved into two or more fragments (Hargrave and Fong, 1977; Towner et al., 1977; Fung and Hubbell, 1978a; Hubbell and Fung, 1979; Albert and Litman, 1978; Hargrave et al., 1982). Initial, mild proteolysis causes the release of about 10% of the amino acids and leaves either two (Hargrave and Fong, 1977) or three (Towner et al., 1977; Albert and Litman, 1978; Zagalsky et al., 1981) major membrane-bound fragments. In conditions which leave two membrane-bound fragments, it has been shown that the larger (F1) comprises about 60% of the original peptide, and the other (F2) some 30%. F1 is shown by dansylation to contain the N-terminal (Hargrave and Fong, 1977), while F2 includes the chromophoric attachment site (Pober and Stryer, 1975; Hargrave et al., 1982). The carboxyl-terminal is located within the cytoplasm, along with the easily digested peptide fragment of about 40 amino acid residues (Hargrave et al., 1982). The cleavage points in papain proteolysis, when 3 fragments are obtained (Sale et al., 1977; Albert and Litman, 1978; Zagalski et al., 1981) indicate that the molecule re-emerges into the cytoplasm in a loop-like fashion. The proteolytically cleaved fractions account for a major fraction of the total molecular weight of rhodopsin, so it is clear that if the molecule is to span the membrane, it must be extremely aspherical and elongate. That this is the case is indicated by energy transfer studies, using fluorescent labels and the intrinsic chromophore (Wu and Stryer, 1972), and by neutron diffraction (Yaeger, 1975) both of which suggest that the molecule is at least 7.5 nm (75 Å) long.

First strong evidence that the molecule also projects into the interior of the disc was obtained by proteolysis and lactoperoxidase catalysed iodination of rhodopsin in a comparative study of native and reconstituted OS membranes (Fung and Hubbell, 1978b). It was also shown that the lectin concanavalin A only binds to the sugar groups of rhodopsin to any great extent after the discs have been ruptured, indicating that these are normally located within the intradiscal space (Rohlich, 1976; Hubbell and Fung, 1979) and therefore unavailable for binding. Since the oligosaccharide attachment sites are located close to the N-terminal of the protein (Hargrave, 1977), it follows that the molecule traverses the membrane an odd number of times.

The location and orientation of the binding site of the chromophore relative to the membrane are also known with some accuracy. For many years it has been known from linear dichroism studies that the chromophoric transition dipole is oriented at an angle of less than 20° to the surface of the disc membrane (Denton, 1959; Liebman, 1962; Kemp; 1973; Harosi, 1975; Chabre and Breton, 1979). This arrangement optimises the effectiveness of the visual pigment for harvesting light incident along the long axis of the rod, and sets limits on the extent to which the region of the protein containing the chromophore is able to undergo major rocking motions in the membrane. The lysine residue to which the retinal is attached lies within a region of hydrophobic amino acids (Hargrave *et al.*, 1982) and is thus likely to be buried within the disc membrane, as is also suggested by fluorescence energy transfer studies (Thomas and Stryer, 1982), and (as mentioned above) the fact that the chromophoric properties of the protein are not lost as a result of limited proteolysis.

Some 50–60% of the rhodopsin molecule is in the form of α-helices (Shichi and Felton, 1974; Stubbs *et al.*, 1976) and that these correspond to trans-membrane sections of the protein is strongly suggested by studies which have shown that they are aligned perpendicular to the membrane (Michel-Villaz *et al.*, 1979; Rothschild *et al.*, 1980).

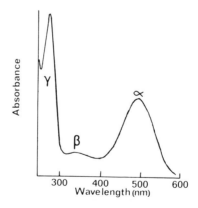

Fig. 6. The spectrum of rhodopsin. The α and β bands are characteristic of the native pigment, while the γ band is due primarily to the aromatic side chains of the protein. The extinction coefficient of the α band is 41 000 (Daemen, 1973). The purity of rhodopsin is routinely estimated either from the ratio of the absorbances obtained at the minimum between the α and β bands and the peak of the α band (which in a good preparation is less than 0.2), or from the relative height of the γ/α bands, which in highly purified rhodopsin is less than 1.6 (Papermaster and Dreyer, 1974). The latter is the better method, since it indicates more clearly the presence of any contaminating proteins (including opsin, the bleached visual pigment).

Given the current levels of knowledge of the structure of rhodopsin, it is possible to construct models of at least parts of the molecule. Hargrave *et al.* (1982) have put forward a model of the F2 fraction of the molecule, based on a prediction algorithm which was shown to be applicable to bacterio-rhodopsin, in which the membrane-bound component of this fragment forms a loop which completely spans the membrane. Other, less firmly founded models have been proposed for the structure: by analogy with the proton pumping pigment bacteriorhodopsin, whose structure has been more fully established, it seems likely that the molecule may span the membrane as many as seven times (Henderson and Unwin, 1975). Hubbell and Fung (1979) have proposed a model consisting of five helices spanning the membrane, and point out that such a structure would have a sufficiently large central core to permit the passage of small ions. An impression of the conformation of the molecule in relation to the membrane is given in Fig. 4, which includes those features which are fairly firmly established.

It is also known that the rhodopsin molecule can undergo multiple phosphorylation (Kuhn and Dreyer, 1972) during its photolysis. Seven of the phosphorylation sites have been shown to lie close to the C-terminal (Fig. 6), since they are released with the initial rapid and mild proteolysis which removes the small cytoplasmic section of the peptide chain (Hargrave *et al.*, 1980). The stoichiometry and kinetics of the phosphorylation will be considered more fully in Section III.B.

B. ROD OUTER SEGMENT MEMBRANES

1. *Membrane Composition*

Rhodopsin forms the largest single component of the OS membrane, constituting up to 40% of its dry weight (Daemen, 1973). Its concentration, which is about 3 mM (Liebman, 1962), far exceeds those of the many other proteins present in OS membranes, and comprises up to about 90% of the total OS protein (Bownds *et al.*, 1971; Papermaster and Dreyer, 1974). The exact proportion found depends upon the conditions used for the analysis, since there are several polypeptides which are only peripherally bound to the membrane and are released from it in the dark, or in low ionic strength buffers (Godchaux and Zimmerman, 1979a; Kuhn, 1980a,b): these proteins, and their interaction with rhodopsin, are considered in more detail in Section 3 below. In circumstances where these proteins have been stripped from the membrane, there remains a high molecular weight protein comprising only about 2% of the total (Papermaster and Dreyer, 1974). The picture is complicated, too, by the observation that SDS-PAGE of frog OS shows multiple forms of rhodopsin, each of similar size (Molday and Molday, 1979; Fatt, 1981). This has not been found to be so in bovine OS, though Uhl *et al.*

(1979a) have suggested that the opsin band contains a considerable amount of a protein other than opsin. On the basis of X-ray diffraction data (Chabre, 1975), and the pattern of aggregation obtained in gluteraldehyde cross-linking experiments (Downer, 1982), it appears that in the dark-adapted OS rhodopsin is monomeric, though the possibility that it may in part be oligomeric has not yet been excluded.

The disc membrane has a high proportion (over 80%) of phospholipids and a low concentration (< 10%) of cholesterol (Daemen, 1973; Anderson and Andrews, 1982). More than 95% of the total phospholipid content is accounted for by five lipids: phosphatidylcholine (PC) (40%); phosphatidyl-ethanolamine (PE) (38%); phosphatidylserine (PS) (13%); sphingomyelin (3%), and phosphatidyl-inositol (PI) (2%) (Daemen, 1973). A high proportion of the fatty acids is unsaturated (the predominant species being 22 : 6). The membrane is therefore expected to be unusually fluid, and also very susceptible to lipid peroxidation. Interestingly, it has been shown that it contains a relatively high concentration of the antioxidant tocopherol (Daemen, 1973; Anderson and Andrews, 1982). The plasma membrane is reported to have a higher proportion of saturated fatty acids (Kamps et al., 1982). There are about 65 lipid molecules for every rhodopsin molecule in the disc membrane (Daemen, 1973; Nemes et al., 1979), and, from chemical labelling studies, the lipids have been reported by some to be asymmetrically arranged, though there is poor agreement as to their quantitative distribution. In the majority of studies, PE and PS have been found to be preferentially concentrated in the outer leaf of the bilayer, and PC on the inner surface (Smith et al., 1977; Crain et al., 1978; Miljanich et al., 1981). However, Drenthe et al. (1980) report a virtually symmetric distribution. A reconstruction of the thermal behaviour of the OS bilayer from model monolayer phospholipid dispersions, using parinaric acid as a fluorescent probe, is consistent with the asymmetric distribution, with the cholesterol in the inner leaf (Sklar and Dratz, 1982).

2. Mobility of Rhodopsin Within the Membrane

Given the ratio of rhodopsin to lipid molecules, it is possible to estimate the proportion of the membrane surface occupied by protein and lipid respectively. Calculations indicate that up to 25% of the area is occupied by rhodopsin and other proteins (Uhl and Abrahamson, 1981), so the visualization of the membrane as a thinly populated "sea" of lipid is inappropriate. Nevertheless there is good evidence that the rhodopsin molecules are highly mobile: Cone (1972) showed, by following the relaxation of the transient photodichroism induced by flash photolysis of the pigment with polarized light, that they rotate in the plane of the membrane with a correlation time of about 20 μs, and estimated that the viscosity of the membrane was of the order of 2 *P*. Saturation transfer EPR experiments have also yielded comparable rotational correlation times for rhodopsin (Baroin *et al.*, 1977; Kusumi *et al.*, 1978). Increasing the viscosity of the solution (by adding sucrose) in which the measurements were made decreased the rate of rotation, thus providing a further indication that a large part of the protein protrudes from the membrane. The optical studies (Cone, 1972) showed that treatment of the OS with gluteraldehyde led to rotational immobilization of the rhodopsin, a result which was confirmed by ST-ESR (Baroin *et al.*, 1979). However, a second ST-ESR study showed only a two-fold increase in the rotational correlation time, which was interpreted as evidence that a high degree of internal mobility within the molecule gives rise to the residual motion (Kusumi *et al.*, 1978).

Not only is rhodopsin free to rotate within the plane of the disc membrane, it is also able to diffuse across the surface of the disc, as was shown by the microspectrophotometric studies of Liebman and Entine (1974) and Poo and Cone (1974). In both cases the lateral diffusion was determined from the time-course of the absorbance recovery observed within a part of the membrane following a local photobleaching flash, and found to be consistent with the viscosity obtained from the rotational diffusion time constant (Cone, 1972). However, in at least one species, the mudpuppy, the recovery of absorptance has been found to be incomplete, suggesting that a part of the visual pigment population is immobile (Drzymala *et al.*, 1980): whether these molecules are randomly distributed across the disc membrane or are, for example, concentrated in regions near the disc edge is not established.

Thus the visual pigment molecules experience an environment which is highly fluid in two dimensions, yet effectively rather rigid in the third, since the protein is unable to tumble.

3. *Rhodopsin–Lipid Interactions*

The degree and extent of interaction of rhodopsin with the lipid bilayer membrane has been the subject of much study. It has been known for some time that the properties of rhodopsin change as the protein is delipidated in some detergents (Heller, 1968a; Zorn and Futterman, 1971) and that the rates of the reactions which it undergoes during bleaching depend upon the presence of lipid (Applebury *et al.*, 1974). The nature of the interaction between the protein and the phospholipids in both OS membranes and reconstituted systems has been probed by several methods, including ESR, NMR and fluorescence, and there has been some dispute over the interpretation of the results obtained. Controversy has centred on the question of the existence of a stoichiometric immobilized "boundary" fraction of the lipid associated with, and forming a shell round, the protein molecule. As has been pointed out by Chapman *et al.* (1979), some of the conflict has arisen from the differences in the time domains probed by the various techniques. Thus, conventional EPR spectroscopy cannot detect motion with a correlation time slower than 10^{-9} s, while for ^{1}H-NMR the corresponding period is 10^{-4} s.

Indeed, from early spin-label ESR measurements on OS membranes, using TEMPO (2,2,6,6-tetramethylpiperidin-1-oxyl), a probe which is soluble only in the fluid membrane fraction, it was suggested that up to half of the lipids were in a rigid layer surrounding the rhodopsin (Pontus and Delmelle, 1975). Other spin-label studies using freely diffusing spin labels have also been interpreted as indicating the presence of immobilized lipid in native OS membranes (Watts et al., 1979). On the other hand, the ST-ESR data on both OS membranes (Favre et al., 1979; Baroin et al., 1979) and reconstituted systems (Davoust et al., 1980) have been interpreted as being inconsistent with such a picture, since the mobility of the boundary lipids was much greater than that of the protein molecule. Favre et al. (1979) found that where rhodopsin aggregation was induced, either as a result of delipidation or by prolonged bleaching exposures, greatly immobilized lipid was observed; they concluded that the use of such conditions so seriously changed the results as to invalidate them for the analysis of native OS membranes. More recently, it has been generally agreed that ESR data indicate that rhodopsin interacts with the lipids in such a way as to reduce by a fairly small amount the mobility of those molecules in contact with it, but that these are in rapid diffusible exchange with the bulk of the membrane (Watts et al., 1981, 1982).

Davoust et al. (1980) and Kusumi et al. (1980) have used reconstituted systems containing purified rhodopsin to determine the effects of changing protein–lipid ratios as a function of temperature. They found that at a given temperature a relatively small reduction in the lipid phase (20%) led to a significant increase in the rhodopsin rotational correlation time; they attribute this to the partial aggregation of the protein. Temperature variations led to broad phase transitions, even in conditions where the lipid–protein ratio was high, suggesting that the rhodopsin molecules do not simply become immobile as the lipids undergo phase transition. Since the observed progressive increase in protein restriction began at temperatures above that of the lipid phase transition, it was attributed to decreasing solubility of the protein in the lipid, an interpretation which is consistent with the distribution patterns of rhodopsin observed in freeze fracture studies (Chen and Hubbell, 1973). Davoust et al. (1980) concluded that immobilization of lipid was associated with molecules which were trapped between proteins, rather than the existence of boundary layers.

Differences between the ^{1}H-NMR spectra of sonicated OS membrane vesicles, which have a broad unresolved component, and those of vesicles of total extracts of OS phospholipids, which do not, were also originally attributed to a population of slowly exchanging lipids (Brown *et al.*, 1977b). Brown *et al.* (1982) have since concluded that the broad contribution to the OS spectrum is more likely to be due to incomplete motional averaging as a result of the relatively large and possibly polydispersed vesicles used. Similarly, the non-exponential data obtained from ^{1}H-NMR spin-lattice relaxation time measurements, which were originally interpreted as evidence of immobilized boundary lipids (Brown *et al.*, 1977c) have been shown to be due to the presence of residual water: it is not observed in lyophilized preparations (Brown *et al.*, 1982). Zumbulyadis and O'Brien (1979), using phospholipid vesicles, also concluded that the lipids in contact with rhodopsin are in fast exchange with those in the bulk of the membrane.

^{13}C-NMR data on rhodopsin–lipid interactions are conflicting: those of Zumbulyadis and O'Brien (1979) suggest that rhodopsin does decrease the motion of adjacent lipids, while Brown *et al.* (1982) suggest that this is an artefactual result of oxidative degradation.

There is also disagreement over the extent to which the presence of rhodopsin plays a structural role in the organization of the membrane: on the one hand it is claimed that without the protein the phospholipids do not have a bilayer configuration (De Grip *et al.*, 1979), while a second study (using, like the first, ^{31}P-NMR) finds that there is little difference between the membranes with or without the rhodopsin present (Deese *et al.*, 1981).

Thus, although there are still areas of conflict, the picture which is emerging of the OS membrane is one in which the interactions between rhodopsin and the lipids are highly dynamic, occurring on a time-scale (about 10^{-7} s) which is much faster than that of rhodopsin rotation, so that the notion that there is some stoichiometric number of lipid molecules in an immobilized sheath surrounding the rhodopsin is not supported (Devaux, 1982; Marsh *et al.*, 1982).

III. Dynamic Changes Induced by Light

A. RHODOPSIN PHOTOLYSIS

The absorption of a quantum of light by the rhodopsin molecule is rather efficient in leading to photolysis: the quantum yield of the photoreaction is nearly 0.7 (Dartnall, 1968). The photolysis involves a complex sequence of reactions with rate constants which differ by some 14 orders of magnitude (Fig. 7). Only the initial reaction is photochemical, all the subsequent ones being thermal. Knowledge of the reactions, which are identified by the spectral changes which accompany them, has largely been established from spectrophotometric studies at a variety of temperatures, and by high speed flash photolysis. While this has been an extremely profitable way of characterizing the photolytic sequence, until relatively recently it has tended to divert attention from reactions which may occur but not involve spectral changes.

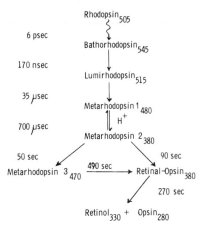

$Rhodopsin_{505}$

6 psec

$Bathorhodopsin_{545}$

170 nsec

$Lumirhodopsin_{515}$

35 μsec

$Metarhodopsin\ 1_{480}$

H^+

700 μsec

$Metarhodopsin\ 2_{380}$

50 sec   90 sec

490 sec

$Metarhodopsin\ 3_{470}$ → $Retinal\text{-}Opsin_{380}$

270 sec

$Retinol_{330}$ + $Opsin_{280}$

FIG. 7. The photolysis sequence of vertebrate rhodopsin in the intact isolated retina, determined spectrophotometrically in conditions where a large fraction ($> 10\%$) of the pigment is bleached. The wavelength at which each species absorbs maximally varies slightly between species; those shown are for the frog. The reaction half lives are approximate. The reaction scheme for the decay of metarhodopsin 2 is one of several which have been shown to be kinetically compatible with the observed spectral changes (Ernst *et al.*, 1978).

While there is broad agreement concerning the spectrally identifiable products, interpretations of the kinetics of several of the steps vary. For the present purpose attention will be centred on the reactions which are known to occur in intact membranes at or near to physiological temperatures. These reactions, illustrated in Fig. 7, may be broadly broken into two groups: those which are fast enough to lead to the primary triggering of the rod's response, and those which are much too slow to play any direct role in transduction. This break occurs at the formation of metarhodopsin 2 (meta 2), which is incidentally the first stage of the photolytic sequence which is accompanied by "bleaching" (the formation of the UV absorbing product).

1. *The Primary Reaction: Formation of Bathorhodopsin*

Although much attention has been focused on the mechanism of the initial process in rhodopsin bleaching, there are still unresolved questions concerning the reaction. Picosecond laser flash photolysis experiments have shown that, in addition to its high quantum efficiency, the reaction is extremely rapid, with the formation of the photoproduct occurring within less than 3–6 ps of absorption of the quantum (Busch *et al.*, 1972; Green *et al.*,

1977). There is some question as to the first spectrally identifiable product: at very low temperatures it appears that the formation of bathorhodopsin is accompanied or preceded by a second species, hypsorhodopsin, which absorbs maximally at about 430 nm (Yoshizawa, 1972). While evidence of the intermediate has been obtained by picosecond absorption spectroscopy at 77°K (Horiuchi *et al.*, 1978), the species is not seen in most studies at room temperature (Applebury and Rentzepis, 1982; Kobayashi and Nagakura, 1982). There is considerable evidence that the primary reaction involves the isomerization of the chromophore from the 11-*cis* to a distorted all-*trans* form, as initially proposed by Wald and his co-workers (Wald, 1968). However, a study of the temperature dependence of the kinetics of the reaction indicated that the transition occurs much more rapidly at very low temperatures than would be expected for isomerization, being complete in less than 40 ps at 4°K (Peters *et al.*, 1977). These authors have therefore favoured the proposal (Thompson, 1975) that the initial step involves a proton transloca-tion between the chromophore and an amino acid side group (probably an imidazole) of the protein. Peters *et al.* (1977) also showed that the rate of formation of bathorhodopsin is very sensitive to proton–deuterium exchange. Since the exchangeable proton on the chromophore is expected to be that involved in protonating the Schiff base linkage, they suggested that, at least at temperatures below 77°K, proton translocation may be the initial step.

Evidence has been put forward against this proposal, based on the photo-conversions of rhodopsin and bathorhodopsin at 77°K (Mao *et al.*, 1980). Bathorhodopsin is itself photosensitive: upon absorption of a quantum it may undergo reaction to form either rhodopsin or isorhodopsin, an analogue visual pigment (absorbing maximally at about 486 nm) in which the chromo-phore is 9-*cis* retinal (Bridges, 1961; Yoshizawa and Wald, 1963). The experiments of Mao *et al.* (1980), which showed that the route between the two visual pigments had bathorhodopsin as a common intermediate, provide strong evidence that *cis-trans* isomerization does occur, for it is otherwise difficult to account for the isomeric changes between the two pigments.

Further support to the notion that the initial step involves isomerization is provided by resonance Raman studies, which show that bathorhodopsin exhibits the characteristics of a distorted protonated all-*trans* retinal Schiff base (Eyring and Mathies, 1979) and that these appear within 30 ps after flash photolysis of either rhodopsin or isorhodopsin (Hayward *et al.*, 1981). Similar conclusions are reached from analyses of the reaction which assume the location of charges on the protein close to the chromophore (Honig *et al.*, 1979; Birge and Hubbard, 1980; Birge, 1981). Semiempirical molecular dynamics calculations (Birge and Hubbard, 1980; Birge, 1981) theoretically predict that the isomerization could occur in ∼ 2 ps, and with a quantum yield of ⩾ 0.6.

2. The Rapid Thermal Reactions

As shown in Fig. 7, there are three thermal reactions in the photolytic sequence of rhodopsin which are fast enough to play a role in transduction. The first of these is the conversion of bathorhodopsin to lumirhodopsin, whose activation parameters are small (Ostroy, 1977). This transition is extremely rapid in solution, with a half life of about 40 ns (Peters et al., 1977), although less so—by a factor of about 4—in the intact rod (Cone, 1972; Catt et al., 1983).

The transition from lumi to meta 1, which occurs in less than about 100 microseconds (Cone, 1972; Ostroy, 1977), has been investigated in a variety of conditions with rather variable results, suggesting that detailed comparison of the various sets of data may not be justified. For example, there have been suggestions that the reaction is first order (Cone, 1972), second order (Rapp, 1979), or involves two or more parallel first order processes (Abrahamson and Wiesenfeld, 1972; Stewart et al., 1977; Hofmann et al., 1978). However, it appears that the reaction involves no major alteration in the configuration of the protein (Ostroy, 1977). The reaction does involve a charge displacement which gives rise to a component of the fast photovoltage, or early receptor potential (ERP), which is the first electrical activity displayed by rods following light absorption (Arden, 1969). The properties of the ERP are so dissimilar from those of the rod's operating characteristics that it is widely assumed that the ERP has no functional significance. However, the ERP has proved a useful probe of reconstituted membrane systems containing rhodopsin (Montal et al., 1981; Skulachev, 1982).

It is in fact noteworthy that the lack of influence of the medium surrounding the visual pigment molecule is such that the reactions up to and including the formation of meta 1 occur in preparations which have been fully dried (Kimbel et al., 1970) or have abnormal lipid environments (Applebury et al., 1974; O'Brien et al., 1977).

In marked contrast to the preceding reactions, the formation of metarhodopsin 2 involves intermolecular processes, requires the presence of water and is sensitive to pH (Matthews et al., 1963; Baumann and Zeppenfeld, 1981) and lipid environment (Applebury et al., 1974). One or more proton is taken up for each protein molecule undergoing the reaction (Emrich, 1971; Bennett, 1978; Kaupp et al., 1981a,b) though the Schiff base linkage becomes deprotonated, as indicated by the large hypsochromic shift in absorption which accompanies it (Fig. 7) and by Raman studies (Lewis et al., 1973). It has been suggested that the Schiff base linkage is also cleaved (Allan and Cooper, 1980), though this has not been confirmed (Pande et al., 1982).

Several lines of evidence suggest that major conformational changes occur in the protein at this stage. Thus, the reaction has an energy of activation of about 33 kcal/mol and an associated entropy of activation of about 70 cal/deg. mol (Abrahamson, 1973; Ostroy, 1977). The chromophore in meta 2 is readily available to attack by hydroxylamine (Bridges, 1962) and NaBH$_4$ (Bownds, 1967), indicating that it becomes more exposed to the cytoplasm. The number of reactive sulphydryls also increases (Regan et al., 1978), again suggesting that there is an unfolding of the protein. The rate at which the C-terminal of the molecule is cleaved by thermolysin increases with the formation of meta 2, indicating that this region of the protein becomes more exposed (Kuhn et al., 1982).

There are several unusual features about the meta 1 – 2 reaction which have led to a variety of descriptions of it. In early studies on solubilized preparations of rhodopsin it was widely recognized that the reaction was an equilibrium, being driven in the forward direction by increased temperature and acidity (Matthews et al., 1963; Ostroy et al., 1966), and there was evidence that there were several parallel reaction pathways (Linschitz et al., 1957; Abrahamson and Wiesenfeld, 1972). In OS preparations the reaction has frequently been described as a single first order process (Kimbel et al., 1970; Rapp et al., 1970; Baumann, 1976; Rapp, 1979; Attwood and Gut-freund, 1980). However, it has been suggested that there are two parallel first order reaction paths involved in isolated bovine disc preparations (Hofmann et al., 1978).

The kinetic complexities of the reaction have led to suggestions that there are in fact two isochromic forms of metarhodopsins 1 and 2. Williams et al. (1974) proposed two forms of meta 1, one of which was able to photo-regenerate rhodopsin, while the other was not. Bennett (1978) studied the reaction in bovine discs at 3°C, and concluded that there are protonated and unprotonated forms of both meta 1 and 2 and that it is the state of protonation of the chromophore which determines the spectral characteristics. When the changes in the light-scattering properties of an OS preparation were used to monitor the reaction, Uhl et al. (1978) found that a scheme in which meta 1 decayed irreversibly to meta 2, and there was then an equilibrium reaction between meta 2 and an isochromic form of meta 1 best described their data. More recently, Baumann and Zeppenfeld (1981) have proposed that in frog OS vesicles the transition is best described as an initial rapid, non pH dependent equilibrium meta 1 to 2 reaction, accompanied by a pH sensitive equilibrium between isochromic forms of meta 2. The characteristics of the reaction have also been shown to change with the extent of bleach in intact OS: for flashes which photolyse less than 10% of the rhodopsin, a higher ratio of meta 2 v. meta 1 is found than is the case with more intense flashes (Emeis and Hofmann, 1981). This bleaching range is the same as that over which discs have been shown to exhibit rapid light-induced changes in their scattering properties resulting from binding of a GTPase protein (see below) to the membrane (Kuhn *et al.*, 1981). Experiments in which the stoichiometry of the GTPase and rhodopsin were varied strongly suggest that it is the binding of the GTPase which leads to displacement of the meta 1 to meta 2 equilibrium towards meta 2 (Emeis *et al.*, 1982).

The existence of isochromic forms of the intermediates has been extended to the notion that there are two forms of rhodopsin itself. Hofmann *et al.* (1978) point out that several apparently anomalous observations in the literature would be resolved if this were the case, and Molday and Molday (1979) and Fatt (1981) have shown that SDS-PAGE separations of frog OS proteins indicate that they contain more than one species of rhodopsin.

3. *The Slow Reactions*

None of the spectrally observable reactions following the formation of meta 2 occurs fast enough to play any direct role in triggering the transduction process. Nevertheless there has been considerable interest in them, since it has been suggested that they may reflect some regulatory role of rhodopsin in controlling the sensitivity of the receptor, i.e. its light and dark adaptation (Donner and Reuter, 1967; Ernst and Kemp, 1972; Donner and Hemilä, 1979).

The disappearance of meta 2 is a complex reaction, for which several routes have been proposed (Baumann, 1972; Ernst and Kemp, 1972; Brin and Ripps, 1977; Ernst *et al.*, 1978; Blazynski and Ostroy, 1981). Thus, although there is general agreement that meta 2 decays to meta 3 and to retinal, it is uncertain whether the meta 3 is in equilibrium with meta 2 (Chabre and Breton, 1979; Baumann and Zeppenfeld, 1981), or itself decays also, either to retinal (Baumann, 1972, Ernst *et al.*, 1978; Blazynski and Ostroy, 1981) or to retinol (Cone and Cobbs, 1969). As is the case for the formation of meta 2, the picture is further complicated by the suggestion that the reaction sequence depends on the fraction of rhodopsin bleached: Donner and Hemiliä (1975) found that when the light exposure caused less than 5% of the pigment to photolyse, no meta 3 could be detected.

It has been shown that the binding site becomes available as meta 2 decays (Rotmans *et al.*, 1974), which indicates that the all-*trans* retinal migrates at this stage, though polarized absorption spectra of intact rods show that the chromophore in meta 3 and retinal–opsin maintains approximately the same orientation as in rhodopsin (Ernst and Kemp, 1978). Since the chromophore does not appear to become isotropically distributed, nor reorientates to an angle greater than 50° to the plane of the membrane, as occurs for retinol (Wald, 1968) it presumably remains rather specifically bound (Van Breugel *et al.*, 1979).

In the intact retina, the retinal produced as a result of rhodopsin photolysis is enzymically reduced to retinol, at least when a substantial fraction of the pigment is bleached (Wald, 1968; Cone and Cobbs, 1969). This reaction is not seen in many of the OS preparations used for studies of rhodopsin photolysis because the necessary coenzyme, NADPH (Futterman, 1963), is washed out during the purification of the preparation. It is observed, however, when precautions are taken to ensure that the OS are maintained osmotically intact (Schnetkamp et al., 1979).

4. Rhodopsin Regeneration

The apparatus necessary for rhodopsin regeneration to occur is lost when the retina is separated from the pigment epithelial cells. However, rhodopsin can be regenerated by adding exogenous 11-cis retinal to preparations consisting of intact isolated retinas (Pepperberg et al., 1976), OS membranes (DeGrip et al., 1972), reconstituted membranes (Darszon et al., 1979) or micellar solution. The choice of solubilising agent is important, since no regeneration is obtained in several detergent extracts (DeGrip, 1982).

B. OTHER LIGHT-MODULATED PROPERTIES

While spectrophotometric studies of rhodopsin photolysis have led to much insight into the behaviour of the molecule, the search for the mechanism by which it triggers transduction has led to increasing attention being paid to the other light-induced changes in the OS. In particular, the last decade has seen a great broadening of knowledge about the biochemical changes within the OS which accompany rhodopsin photolysis. Several of the observed changes in enzymic activities and the concentrations of intracellular components have been implicated in suggested mechanisms linking rhodopsin bleaching to rod transduction or adaptation. The system which has received the most attention is that which modulates the concentrations of cyclic nucleotides, at least partly because of their known involvement in the mechanism of hormone action (Greengard, 1978).

1. cGMP Phosphodiesterase (PDE) Activity

The intracellular concentration of cGMP in rod OS in the dark is about 10–50 μM, an unusually high level (Farber and Lolley, 1974; Lolley and Farber, 1976; Orr et al., 1976; Berger et al., 1980, Kilbride, 1980). Illumination causes this to drop, by more than 50% (Ferendelli and Cohen, 1976; Woodruff et al., 1977; Cohen et al., 1978; Kilbride, 1980). It has been shown that the decrease can take place extremely rapidly, at least in in vitro conditions, with a $t_{\frac{1}{2}}$ of less than 200 ms. (Woodruff et al., 1977; Yee and Liebman, 1978; Woodruff and Bownds, 1979). It has therefore been suggested that the reaction may play a role in transduction (Farber et al., 1978; Liebman and Pugh, 1979; Hubbell and Bownds, 1979). Experiments on intact retinas, on the other hand, have yielded much lower estimates of the rate at which the reaction occurs and imply that it is too slow to be consistent with its involvement in the mechanism of transduction (Kilbride and Ebrey, 1979; Ebrey et al., 1981). However, these experiments suffer the disadvantage that the cellular origins of the cGMP being measured are less precisely known, so that changes within the OS may be masked by other retinal changes.

The extent of the reduction depends on the intensity of light stimulation and has been shown to be half saturated when fewer than 1000 of the rhodopsin molecules in a rod are photolysed (Woodruff et al., 1977). Thus, the range of light intensities over which the extent of the observed change varies is higher than that of the receptor's response by a factor of less than 50.

The magnitude of the light-induced change also depends upon the experimental conditions in which it is studied: in media containing very low levels of Ca^{2+} (10^{-9} M) cGMP levels in dark adapted ROS rise tenfold, and the fall is correspondingly larger (Kilbride, 1980). The suggestion that the rapid reaction might therefore occur only at low Ca levels was shown not to be the case in isolated OS, where the cGMP decrease is substantially complete within 1 second at Ca concentrations ranging from 10^{-9} to 10^{-3} M (Polans et al., 1981).

The PDE which is responsible for the changes in cGMP concentration has been purified and characterized (Miki et al., 1975; Baehr et al., 1979; Kuhn, 1981). It is loosely bound to the membrane and is readily dissociated from it by reduction in ionic strength and the removal of Mg^{2+} ions (Miki et al., 1975; Wheeler et al., 1977; Bignetti et al., 1978; Baehr et al., 1979). Unlike some of the proteins which are peripherally attached to the membrane, the tightness with which it is bound does not change when the OS are exposed to light (Baehr et al., 1979; Kuhn, 1980a,b). The PDE will also bind to pure PC–PE lipid bilayers in isotonic media, which suggests that it adheres to the lipids rather than to rhodopsin in the native membrane (O'Brien, 1982b). Estimates of its abundance in the OS relative to that of rhodopsin vary from about 0.5–2.0% (Baehr et al., 1979; Kuhn, 1981).

The isolated PDE system consists of two large subunits, with molecular weights of about 88 000 and 84 000 (Wheeler and Bitensky, 1977; Baehr et al., 1979; Hurley and Stryer, 1982) and a third small (molecular weight about 11 000) polypeptide (Kuhn, 1981; Hurley and Stryer, 1982). The PDE is inactive in the absence of GTP and light activated OS membranes (Wheeler et al., 1977; Wheeler and Bitensky, 1977; Yee and Liebman, 1978), but can be fully activated by limited proteolysis (Miki et al., 1975). The relative inactivity of the enzyme in dark-adapted OS membranes has led to the suggestion that light and limited proteolysis activation arise as the result of the removal of an inhibitory constraint (Hurley, 1980), and Hurley and Stryer (1982) have shown that the small subunit of the PDE has this regulatory role.

Hydrolysis of the GTP is not a requirement for PDE activation, as has been demonstrated in studies where guanosine 5'-(β,γ-imido)triphosphate, a non-hydrolysable analogue have been used (Godchaux and Zimmerman, 1979b; Liebman and Pugh, 1980). Estimates of the K_m value of the purified PDE are 70 μM for the frog OS enzyme (Miki *et al.*, 1975) and 150 μM for the cow (Baehr *et al.*, 1979).

Liebman and his colleagues (Yee and Liebman, 1978; Liebman and Pugh, 1979; Liebman and Pugh, 1980, 1981, 1982) have made extensive use of a rapid monitoring technique which has enabled them to follow the kinetic course of the changes in cGMP concentration in some detail. The method uses either a fast pH electrode (Yee and Liebman, 1978) or a pH indicator dye (Liebman and Evanczuk, 1982) to monitor the release of H^+ which accompanies the hydrolysis of cGMP. Using this technique, they have shown that the change in cGMP concentration induced by a flash which bleaches $\sim 10\%$ of the pigment occurs with a delay of 50 ms, a time-scale similar to, but slower than that of meta 2 formation (Liebman and Pugh, 1981).

Thus, both the speed and the apparent amplification by the PDE observed in *in vitro* conditions are extremely high: photoactivation of a single rhodopsin molecule can lead to the hydrolysis of up to 10^6 cGMP hydrolysed per second (Yee and Liebman, 1978; Liebman and Pugh, 1979, 1981). Such a rate of activation appears to be impossible without some intermediary catalytic step. It was originally proposed by Liebman and Pugh (1979) that a photolysis product of rhodopsin was the catalyst, activating up to 500 PDE molecules by causing them to bind a GTP molecule. The activation was postulated to result from interaction between R* and the PDE molecules as a result of the lateral diffusion of the proteins across the membrane. More recent evidence indicates that the activation of PDE by rhodopsin involves a second enzyme system which binds GTP (Fung and Stryer, 1980; Fung *et al.*, 1981), though the lateral diffusion of rhodopsin may still play a major role in the activating of this latter protein (Liebman *et al.*, 1982).

The light-activated PDE activity has been shown to be strongly regulated by ATP (Liebman and Pugh, 1980, 1982; Kawamara and Bownds, 1981). The initial rate of cGMP hydrolysis is little changed by the presence of 5 μM ATP, but the time over which the reaction continues is much reduced, so that the overall extent of hydrolysis is reduced (Liebman and Pugh, 1980). One mM ATP causes the range over which light modulates the extent of the hydrolysis to undergo a threefold shift to higher intensities (Liebman and Pugh, 1982).

2. *GTP-binding Protein, or GTPase*

The requirement of both light and GTP for transient activation of the PDE system led to studies which have shown that a second peripheral OS membrane protein, originally described as a GTPase, plays an intermediary role in the light activated modulation of cGMP. As for the PDE, the spectral dependence of the light activation of the protein matches the rhodopsin absorption spectrum (Wheeler *et al.*, 1977). The extent of activation by light is very high: Fung and Stryer (1980) estimated that the photolysis of a single rhodopsin leads to the release of up to 500 GDP molecules, while Liebman and Pugh (1982) conclude that the number is several thousands. The reaction involves the displacement of bound GDP by GTP and does not require hydrolysis of the latter, as has been shown by the use of non-hydrolysable analogues (Fung *et al.*, 1981; Stryer *et al.*, 1981). The K_m of the GTPase is about 0.5 μM, a value close to the concentration of GTP required for half-maximal activation of the PDE (Wheeler *et al.*, 1977; Kuhn, 1981). The maximum rate of GTP hydrolysis by the enzyme is slow, at about 1 mole of phosphate released per mole of GTPase per minute (Wheeler *et al.*, 1977; Caretta *et al.*, 1979; Kuhn 1981), and it is the GTP/GDP exchange properties which lead to the rapid light activated decrease in GTP in the OS.

The properties of the GTPase described here, and its association with rhodopsin and the transduction process, have led to it being given a variety of names by different authors. It is sometimes referred to as the GTP-binding protein (Godchaux and Zimmerman, 1979b), transducin (Fung *et al.*, 1981), Γ (Liebman and Pugh, 1979) and the G- or N-protein (Abood *et al.*, 1982; Somers and Shichi, 1982), these last two terms being by analogy with a protein known to be involved in hormonal activation of adenylate cyclase (Rodbell, 1980). In fact, the homology between the proteins in the OS and the cyclase system has been demonstrated by their functional interchangeability (Bitensky *et al.*, 1982), as well as similarities in their inhibition by cholera toxin (Abood *et al.*, 1982). For simplicity, it will be referred to as the G-protein, or G, here.

Like the PDE, the G-protein is only peripherally bound to the disc membrane in dark-adapted OS and can be released by reducing the ionic strength of the bathing medium (Bignetti *et al.*, 1978; Kuhn, 1980a). However, in contrast to the PDE, it becomes much more tightly attached following illumination. This light-modulated binding is also dependent on the absence of GTP: if the light-adapted membranes are washed with GTP the protein is released (Kuhn, 1980b).

This property of the G-protein has been of value in its purification, which has been carried out in several laboratories (Godchaux and Zimmerman, 1979b; Kuhn, 1980a,b; Fung *et al.*, 1981). Stoichiometric studies indicate that the rhodopsin : G-protein ratio in the OS is between 16 and 10 : 1 (Godchaux and Zimmerman, 1979a; Kuhn, 1980a, 1981; Baehr *et al.*, 1982). The G- protein is found to consist of three polypeptides whose molecular weights are approximately 37 000, 35 000 and 6000 (Fung *et al.*, 1981; Kuhn,

1981). The highest molecular weight unit α has been shown by Fung *et al.* (1981) to be the one which binds GDP in the dark and exchanges it with GTP upon light activation of the OS membranes. The factors affecting the ease with which the subunits may be solubilized from the OS membrane also differ. The original observations of the ability of GTP to bring about dissociation of the GTPase from the membrane were carried out at low ionic strengths, and it has been questioned whether the process occurs in more physiological conditions (Godchaux and Zimmerman, 1979b; Kuhn, 1980b). In the presence of 1 mM GTP and 2 mM ATP, that is concentrations similar to those in OS (Biernbaum and Bownds, 1979), in a medium whose ionic composition matches that of the rod (Robinson and Hagins, 1979), all three sub-units are bound to the membrane in darkness, and the G-protein becomes only partially soluble in light. It appears that the α sub-unit is preferentially solubilized, suggesting that it is this polypeptide which has light and GTP dependent binding properties, while the light dependent change in solubility of the 35 000 subunit is sensitive to ionic strength (Kuhn, 1981).

The solubilized protein is inactive, and its activation requires the presence of OS membranes, light and an ionic strength sufficient to induce modulation of its binding to the membrane (Kuhn, 1981). Like the PDE, the G-protein adheres to lipid bilayer membranes, suggesting that it is not necessary for it to be bound to rhodopsin (O'Brien, 1982b). However, the light-activated increase in binding observed in the absence of GTP is almost certainly due to formation of a complex between the G-protein and the rhodopsin molecule, since it has been shown that the OS can be replaced by reconstituted membranes consisting of phosphatidylcholine and rhodopsin purified by chromatography (Stryer *et al.*, 1981). Activation of the G-protein has been shown to be dependent on the structural integrity of the rhodopsin in the membranes, and it is a photoactivated form of the pigment molecule which binds it (Kuhn and Hargrave, 1981). When the rhodopsin molecule is extensively proteolysed, to leave only the F1–F2 core, the G-protein binding is lost. Limited proteolysis to remove only the carboxyl terminal peptide, on the other hand, does not inhibit binding (Kuhn and Hargrave, 1981).

Measurements of light-induced changes in the turbidity properties of OS have proved fruitful for probing the interactions of rhodopsin and other molecules (Uhl and Abrahamson, 1981). Recent studies have convincingly coupled the optical effects to the interactions between rhodopsin and the G-protein events, and have enabled the method to be used to probe the kinetics and stoichiometry of the process (Kuhn *et al.*, 1981; Bennett, 1982). In the absence of GTP, the turbidity change, termed the binding signal, increases linearly with the extent of rhodopsin photolysed at low levels and saturates at about the 10% bleaching level. Since this is near to the G-protein : rhodopsin stoichiometric ratio, it suggests that the two species form a 1 : 1 binding complex. The time course of the rapid phase of the binding signal is reported to be slower than that of metarhodopsin 2 formation. In the presence of GTP, a slightly delayed transient of opposing sign is observed (the dissociation signal), as the GTP–GDP exchange occurs and the G–GTP–protein dissociates from rhodopsin. This signal also saturates with increasing light intensity, but at much lower levels (less than 0.5% rhodopsin bleached) than does the binding signal. Bennett (1982) has also monitored the H^+ release which accompanies cGMP hydrolysis and has shown that it follows the dissociation signal with little delay.

In addition to the light-activated decrease in GTP by exchange with bound GDP, there may be a second light-activated GTPase operating in the OS. Biernbaum and Bownds (1979) and Robinson and Hagins (1979) have shown that the light dependent change in intrinsic GTP levels in intact frog OS also depends on Ca concentration, unlike the activity described above (Kuhn, 1981), and that the range of light intensities over which its activation varies is also much greater, spanning more than 5 log units (Bownds, 1981). The time-course of the reaction, which occurs over several seconds, has not been linked with any specific reaction in the rhodopsin photolysis sequence.

3. ATPases

A rapid change in OS cytosol ATP and ADP levels has recently been reported by Zuckerman et al. (1982). The reaction, like that involving the G-protein, is highly amplified at low light levels, and at sufficiently high light levels occurs within 250 ms. The characteristics of the reaction appear to be inconsistent with a light-activated hydrolysis and suggest that the process involves the exchange of bound ADP with ATP. The similarities to the GTPase exchange do not extend to the behaviour in media of low ionic strength. In conditions where the GTPase would be solubilized from the membrane and inactivated, ATPase activity is increased. It therefore appears that the two systems are distinct from one another. The time-course of the reaction is described as being directly linked to the formation of metarhodopsin 2 (Zuckerman et al., 1982), though the evidence for this is not compelling.

A further ATPase activity has been reported to be triggered by light (Uhl et al., 1979a,b; Borys et al., 1982). The enzyme is enabled by the presence of Mg^{2+}, and is active in the dark. In OS preparations where the ATPase is active, light induces a rapid ($\leqslant 50$ ms) transient change in the light-scattering properties of the discs. It has been suggested that the ATPase causes the uptake of H^+ and anions into the discs in the dark, and that light causes a rapid efflux of H^+ ions without accompanying anions, and further enhances the ATPase activity (Borys et al., 1982).

Transient light-activated phosphorylation by ATP of a large membrane bound protein in toad OS has been reported by Thacher (1978, 1982). The activity, which is Mg^{2+} dependent, is stimulated by light by a factor of two. The function of this ATPase activity is unknown, though it has been suggested that it may mediate ion transport across OS membranes (Thacher, 1982).

4. Protein Dephosphorylation

A light-dependent dephosphorylation of two extrinsic membrane proteins with molecular weights of 13 000 and 12 000 has been described (Polans et al., 1979). The reaction, which is activated over a similar range of intensities to that observed for the cGMP changes, is observed only in ROS which remain attached to the retina. The phosphorylation of the two proteins, which occurs in the dark, is cGMP dependent and is inhibited by Ca (Hermolin et al., 1982).

5. Rhodopsin Phosphorylation

As indicated earlier (Section II.A.4), it is known that rhodopsin in OS becomes phosphorylated during its photolysis (Kuhn and Dreyer, 1972). The rhodopsin kinase is loosely bound to the disc membranes in the dark and tightly bound upon illumination (Frank and Buzney, 1975; Kuhn, 1978). The molecular weight of the protein has been estimated as 53 000 (Shichi and Somers, 1978) and 68 000 (Kuhn, 1978). More recently it has been suggested that it consists of two polypeptides, of which only the 68 000 dalton species undergoes light-dependent changes in the tightness with which it is bound to the membrane (Kuhn, 1981). The kinase transfers the phosphate of either ATP or GTP to the rhodopsin (Chader et al., 1975; Miller and Paulsen, 1975; Weller et al., 1976). The reaction has been shown to occur in vivo, and to be followed by a slow dephosphorylation process (Kuhn, 1974).

There is general agreement that the kinase is not itself activated by light (Kuhn and Dreyer, 1972; Weller et al., 1975) but there is controversy over the exact pattern of the kinase activity. Thus, most studies have found that only those rhodopsin molecules which have been photolysed can act as substrate (e.g. Kuhn and Dreyer, 1972; Chader et al., 1976). However, it has been reported that when only very small fractions of the pigment are photolysed, unbleached rhodopsin molecules may also be phosphorylated (Bownds et al., 1972; Miller et al., 1977). The picture is further complicated by the suggestion that only the rhodopsin in the plasma membrane and in the newly formed discs at the base of the OS are phosphorylated (Shichi and Somers, 1978). However, this result has not been confirmed by a more recent study (Wilden and Kuhn, 1982).

The stoichiometry of phosphorylation is also complex: it appears that the amount of phosphate incorporated per rhodopsin molecule is highest when little of the pigment is bleached, at least in frogs (Bownds *et al.*, 1972; Miller and Paulsen, 1975; Hermolin *et al.*, 1982). In bovine rods the degree of phosphorylation has been found to be linearly proportional to the extent of bleaching, though it has been shown that the extent of phosphorylation varies among the population of bleached molecules (Shichi *et al.*, 1974; Kuhn and McDowell, 1977; Sale *et al.*, 1978; Shichi and Somers, 1978). The highest level of incorporation observed for the larger bleaches averaged seven phosphates per rhodopsin molecule (Wilden and Kuhn, 1982), though the population was always heterogeneous, the stoichiometric ratio varying from a maximum of 9 : 1 to a total lack of phosphate incorporation. Most studies have hitherto found lower values (Miller and Paulsen, 1975; Kuhn and McDowell, 1977). The extent of phosphorylation depends on Ca levels (Chader *et al.*, 1980; Hermolin *et al.*, 1982); increasing the concentration of Ca from nM to mM reduces the phosphate incorporated by more than twofold.

Most, if not all, of the phosphorylation sites on the pigment molecule are known to be situated near to the carboxyl terminal, since they are readily removed by limited proteolysis (Virmaux *et al.*, 1975; Hargrave and Fong, 1977). More detailed analysis of the peptide fragments indicates that all 7 of the serines and threonines present in the sequence 6 to 15 are capable of becoming phosphorylated (Hargrave *et al.*, 1982).

The species of bleached rhodopsin which acts as substrate to the kinase is uncertain, though several attempts have been made to discover whether the ability of the molecule to become phosphorylated follows a similar time course to the appearance or decay of one or more of the photoproducts. It has been reported that the decay of meta 2 is too fast to be correlated with decay of kinase activity and that of meta 3 is too slow (Kuhn, 1978). While Miller *et al.* (1977) found that there was similarity between the decay of meta 3 and kinase activity in frog, the two processes varied differently with temperature. It thus appears that the phosphorylation of rhodopsin may be "spectrally silent", which is consistent with wide separation of the phosphate binding sites from that of the chromophore.

The dephosphorylation of rhodopsin which accompanies the regeneration of the visual pigment *in vivo* (Kuhn, 1974) is usually lost in preparations of isolated OS (Bownds *et al.*, 1972; Frank *et al.*, 1973). Its absence does not prevent the bleached pigment from being regenerated when exogenous chromophore is added (Miller and Paulsen, 1975), which again suggests that the phosphorylation has no direct effect on the chromophoric binding site.

6. *Modulation of OS Ca Levels*

Estimates of the concentration of Ca within OS range from 0.1 to 12 Ca/rhodopsin, though most determinations lie in the range between 1 and 3

Ca/rhodopsin (Kaupp and Schnetkamp, 1982), equivalent to a concentration of 3–10 mM. The major fraction of this is located within the discs (Schnetkamp, 1979; Kaupp et al., 1979). The level of free Ca in the cytoplasm is almost certainly much lower than this and is likely to be of the order of 10^{-7}–10^{-6} M (Fain and Lisman, 1981; Kaupp and Schnetkamp, 1982). Following the original suggestion that the transmitter which carries the information from the light-activated rhodopsin to the plasma membrane is Ca which is sequestered in the discs in darkness (Yoshikami and Hagins, 1971; Hagins, 1972), there have been many attempts to determine whether the concentration of the ion within the OS is indeed modulated by light. Quantitative analyses of the required change show that this must be substantial in order to provide the observed amplification (Cone, 1973; Yoshikami and Hagins, 1973; Hagins and Yoshikami, 1974; Hagins and Yoshikami, 1977; Uhl and Abrahamson, 1981); estimates range between 20 and 100 000 Ca ions released as a result of bleaching a single rhodopsin molecule. While the analysis which produces the highest estimate (Uhl and Abrahamson, 1981) is based on the inclusion of several assumptions concerning the buffering capacity of the nucleotides present in the OS, it is plausibly argued that previous calculations underestimate the required change.

Almost all of the studies carried out have failed to demonstrate such a release of Ca (or the re-uptake which should occur in darkness). These have ranged from experiments on OS fragments (Smith et al., 1977; Szuts and Cone, 1977; Liebman, 1974, 1978) to intact receptors (Szuts and Cone, 1977; Szuts, 1980; 1981). A wide variety of techniques has also been used, including atomic absorption spectroscopy (Hendriks et al., 1974; Liebman, 1974; Szuts and Cone, 1977), ^{45}Ca labelling (Bownds et al., 1971; Weller et al., 1975; Hemminki, 1975; Smith et al., 1977; Szuts, 1980, 1981), and flash spectrophotometry, using a Ca sensitive dye as indicator (Kaupp and Junge, 1977; Hubbell and Bownds, 1979; Kaupp et al., 1979; Smith and Bauer, 1979; Noll et al., 1979; Kaupp et al., 1981a,b). There has been a correspondingly wide diversity of results, but in general no satisfactory demonstration of a sufficiently large release to fulfil the requirements of internal transmitter, at least in the model originally proposed (Yoshikami and Hagins, 1971). Thus, in most cases the stoichiometry of release of Ca has been about 1 per rhodopsin bleached (Noll et al., 1979). The major exception, where up to 90 Ca were released for each pigment molecule bleached, was in an artificial rhodopsin–phospholipid recombinant system containing a very low concentration of rhodopsin (O'Brien, 1979).

The kinetics of the Ca^{2+} change induced by light are reported to be only slightly less rapid than those of meta 2 formation in sonicated OS preparations to which the ionophore A23187 had been added (Kaupp et al., 1979); in intact OS they were slower by two orders of magnitude (Kaupp et al.,

1981b). In both cases the release of each Ca ion was accompanied by uptake of about $3H^+$ (Kaupp et al., 1981a,b).

Many of the studies of Ca modulation suffered the disadvantage that they involved severe disruption of the OS biochemical apparatus in the purification of the membranes. Even when carried out on "intact" OS, conditions were often used which have been shown to have caused the preparation to become leaky and to lose osmotic integrity (Yoshikami et al., 1974; Schnetkamp et al., 1979), so that the necessary apparatus for the Ca translocation may be lost. In addition, where changes in Ca have been observed, there has been doubt as to whether they can be attributed to contamination of the OS membrane preparation by other subcellular organelles (Szuts, 1980). Nevertheless, even in the intact isolated retina Szuts (1981) found that incubation with ^{45}Ca resulted in only a very slow accumulation of the radiolabel within the discs, the extent of which was virtually unchanged when the retinas were repeatedly light stimulated. Szuts (1981) concluded that if there is a light sensitive pool of Ca then it is located in bound form within the cytoplasm.

There is, however, evidence of a transient light-activated release of Ca from the photoreceptors of the intact isolated retina into the extracellular space. Yoshikami et al. (1980), using a Ca selective microelectrode positioned close to the OS, detected a transient extracellular increase in Ca from rat photoreceptors. In bathing media containing 0.1 mM or less Ca, the time course of the increase was similar to that of the receptors' electrical response. A similar, but larger, Ca release has also been found to occur from the receptors in the isolated toad retina by Gold and Korenbrot (1980a,b,c, 1981), who showed that the action spectrum of the process was consistent with the spectral properties of the rhodopsin-containing rods. Like the light-activated change in cGMP, both the rate and extent of the efflux depend on the stimulus intensity (Yoshikami et al., 1980; Gold and Korenbrot, 1981). However, in the toad, the range of light intensities over which the efflux is modulated is very wide, with increasing amounts of Ca being released until the light intensity has been increased by more than 4 log units. The size of the effect is larger in the toad than in the rat retina. In the rat, 800 or so Ca were estimated to be released for each rhodopsin photolysed (Yoshikami et al., 1980), while in the toad the corresponding figure is 20 000 (Gold and Korenbrot, 1981). The maximal Ca release observed from toad receptors resulting from a single light flash amounts to about 10^7 Ca per rod, which constitutes about 1 % of the total Ca content of the rod (Hagins and Yoshikami, 1975; Szuts and Cone, 1977). The latency between light stimulation and the onset of the increase in extracellular Ca is estimated to be less than 200 ms. Gold and Korenbrot (1981) conclude, from experiments in which the electrophysiological response was measured in a variety of bathing media, that the extracellular increase results from a light stimulated active Ca efflux (possibly primarily through a Na–Ca exchange process), rather than as a passive consequence of closure of Na channels through which the Ca ions had been entering the OS in the dark.

IV. Transduction Mechanisms

A. OUTLINE OF REQUIREMENTS OF MODELS

Several authors have reviewed the requirements which any proposed mechanism for the transfer of information from rhodopsin to the outer membrane of the OS must meet, and have made critical appraisals of existing hypotheses (Hagins, 1972; Cone, 1973; Hubbell and Bownds, 1979; Miller, 1981; Uhl and Abrahamson, 1981; Liebman and Pugh, 1981, 1982; Fatt, 1982; O'Brien, 1982b; Saibil, 1982; Skulachev, 1982). The criteria, which are based on many electrophysiological studies of the rod's response in a variety of conditions will therefore only be dealt with in summary.

Central to any mechanism must be the known speed and sensitivity (gain) of the photoreceptor. Additionally, however, any model must be able to generate the signal in a way that mimics the characteristics of the rod photoresponse to a variety of light stimuli. From the properties outlined in Section I.C, it follows that:

(a) Absorption of a single photon is sufficient to cause a rod to respond (Yau *et al.*, 1977);

(b) Modulation of some intracellular constituent is needed to cause the observed reduction in the conductance of the plasma membrane, since the absorption of a quantum by rhodopsin almost always occurs in the disc, rather than in the plasma membrane;

(c) Bleaching of a single rhodopsin reduces the conductance of the plasma membrane by about 3 % (Penn and Hagins, 1972; Baylor *et al.*, 1979a);

(d) It has been estimated that there are at least 3600 light modulatable conductance channels in the plasma membrane (Yoshikami and Hagins, 1973), so that a single photon presumably causes the closure of at least 100 of these;

(e) The manner in which the modulation occurs must be consistent with the observation that the decrease in the plasma conductance in response to a single quantum absorbed does not vary (Baylor *et al.*, 1979b), as would be expected if it were the result of a simple diffusion limited process;

(f) The time course of the response is complex, with an S-shaped delay which shortens with increasing light intensity (Fig. 1). The delay can be described kinetically as a sequence of 4 or more first order processes

(Penn and Hagins, 1972; Baylor *et al.*, 1979a): these do not correspond to spectrally observed transitions in rhodopsin photolysis;

(g) Light absorption by rhodopsin in only a few adjacent discs leads to changes in the plasma membrane near those discs, rather than a spread of excitation throughout the length of the membrane (McNaughton *et al.*, 1980).

As pointed out earlier, all these criteria strongly imply that there is an internal messenger within the OS which carries the information triggered by rhodopsin activation to the plasma membrane. Hypotheses for the mechanism by which the process occurs have centred either on rhodopsin acting as an initiator of enzymes which amplify and relay the information to the plasma membrane, or acting as a gating protein and releasing large numbers of a sequestered transmitter particle.

B. THE cGMP HYPOTHESIS

The rate and extent of the modulation of cGMP levels resulting from rhodopsin photolysis *in vitro* has led to the suggestion that this nucleotide plays a primary role in the transduction mechanism of the rod (e.g., Liebman and Pugh, 1979, 1981; Fung and Stryer, 1980; Miller and Nicol, 1979, 1981). This has appeared to be reinforced by electrophysiological data which show that changing the cGMP levels within rod OS produces changes in the plasma membrane permeability and the rod response characteristics consistent with such a hypothesis (Miller and Nicol, 1979, 1981).

The reaction sequence by which the absorption of light by rhodopsin is thought to bring about the fall in cGMP levels is illustrated in Fig. 8. Two representations are shown; one (Fig. 8a) to emphasize the potential amplification of the system, and the other (Fig. 8b) the way in which rhodopsin and the disc membrane are involved. The steps involved have been deduced as a result of the many studies characterizing the proteins, and the light-invoked changes in their activities described above.

Thus, the process occurs in two stages, each of which amplifies the unitary effect of the quantal absorption. In the dark, the G-protein (G) is bound to GDP and is in the inactive form (Fung *et al.*, 1981). The photolysis of rhodopsin leads to the formation of an activated product (R*), which interacts with a G-protein molecule and causes exchange of the bound GDP by GTP to occur (Fig. 8a). The G-protein dissociates from R*, and itself separates into its subunits. The largest of these (α) leaves the membrane and diffuses in the cytoplasm (Fig. 8b). While the rhodopsin is still in its active form (R*) this process can be repeated to an extent which is limited by the availability of G–GDP molecules. Thus the absorption of a single quantum has been

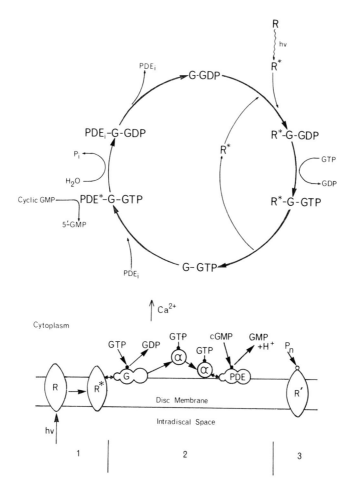

FIG. 8. Diagrammatic representations of the reactions triggered by rhodopsin bleaching. (A) The cyclic nature of the postulated sequence: absorption of light by rhodopsin (R) leads to formation of an active species R^*, which triggers GTP/GDP exchange on the GTP-binding protein (G). The activated species of G (G-GTP) in turn causes disinhibition of the cGMP phosphodiesterase (PDE_i). The PDE (PDE^*-G-GTP) is no longer activated when the GTPase activity of G hydrolyses the bound GTP to GDP. Derived from Fung and Stryer (1980). (B) The reactions in relation to the disc membrane. Section 1 shows the light triggered activation ($R - R^*$) of the peripherally bound trimeric G-protein. Following the GTP/GDP exchange the α sub-unit of G dissociates from the membrane and subsequently interacts with and activates the peripherally bound trimeric PDE (section 2). The phosphorylation of rhodopsin is shown in section 3. The arrow indicates the increase in Ca concentration in the cytoplasm which appears to coincide temporarilly with the events in sections 1 and 2.

estimated to lead to the activation of between 100 and 7000 G-protein molecules (Fung and Stryer, 1980; Fung *et al.*, 1981; Bennett, 1982; Liebman and
Pugh, 1982). While the G-protein molecules could, in principle, originate
from either the disc containing the photolysed rhodopsin molecule or an
adjacent one, it has been calculated that the observed activation rate is
consistent with mediation by lateral diffusion within the same membrane,
rather than by "solution-hopping" (Liebman *et al.*, 1982).

Each of the G–GTP molecules released from the membrane can interact
with the PDE molecules, either on the surface of the same disc (as in Fig. 8b),
or of an adjacent one. Binding of the G–GTP to the PDE leads to the removal
of inhibition, probably as a result of changing the interaction between the
subunits. The active PDE–G–GTP complex catalyses the hydrolysis of cGMP.
As a result of the coupling of the two activities into a cascade the overall
rate of cGMP hydrolysis resulting from the activation of a single rhodopsin
molecule could reach some 10^6 per second. In the scheme shown in Fig. 8a,
the activity of the PDE–G–GTP complex is lost when the GTP bound to it is
hydrolysed to GDP as a result of the GTPase activity. The complex then
dissociates and the PDE returns to an inhibited configuration. The tightness
with which the G–GTP molecule is bound to the PDE is not universally
agreed: Liebman and Pugh (1982) find that the amount of G activated
exceeds that of the PDE by up to a factor of 10 and therefore suggest that
there is rapid binding and dissociation between the two systems, with an
incomplete approach to equilibrium, rather than continuous binding throughout the lifetime of PDE–G–GTP.

In the scheme outlined in Fig. 8a, the cascade could continue until the
supply of either the GTP or the cGMP present in the OS was exhausted, for
there would be a continual replenishment of the cycle by R* molecules. Thus,
a regulatory mechanism is also necessary. There is evidence that the photolysed rhodopsin molecule remains active only transiently, and the question
of the identity of R* arises. It has been reported that the onset of activation
of the hydrolysis of cGMP parallels the time course of formation of meta 2
(Liebman and Pugh, 1981). Kuhn *et al.* (1981) state that the turbidity
measurements attributed to the binding of GTP to the G-protein occurs
more slowly than the formation of meta 2. Given the complexities of both
the kinetics of formation of meta 2 and the dependence of the turbidity
changes associated with binding of R* to the G-protein and the subsequent
dissociation signal, it is clearly difficult to make too definite a comparison
with this spectrally defined photolysis intermediate. This is reinforced by the
variety of conditions in which the processes have been measured: quantitative
determinations of the kinetics of meta 2 formation are difficult in circumstances where less than about 10% of the rhodopsin is bleached. This is in
the uppermost range of G-protein : rhodopsin stoichiometry, so that most

determinations of the binding activity and subsequent cGMP hydrolysis have been measured at lower bleaching levels. Moreover, there may be a membrane diffusional rate-limited delay in GTP–GDP exchange caused by the need for the R* to sweep out the membrane surface for the G-protein molecules with which it successively interacts (Liebman and Pugh, 1981, 1982). However, it is clear that even if total identity with the spectral characterization of meta 2 cannot be made at present, the kinetics of its formation appear to be similar to those of R*.

It is also not universally agreed that regulation of the PDE activity results from inactivation of R*, although there is evidence to support this. Clearly, a conformational change on the pigment protein molecule which accompanies the interconversion of meta 2 to the long-lived photoproducts could lead to inactivation, as could a possibly spectrally silent reaction occurring in a region of the molecule remote from the chromophoric binding site. On the basis of the observed regulation of the cGMP change by ATP, Liebman and Pugh (1980, 1981, 1982) suggest that the inactivation results from phosphorylation of R* by the peripherally bound rhodopsin kinase (as indicated in Fig. 8b). Most estimates of the rate at which the phosphorylation occurs indicate that it is too slow for the reaction to play any direct role in transduction, its $t_{\frac{1}{2}}$ being of the order of 0.4 to 2 minutes (Miller et al., 1975; Kuhn and Bader, 1976; Hermolin et al., 1982), and it has therefore more frequently been associated with a possible role in receptor adaptation. However, there is evidence that the time course of the reaction may be accelerated when only small fractions of rhodopsin are bleached (Wilden and Kuhn, 1982; but cf. Hermolin et al., 1982), and it may be that in these circumstances the reaction is fast enough to act as the deactivation mechanism of R*. The proposal is also supported by the observation that preparations containing partially phosphorylated rhodopsin lead to a reduced level of PDE activation (Polans et al., 1979), and that the light-induced binding of ATP to the OS membrane occurs with a similar time course (Zuckerman et al., 1982).

On the other hand, the regulation of the PDE activity could occur as the result of ATP-mediated phosphorylation of other proteins: Hermolin et al. (1982) report that there are conditions in which rhodopsin phosphorylation can be almost completely suppressed, yet both the activation of the PDE and its regulation by ATP remain unchanged.

Although there is controversy over the mechanisms by which the reactions occur, it appears that the hydrolysis of cGMP could, in principle, fulfil the requirements of a transduction mechanism. The coupling of two enzyme reactions to form a cascade provides the very high amplification necessary to provide sufficient sensitivity and accounts for at least two of the delays which contribute to the latency of the response. It is also consistent with the relative constancy observed between the responses to individual photons

(Liebman and Pugh, 1981), and the diffusional rate expected for the local reduction of the cGMP levels is consistent with the longitudinal spread of excitation along the plasma membrane (Miller, 1981; Yau et al., 1981).

Nevertheless, this ability to postulate at a molecular level the way in which the bleaching of a single rhodopsin molecule can lead to a large and rapid change in the intracellular concentration of cGMP cannot be considered to be sufficient to justify the assumption that cGMP is in fact the transmitter. For example, the finding that the kinetics of the reactions in situ are much slower than those of isolated OS preparations has yet to be resolved. Further, as Liebman and Pugh (1981) have pointed out, the extent of the cGMP modulation is much reduced by the presence of ATP at levels similar to those in the rod, and the amplification of the process in situ may therefore be insufficient to account for the observed sensitivity.

Doubts have also been cast by recent electrophysiological experiments (Woodruff et al., 1982), which show that although changing the extracellular concentration of Ca affects both the levels of cGMP within the OS and the permeability of the plasma membrane, there are major quantitative dissimilarities in the way in which each of them varies: again, the modulation of cGMP is found to be both too small and too slow to be consistent with the hypothesis that it directly induces the membrane permeability changes. Experiments in which 8-bromo-cGMP, an analogue of cGMP which is normally resistant to PDE hydrolysis, was iontophoretically injected into a rod (Waloga and Bitensky, 1981) were reported to yield results similar to those obtained with cGMP itself (Miller and Nicol, 1979, 1981). On the assumption that the analogue is resistant to hydrolysis by the OS PDE, Waloga and Bitensky suggested that the results of Miller and Nicol do not demonstrate a causal link between the light modulation of cGMP concentration and changes in permeability of the plasma membrane.

The effects of NH_2OH on the photoresponse also appear to be inconsistent with elements of the hypothesis. Isolated retinas bathed in a medium containing $\sim mM$ concentrations of NH_2OH show little change in their electrophysiological properties, though the kinetics of the rhodopsin photolysis sequence are greatly altered (Brin and Ripps, 1977), the slow reactions occurring much faster (Catt et al., 1982). However, in the dark-adapted rat retina the time course of the rod photoresponse to light flashes too dim to cause saturation is slower, both to reach its maximum, and to decay, than normal (Ernst and Kemp, in preparation). The maximum, saturated, response amplitude (V_{max}) in the presence of NH_2OH is within the normal range, though the operating range is shifted to lower light levels, by ~ 0.4 log units. In contrast, it has been found that the rates of activation and switch-off of the PDE in isolated disc preparations are increased by the addition of low concentrations of NH_2OH in a manner that parallels the changes induced in the kinetics of meta 2 (Liebman and Pugh, 1982; Parkes and Liebman, 1982; Buzdygon and Liebman, 1982), which are similar to those found in the isolated retina (Catt *et al.*, 1982).

C. THE Ca HYPOTHESIS

The Ca transmitter hypothesis has been tested electrophysiologically in many studies, and several lines of evidence from these experiments are consistent with the possibility that Ca plays a role in transduction or adaptation (Yoshikami and Hagins, 1973; Hagins and Yoshikami 1974, 1977; Brown *et al.*, 1977a). The properties of rhodopsin and the disc membrane are not incompatible with the idea of transmitter release from within the disc. Thus, rhodopsin spans the membrane, so it could act as a pore, and there is a rapid transient increase in the conductance of the disc membrane following illumination of intact OS (Falk and Fatt, 1973a,b; Fatt and Falk, 1977). However, there have been areas of discrepancy between the experimental observations and the hypothesis, at least in its original form. Chief of these has been the repeated failure to find either an accumulation of exchangeable Ca in the discs (see above), or a sufficiently large movement of the ion across the membrane following light stimulation. However, it has been pointed out that this may reflect the extent to which the membranes have been damaged (Szuts, 1981; Fatt, 1982), and it is notable that large transient Ca release has been observed from receptors in intact isolated retina preparations (Yoshikami *et al.*, 1980; Gold and Korenbrot, 1981). Recent preliminary reports that a substantial transient Ca release can also be detected from the isolated frog rod, so long as the inner segment of the cell remains attached (Biernbaum and Bownds, 1982), and from bovine OS (Ebrey and Suh, 1982), in appropriate conditions (addition of exogenous GTP), further suggest that extensive light-induced changes in Ca do occur.

A second problem is that of accounting for the delay in the onset of the responses. On the basis that the Ca is freely diffusible, the time course of the rising phase of the response would be expected to be much faster than is the case. The widely held explanation of the response latency has been that there is a series of (unspecified) delay reactions before the release of the Ca is initiated (e.g. Arden, 1969; Penn and Hagins, 1972; Baylor *et al.*, 1979a). However, it has been argued that since it is to be expected that much of the Ca would be interacting with the large number of fixed negative charges on the surface of the disc membrane, it is inappropriate to assume that the ion is freely diffusible (McLaughlin and Brown, 1981). On the basis of binding studies of Ca to lipid bilayers containing PS, PE and PC (see section II.2), they calculate that there are 10–100 times as many Ca bound to the disc membrane as there are free ions in the cytoplasm, and that this will reduce the effective diffusion coefficient, to a value consistent with the Ca hypothesis. The very small variation in both the time course and amplitude observed for responses to the absorption of a single quantum by the rod is also inconsistent with a mechanism which presupposes that the response simply reflects the amount of transmitter released from the disc into the cytoplasm and the subsequent closure by it of channels in the plasma membrane. Such a system would show first order kinetics, so that the duration of pore closure (and thus the response) would follow an exponential pattern quite unlike the one observed (Liebman and Pugh, 1981). It is therefore necessary to postulate that Ca release from the disc continues for a longer period than rising phase of the photoresponse (Fatt, 1982).

D. COMBINATION HYPOTHESES

The unresolved questions posed by the cGMP and Ca hypotheses have led to an increasing tendency for models which combine elements of each to be put forward (e.g. Bownds, 1980; Lipton and Dowling, 1981; Miller, 1981; Fatt, 1982; O'Brien, 1982b; Saibil, 1982; Skulachev, 1982). Two patterns of modelling are emerging: one which presumes that the primary role of rhodopsin is to initiate the cGMP modulation, and that this leads to a change in Ca which then acts to close the Na channels in the plasma membrane; and another in which rhodopsin directly modulates both Ca and cGMP in parallel and the two processes are interlinked, possibly in a regulatory manner.

In these schemes, the linking mechanisms proposed remain largely hypothetical, though it has been shown that the modulation of cGMP and Ca are closely inter-related. Thus, changes in cGMP cause the release of tightly bound Ca from discs (Cavaggioni and Sorbi, 1980), while Ca at levels around 0.1 mM has a modulatory effect on the cGMP-sensitive dephosphorylation of two small proteins associated with the OS membrane (Hermolin *et al.*, 1982) and on the guanylate cyclase (Fleischman, 1981) which is presumed to replenish the cGMP following its light-induced hydrolysis.

In a model recently put forward by Fatt (1982), the proposed mechanism is similar to that originally postulated for Ca, with the initial action of light being the release of Ca from the interior of the disc, as a result of transient oligomerization of the light activated rhodopsin. The role of the cGMP cascade is envisaged to be the modulation of Ca binding sites on the disc membrane (on a postulated cGMP dependent Ca-binding protein), so that the enzyme reactions are deemed to be primarily responsible for the falling phase of the response. Sklachev (1982), on the other hand, has proposed a dualistic scheme: he suggests that the cGMP system is responsible for the response of the rod to dim light stimuli, while Ca release, which he postulates is the result of charge translocation in the disc membrane, produces the more rapid, but less sensitive, response to bright stimuli.

Clearly, the last decade has seen a massive unveiling of the properties of rhodopsin, the membrane in which it is embedded, and the proteins with which its bleaching is inter-related. Experimental evidence of a direct link between the visual pigment molecule and the cyclic nucleotide system marked the beginning of a molecular understanding of at least one way in which rhodopsin triggers changes within the rod: it remains to unravel the full intricacies of the machinery of transduction and adaptation. What has become apparent is that the early assumptions of a single, simple information pathway were too optimistic. With the increasingly complex biochemical machinery which is being shown to exist within the outer segment, the possibilities become wider, and the experimental criteria by which to judge the role of rhodopsin, both in transduction and adaptation, more rigorous.

References

Abood, M. E., Hurley, J. B., Pappone, M. C., Bourne, H. R. and Stryer, L. (1982). *J. Biol. Chem.* **257**, 10540.

Abrahamson, E. (1973). *In* "Biochemistry and Physiology of Visual Pigments" (ed. H. Langer) p. 47. Springer-Verlag, Berlin.

Abrahamson, E. W. and Wiesenfeld, J. R. (1972). *In* "Handbook of Sensory Physiology, VII/I Photochemistry of the visual pigments (ed. H. J. A. Dartnall) p. 69, Springer-Verlag, Berlin.

Abrahamson, E. W., Shriver, J., Fager, R., Matescu G. and Torchia, D. (1977). *Nature* **270**, 271.

Albert, A. D. and Litman, B. J. (1978). *Biochemistry* **17**, 3893.

Allan, A. E. and Cooper, A. (1980). *FEBS Lett.* **119**, 238.

Anderson, R. E. and Andrews, J. (1982). *In* "Visual Cells in Evolution" (ed. J. A. Westfall) **1**, Raven, New York.

Applebury, M. L. and Rentzepis, P. M. (1982). *Methods in Enzymol.* **81**, 354.

Applebury, M. L., Zuckerman, D. M., Lamola, A. A. and Jovin, T. M. (1974). *Biochemistry* **13**, 3448.

Arden, G. B. (1969). *Prog. Biophys. Mol. Biol.* **19**, 373.

Attwood, P. V. and Gutfreund, H. (1980). *FEBS Lett.* **119**, 323.

Baehr, W., Devlin, M. J. and Applebury, M. L. (1979). *J. Biol. Chem.* **254**, 11669.

Baehr, W., Morita, E. A., Swanson, R. J. and Applebury, M. L. (1982). *J. Biol. Chem.* **257**, 6452.

Baroin, A. Thomas, D. D., Osborne, B. and Devaux, P. F. (1977). *Biochem. Biophys. Res. Commun.* **78**, 442.

Baroin, A., Bienvenue, A. and Devaux, P. F. (1979). *Biochemistry* **18**, 1151.

Basinger, S., Bok, D. and Hall, M. O. (1976). *J. Cell Biol.* **69**, 29.

Bastian, B. L. and Fain, G. L. (1979). *J. Physiol.* **207**, 77.

Baumann, C. (1972). *J. Physiol.* **222**, 643.

Baumann, C. (1976). *J. Physiol.* **259**, 357.

Baumann, C. and Zeppenfeld, W. (1981). *J. Physiol.* **317**, 347.

Baylor, D. A. and Fuortes, M. G. F. (1970). *J. Physiol.* **207**, 77, 93
Baylor, D. A., Lamb, T. D. and Yau, K-W (1979a). *J. Physiol.* **288**, 589.
Baylor, D. A., Lamb, T. D. and Yau, K-W (1979b). *J. Physiol.* **288**, 613
Bennett, N. (1978). *Biochem. Biophys. Res. Commun.* **83**, 457.
Bennett, N. (1982). *Eur. J. Biochem.* **123**, 133
Berger, S. J., De Vries, G. W., Carter, J. G., Schulz, D. W., Passonneau, P. N., Lowry, O. H. and Ferendelli, J. A. (1980). *J. Biol. Chem.* **255**, 3128.
Biernbaum, M. S. and Bownds, M. D. (1979). *J. Gen. Physiol.* **74**, 649.
Biernbaum, M. S. and Bownds, M. D. (1982). *Invest. Ophthalmol. Visual Sci.* **22**, Supp. 187.
Bignetti, E., Cavaggioni, A. and Sorbi, R. T. (1978). *J. Physiol* **279**, 55.
Birge, R. R. (1981). *Ann. Rev. Biophys. Bioeng.* **10**, 315.
Birge, R. R. and Hubbard, L. M. (1980). *J. Am. Chem. Soc.* **102**, 2195.
Bitensky, M. W., Wheeler, M. A., Rasenick, M. M., Yamazaki, A., Stein, P. J., Halliday, K. R. and Wheeler, G. L. (1982). *Proc. Natl Acad. Sci. USA* **79**, 3408.
Blasie, J. K., Worthington, C. R. and Dewey, M. (1969). *J. Molec. Biol.* **39**, 407.
Blatchly, R. A., Carriker, J. D., Balogh-Nair, V. and Nakauishi, K. (1980). *J. Am. Chem. Soc.* **102**, 2495.
Blazynski, C. and Ostroy, S. E. (1981). *Vision Res.* **21**, 833.
Borys, T., Uhl, R. and Abrahamson, E. W. (1982). *Invest. Ophthalmol. Visual Sci.* **22**, Supp. 43.
Bownds, M. D. (1967). *Nature* **216**, 1178.
Bownds, M. D. (1980). *Photochem. Photobiol.* **32**, 487.
Bownds, M. D. (1981). *Current Topics in Membranes and Transport* **15**, 203.
Bownds, M. D. and Wald, G. (1965). *Nature* **205**, 264.
Bownds, D., Dawes, J., Miller, J. and Stahlman, M. (1972). *Nature New Biol.* **237**, 125.
Bownds, D., Gordon-Walker, A., Gaide-Huguenin, A. and Robinson, W. (1971). *J. Gen. Physiol.* **58**, 225.
Bridges, C. D. B. (1961). *Biochem. J.* **79**, 135.
Bridges, C. D. B. (1962). *Vision Res.* **2**, 215.
Bridges, C. D. B. (1972). *In* "Handbook of Sensory Physiology, VII/I, Photochemistry of Vision" (ed. H. Dartnall), p. 417. Springer-Verlag, Berlin.
Brin, K. P. and Ripps, H. (1977). *J. Gen. Physiol.* **69**, 97.
Brown, J. E., Coles, J. A. and Pinto, L. H. (1977a). *J. Physiol.* **269**, 707.
Brown, M. F., Miljanich, G. P. and Dratz, E. A. (1977b). *Biochemistry*, **16**, 2640.
Brown, M. F., Miljanich, G. P. and Dratz, E. A. (1977c). *Proc. Natl Acad. Sci. USA* **74**, 1978.
Brown, M. F., Deese, A. J. and Dratz, E. A. (1982). *Methods in Enzymology* **81**, 709.
Busch, C. E. Applebury, M., Lamola, A. and Rentzepis, P. M. (1972). *Proc. Natl Acad. Sci. USA* **69**, 2802.
Buzdygon, B. and Liebman, P. A. (1982). *Invest. Ophthalmol. Visual Sci.* **22**, Supp., 44.
Carretta, A., Cavaggioni, A. and Sorbi, R. T. (1979). *Biochim. Biophys. Acta.* **583**, 1.
Cavaggioni, A. and Sorbi, R. T. (1980). *Proc. Natl Acad. Sci. USA.* **78**, 3964.
Catt, M., Ernst, W. and Kemp, C. M. (1982). *Biochem. Soc. Trans.* **10**, 343.
Catt, M., Ernst, W. and Kemp, C. M. (1983). *Vision Res.*, in press.
Chabre, M. (1975). *Biochim. Biophys. Acta* **382**, 322.
Chabre, M. and Breton, J. (1979). *Vision Res.* **15**, 985.

Chader, G. J., Fletcher, R. T. and Krishna, G. (1975). *Biochem. Biophys. Res. Comm.* **64**, 535.

Chader, G. J., Fletcher, R. T., O'Brien, P. J. and Krishna, G. (1976). *Biochemistry* **15**, 1615.

Chader, G. J., Fletcher, R. T., Russell, P. and Krishna, G. (1980). *Biochemistry* **19**, 2634.

Chapman, D., Gomez-Fernandez, J. C. and Goni, F. M. (1979). *FEBS Lett.* **98**, 211.

Chen, Y. S. and Hubbell, W. L. (1973). *Exp. Eye Res.* **17**, 517.

Chen, Y. S. and Hubbell, W. L. (1978). *Membrane Biochem.* **1**, 107.

Cohen, A. I., Hall, I. A. and Ferrendelli, J. A. (1978). *J. Gen. Physiol.* **71**, 595.

Cone, R. A. (1972). *Nature New Biol.* **236**, 39.

Cone, R. A. (1973). *In* "Biochemistry and Physiology of Visual Pigments" (ed. H. Langer), p. 475. Springer-Verlag, Berlin.

Cone, R. A. and Cobbs, W. H. (1969). *Nature* **221**, 820.

Crain, R. C., Marinetti, G. V. and O'Brien, D. F. (1978). *Biochemistry* **17**, 4186.

Crescitelli, F., Mommaerts, W. F. H., and Shaw, T. I. (1966). *Proc. Natl Acad. Sci. USA* **56**, 1729.

Crouch, R., Purvin, V., Nakanishi, K. and Ebrey, T. G. (1975). *Proc. Natl Acad. Sci. USA* **72**, 1538.

Daemen, F. J. M. (1973). *Biochim. Biophys. Acta* **300**, 255.

Daemen, F. J. M. (1978). *Nature* **276**, 847.

Daemen, F. J. M., De Grip, W. J. and Jansenn, P. A. A. (1972). *Biochim. Biophys. Acta* **271**, 419.

Darszon, A., Blair, L. and Montal, M. (1979). *FEBS Lett.* **107**, 213.

Dartnall, H. J. A. (1968). *Vision Res.* **8**, 339.

Davoust, J., Bienvenue, A., Fellman, P. and Devaux, P. F. (1980). *Biochim. Biophys. Acta.* **596**, 28.

Deese, A. J., Dratz, E. A. and Brown, M. F. (1981). *FEBS Lett.* **124**, 93.

De Grip, W. J. (1982). *Methods in Enzymol.* **81**, 256.

De Grip, W. J., Daemen, F. J. M. and Bonting, S. L. (1972). *Vision Res.* **12**, 694.

De Grip, W. J., Van De Laar, G. L. M., Daemen, F. J. M. and Bonting, S. L. (1973). *Biochim. Biophys. Acta* **325**, 315.

De Grip, W. J., Liu. R. S. H., Rammurthy, V. and Asato, A. (1974). *Nature* **262**, 416.

De Grip, W. J., Drenthe, E. H. S., Van Echteld, C. J. A., De Kruijff, B. and Verkleij, A. J. (1979). *Biochim. Biophys. Acta* **558**, 330.

Denton, E. J. (1959). *Proc. Roy. Soc. B* **150**, 78.

Devaux, P. F. (1982). *Methods in Enzymol.* **81**, 703.

Donner, K. O. and Reuter, T. (1967). *Vision Res.* **7**, 17.

Donner, K. O. and Hemiliä, S. (1975). *Vision Res.* **15**, 985.

Donner, K. O. and Hemila, S. (1979). *J. Physiol.* **287**, 92.

Downer, N. W. (1982). *Biophys. J.* **37**, 86a.

Drenthe, E. H. S., Klompmakers, A. A., Bonting, S. L. and Daemen, F. J. M. (1980). *Biochim. Biophys. Acta* **603**, 130.

Drzymala, R., Weiner, H. and Liebman, P. A. (1980). *Fed. Proc.* **39**, 2137.

Ebrey, T. G. and Honig, B. (1975). *Quart. Rev. Biophys.* **8**, 129.

Ebrey, T. G., Govindjee, R., Honig, B., Pollock, E., Chan, W., Crouch, R., Yudd, A. and Nakanishi, K. (1975). *Biochemistry* **14**, 3933.

Ebrey, T. C., Kilbride, P., Hurley, J. B., Calhoon, R. and Tsuda, M. (1981). *Current Topics in Membranes and Transport* **15**, 121.

Ebrey, T. G. and Suh, C. K. (1982). *Biophys. J.* **37**, 84a.
Emeis, D. and Hofmann, K. P. (1981). *FEBS Lett.* **136**, 201.
Emeis, D., Kuhn, H., Reichert, J. and Hofmann, K. P. (1982). *FEBS Lett.* **143**, 29.
Emrich, H. M. (1971). *Naturforsch.* **26b**, 352.
Ernst, W. and Kemp, C. M. (1972). *Vision Res.* **12**, 1937.
Ernst, W. and Kemp, C. M. (1978). *Exp. Eye Res.* **27**, 101–116.
Ernst, W., Kemp, C. M. and White, H. A. (1978). *Exp. Eye Res.* **26**, 337.
Eyring, G. and Mathies, R. (1979). *Proc. Natl Acad. Sci. USA* **76**, 33
Fager, R. S., Gentilcore, P. C. and Abrahamson, E. W. (1978). *Vision Res.* **18**, 483
Fain, G. L. (1976). *J. Physiol.* **172**, 239.
Fain, G. L. and Lisman, J. E. (1981). *Prog. Phys. Mol. Biol.* **37**, 91.
Falk, G. and Fatt, P. (1973a). *J. Physiol.* **229**, 185.
Falk, G. and Fatt, P. (1973b). *J. Physiol.* **229**, 221.
Farber, D. B., Brown, B. M. and Lolley, R. N. (1978). *Vision Res.* **18**, 497.
Farber, D. B., and Lolley, R. N. (1974). *Science* **186**, 449.
Fatt, P. (1981). *Exp. Eye Res.* **33**, 31.
Fatt, P. (1982). *FEBS Lett.* **149**, 159.
Fatt, P. and Falk, G. (1977). *In* "Vertebrate Photoreception" (eds H. B. Barlow and P. Fatt) p. 77. Academic Press, London.
Favre, E., Baroin, A., Bienvenue, A. and Devaux, P. F. (1979). *Biochemistry* **18**, 1156.
Ferendelli, J. A. and Cohen, A. I. (1976). *Biochem. Biophys. Res. Commun.* **73**, 421.
Findlay, J. B. C., Brett, M. and Pappin, D. J. C. (1981). *Nature* **293**, 314.
Fleischmann, D. (1981). *Current Topics in Membranes and Transport* **15**, 109.
Frank, R. N. and Buzney, S. M. (1975). *Biochemistry* **14**, 5110.
Frank, R. N., Cavanaugh, H. D. and Kenyon, D. R. (1973). *J. Biol. Chem.* **248**, 596.
Frank, R. N. and Rodbard, D. (1975). *Arch. Biochem., Biophys.* **171**, 1.
Fukuda, M. N., Papermaster, D. P. and Hargrave, P. A. (1979). *J. Biol. Chem.* **254**, 8201.
Fung, B. K.-K. and Hubbell, W. L. (1978a). *Biochemistry* **17**, 4396.
Fung, B. K.-K. and Hubbell, W. L. (1978b). *Biochemistry* **17**, 4403.
Fung, B. K.-K., Hurley, J. B. and Stryer, L. (1981). *Proc. Natl Acad. Sci. USA* **78**, 152.
Fung, B. K.-K. and Stryer, L. (1980). *Proc. Natl Acad. Sci. USA* **77**, 2500.
Futterman, S. (1963). *J. Biol. Chem.* **238**, 1145.
Gilardi, R., Karle, I. L., Karle, J. and Sperling, W. (1971). *Nature* **22**, 187.
Godchaux, W. and Zimmerman, W. F. (1979a). *Exp. Eye Res.* **28**, 483.
Godchaux, W. and Zimmerman, W. F. (1979b). *J. Biol. Chem.* **254**, 7874.
Gold, G. H. and Korenbrot, J. I. (1980a). *Invest. Ophthalmol. Visual Sci.*, **19**, Supp. 281.
Gold, G. H. and Korenbrot, J. I. (1980b). *Fed. Proc. Fed. Am. Soc. Exp. Biol.* **39**, 1814.
Gold, G. H. and Korenbrot, J. I. (1980c). *Proc. Natl Acad. Sci. USA* **77**, 5557.
Gold, G. H. and Korenbrot, J. I. (1981). *Current Topics in Membranes and Transport*, **15**, 307.
Grabowski, S. R. and Pak, W. L. (1975). *J. Physiol.* **247**, 363.
Green, B., Monger, T., Alfano, R., Aton, B. and Callender, R. (1977). *Nature* **269**, 179.
Greengard, P. (1978). *Science*, **199**, 146.
Hagins, W. A. (1972). *Ann. Rev. Biophys. Bioeng.* **1**, 131.
Hagins, W. A. and Yoshikami, S. (1974). *Exp. Eye Res.* **18**, 299.

Hagins, W. A. and Yoshikami, S. (1975). *Ann. N. Y. Acad. Sci.* **264**, 314.
Hagins, W. A. and Yoshikami, S. (1977). *In* "Vertebrate Photoreception" (eds H. B. Barlow and P. Fatt) p. 97. Academic Press, London.
Hargrave, P. A. (1977). *Biochim. Biophys. Acta* **429**, 83.
Hargrave, P. A. and Fong, S. L. (1977). *J. Supramol. Struct.* **6**, 559.
Hargrave, P. A., Fong, S. L., McDowell, J. H., Mas, M. T., Curtis, D. R., Wang, J. K., Juszczak, E. and Smith, D. P. (1980). *Neurochem. Int.* **1**, 231.
Hargrave, P. A., Mcdowell, J. M., Siemialkowski, E. C., Juszczak, E., Fong, S. L., Kuhn, H., Wang, J. K., Curtis, D. R., Rao, J. K. M., Argos, P. and Feldmann, R. (1982). *Vision Res.* **2**, 1429.
Harosi, F. (1975). *J. Gen. Physiol.* **66**, 357.
Hayward, G., Carlsen, W., Siegmai, A. and Stryer, L. (1981). *Science* **211**, 942.
Heller J. (1968a). *Biochemistry* **7**, 2906.
Heller J. (1968b). *Biochemistry* **7**, 2914.
Heller, J. and Lawrence, M. A. (1970). *Biochemistry* **9**, 864.
Hemminki, K. (1975). *Vision Res.* **15**, 69.
Henderson, R. and Unwin, P. N. T. (1975). *Nature* **275**, 28.
Hendriks, T., Daemen, F. J. M. and Bonting, S. L. (1974). *Biochim. Biophys. Acta* **345**, 468.
Hermolin, J., Karell, M. A., Hamm, H. E. and Bownds, M. D. (1982). *J. Gen. Physiol.* **79**, 633.
Hofmann, W., Siebert, F., Hofmann, K. P. and Dreutz, W. (1978). *Biochem. Biophys. Acta* **503**, 450.
Honig, B., Greenberg, A. D., Dinur, U. and Ebrey, T. G. (1976). *Biochemistry* **15**, 4593.
Honig, B., Dinur, U., Nakanishi, K., Balogh-Nair, V., Gawinowicz, M. A., Arnaboldi, M. and Motto, M. G. (1979). *J. Am. Chem. Soc.* **101**, 7084.
Horiuchi, S., Tokunaga, F. and Yoshizohwa, T. (1978). *Biochim. Biophys. Acta.* **503**, 402.
Hubbell, W. L. and Bownds, M. D. (1979). *Ann. Rev. Neurosci.* **2**, 17.
Hubbell, W. L. and Fung, B. K. K. (1979). *In* "Membrane Transduction Mechanisms" (eds. R. A. Cone and J. E. Dowling), p. 17. Raven, New York.
Hurley, J. B. (1980). *Biochem. Biophys. Res. Comm.* **92**, 505.
Hurley, J. B. and Stryer, L. (1982). *J. Biol. Chem.* **257**, 11094.
Kamps, K. M. P., De Grip, W. J. and Daemen, F. J. M. (1982). *Biochim. Biophys. Acta* **687**, 298.
Kaupp, U. B. and Junge, W. (1977). *FEBS Lett.* **81**, 229.
Kaupp, U. B., Schnetkamp, P. P. M. and Junge, W. (1979). *Biochim. Biophys. Acta* **552**, 390.
Kaupp, U. B., Schnetkamp, P. P. M. and Junge, W. (1981a). *Biochemistry* **20**, 5500.
Kaupp, U. B., Schnetkamp, P. P. M. and Junge, W. (1981b). *Biochemistry* **20**, 5511.
Kaupp, U. B. and Schnetkamp, P. P. M. (1982). *Cell Calcium* **3**, 83.
Kawamara, S. and Bownds, M. D. (1981). *J. Gen. Physiol.* **77**, 571.
Kemp, C. M. (1973). *In* "Biochemistry and Physiology of Visual Pigments (ed. H. Langer), p. 307. Springer-Verlag, Berlin.
Kilbride, P. (1980). *J. Gen. Physiol.* **75**, 457.
Kilbride, P. and Ebrey, T. G. (1979). *J. Gen. Physiol.* **74**, 415.
Kimbel, R., Poincelot, R. P. and Abrahamson, E. W. (1970). *Biochemistry* **9**, 1871.
Kleinshmidt, J. and Dowling, J. E. (1975). *J. Gen. Physiol.* **66**, 617.
Kobayashi, T. and Nagakura, S. (1982). *Methods in Enzymol.* **81**, 368.
Kuhn, H. (1974). *Nature* **250**, 588.

Kuhn, H. (1978). *Biochemistry* **17**, 4389.

Kuhn, H. (1980a). *Nature* **283**, 587.

Kuhn, H. (1980b). *Neurochem. Int.* **1**, 269.

Kuhn, H. (1981). *Current Topics in Membranes and Transport* **15**, 171.

Kuhn, H. and Dreyer, W. J. (1972). *FEBS Lett.* **20**, 1.

Kuhn, H. and Bader, S. (1976). *Biochim. Biophys. Acta* **428**, 13.

Kuhn, H. and McDowell, J. H. (1977). *Biophys. Struct. Mech.* **3**, 199.

Kuhn, H. and Hargrave, P. A. (1981). *Biochemistry* **20**, 2410.

Kuhn, H., Bennett, N., Michel-Villaz, M. and Chabre, M. (1981). *Proc. Natl Acad. Sci. USA* **78**, 6873.

Kuhn, H., Mommertz, O. and Hargrave, P. A. (1982). *Biochim. Biophys. Acta* **679**, 95.

Kusumi, A., Ohuishi, S., Ito, T. and Yoshizawa, T. (1978). *Biochim. Biophys. Acta* **507**, 539.

Kusumi, A., Sakate, T., Yoshizawa, T. and Ohnishi, S. (1980). *J. Biochem. (Tokyo)* **88**, 1103.

Lewis, A., Fager, R. and Abrahamson, E. (1973). *J. Raman Spectrosc.* **1**, 465.

Liang, C. J., Yamashita, K., Muellenberg, G. G., Shichi, H. and Kobata, A. (1979). *J. Biol. Chem.* **254**, 6414.

Liebman, P. A. (1962). *Biophys J.* **2**, 161.

Liebman, P. A. (1974). *Invest. Ophthalmol.* **13**, 700.

Liebman, P. A. (1978). *Ann. N. Y. Acad. Sci.* **307**, 642.

Liebman, P. A. and Entine, G. (1974). *Science* **185**, 457.

Liebman, P. A. and Pugh, E. N. (1979). *Vision Res.* **19**, 375.

Liebman, P. A. and Pugh, E. N. (1980). *Nature* **287**, 734.

Liebman, P. A. and Pugh, E. N. (1981). *Current Topics in Membranes and Transport* **15**, 157.

Liebman, P. A. and Evanczuk, A. T. (1982). *Methods in Enzymol.* **81**, 532.

Liebman, P. A. and Pugh, E. N. (1982). *Vision Res.* **22**, 1475.

Liebman, P. A., Sitaramayya, A. and Pugh, E. N. (1982). *Biophys. J.* **7**, 88a.

Linschitz, H., Wulff, V. J., R. G. and Abrahamson, E. W. (1957). *Arch. Biochem.* **68**, 233.

Lipton, S. A. and Dowling, J. E. (1981). *Current Topics in Membranes and Transport* **15**, 381.

Lolley, R. N. and Faber, D. B. (1976). *Exp. Eye Res.* **22**, 477.

Mao, B., Ebrey, T. G., and Crouch, R. (1980). *Biophys. J.* **29**, 247.

Marsh, D., Watts, A., Knowles, P. F., Pates, R. D., Uhl, R. and Esman, M. (1982). *Biophys. J.* **37**, 265.

Mathies, R., Oseroff, A. R. and Stryer, L. (1976). *Proc. Natl Acad. Sci. USA* **73**, 1.

Mathies, R., Freedman, T. D. and Stryer, L. (1977). *J. Mol. Biol.* **109**, 367.

Matsumoto, H. and Yoshizawa, T. (1975). *Nature* **228**, 523.

Matsumoto, H. and Yoshizawa, T. (1978). *Vision Res.* **18**, 607.

Matsumoto, H., Asato, A. E., Denny, M., Beretz, B., Yen, Y. P., Tong, D. and Liu, R. S. H. (1980). *Biochemistry* **19**, 4589.

Matthews, R., Hubbard, R., Brown, P. and Wald, G. (1963). *J. Gen. Physiol* **47**, 215.

McLaughlin, S. and Brown, J. E. (1981). *J. Gen. Physiol.* **77**, 475.

McNaughton, P. A., Yau, K. W. and Lamb, T. D. (1980). *Nature* **283**, 85.

Michel-Villaz, M., Saibil, H. R. and Chabre, M. (1979). *Proc. Natl Acad. Sci. USA* **76**, 4405.

Miki, N., Keirns, J. J., Marcus, F. R., Freeman, J. and Bitensky, M. W. (1975). *Proc. Natl Acad. Sci. USA* **76**, 3820.

Miljanich, G. P., Nemes, P. P., White, D. L. and Dratz, E. A. (1981). *J. Mem. Biol.* **60**, 249.

Miller, J. A., Brodie, A. E. and Bownds, M. D. (1975). *FEBS Lett.* **59**, 20.

Miller, J. A. and Paulsen, R. (1975). *J. Biol. Chem.* **250**, 4427.

Miller, J. A., Paulsen, R. and Bownds, M. D. (1977). *Biochemistry* **16**, 2633.

Miller, W. H. (1981). *Current Topics in Membranes and Transport* **15**, 441.

Miller, W. H. and Nicol, G. D. (1979). *Nature* **280**, 64.

Miller, W. H. and Nicol, G. D. (1981). *Current Topics in Membranes and Transport* **15**, 417.

Molday, R. S. and Molday, L. L. (1979). *J. Biol. Chem.* **254**, 4653.

Montal, M., Darszon, A. and Schindler, H. (1981). *Quart. Rev. Biophys.* **14**, 1.

Mullen, E. and Akhtar, M. (1981). *FEBS Lett.* **132**, 261.

Nemes, P. P., Miljanich, G. P., White, D. L. and Dratz, E. A. (1979). *Biochemistry* **19**, 2067.

Noll, G., Stieve, H. and Winterhager, J. (1979). *Biophys. Struct. Mech.* **5**, 43.

Nunn, B. J. and Baylor, D. A. (1982). *Nature* **299**, 726.

O'Brien, D. F. (1979). *Photochem. Photobiol.* **29**, 679.

O'Brien, D. F. (1982a) *Methods in Enzymol.* **81**, 378.

O'Brien, D. F. (1982b). *Science* **218**, 961.

O'Brien, D. F., Costa, L. F. and Ott, R. A. (1977). *Biochemistry* **16**, 1295.

Orr, H. T., Lowry, O. H., Cohen, A. I. and Ferrendelli, J. A. (1976). *Proc. Natl Acad. Sci. USA* **73**, 4442.

Oseroff, A. R. and Callender, R. H. (1974). *Biochemistry* **13**, 4243.

Ostroy, S. E. (1977). *Biochim. Biophys. Acta* **463**, 91.

Ostroy, S. E., Erhardt, F. and Abrahamson, E. W. (1966). *Biochim. Biophys. Acta* **112**, 265.

Ovchinnikov, Y. A., Abdulaev, N. G., Feigina, M. Y., Artamonov, I. D., Zolotarev, A. S., Kostina, M. B., Bogachuk, A. S., Miroshnikov, A. I., Martinov, V. I. and Kudelin, A. B. (1982). *Biorg. Khim.* **8**, 1011.

Pande, J., Pande, A. and Callender, R. H. (1982). *Photochem. Photobiol.* **36**, 107.

Papermaster, D. S. and Dreyer, W. J. (1974). *Biochemistry* **13**, 2438.

Papermaster, D. S., Schneider, B. G., Zorn, M. A. and Kraehenbuhl, J. P. (1978). *J. Cell Biol.* **78**, 415.

Parkes, J. H. and Liebman, P. A. (1982). *Invest. Ophthalmol. Visual Sci.* **22**, Supp, 44.

Pellicone, C., Bouillon, P. and Virmaux, N. (1980). *C.R. Hebd. Séances Acad. Sci. Paris. D* **290**, 567.

Penn, P. D. and Hagins, W. A. (1972). *Biophys. J.* **12**, 1073.

Pepperberg, D. R., Lurie, M., Brown, P. K. and Dowling, J. E. (1976). *Science* **191**, 394.

Peters, K., Applebury, M. and Rentzepis, P. M. (1977). *Proc. Natl Acad. Sci. USA* **74**, 3119.

Pitt, G. A. J., Collins, F. D., Morton, R. A. and Stok, P. (1955). *Biochem. J.* **59**, 122.

Plantner, J. J. and Kean, E. L. (1976). *J. Biol. Chem.* **251**, 1548.

Pober, J. S. and Stryer, L. (1975). *J. Mol. Biol.* **95**, 477.

Polans, A. S., Hermolin, J. and Bownds, M. D. (1979). *J. Gen. Physiol.* **74**, 595.

Polans, A. S., Kawamura, S. and Bownds, M. D. (1981). *J. Gen. Physiol.* **77**, 41.

Pontus, M. and Delmelle, M. (1975). *Biochim. Biophys. Acta* **401**, 221.

Poo, M. M. and Cone, R. A. (1974). *Nature* **247**, 438.

Rafferty, C. N., Muellenberg, C. G. and Shiehi, H. (1980). *Biochemistry* **19**, 2145.

Rapp, J. (1979). *Vision Res.* **19**, 137.
Rapp, J., Wiesenfeld, J. R. and Abrahamson, E. W. (1970). *Biochim. Biophys. Acta* **201**, 119.
Regan, C. M., De Grip, W. J., Daemen, F. J. M. and Bontig, S. L. (1978). *Biochim. Biophys. Acta* **537**, 145.
Robinson, W. E. and Hagins, W. A. (1979). *Nature* **280**, 398.
Robinson, W. E., Gordon-Walker, A. and Bownds, M. D. (1972). *Nature New Biol.* **235**, 112.
Rodbell, M. (1980). *Nature* **284**, 17.
Rohlich, P. (1976). *Nature* **263**, 789.
Rothschild, K. J., Sanches, R., Hsiao, T. L. and Clark, N. A. (1980). *Biophys. J.* **31**, 53.
Rotmans, J. P., Daemen, F. J. M. and Bonting, S. L. (1974). *Biochim. Biophys. Acta.* **357**, 151.
Saari, J. C. (1974). *J. Cell. Biol.* **63**, 480.
Saibil, H. (1982). *Nature* **297**, 106.
Saibil, H., Chabre, M. and Worcester, D. (1976). *Nature* **262**, 266.
Sale, G. J., Towner, P. and Akhtar, M. (1977). *Biochemistry* **16**, 5641.
Sale, G. J., Towner, P. and Akhtar, M. (1978). *Biochem. J.* **175**, 421.
Schnetkamp, P. P. M. (1979). *Biochim. Biophys. Acta* **554**, 441.
Schnetkamp, P. P. M. (1980). *Biochim. Biophys. Acta* **598**, 66.
Schnetkamp, P. P. M., Klompmakers, A. A. and Daemen, F. J. M. (1979). *Biochim. Biophys. Acta* **552**, 379.
Schwartz, E. A. (1973). *J. Physiol.* **232**, 503.
Schwartz, E. A. (1975). *J. Physiol.* **246**, 617.
Sen, R., Carriker, J. D., Balogh-Nair, V. and Nakanishi, K. (1982). *J. Am. Chem. Soc.* **104**, 3214.
Shichi, H. (1971). *Photochem. Photobiol.* **13**, 499.
Shichi, H. and Felton, E (1974). *J. Supramol. Struct.* **2**, 7.
Shichi, H. and Somers, R. L. (1978). *J. Biol. Chem.* **253**, 7040.
Shichi, H., Somers, R. L. and O'Brien, P. J. (1974). *Biochem. Biophys. Res. Commun.* **61**, 217.
Shriver, J. W., Mateescu, G. D. and Abrahamson, E. W. (1979). **18**, 4787.
Siebert, F. and Mantele, W. (1980). *Biophys. Struct. Mech.* **6**, 147.
Sklar, L. A. and Dratz, E. A. (1982). *Methods in Enzymol.* **81**, 685.
Skulachev, V. (1982). *FEBS Lett.* **146**, 244.
Smith, H. G., Fager, R. S. and Litman, B. J. (1977). *Biochemistry* **16**, 1399.
Smith, H. G. and Bauer, P. J. (1979). *Biochemistry* **18**, 5067.
Somers, R. L. and Shichi, H. (1982). *Invest. Ophthalmol. Visual Sci.* **22**, Supp., 43.
Stewart, J. G., Baker, B. N. and Williams, T. P. (1977). *Biophys. Struct. Mech.* **3**, 19.
Stryer, L., Hurley, J. B. and Fung, B. K. K. (1981). *Current Topics in Membranes and Transport* **15**, 93.
Stubbs, G. W., Smith, H. G. and Litman, B. J. (1976). *Biochim. Biophys. Acta* **426**, 46.
Szuts, E. Z. (1980). *J. Gen. Physiol.* **76**, 253.
Szuts, E. Z. (1981). *Current Topics in Membranes and Transport* **15**, 291.
Szuts, E. Z. and Cone, R. A. (1977). *Biochim. Biophys. Acta* **468**, 194.
Thacher, S. M. (1978). *Biochemistry* **17**, 3005.
Thacher, S. M. (1982). *Biochim. Biophys. Acta* **648**, 199.

Thomas, D. D. and Stryer, L. (1982). *J. Mol. Biol.*

Thompson, A. J. (1975). *Nature* **254**, 178.

Towner, P., Sale, G. J. and Akhtar, M. (1977). *FEBS Lett.* **76**, 51.

Towner, P., Gaertner, W., Walckhott, B., Oesterhelt, D. and Hopf, H. (1981). *Eur. J. Biochem.* **117**, 353.

Toyoda, J., Nosaki, H. and Tomita, T. (1969). *Vision Res.* **9**, 453.

Trayhurn, P., Mandel, P. and Virmaux, N. (1974). *FEBS Lett.* **38**, 351.

Tsunasawa, S., Narita, K. and Shichi, H. (1980). *Biochim. Biophys. Acta* **624**, 218.

Uhl, R. and Abrahamson, E. W. (1981). *Chem. Rev.* **81**. 291.

Uhl, R., Borys, T., Semple, N., Pasternak, J. and Abrahamson, E. W. (1979a). *Biochem. Biophys. Res. Commun.* **90**, 58.

Uhl, R., Borys, T. and Abrahamson, E. W. (1979b). *FEBS Lett.* **107**, 317.

Uhl, R., Hofmann, K. P. and Dreutz, W. (1978). *Biochemistry* **17**, 5347.

Van Breugel, P. J. G. M., Bovee-Geurts, P. H. M., Bonting, S. L. and Daemen, F. J. M. (1979). *Biochim. Biophys. Acta* **557**, 188.

Virmaux, N., Weller, M., Mandel, P. and Trayhurn, P. (1975). *FEBS Lett.* **53**, 320.

Wald, G. (1968). *Nature* **219**, 800.

Waloga, G. and Bitensky, M. W. (1981). *Invest. Ophthalmol. Visual Sci.* **20** Supp. 233.

Watts, A., Volotovski, I. D. and Marsh, D. (1979). *Biochemistry* **18**, 5006.

Watts, A., Davoust, J., Marsh, D. and Devaux, P. F. (1981). *Biochim. Biophys. Acta* **643**, 673.

Watts, A., Volotovski, I. D., Pater, R. and Marsh, D. (1982). *Biophys. J.* **37**, 93.

Weller, M., Virmaux, N. and Mandel, P. (1975). *Nature* **256**, 68.

Weller, M., Virmaux, N. and Mandel, P. (1976). *Exp. Eye Res.* **23**, 65.

Wheeler, G. L. and Bitensky, M. W. (1977). *Proc. Natl Acad. Sci. USA* **74**, 4238.

Wheeler, G. L., Matuo, Y. and Bitensky, M. W. (1977). *Nature* **269**, 822.

Wilden, V. and Kuhn, H. (1982). *Biochemistry* **21**, 3014.

Williams, T. P., Baker, B. N. and McDowell, J. H. (1974). *Exp. Eye Res.* **18**, 69.

Woodruff, M. L. and Bownds, M. D. (1979). *J. Gen. Physiol.* **73**, 629.

Woodruff, M. L., Bownds, M. D., Green, S. H., Morrisey, J. L. and Shedlovsky, A. (1977). *J. Gen. Physiol.* **667**.

Woodruff, M. L., Fain G. and Bastian (1982). *J. Gen. Physiol.* **80**, 537.

Wu, C. W. and Stryer, L. (1972). *Proc. Natl Acad. Sci. USA* **69**, 1104.

Yeager, M. J. (1975). *Brookhaven Symp. Biol.* **27**, III 3.

Yau, K. W., Lamb, T. D. and Baylor, D. D. (1977). *Nature* **269**, 78.

Yau, K. W., Lamb, T. D. and McNaughton, P. A. (1981). *Current Topics in Membranes and Transport* **15**, 19.

Yee, R. and Liebman, P. A. (1978). *J. Biol. Chem.* **253**, 8902.

Yoshikami, S. and Hagins, W. A. (1971). *Biophys. J.* **11**, 47a.

Yoshikami, S. and Hagins, W. A. (1973). *In* "Biochemistry and Physiology of Visual Pigments" (ed. H. Langer), p. 245. Springer-Verlag, Berlin.

Yoshikami, S., George, J. S. and Hagins, W. A. (1980). *Nature* **286**,

Yoshikami, S., Robinson, W. E. and Hagins, W. A. (1974). *Science* **185**, 1176.

Yoshizawa, T. (1972). *In* "Handbook of Sensory Physiology. VII/1 Photochemistry of Vision" (ed. H. J. A. Dartnall), p. 146. Springer-Verlag. Berlin.

Yoshizawa, T. and Wald, G. (1963). *Nature* **197**, 1279.

Young, R. W. (1967). *J. Cell Biol.* **33**, 61.

Young, R. W. (1971). *J. Cell Biol.* **49**, 303.

Young, R. W. and Droz, B. (1968). *J. Cell Biol.* **39**, 169.
Young, R. W. and Bok, D. (1969). *J. Cell. Biol.* **42**, 392.
Zagalski, P. F., Virmaux, N. and Mandel, P. (1981). *Exp. Eye Res.* **32**, 627.
Zorn, M. and Futterman, S. (1971). *J. Biol. Chem.* **246**, 881.
Zuckerman, R., Schmidt, G. J. and Dacko, S. M. (1982). *Proc. Natl Acad. Sci. USA* **79**, 6414.
Zumbulyadis, N. and O'Brien, D. F. (1979). *Biochemistry* **18**, 5427.

Chapter 5

Bacteriorhodopsin Topography in Purple Membrane

Yu A. Ovchinnikov, N. G. Abdulaev and A. V. Kiselev

Shemyakin Institute of Bioorganic Chemistry,
U.S.S.R. Academy of Sciences, 117988 Moscow, U.S.S.R.

I. Introduction

The study of the spatial organization of membrane proteins is an important step in elucidating their functions. Recently published data shed light on the topography of cytochrome c oxidase (Azzi and Casey, 1979), glycophorine (Tomita and Marchesi, 1975), "band 3"—the main protein of erythrocyte membrane (Steck, 1978; Cabantchik *et al.*, 1978), bovine rhodopsin (Dratz *et al.*, 1979; Ovchinnikov, 1982), cytochrome b_5 (Enoch *et al.*, 1979) and other proteins. A great variety of reagents selectively interacting with membrane surfaces as well as sites of protein molecules located in the lipid matrix of the membrane were used in these studies; they were designated as impenetrating and penetrating membrane reagents, respectively. Immunochemical methods for obtaining monoclonal antibodies to various sites of protein molecules furnished these investigations with accuracy and elegance (Etemadi, 1980).

At present bacteriorhodopsin is one of the most extensively studied membrane proteins. Almost all modern methods of physics, chemistry and biology have been used to obtain information on its structure and function. Henderson and Unwin (1975) and later Hayward and Stroud (1981) determined the molecular structure of bacteriorhodopsin by combination of electron-microscopy and diffraction methods. The protein molecule was shown to be composed of 7 α-helical rods spanning the membrane width perpendicularly to its surface. However, even the three-dimensional picture of high resolution (3.7 Å) did not ascertain which groups are involved in the process of proton translocation across the membrane. Apparently, knowledge of the primary structure and peculiarities of the polypeptide chain folding could provide a new insight into understanding of the mechanism of proton pumping by bacteriorhodopsin.

Determination of the complete amino acid sequence of bacteriorhodopsin was an important step in studying its three-dimensional structure and mechanism of functioning. The protein consists of 248 amino acid residues (Ovchinnikov et al., 1978, 1979; Khorana et al., 1979), 67% of them being of hydrophobic character (Fig. 1).

The present work deals with the study of bacteriorhodopsin topography by means of impenetrating reagents—proteolytic enzymes: papain, carboxypeptidase A, trypsin and α-chymotrypsin. As is well known the purple membrane is a quite dense semi-crystalline formation with more or less rigid, well-defined mutual disposition of its protein and lipid components. One could thus expect that relatively mild treatment of the purple membrane with proteolytic enzymes should lead to splitting only in the regions exposed to the exterior whereas the inaccessible regions of the protein deeply embedded in the membrane would withstand an attack.

Analysing the primary structure of bacteriorhodopsin we found the polypeptide chain regions with clearly defined hydrophilic properties: fragments 33–41, 102–107, 157–166 and the C-terminal region of the protein molecule (231–248). This region makes up 7% of the polypeptide chain and 25% of residues with negatively charged side chains. Hydrophilic regions of the protein molecule were assumed to be on the membrane surface; some of these regions link α-helical rods, and they could be precisely the sites of the action of proteolytic enzymes.

```
                                    5                          10                        15                        20                        25
<Glu Ala Gln Ile Thr Gly Arg Pro Glu Trp Ile Trp Leu Ala Leu Gly Thr Ala Leu Met Gly Leu Gly Thr Leu

                           30                        35                        40                        45                        50
Tyr Phe Leu Val Lys Gly Met Gly Val Ser Asp Pro Asp Ala Lys Phe Tyr Ala Ile Thr Thr Leu Val Pro

                           55                        60                        65                        70                        75
Ala Ile Ala Phe Thr Met Tyr Leu Ser Met Leu Leu Gly Tyr Gly Leu Thr Met Val Pro Phe Gly Gly Glu Gln

                           80                        85                        90                        95                        100
Asn Pro Ile Tyr Trp Ala Arg Tyr Ala Asp Trp Leu Phe Thr Thr Pro Leu Leu Leu Asp Leu Ala Leu Leu

                           105                       110                       115                       120                       125
Val Asp Ala Asp Gln Gly Thr Ile Leu Ala Leu Val Gly Ala Asp Gly Ile Met Ile Gly Thr Gly Leu Val Gly

                           130                       135                       140                       145                       150
Ala Leu Thr Lys Val Tyr Ser Tyr Arg Phe Val Trp Trp Ala Ile Ser Thr Ala Ala Met Leu Tyr Ile Leu Tyr

                           155                       160                       165                       170                       175
Val Leu Phe Phe Gly Phe Thr Ser Lys Ala Glu Ser Met Arg Pro Glu Val Ala Ser Thr Phe Lys Val Leu Arg

                           180                       185                       190                       195                       200
Asn Val Thr Val Val Leu Trp Ser Ala Tyr Pro Val Val Trp Leu Ile Gly Ser Glu Gly Ala Gly Ile Val Pro

                           205                       210                       215                       220                       225
Leu Asn Ile Glu Thr Leu Leu Phe Met Val Leu Asp Val Ser Ala Lys Val Gly Phe Gly Leu Ile Leu Leu Arg

                           230                       235                       240                       245
Ser Arg Ala Ile Phe Gly Glu Ala Glu Ala Pro Glu Pro Ser Ala Gly Asp Gly Ala Ala Ala Thr Ser
```

FIG. 1. Complete amino acid sequence of bacteriorhodopsin.

II. Proteolysis of Bacteriorhodopsin

A. PAPAIN ACTION

Comprehensive information on the bacteriorhodopsin topography in the purple membrane was obtained by means of papain-protease of wide specificity.

Not all regions in the bacteriorhodopsin polypeptide chain turned out to be equally accessible to the enzyme action. Protease split rapidly the most accessible ones at low enzyme : substrate ratios. At increasing enzyme: substrate ratios as well as the reaction time bacteriorhodopsin was cleaved also in other regions less accessible to the enzyme. Changing the reaction conditions, we succeeded in selective cleavage of the protein in different positions.

Figure 2 shows the kinetics of papain action on bacteriorhodopsin (enzyme/substrate ratio—1 : 10 mg/mg). Judging by SDS electrophoresis in polyacrylamide gel bacteriorhodopsin via intermediate product I was transformed into fragment II, which a few hours later split into fragments III and IV.

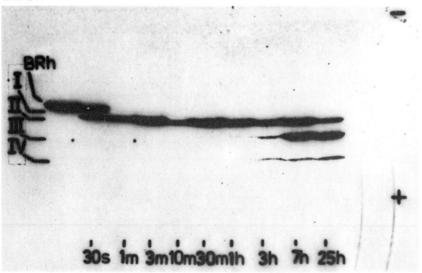

FIG. 2. Kinetics of bacteriorhodopsin cleavage with papain. S, seconds; M, minutes; H, hours.

Ten minutes after incubation the membranes were centrifuged and from the obtained supernatant peptides P-1, P-2, P-3 were isolated by chromatography and electrophoresis on cellulose thin layer plates (Fig. 3). The

amino acid sequences of these peptides were shown to be Asp–Gly–Ala–Ala–
Ala–Thr–Ser (242–248), Ala–Pro–Glu–Pro–Ser–Ala–Gly (235–241) and
Glu–Ala–Glu (232–234), respectively. The C-terminal amino acid residue of
membrane-bound fragment II, glycine, was determined by hydrazinolysis.

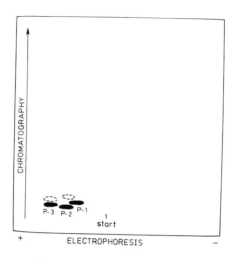

FIG. 3. Micropreparative separation of papain peptides of bacteriorhodopsin
C-terminal region by peptide mapping.

Further cleavage of fragment II (1–231) was accompanied with appearance
of peptides P-4, P-5, P-6 in the reaction medium. The peptides were isolated
from the supernatant by peptide mapping on cellulose thin layer plates
(Fig. 4). Amino acid sequences Leu–Thr (66–67) and Val–Pro–Phe–Gly
(69–72) were established for peptides P-4 and P-5, correspondingly. Pyroglutamic acid was found to be the N-terminal amino acid residue of peptide
P-6. Its amino acid sequence was found to be < Glu–Ala–Gln (1–3).

Table I presents data on structures of the peptides split off from bacteriorhodopsin.

TABLE I

Structures of papain peptides

P-1	Asp–Gly–Ala–Ala–Ala–Thr–Ser
P-2	Ala–Pro–Glu–Pro–Ser–Ala–Gly
P-3	Glu–Ala–Glu
P-4	Leu–Thr
P-5	Val–Pro–Phe–Gly
P-6	< Glu–Ala–Gln

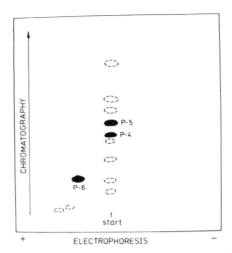

FIG. 4. Micropreparative separation of peptides obtained as result of fragment **II** hydrolysis by peptide mapping.

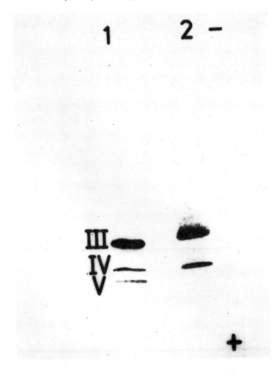

FIG. 5. Electrophoresis of bacteriorhodopsin papain fragments. (1) Enzyme/ bacteriorhodopsin 1 : 10, 7 days. (2) Enzyme/bacteriorhodopsin 1 : 1, 20 hours.

Quantitative cleavage of fragment II into fragments III, IV occurred at enzyme/substrate ratio 1 : 1 (mg/mg) for 20 h (Fig. 5.2).

A longer treatment (7 days) of membranes with papain led to formation of new membrane-bound fragment V as well as fragments III, IV (Fig. 5.1). Glycine, isoleucine and methionine were found to be N-terminal amino acid residues of the formed fragments. The N-terminal amino acid sequences Gly–Glu–Gln–, Ile–Thr–Gly–, Met–Arg–Pro– were identified by a sequence analysis of fragments III, IV, V, respectively. Assignment of amino acids to one of the three given sequences was unequivocal as two of them belonged to known fragments III and IV.

B. CARBOXYPEPTIDASE A AND TRYPSIN ACTION

Treatment of bacteriorhodopsin with carboxypeptidase A for 2 h (enzyme/substrate ratio 1 : 40 mg/mg) cleaved off five amino acids from the protein molecule. The kinetic analysis of this process led to determination of the C-terminal amino acid sequence of bacteriorhodopsin: –Ala–Ala–Ala–Thr–Ser (Fig. 6).

As a result of tryptic digestion bacteriorhodopsin was transformed into a product with lower molecular mass. The reaction product was analysed by SDS electrophoresis in polyacrylamide gel (Fig. 7).

From the supernatant of tryptic hydrolysate a peptide of the amino acid sequence Ala–Gly–Asp–Gly–Ala–Ala–Ala–Thr–Ser (240–248) was isolated

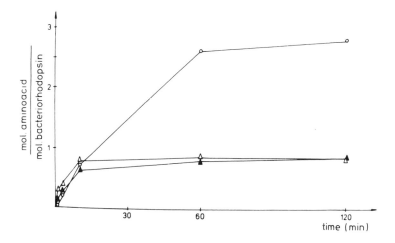

FIG. 6. Kinetics of carboxypeptidase A cleavage of C-terminal amino acids of bateriorhodopsin.

by chromatography on thin layer cellulose. The peptide is the C-terminal region of the protein molecule formed as a result of nonspecific trypsin cleavage of bond Ser–Ala (239–240).

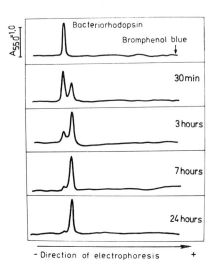

FIG. 7. Kinetics of trypsin cleavage of bacteriorhodopsin. Enzyme/bacteriorhodopsin 1 : 20 (mg/mg).

C. α-CHYMOTRYPSIN ACTION

α-Chymotrypsin treatment of purple membranes led to splitting of bacteriorhodopsin into two large membrane-bound fragments (Fig. 8).

Taking into account that the N-terminal amino acid of bacteriorhodopsin was pyroglutamic acid and α-chymotrypsin cleavage led to appearance of only one N-terminal amino acid—glycine—we determined the cleavage site by simultaneous analysis of the fragments and uncleaved protein. Gly–Gly–Glu–Gln–Asn– (72–76) was shown to be the N-terminal amino acid sequence of one of bacteriorhodopsin chymotryptic fragments. No amino acids and peptides cleaved off as a result of chymotrypsin action on bacteriorhodopsin were found in the supernatant. The C-terminal amino acids (serine and phenylalanine) were determined by hydrazinolysis of the mixture of the protein and its two fragments. Thus it was shown that α-chymotrypsin cleaved the only (71–72) peptide bond in the polypeptide chain of bacteriorhodopsin.

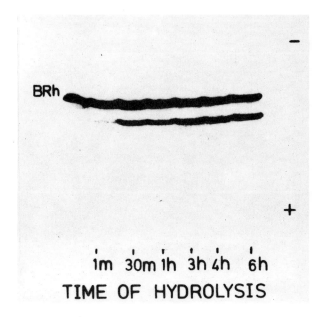

−

BRh

+

ı́m   ³0m ı̇h   ³h ⁴h   ı́6h
TIME OF HYDROLYSIS

FIG. 8. Kinetics of α-chymotrypsin cleavage of bacteriorhodopsin.

D. COMPARATIVE STUDY OF PAPAIN AND α-CHYMOTRYPSIN ACTION ON
BACTERIORHODOPSIN IN PURPLE MEMBRANES AND RIGHT-SIDE-OUT
VESICLES

Determination of the disposition of peptide bond 71–72 and the C-terminal region of the protein relative to the membrane surface was the next step in the study of bacteriorhodopsin topography. Comparative analysis of papain and α-chymotrypsin action on bacteriorhodopsin in purple membranes and vesicles obtained from halophilic bacteria cells by the freeze-thawing method was carried out. It was revealed by the electronmicroscopy technique that over 90% of vesicles were morphologically intact (Fig. 9.1). The study of these vesicles by the freeze-fracture technique showed that convex and con-cave membrane fracture faces contained different amounts of the so-called intramembrane particles (IMP). As the analysis showed 80% of convex fracture faces contained much more IMP than concave ones (Fig. 10). According to Blaurock and Stoeckenius (1971), such distribution of the particles is characteristic of intact cell membranes. Consequently, 80% of vesicles were of the same membrane orientation as in intact cells. It is note-worthy that treatment of vesicles with proteolytic enzymes did not change their morphological integrity (Fig. 9.2). Thus the obtained vesicles were morphologically intact and preserved their membrane orientation that was successfully applied for the study of the bacteriorhodopsin topography.

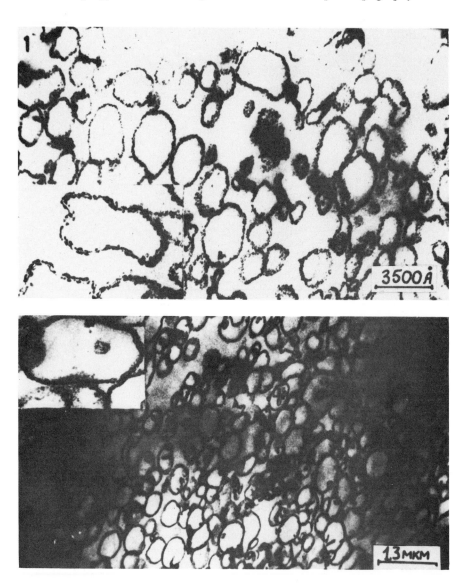

FIG. 9. Ultrathin sections of vesicles. (1) Before papain treatment. (2) After papain treatment.

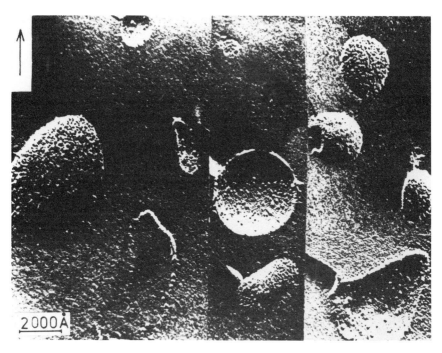

FIG. 10. Structure of freeze-fracture faces of membrane vesicles. Arrow shows direction of shadowing layer.

As the vesicle preparation contained a considerable amount of carotenoids and absorption bands of these compounds and bacteriorhodopsin overlapped, the spectral methods for determination of bacteriorhodopsin concentration were unsatisfactory. Therefore the quantitative estimation of bacteriorhodopsin content in vesicles was carried out by electrophoresis in polyacrylamide gel. With this aim in view, the calibration curve was plotted representing the dependence of the bacteriorhodopsin amount, used for electrophoresis, on the peak area on electrophoregram, obtained after staining and scanning of the gel (Fig. 11). An aliquot of the vesicle suspension was lyophilized, dissolved in the incubation buffer and then applied to the gel. The protein content in vesicles was determined using the calibration curve and the peak area corresponding to bacteriorhodopsin on electrophoregram.

After treatment with proteolytic enzymes, the vesicles were centrifuged and the pellet was washed with water. This pellet turned out to be purple membranes containing bacteriorhodopsin and its fragments as revealed by electrophoresis in the presence of sodium dodecyl sulphate.

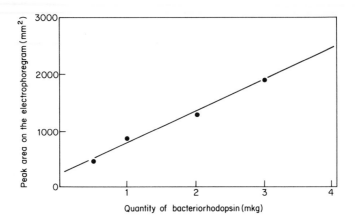

FIG. 11. Calibration curve for determination of bacteriorhodopsin content in vesicles.

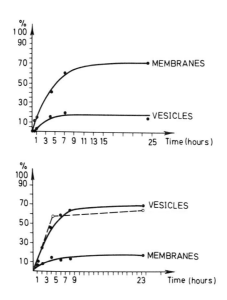

FIG. 12. Comparative study of papain and α-chymotrypsin action on bacterio-rhodopsin of purple membranes and vesicles. (1) Papain cleavage of 17 C-terminal amino acids of bacteriorhodopsin in vesicles and membranes. (2) α-Chymotrypsin cleavage of bond Phe–Gly (71–72) of bacteriorhodopsin in vesicles and membranes.

Figure 12.1 presents data on the comparative study of the papain cleavage of the bacteriorhodopsin C-terminal fragment in vesicles and membranes under the same conditions. The enzyme : bacteriorhodopsin ratio was 1 : 50 (mg/mg). The curves were lined through the points obtained after gel scanning and electrophoregram calculation according to the formula

$$\frac{\text{S-II}}{\text{S-BRh}} \times 100 \; (\%),$$

where S-BRh and S-II were the peak areas corresponding to bacteriorhodopsin and fragment II (bacteriorhodopsin without 17 C-terminal amino acids).

Kinetics of the cleavage of bacteriorhodopsin bond 71–72 in a bacteriorhodopsin molecule with α-chymotrypsin at the enzyme/bacteriorhodopsin ratio 1 : 10 (mg/mg) is shown in Fig. 12.2 for the same preparations. The degree of the protein hydrolysis calculated according to the formula

$$\frac{\text{S-CH1} + \text{S-CH2}}{\text{S-Ch1} + \text{S-Ch2} + \text{S-BRh}} \times 100 \; (\%),$$

where S-BRh, S-Ch1 and S-Ch2 are the peak areas on electrophoregram corresponding to bacteriorhodopsin and its two chymotryptic fragments, was plotted on axis of ordinates.

III. Chromophore Localization in Bacteriorhodopsin Polypeptide Chain

To localize the retinal binding site, the experimental approach based on α-chymotrypsin cleavage of bacteriorhodopsin in purple membranes into fragments 1–71 and 72–248 was used. It turned out that retinyl bacterioopsin obtained after reduction of bacteriorhodopsin with sodium borohydride (1 %, 30 min, pH 9.5) in the light was cleaved with α-chymotrypsin into the same fragments: 1–71 and 72–248. Analysing intensity of fluorescence zones, corresponding to these fragments on electrophoregram, we identified the retinal binding fragment. Figure 13 shows data on distribution of retinyl fluorescence between the fragments of retinyl bacterioopsin. Under these conditions the retinal was bound preeminently to fragment 72–248.

In order to determine which lysine residue of fragment 72–248 is a chromophore-binding site, retinyl bacterioopsin was delipidated and subjected to pronase digestion at the enzyme : substrate ratio 1 : 15 (mg/mg) for 48 h. The obtained hydrolysate was centrifuged, and the fluorescent residue was subsequently washed with 0.25 % and 0.025 % aqueous ammonium. After the last centrifugation the precipitate was dissolved in pyridine and chromatographed on a thin layer silica gel plate; as a result ε-N-retinyl derivative of H–Ala–Lys–OH was isolated. The structure of the isolated peptide was determined by the Edman degradation and amino acid analysis. Since bond Ala–Lys (215–216) occurs only once in fragment 72–248, the presented data testify that Lys-216 is the retinal-binding site in the protein.

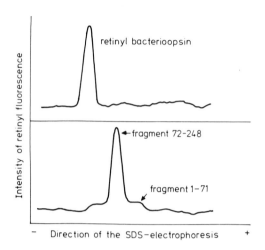

FIG. 13. Localization of retinyl label in chymotryptic fragments of bacteriorhodopsin.

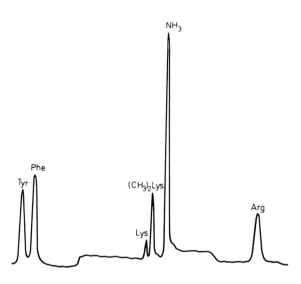

FIG. 14. Chromatogram fragment of methylated bacteriorhodopsin hydrolysate.

In a separate experiment we prepared an ε-N-dimethyl analogue of bacteriorhodopsin. The reductive methylation of the protein was carried out using formaldehyde (CH_2O) and sodium cyanoborohydride ($NaCNBH_3$) in the dark in 0.5 M borate buffer (pH 8.5); as follows from the amino acid analysis (Fig. 14) six of seven Lys residues of the protein molecule were dimethylated. Similarly to unmodified protein more than 95% of retinyl fluorescence appeared to be provided by fragment 72–248. Methylated bacteriorhodopsin was released from lipids and retinal with organic solvents and the only unmodified Lys residue was methylated with [$^{14}C]H_2O$ and $NaCNBH_3$. The completeness of this reaction was followed by amino acid analysis. The obtained radioactively labelled opsin was cleaved with cyanogen bromide at methionine residues and the mixture of resultant peptides was subjected to the automated Edman degradation. More than 92% of radioactivity was cleaved at the seventh cycle that corresponded to Lys-216 in the primary structure.

IV. Discussion

The study of the papain action on bacteriorhodopsin along with the information on its primary structure revealed that formation of the fragment II was a result of the cleavage of 17 C-terminal amino acids from the protein molecule.

Thus at least 17 amino acids of the C-terminal region of bacteriorhodopsin cleaved with papain were exposed at the membrane surface (Fig. 15).

The structural analysis of peptides P-4, P-5 and P-6 allowed us to localize two more regions of the bacteriorhodopsin polypeptide chain, 1–4 and 65–73, accessible to the papain action. Fragment III was the region of the polypeptide chain between Gly-73 and Gly-231, fragment IV was located between Ile-4 and Gly-65. It is noteworthy that under the same conditions but in the presence of 0.2–2.3 M NaCl papain did not act on molecule regions 1–4 and 65–73 but cleaved off only 17 C-terminal acids of bacteriorhodopsin. Apparently polypeptide chain regions 1–4 and 65–73 located on the membrane surface could be buried in the lipid matrix and thus are inaccessible to the enzymes at ionic strength increase.

It was found that peptides of the C-terminal region, P-1, P-2 and P-3, were rich in acidic amino acids which made it hydrophilic, whereas peptides P-4 and P-5 of the 65–73 region contained no charged amino acids. Four of six amino acids of these peptides were hydrophobic. Some connections between α-helical segments of the bacteriorhodopsin molecule accessible to proteolytic enzymes did not seem to be hydrophilic.

In the very beginning of these studies we believed that bacteriorhodopsin amino acid residues exposed outside the membrane should participate in its functioning, binding and releasing hydrogen ions at the opposite sides of the membrane. However, removal of regions 1–3, 66–72 and 232–248 from the polypeptide chain did not change its ability to translocate protons across the membrane (Abdulaev *et al.*, 1978).

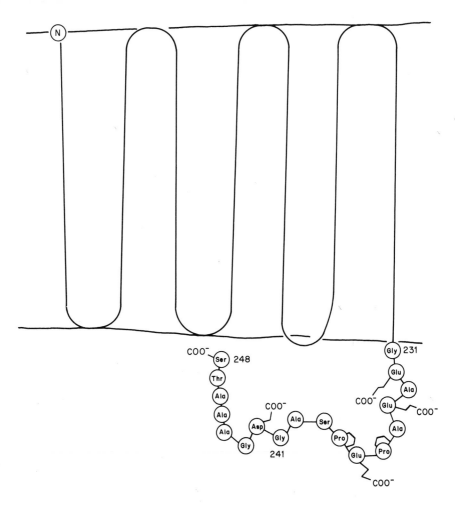

FIG. 15. Topography of C-terminal region of bacteriorhodopsin.

The data on the N-terminal amino acid sequences of fragments III, IV and V made it possible to locate one more site of cleavage in the polypeptide chain, bond Ser–Met (162–163). This bond appeared to be the least accessible to the papain action.

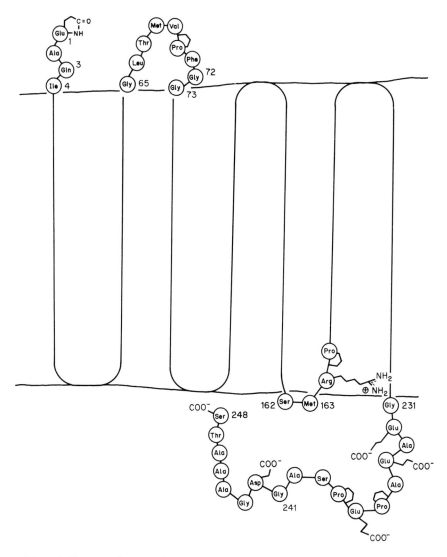

FIG. 16. Topography of regions 1–4, 65–73 and 162–165 of bacteriorhodopsin.

Taking into account that bacteriorhodopsin molecule consists of 248 amino acid residues, and the N- and C-terminal protein regions (at least 20 amino acids) are located outside the membrane as shown in the present work, each of 7 α-helical segments of bacteriorhodopsin (with their connectivities within the framework of Henderson's model) should contain 32 amino acids. Polypeptide chain region 65–73 accessible to papain is at the connection site between segments II and III, i.e. this region and the C-terminus are situated on opposite membrane surfaces. That was experimentally confirmed when analysing the action of proteolytic enzymes on bacteriorhodopsin of vesicles with right-side-out orientation of the membrane (see below). At the same time peptide bond Ser–Met (162–163) should be in the chain region connecting segments V and VI of the protein molecule (Fig. 16).

Upon studying the papain action on bacteriorhodopsin, four regions in the polypeptides chain accessible to this enzyme were found which made possible their location on the membrane surface. It should be noted that attempts to identify the connections between the first and second, third and fourth, fourth and fifth, as well as sixth and seventh α-helical segments of bacteriorhodopsin molecule by papain treatment were unsuccessful. However, this fact does not testify that they are embedded in the membrane. In fact Kimura *et al.* (1982) not only confirmed the data on the out of membrane localization of the C-terminal region of bacteriorhodopsin with the aid of monoclonal antibodies but also demonstrated that connection between protein segments III and IV was exposed outside the membrane.

To study bacteriorhodopsin topography use was also made of such proteolytic enzymes as carboxypeptidase A, trypsin and α-chymotrypsin.

Carboxypeptidase A cleaved off 5 C-terminal amino acid residues from bacteriorhodopsin. Gerber *et al.* (1977) showed that carboxypeptidase A cleaved off four residue –Ala–Ala–Thr–Ser from bacteriorhodopsin of the purple membrane. Cleavage of additional alanine residue in our experiment could be explained by the higher (1 : 40) enzyme : substrate ratio comparing with that (1 : 65) used in the above cited work.

Trypsin was found to act on the bacteriorhodopsin C-terminal region exposed into the aqueous phase. The anomalous specificity of trypsin could be explained by the presence of another protease in the trypsin preparation. However, the same preparation of trypsin did not cleave bond 239–240 in the chymotryptic peptide 231–248. Moreover, increasing ionic strength in the reaction medium (0.2 M NaCl) eliminated the observed effect of non-specific hydrolysis of the protein, whereas trypsin activity under these conditions was completely preserved. Therefore we assumed that the surface charge of the membrane changed the trypsin specificity.

α-Chymotrypsin digested the polypeptide chains of bacteriorhodopsin and retinyl bacterioopsin at the only bond Phe–Gly (71–72). Reduction of the Schiff base did not result in appearance of new regions in the polypeptide chain accessible to α-chymotrypsin. Accessibility of the same regions of the molecule to the action of immunoglobulins and enzymes with different nature of the active site and specificity made the obtained information on the packing of the bacteriorhodopsin polypeptide chain more reliable.

Comparative study of the papain action on bacteriorhodopsin in purple membranes and vesicles showed that the cleavage of the C-terminal region in vesicles proceeded more slowly than the membrane sample. This finding supported the intracellular location of this fragment, which is in a complete accord with the data of Gerber *et al.* (1977). On the other hand, the cleavage of bacteriorhodopsin with α-chymotrypsin in vesicles proceeded more rapidly. Different rates of cleavage are explained most probably by aggregation of membrane preparations in the presence of 30 mM $MgSO_4$ necessary for integrity of vesicles (Blaurock *et al.*, 1976) In fact hydrolysis of bacteriorhodopsin in purple membranes with α-chymotrypsin in 50 mM ammonium–bicarbonate buffer (pH 8.0) at the same enzyme-substrate ratio proceeded more rapidly (dotted lines in Fig. 12.2). In this case the rates of the bacteriorhodopsin cleavage in membranes and vesicles became comparable. The results testify to location of bond 71–72 of the bacteriorhodopsin molecule on the outer membrane surface.

Retinal localization in the polypeptide chain of bacteriorhodopsin was of utmost importance. Different physical methods, such as neutron diffraction (King *et al.*, 1980) or fluorescence analysis (Thomas and Stryer, 1980), provided information on the spatial disposition of the β-ionone ring of the chromophore. However, they could not establish the Lys residue in the polypeptide chain responsible for retinal binding. In our first publication (Ovchinnikov *et al.*, 1977) we assigned the chromophore binding site to amino acid residue Lys-41. Data on the structure of thermolytic retinal-containing peptide Val–Ser–Asp–Pro–Asp–Lys–Lys– isolated by Bridgen and Walker (1976) and on bacteriorhodopsin primary structure favoured this conclusion. Findings on chromophore localization in fragment 72–248, which did not contain Lys-41 (Ovchinnikov *et al.*, 1980) and subsequent studies in our laboratory and others revealed Lys-216 as the principal site of retinyl localization (Bayley *et al.*, 1981; Lemke and Oesterhelt, 1981; Mullen *et al.*, 1981; Rodionov *et al.*, 1981; Katre *et al.*, 1981).

However, it was still unclear if the light fixation of retinal on Lys-216 is an artefact, in other words, a consequence of formation of preferred but not functional aldimine caused by influence of pH and the presence of $NaBH_4$.

This problem was solved when studying retinal localization in bacteriorhodopsin where 6 of 7 lysine ε-amino groups were dimethylated. The modified protein completely preserved its spectral and functional characteristics and the possibility of light-induced transiminization in such a sample was completely excluded. Determination of the Lys-216 residue as the chromophore-binding site in dimethylated bacteriorhodopsin made it possible to elicit the real functional localization of retinal in the protein.

All the above mentioned results underlay the model of spatial organization of the bacteriorhodopsin polypeptide chain in the purple membrane (Ovchinnikov *et al.*, 1979).

Analysis of X-ray and electronmicroscopy data within the framework of the established amino acid sequence (Engelman *et al.*, 1980), immunological approaches (Kimura *et al.*, 1982) and some other methods (Katre and Stroud, 1981) led to a refinement of our model. Figure 17 summarizes the current views of disposition of the bacteriorhodopsin polypeptide chain.

Consideration of the arrangement of amino acid residues of bacteriorhodopsin in the membrane resulted in the following conclusions:

(1) Several of proline residues (e.g. Pro-50, Pro-91, Pro-186) are located well in the membrane and can cause bending of helices.

(2) Some charged groups (Arg-82, Asp-85, Asp-115, Asp-212, Lys-216) are in the membrane width. According to thermodynamic considerations these residues could form ion pairs in membranes. In particular, the mutual disposition of such interactions gives some information concerning the packing of certain segments of the bacteriorhodopsin molecule (Engelman *et al.*, 1980). These residues probably participate in the formation of pathways for charge transfer across the membrane.

(3) The retinal binding site (Lys-216) is located in the seventh segment close to the centre of the membrane.

(4) The majority of basic amino acids (Arg-7, Lys-30, Lys-40, Lys-41, Lys-129, Lys-159, Arg-164, Lys-172, Arg-175, Arg-225, Arg-227) are in the vicinity of the membrane surface and can stabilize the lipoprotein complex interacting with negatively charged phosphate and sulphate and sulphate lipid groups and acidic amino acids.

(5) Most of negatively charged carboxyl groups are located on the cytoplasmic surface of the membrane, whereas practically all tryptophan and many tyrosine residues are concentrated on the outer surfaces of the membrane that should restrict the conformational mobility of bacteriorhodopsin molecule segments due to so-called stacking interactions. Probably the structural asymmetry of bacteriorhodopsin in the membrane is the molecular basis of functional asymmetry—the vector transport of protons across the membrane.

V. Conclusions

The study of proteolysis of bacteriorhodopsin in membranes resulted in finding four regions of its polypeptide chain 1–4, 65–73, 162–163, 231–248, accessible to enzyme action which made possible their localization on the membrane surface.

The comparative investigation of the action of proteolytic enzymes on bacteriorhodopsin in purple membranes and in vesicles with right-side-out orientation of the membrane testified to location of bond 71–72 on the outer surface and the C-terminal region on the cytoplasmic one.

Analysis of the chemical structure of outer and intramembrane regions of the bacteriorhodopsin molecule revealed some details of the spatial organization of its polypeptide chain.

Amino acid residue Lys-216 was identified as the functional retinal-binding site.

References

Abdulaev, N. G., Feigina, M. Yu., Kiselev, A. V., Ovchinnikov, Yu. A., Drachev, L. A., Kaulen, A. D., Khitrina, L. V. and Skulachev, V. P. (1978). *FEBS Lett.* **90**, 190–194.

Azzi, A. and Casey, R. P. (1979). *Mol. Cell Biochem.* **28**, 169–184.

Bayley, H., Huang, K.-S., Radhakrishnan, R., Ross, A. H., Takagaki, Y. and Khorana, H. G. (1981). *Proc. Natl Acad. Sci. USA* **78**, 2225–2229.

Blaurock, A. E. and Stoeckenius, W. (1971). *Nature New Biol.* **233**, 152–155.

Blaurock, A. E., Stoeckenius, W. and Oesterhelt, D. (1976). *J. Cell Biol.* **71**, 1–22.

Bridgen, J. and Walker, I. D. (1976). *Biochemistry* **15**, 792–798.

Cabantchik, Z. I., Knauf, P. A. and Rothstein, A. (1978). *Biochim. Biophys. Acta* **515**, 239–302.

Dratz, E. A., Milianich, G. P., Hemes, P. P., Gaw, J. E. and Schwarts, S. (1979). *Photochem. Photobiol.* **29**, 661–670.

Engelman, D. H., Henderson, R., McLachlan, A. D. and Wallace, B. A. (1980). *Proc. Natl Acad. Sci. USA* **77**, 2023–2027.

Etemadi, A.-H. (1980). *Biochim. Biophys. Acta* **604**, 347–422.

Enoch, H. G., Fleming, P. J. and Strittmatter, P. (1979). *J. Biol. Chem.* **254**, 6483–6488.

Gerber, G. E., Gray, C. P., Wildenauer, D. and Khorana, H. G. (1977). *Proc. Natl Acad. Sci. USA* **74**, 5426–5430.

Hayward, S. B. and Stroud, R. M. (1981). *J. Mol. Biol.* **151**, 491–517.

Henderson, R. and Unwin, P. N. T. (1975). *Nature* **257**, 28–31.

Katre, N. V. and Stroud, R. M. (1981). *FEBS Lett.* **136**, 170–174.

Katre, N. V., Wolber, P. K., Stoeckenius, W. and Stroud, R. M. (1981). *Proc. Natl Acad. Sci. USA* **78**, 4068–4072.

Khorana, H. G., Gerber, G. E., Herlihy, W. C., Gray, C. P., Anderegg, R. J., Nihei, K. and Biemann, K. (1979). *Proc. Natl Acad. Sci. USA* **76**, 5046–5050.

Kimura, K., Mason, T. L. and Khorana, H. G. (1982). *J. Biol. Chem.* **257**, 2859–2867.

King, G. I., Mowery, P. C., Stoeckenius, W., Crespi, H. L. and Schoenborn, B. P. (1980). *Proc. Natl Acad. Sci. USA* **77**, 4726–4730.

Lemke, H. D. and Oesterhelt, D. (1981). *Eur. J. Biochem.* **115**, 595–604.

Lemke, H. D. and Oesterhelt, D. (1981). *FEBS Lett.* **128**, 255–260.

Mullen, E., Johnson, A. H. and Akhtar, M. (1981). *FEBS Lett.* **30**, 187–193.

Ovchinnikov, Yu. A. (1982). *FEBS Lett.* **148**, 179–191.

Ovchinnikov, Yu. A., Abdulaev, N. G., Feigina M. Yu., Kiselev, A. V. and Lobanov, N. A. (1977). *FEBS Lett.* **84**, 1–4.

Ovchinnikov, Yu. A., Abdulaev, N. G., Feigina, M. Yu., Kiselev, A. V., Lobanov, N. A. and Nasimov, I. V. (1978). *Bioorg. Khim.* **4**, 1573–1574.

Ovchinnikov, Yu. A., Abdulaev, N. G., Feigina, M. Yu., Kiselev, A. V. and Lobanov, N. A. (1979). *FEBS Lett.* **100**, 219–224.

Ovchinnikov, Yu. A., Abdulaev, N. G., Tsetlin, V. I., Kiselev, A. V. and Zakis, V. I. (1980). *Bioorg. Khim.* **6**, 4427–4429.
Rodionov, A. V., Bairamashvili, D. I., Kudelin, A. B., Feigina, M. Yu., Shkrob, A. M. and Ovchinnikov, Yu. A. (1981). *Bioorg. Khim.* **7**, 1328–1334.
Steck, T. L. (1978). *J. Supramol. Struct.* **8**, 311–324.
Thomas, D. D. and Stryer, L. (1980). *Fed Proc.* **39**, 1847.
Tomita, M. and Marchesi, V. T. (1975). *Proc. Natl Acad. Sci. USA* **72**, 2964–2968.

Note Added in Proof

Two groups have now presented complete amino acid sequences for bovine rhodopsin (Ovchinnikov, 1982; Hargrave *et al.*, 1983). Each finds that the molecule consists of 348 amino acid residues (Fig. 18), and each has proposed model structures in which the molecule spans the membrane seven times.

```
Ac
Met.Asn.Gly.Thr.Glu.Gly.Pro.Asn.Phe.Tyr.Val.Pro.Phe.Ser.Asn.Lys.Thr.Gly.Val.Val
1                                 10                                           20
Arg.Ser.Pro.Phe.Glu.Ala.Pro.Gln.Tyr.Tyr.Leu.Ala.Glu.Pro.Trp.Gln.Phe.Ser.Met.Leu
21                                30                                           40
Ala.Ala.Tyr.Met.Phe.Leu.Leu.Ile.Met.Leu.Gly.Phe.Pro.Ile.Asn.Phe.Leu.Thr.Leu.Tyr
41                                50                                           60
Val.Thr.Val.Gln.His.Lys.Lys.Leu.Arg.Thr.Pro.Leu.Asn.Tyr.Ile.Leu.Leu.Asn.Leu.Ala
61                                70                                           80
Val.Ala.Asp.Leu.Phe.Met.Val.Phe.Gly.Gly.Phe.Thr.Thr.Thr.Leu.Tyr.Thr.Ser.Leu.His
81                                90                                          100
Gly.Tyr.Phe.Val.Phe.Gly.Pro.Thr.Gly.Cys.Asn.Leu.Glu.Gly.Phe.Phe.Ala.Thr.Leu.Gly
101                               110                                         120
Gly.Glu.Ile.Ala.Leu.Trp.Ser.Leu.Val.Val.Leu.Ala.Ile.Glu.Arg.Tyr.Val.Val.Val.Cys
121                               130                                         140
Lys.Pro.Met.Ser.Asn.Phe.Arg.Phe.Gly.Glu.Asn.His.Ala.Ile.Met.Gly.Val.Ala.Phe.Thr
141                               150                                         160
Trp.Val.Met.Ala.Leu.Ala.Cys.Ala.Ala.Pro.Pro.Leu.Val.Gly.Trp.Ser.Arg.Tyr.Ile.Pro
161                               170                                         180
Glu.Gly.Met.Gln.Cys.Ser.Cys.Gly.Ile.Asp.Tyr.Tyr.Thr.Pro.His.Glu.Glu.Thr.Asn.Asn
181                               190                                         200
Glu.Ser.Phe.Val.Ile.Tyr.Met.Phe.Val.Val.His.Phe.Ile.Ile.Pro.Leu.Ile.Val.Ile.Phe
201                               210                                         220
Phe.Cys.Tyr.Gly.Gln.Leu.Val.Phe.Thr.Val.Lys.Glu.Ala.Ala.Ala.Gln.Gln.Gln.Glu.Ser
221                               230                                         240
Ala.Thr.Thr.Gln.Lys.Ala.Glu.Lys.Glu.Val.Thr.Arg.Met.Val.Ile.Ile.Met.Val.Ile.Ala
241                               250                                         260
Phe.Leu.Ile.Cys.Trp.Leu.Pro.Tyr.Ala.Gly.Val.Ala.Phe.Tyr.Ile.Phe.Thr.His.Gln.Gly
261                               270                                         280
Ser.Asp.Phe.Gly.Pro.Ile.Phe.Met.Thr.Ile.Pro.Ala.Phe.Phe.Ala.Lys.Thr.Ser.Ala.Val
281                               290                                         300
Tyr.Asn.Pro.Val.Ile.Tyr.Ile.Met.Met.Asn.Lys.Gln.Phe.Arg.Asn.Cys.Met.Val.Thr.Thr
301                               310                                         320
Leu.Cys.Cys.Gly.Lys.Asn.Pro.Leu.Gly.Asp.Asp.Glu.Ala.Ser.Thr.Thr.Val.Ser.Lys.Thr
321                               330                                         340
Glu.Thr.Ser.Gln.Val.Ala.Pro.Ala.OH
341                       348
```

FIG. 18. The amino acid sequence of bovine rhodopsin.

High resolution solid-state ^{13}C NMR has been used to study rhodopsin–lipid interactions in native and artificial membranes: the results suggest that the presence of rhodopsin reduces the rates, but not the amplitudes, of ultra-high frequency lipid motions in the membrane (Sefcik *et al.*, 1983). The environments in the retinal and reconstituted PC membranes containing rhodopsin have also been probed using ^{31}P NMR (Albert and Yeagle, 1983). The spectra obtained indicate that there are two phospholipid head-group domains in both cases, and it is suggested that these may represent unperturbed phospholipids and those (about 23 per rhodopsin) whose head-groups are motionally restricted by the protein.

A detailed description of the phospholipid molecular species of frog disk membranes has been provided by Wiegand and Anderson (1983), who report that the saturated fatty acids are predominantly located on postition-1 and polyunsaturates on position-2.

Further evidence that 11-*cis* to all-*trans* isomerization of retinal derivative is the primary step in the photolysis of rhodopsin has been provided by picosecond kinetic absorption and fluorescence measurements on an analogue pigment containing a retinal derivative having a fixed 11-*cis* geometry (Buchert *et al.*, 1983). Data from this non-bleachable pigment also strongly suggest that isomerization in rhodopsin must occur on a picosecond timescale.

The bleaching of rhodopsin in pure phosphatidylserine (PS) membranes has been compared with that in membranes reconstituted from endogenous lipids (De Grip *et al.*, 1983). When the lipid/protein ratio is reduced to below 30, the metarhodopsin 1–2 transition in endogenous membranes is greatly altered, whereas it is only slightly affected in the PS membranes. This is interpreted as evidence that the PS membranes minimize protein–protein interaction, so that the rhodopsin remains dispersed even at low lipid volumes.

There has also been progress in defining more clearly the nature of the factors linking the excitation of rhodopsin and the hydrolysis of cGMP. Thus, there is increasing evidence that the bleaching product of rhodopsin which activates GTP/GDP exchange by the G protein is metarhodopsin 2 (Pfister *et al.*, 1983), and that for weak bleaches the phosphorylation of metarhodopsin can occur fast enough to account for the observed cGMP phosphodiesterase quench times (Sitaramayya and Liebman, 1983).

The G-protein which serves to couple the activated rhodopsin to the phosphodiesterase has also been more fully characterized: Fung (1983) has separated the molecule into two subunits (α and β,γ) and shown that neither displays significant GTPase activity, GTP/GDP exchange or ability to bind to rhodopsin. However, all the activities are restored if the subunits are recombined and analysis of the GTPase activity as a function of the stoichiometries of the two subunits strongly suggests that the β,γ subunit activates the α subunit by enabling it to bind to rhodopsin and dissociates from it during GTP hydrolysis. Limited trypsin proteolysis of the α subunit removes the ability of the reconstituted G-protein to interact with rhodopsin, while the accessibility of the GDP binding site to proteolysis depends upon the presence or absence of GDP (Fung and Nash, 1983). It thus appears that the rhodopsin and GDP binding sites are topologically distinct, and that a large conformational change accompanies conversion of bound GDP to GTP.

Debate over the mechanism of transduction has continued, with the case for the cGMP hypothesis being reviewed by Miller (1983a,b). Further support for the hypothesis came from experiments which showed that pressure-injecting the purified G-protein bound to the hydrolysis resistant GTP analogue, guanylyl imidodiphosphate (p(NH)ppG), or the partially purified PDE into toad rods causes a reversible hyperpolarization of the plasma membrane (Clack *et al.*, 1983). The case for Ca has also been fortified by the demonstration that a sufficiently large store, and rapid light-stimulated efflux, of this cation occurs in preparations of ROS membranes suspended in media containing high-energy phosphate esters and with an electrolytic composition similar to that of the living rod cell (George and Hagins, 1983).

A further hypothesis which combines roles for both cyclic GMP and Ca has been proposed, based on the observation that protons block the dark current of isolated frog rods, and that the site of their action is inside the cell (Mueller and Pugh, 1983; Liebman et al., 1984). In this model it is suggested that the protons which are produced by cyclic GMP hydrolysis following light absorption undergo exchange with Ca bound to or contained within disk membranes, and that the Ca released causes the Na conductance decrease of the plasma membrane.

References

Albert, A. D. and Yeagle, P. L. (1983). *Proc. Natl Acad. Sci. USA* **80**, 7188.

Buchert, J., Stefanic, V., Doukas, A. G., Alfano, R. R., Callender, R. H., Pande, J., Akita, H., Balogh-Nair, V. and Nakanishi, K. (1983). *Biophys. J.* **43**, 279.

Clack, J. W., Oakley, B. and Stein, P. J. (1983). *Nature* **305**, 50.

De Grip, W. J., Olive, J. and Bovee-Geurts, P. H. M. (1983). *Biochim. Biophys. Acta* **734**, 168.

Fung, B. K.-K. (1983). *J. Biol. Chem.* **258**, 10495.

Fung, B. K.-K. and Nash, C. R. (1983). *J. Biol. Chem.* **258**, 10503.

George, J. S. and Hagins, W. A. (1983). *Nature* **303**, 344.

Hargrave, P. A., McDowell, J. H., Curtis, D. R., Wang, J. K., Juszczak, E., Fong, S. L., Rao, J. K. M. and Argos, P. (1983). *Biophys. Struct. Mech.* **9**, 235.

Liebman, P. A., Mueller, P. and Pugh, E. N. (1984). *J. Physiol.* **347**, 85.

Miller, W. H. (1983a). *Adv. Cyclic Nucleotide Res.* **15**, 495.

Miller, W. H. (1983b). *Trends in Physiol. Sci.* **4**, 1253.

Mueller, P. and Pugh, E. N. (1983). *Proc. Natl Acad. Sci. USA* **80**, 1892.

Ovchinnikov, Y. A. (1982). *FEBS Lett.* **148**, 179.

Pfister, C., Kuhn, H. and Chabre, M. (1983). *Eur. J. Biochem.* **136**, 489.

Sefcik, M. D., Schaefer, J., Stejskal, E. O., McKay, R. A., Ellena, J. F., Dodd, S. W. and Brown, M. F. (1983). *Biochem. Biophys. Res. Commun.* **114**, 1048.

Sitaramayya, A. and Liebman, P. A. (1983). *J. Biol. Chem.* **258**, 12106.

Wiegand, R. D. and Anderson, R. E. (1983). *Exp. Eye Res.* **37**, 159.

Chapter 6

Acetylcholine Receptor Structure and Function

ROBERT M. STROUD

*Department of Biochemistry and Biophysics,
University of California School of Medicine, San Francisco,
California 94143, U.S.A.*

I. Introduction

Nicotinic acetylcholine receptors (AcChR) found in skeletal muscle or electric tissue translate the binding of a neurotransmitter, acetylcholine (AcCh) or other agonists, into a rapid increase, and subsequent decrease in the permeability of the endplate membrane to passage of cations. The physiological effect is to temporarily short-circuit, or depolarize, the endplate, a response which is translated into muscular contraction in the case of neuromuscular junctions, or potentiation of electric tissue in the stacked asymmetric cells of organs in *Torpedo* (a marine elasmobranch) or *Electrophorus* (a freshwater teleost). The cells in these electric species are innervated on one side only, and the electric discharge, which is used to stun or kill their prey or for defence, is generated as the sum of many thousands of ~ 70 mV elementary discharges of the oriented rostrocaudal stacks of

BIOLOGICAL MEMBRANES Vol. 5
ISBN 0 12 168546 2

electrocytes (Bennett, 1961; Bennett, 1970). The rich supply of AcChR in electric organs of *Torpedo* or *Electrophorus* permitted a much more extensive analysis of the structure, amino acid and gene sequence, and function of this cell surface receptor than has so far been possible for any other eukaryotic cell surface receptor.

These AcChR species have been studied extensively at the electrophysiological, the cellular, the kinetic, the biochemical and the structural levels (for reviews see Rang, 1974; Heidmann and Changeux, 1978; Fambrough, 1979; Barrantes, 1979; Karlin, 1980).

Acetylcholine binding to AcChR leads to opening of a cation selective channel across the postsynaptic membrane, a channel shown to be large enough to pass isopropylammonium ion (Furukawa and Furukawa, 1959) or dimethyl diethanolammonium ion (Maeno *et al.*, 1977). Therefore, the "open" channel must be about 0.72 nm (7.2 Å) across throughout the transmembrane length. Within about 1 millisecond, a time depending on the nature of the agonist, temperature, and membrane polarization, the channel is closed with a second, slower rate (Takeuchi and Takeuchi, 1960; Magleby and Stevens, 1972a, Magleby and Stevens, 1972b; Katz and Miledi, 1972; Anderson and Stevens, 1973; Stevens, 1977). Under physiological conditions, each receptor activation can lead to passage of about 10^4 sodium ions (Neher and Sakmann, 1976; Neher and Steinbach, 1978), which serve to depolarize the endplate membrane from its normal -65 mV potential inside the cell to about -5 mV.

With a much slower rate the agonist-bound AcChR undergoes a transition to a third "desensitized" state (Katz and Thesleff, 1957). At neuromuscular junctions desensitization is correlated with a decrease in the amplitude of ion conductivity response to applied agonist, with a half-time (dependent on agonist type and concentration) of several seconds to minutes, an effect which is reversed by removal of agonist. The rate of recovery is independent of agonist type and typically occurs in seconds to minutes (reviewed in Heidmann and Changeux, 1978; Karlin, 1980). Even in the constant presence of desensitizing amounts of AcCh however, individual AcChR molecules partially revive and can trigger clusters of opening events prior to returning to the deeply desensitized state (Sakmann *et al.*, 1980). This seems to imply a fourth structural state of AcChR. Physiologically, desensitization is correlated with transition to a form of AcChR which shows higher affinity (typically 200-400 times higher) for agonists such as carbamylcholine or acetylcholine. The concept of a desensitized state also agrees well with *in vitro* data currently available for subsynaptic membrane fragments from *Torpedo*. In spite of the higher affinity, acetylcholine diffuses away from the receptor *without* inducing a second channel opening, and thus it is not a microscopic reverse of AcCh binding. This neurotransmitter is then hydro-lysed by acetylcholine esterase, an enzyme located on the basal lamina between the pre- and post-synaptic membrane (Hall and Kelly, 1971).

II. Characterization of AcChR-rich Membranes

AcChR membranes are prepared from *Torpedo californica* or *Electrophorus electricus* by homogenization and differential centrifugation (Elliott *et al.*, 1978). Specific activity is generally 3–6 nmol α-bungarotoxin per mg of protein (or about 1–1.5 toxin binding sites per AcChR equivalent), at this stage. A further purification step due to Neubig *et al.* (1979) involves treat-ment with NaOH at pH 11 for one hour. This removes several of the major protein species (a doublet around 43 kD, another around 55 kD, and one of 90 kD) without loss of functionality (Neubig *et al.*, 1979; Moore *et al.*, 1979) as judged from SDS polyacrylamide gels, and agonist-induced ion-fluxing ability.

At this stage, the membranes are predominantly right-side-out vesicles densely packed with AcChR molecules, clearly separated from membranes containing acetylcholine esterase. The AcChR binds 2 molecules of [^{125}I]-bungarotoxin per 250 000 daltons of protein (Karlin, 1980).

AcChR protein can be solubilized in detergents such as Triton X-100 or deoxycholate, and has been further purified using affinity chromatographic procedures. Under these conditions, the subunits of AcChR remain tightly associated together (they can be dissociated in SDS). The fluxing ability of AcChR in response to ligand binding can be restored in membranes recon-stituted from protein solubilized in the presence of lipids (Epstein and Racker, 1978; Lindstrom *et al.*, 1980; Sobel *et al.*, 1980; Heidmann *et al.*, 1980), or octylglucoside-solubilized, and affinity purified AcChR (Gonzalez-Ros *et al.*, 1980). Structural analysis has centred on the membrane vesicles generated from homogenized organ which do display ligand induced permeability response, and only recently has it included studies of octylglucoside-solubilized, or reconstituted material (Kistler *et al.*, 1982).

Membranes prepared with 10 mM *N*-ethyl maleimide (NEM) preserve the native dimer, sedimenting as a 13 S particle of AcChR cross-linked through the δ subunits, while omission of NEM, or subsequent treatment with a disulphide reducing agent such as β-mercaptoethanol leads to 9 S monomer containing membranes (Chang and Bock, 1977; Hamilton *et al.*, 1977). No specific functions have yet been assigned to the dimerization of AcChR *in vivo*; full functionality (binding of ligands, toxins, or ligand induced perme-ability) is retained by membrane preparations containing only monomeric AcChR (Anholt *et al.*, 1980).

```
                                          -1
α  MILCSYWHVCLVLLLFSCCGLVLC
β  MENVRRMALGLVVMMALALSGVGA
γ        MVLTLLLIICLALEVRS
δ  MGNIHFVYLLISCLYYSGCS   G

   1
   SEHETRLVANLL  EN YNKVIRPVEHHTHFVDITVGLQLIQLISVDEVNQIVETNV
   SVMEDTLLSVLF  ET YNPKVRPAQTVGDKVTVRVGLTLTNLLILNEKIEEMTTNV
   ENEEGRLIEKLL  GD YDKRIIPAKTLDHIIDVTLKLTLTNLISLNEKEEALTTNV
   VNEEERLINDLLIVNKYNKHVRPVKHNNEVVNIALSLTLSNLISLKETDETLTSNV

                                             100
   RLRQQWIDVRLRWNPADYGGIKKIRLPSDDVWLPDLVLYNNADGDFAIVHMTKLLL
   FLNLAWTDYRLQWDPAAYEGIKDLRIPSSDVWQPDIVLMNNNDGSFEITLHVNVLV
   WIEIQWNDYRLSWNTSEYEGIDLVRIPSELLWLPDVVLENNVDGQFEVAYYANVLV
   WMDHAWYDHRLTWNASEYSDISILRLPPELVWIPDIVLQNNNDGQYHVAYFCNVLV

   DYTGKIMWTPPAIFKSYCEIIVTHFPFDQQNCTMKLGIWTYDGTKVSIS
   QHTGAVSWQPSAIYRSSCTIKVMYFPFDWQNCTMVFKSYTYDTSEVTLQ    HALD
   YNDGSMYWLPPAIYRSTCPIAVTYFPFDWQNCSLVFRSQTYNAHEVNLQLS
   RPNGYVVTWLPPAIFRSSCPINVLYFPFDWQNCSLKFTALNYDANEITMDLMTDTID

                                  200
   PESDRP      DLSTFMESGEWVMKDYRGWKH  WVYYTCCPD  TPYLDITYHFIM
   AKGEREVKEIVINKDAFTENGQWSIEH KPSRKNW RSD DP  S YEDVTFYLII
   AEEGEAVEWIHIDPEDFTENGEWTIRH RPAKKNYNWQLTKDD TDFQEIIFFLII
   GK DYPIEWIIIDPEAFTENGEWEIIH KPAKKN IYPDKFPNGTNYQDVTFYLII

   QRIPLYFVVNVIIPCLLFSFLTGLVFYLPTDSG EKMTLSISVLLSLTVFLLVIVE
   QRKPLFYIVYTIIPCILISILAILVFYLPPDAG EKMSLSISALLAVTVFLLLLAD
   QRKPLFYIINIIAPCVLISSLVVLVYFLPAQAGGQKCTLSISVLLAQTIFLFLIAQ
   RRKPLFYVINFITPCVLISFLASLAFYLPAESG EKMSTAISVLLAQAVFLLLTSQ
```

FIG. 1. Sequences of the four subunits of AcChR, aligned as described by Finer-Moore and Stroud (1984), and see Noda et al. (1983).

The 9 S monomeric AcChR from *T. californica* contains five subunits of stoichiometry: $\alpha_2\beta\gamma\delta$ (Reynolds and Karlin, 1978; Lindstrom *et al.*, 1979; Raftery *et al.*, 1980). Upon SDS gel electrophoresis, protein in these vesicles is contained in four major bands characteristic of the α (~ 40 kD) β (~ 50 kD), γ (~ 60 kD) and δ (65 kD) subunits of AcChR and additional 43 kD, ~ 55 kD and 90 kD components not apparently required for AcChR activity. The latter species can be removed completely by treatment with NaOH at pH 11 and are therefore regarded as peripheral membrane proteins, not part of AcChR. The 43 kD component was shown *not* to be actin (Strader *et al.*, 1980).

The cDNA sequences of all four subunit types have been determined (Sumikawa *et al.*, 1982; Noda *et al.*, 1982, 1983; Claudio *et al.*, 1983). The sequences are all highly homologous, suggesting a common tertiary structure to all five of the subunits (Fig. 1). Each sequence contains four very hydrophobic sequences thought to correspond to transmembrane spanning regions (Noda *et al.*, 1982; Claudio *et al.*, 1983). The molecular weights of the amino acid sequence corresponding to the coding sequences alone add up to 268 000 daltons, not counting the polysaccharides which may be about 18 000 daltons (Anderson and Blobel, 1981). It is not yet proven that the entire sequence is present in the mature protein, which could account for some of the difference between this value and electrophoretic mobilities, or estimates of the total molecular weight as 250 000 by detergent matched sedimentation (Reynolds and Karlin, 1978).

One of the α subunits can be labelled by the affinity label MBTA or with bromoacetylcholine, which compete stoichiometrically with agonist or neurotoxins after reduction of disulphide bonds with dithiothreitol (Damle and Karlin, 1978; Damle *et al.*, 1978). Prior incubation with neurotoxins such as α-bungarotoxin or cobrotoxin (2 sites bound per AcChR) inhibit MBTA binding. Thus the two α subunits contain binding sites for agonists and snake neurotoxins. However, physiologically, a single conducting event involves release of millimolar concentrations of AcCh into the synapse. Most measurements of agonist affinity for AcChR give micromolar dissociation constants ($K_d \sim 10^{-6}$ M) which may measure binding to the desensitized, high affinity state of AcChR. Dunn and Raftery (1982) show that there are additional effects produced by much higher concentrations of agonist, in the mM range, and suggest that these ones, rather than the ones that compete with neurotoxins, could be the effective activator sites. The Hill coefficient for agonist induced permeability response is 1.96 ± 0.06; thus at least two agonists are required for channel opening (Neubig and Cohen, 1980).

III. Structural Characterization:
AcChR is a Funnel-shaped, Transmembrane Protein

X-ray diffraction studies of AcChR membranes led to an electron density profile at 0.65 nm (6.5 Å) resolution through the membrane plane, which shows that the protein extends by 5.5 nm (55 Å) on one side of the membrane, and 1.5 nm (15 Å) on the other side (Ross *et al.*, 1977; Klymkowsky and Stroud, 1979) (Fig. 2). It relied on development of a Fourier refinement scheme for such analysis from continuous X-ray scattering patterns (Stroud and Agard, 1979). The profile could be resolved into lipid, protein, and solvent to give the 0.65 nm (6.5 Å) resolution cylindrically averaged structure shown in Fig. 1 (Ross *et al.*, 1977; Klymkowsky and Stroud, 1979). Subsequently, the funnel-shaped structure was visualized directly in electron micrographs at the edge of a folded-over vesicle (Fig. 2a,b). The AcChR molecules were unambiguously identified by immunoelectronmicroscopy using anti-receptor antibodies coupled to gold beads (Klymkowsky and Stroud, 1979). Similar experiments, in which AcChR was first treated with α-bungarotoxin and then with anti-toxin antibodies attached to gold spheres, established that the extension of about 5.5 nm (55 Å) was on the extracellular, or synaptic side of the membrane (Klymkowsky and Stroud, 1979) (Fig. 3). The small extension and relatively flat surface on the cytoplasmic side makes direct observation of this surface alone difficult by electronmicroscopy; however, folded-over membranes are sometimes observed in which the two lipid

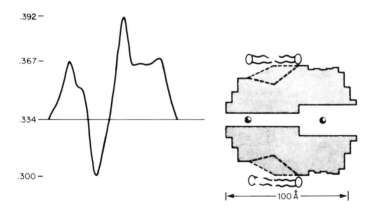

FIG. 2. A 6.5 Å resolution cylindrically averaged structure for AcChR obtained from solution of the continuous X-ray scattering profile of AcChR membranes is reprinted with permission of the *Journal of Molecular Biology*; the measured density of receptors and the in-plane equatorial diffraction maxima which show a particle diameter of 85 Å are implicit.

bilayers are in close apposition. The extracellular head of AcChR is clearly visible on each side of the fold, and the maximum width of the folded double membranes is about 20 nm (200 Å), which corresponds well with twice the thickness of 11 nm (110 Å) determined by X-ray diffraction (Ross *et al.*, 1977). The central double bilayer thickness of about 8.6 nm (86 Å) is close to twice the 4.0 nm (40 Å) distance between phosphatidyl head groups on either side of the single bilayer (Kistler *et al.*, 1982). Thus the projection of the AcChR on the intracellular surface is about 1–1.5 nm (10–15 Å) perpendicular to the membrane plane, in agreement with the X-ray profile structure shown in Fig. 2. The transmembrane nature of AcChR has also been shown immuno-logically, since some anti-AcChR antibodies (coupled to ferritin) bind to the cytoplasmic surface (Tarrab-Hazdai *et al.*, 1978; Strader *et al.*, 1979).

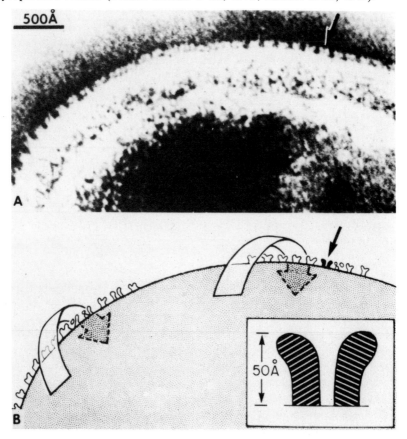

FIG. 3. Negative stain micrographs in which (a) AcChR molecules are apparent at the edge of an AcChR vesicle, as schematized in (b). Reprinted from Klymkowsky and Stroud (1979) with permission from *Journal of Molecular Biology*.

IV. Subunit Organization and the Ion Channel

X-ray diffraction patterns reveal a sharp peak corresponding to a spacing of 5.1 Å perpendicular to the membrane plane (Ross et al., 1977). The sharpness of this reflection indicates that on the average approximately 16 repetitions of structures with a basic repeat unit of 0.51 nm (5.1 Å) are oriented perpendicular to the membrane. The only known secondary structure to give this spacing is the α-helix, and circular dichroism suggests that up to 34% of the ∼250 000 dalton oligomeric AcChR molecule is α-helix (Moore et al., 1974). Thus, α-helices of up to 8.0 nm (80 Å) length are oriented perpendicular to the membrane (Ross et al., 1977). Such helical structures are probably desirable for the cotranslational insertion of polypeptides into membranes during biosynthesis, since the energetic cost of burying unpaired charges or unbonded hydrogen bonding groups within a hydrophobic environs is prohibitive. Helical secondary structures provide intramolecular satisfaction of all main chain hydrogen bonding groups during insertion. The length of the oriented α-helices was the first of several pieces of evidence which demonstrate that the subunits of AcChR are themselves each elongated perpendicular to the membrane plane.

Using proteases as probes of exposure to the aqueous environment, it is clear that all five subunits are exposed to the aqueous phase (Klymkowsky et al., 1980). With trypsin, the initial cleavage of AcChR occurs very rapidly. Within one minute, three new bands are generated: within only 15 minutes, 10 μM trypsin cleaves all subunits. Yet after 72 hours of treatment with 10 μM trypsin (4°C or 37°C), further digestion proceeds relatively slowly through discrete, intermediate-sized bands (a major one is of 45 kD), to a single 25 kD maximum-sized chain, with small fragments in the ultimate digest pattern. Thus all subunits are accessible to trypsin, and also to papain (Lindstrom et al., 1980) in the predominantly right-side-out membrane vesicles, and it has since been shown that all subunits are accessible to trypsin on the intracellular as well as extracellular side of the membrane (Strader and Raftery, 1980; Wennogle and Changeux, 1980). Thus each of the subunits span the bilayer.

The proteolysed AcChR remains a 9 S species, binds ligands, and is capable of agonist-induced ion fluxing (Klymkowsky et al., 1980; Lindstrom et al., 1980; Strader and Raftery, 1980). Thus most of the split chains remain tightly associated in a functional complex.

In the presence of detergent which solubilizes the monomeric AcChR, trypsin is much more rapid in producing the same digest pattern (Klymkowsky et al., 1980), one which produces a single 27 kD major band with only small chains at the dye front (Bartfeld and Fuchs, 1979). This raises the idea that all five subunits have a common structurally homologous folded domain, a common motif which presents loose-chain susceptible sites on either side of the membrane. More recently, homology throughout the complete amino acid sequences of all four subunit types has been demonstrated in *Torpedo californica* AcChR (Raftery *et al.*, 1980).

The cDNA sequences are highly homologous in hydrophobicity plots and align such that the N-termini are in perfect register; the C-termini are of slightly different length, but align within 17 residues (Noda *et al.*, 1983) (see Fig. 1). Thus internal deletions of presumptive loops rather than simply shorter chains can account for the differences in molecular weight (Noda *et al.*, 1982; 1983; Claudio *et al.*, 1983). Using lipophilic nitrenes, Martinez and co-workers (Sator *et al.*, 1979) and Bercovici and Gitler (1978) present strong evidence for exposure of at least the α, β and γ subunits to lipid within the bilayer. There is weak immunocross-reactivity between α and β (Tzartos and Lindstrom, 1979), and between γ and δ subunits (Tzartos and Lindstrom, 1980; Lindstrom *et al.*, 1976; Claudio and Raftery, 1979). In addition, reduction with DTT increases reactivity of all subunits to [^{14}C]-*N*-ethyl maleimide (Hamilton *et al.*, 1979). All except δ can be iodinated from the aqueous phase (Hartig and Raftery, 1977, 1979), again indicating aqueous exposure. Finally, all subunits are glycosylated (Karlin *et al.*, 1975; Vandlen *et al.*, 1979) which clearly indicates that all have an extracellular exposure. There is thus considerable evidence that the subunits are structurally, as well as sequentially, homologous.

The central well can be stained with uranyl or phosphotungstate negative stains throughout the entire length of AcChR (11.0 nm ± 1; 110 Å ± 10) and more densely than any other region on the membrane by a stain thickness equivalent to 11.4 nm (114 Å) (Kistler *et al.*, 1982). It is therefore an ionophoretic channel large enough to retain these ions. Uranyl ions are known to strongly potentiate the effect of AcCh (Reader *et al.*, 1973); thus it is not surprising that the channel seems to be stained throughout its length. The current evidence demonstrates that subunits elongated perpendicular to the membrane plane are arranged as somewhat homologous structural units, much as staves around the central ion channel.

V. Three-dimensional Structure of AcChR

Highly ordered helical or tubular structures can be "annealed" into the membranes to yield reproducible, ordered structures (Fig. 4). The importance of these structures is two-fold. First, a single image contains all particle views, from side view to top view in the helical array, and therefore contains all information for a 3-dimensional image reconstruction. Secondly, the highly regular arrays provide the means for locating subunits of AcChR, after staining with Fab fragments of subunit specific antibodies (Kistler *et al.*, 1982). In addition, binding of Fab fragments to the cytoplasmic side can be compared with binding to the extracellular side in order to quantitate changes which occur upon agonist or effector binding or upon desensitization, channel opening etc.

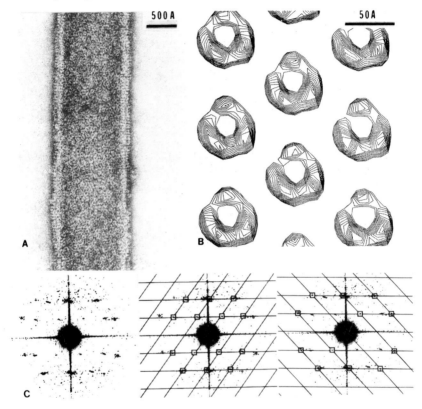

FIG. 4. Electron micrograph of a negatively stained tube membrane array is presented alongside the computer filtration of the surface lattice which is in contact with the support film. Two major, and a minor protein peak and a stain-filled groove surround the central depression in the anti-clockwise sense viewed from the cytoplasmic side. From results of Kistler and Stroud (1981).

Computer filtering of the arrays in tubular structures show that the pattern of stain-excluding regions within the funnel-shaped particles is very reproducible (Kistler and Stroud, 1981) (Fig. 3). Always the two high peaks, the smaller peak, and the deep valley appear anticlockwise as viewed from the cytoplasmic side. These images are similar to the images generated from planar membrane arrays (Ross *et al.*, 1977). A limited ordered arrangement of AcChR (double rows) has also been reported for quick-frozen and deep-etched membrane preparations, frozen on a time scale which does not allow molecular diffusion during preparation (Heuser and Salpeter, 1979). The cell dimensions (or inter-particle distances) in planar arrays are $a = 8.8 \pm 0.5$ nm $(88 \pm 5 \text{ Å})$; $c = 8.8 \pm 0.5$ nm $(88 \pm 5 \text{ Å})$ and $\gamma = 118° \pm 2°$, while for the tube membranes the surface lattice has $a = 9.2 \pm 0.22$ nm $(92 \pm 2.2 \text{ Å})$ parallel to the tube axis, with $b = 8.5 \pm 0.17$ nm $(85 \pm 1.7 \text{ Å})$ and $\gamma = 125° \pm 2°$. Together the X-ray and multiple electronmicroscopy images give the three-dimensional model for AcChR shown in Fig. 5 (Kistler et al., 1982).

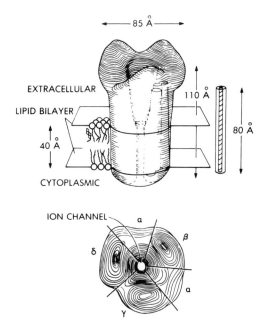

FIG. 5. Three-dimensional model of AcChR at approx. 20–30 Å in-plane, approx. 10 Å perpendicular to the plane resolution (A). Synaptic view of the AcChR molecule with borders arbitrarily drawn between elongated subunits and tentative assignment of subunit types consistent with our cross-linking data (B) and mono-clonal antibody binding. Reprinted with permission from the *Biophysical Journal*, Kistler et al. (1982).

Each "infundibuliform" structure is a single monomeric AcChR (Ross et al., 1977; Kistler et al., 1982). Using the known partial specific volume of protein based on amino acid sequence (Changeux et al., 1975), the volume of this model corresponds to a molecular content of ~277 000 daltons, which is slightly larger than the value 250 000 of Reynolds and Karlin (1978). The computed radius of gyration 4.3 nm (43 Å) is somewhat smaller than the measured value of 4.6 nm (46 Å) (Wise *et al.*, 1979). Volumes of the model on synaptic side, cytoplasmic side, and within the bilayer correspond to 165 000, 32 000 and 72 000 daltons, in fair agreement with models based on sequence topology (see Claudio *et al.*, 1983; Noda *et al.*, 1983; Moore and Stroud, 1983).

Structure of Neurotoxins and Binding to AcChR

Agonists and neurotoxins bind strictly competitively either with each other, or with the affinity label MBTA which binds to one of the two α-subunits (Damle and Karlin, 1978; Damle *et al.*, 1978). The neurotoxins bind with dissociation constants of about 10^{-10} M at 20°C, neutral pH (Weber and Changeux, 1974a; Chicheportiche *et al.*, 1975) and are sufficiently large as to affect the X-ray or electron optical patterns. However, while the dissociation constants indicate exceedingly tight binding, the on-rates for neurotoxin-receptor binding are extremely slow: The half-times, $t_{\frac{1}{2}}$, are typically in the range of 1–3 minutes (Weber and Changeux, 1974b, and these rates (but *not* the off-rates) become very slow below 11°C (Lester, 1971).

The structures of three post-synaptic snake neurotoxins have been determined and help to explain how they bind to AcChR. Erabutoxin, a "short" neurotoxin, was solved by Low and colleagues (1976, 1979) and also by Tsernoglou *et al.* (1976, 1978). α-Cobratoxin structure was solved by Walkinshaw *et al.* (1980). The structure of α-bungarotoxin, a second "long" toxin, is also solved (Mebs, *et al.*, 1972; Agard and Stroud, 1982; Stroud 1981; Agard *et al.*, 1983) (Fig. 6). While all three structures are clearly evolutionary cousins of each other, the comparison between bungarotoxin and erabutoxin illustrated in Fig. 6 reveals differences in the secondary and tertiary folded structure. In the case of α-bungarotoxin, there are only four hydrogen bonds associated with secondary structure of the molecule. This compares with up to 70% of the molecule for erabutoxin (Low *et al.*, 1976). The lack of any precisely determined secondary or tertiary structure may thus explain the extraordinary stability of the neurotoxin molecules (Karlsson, 1979). Furthermore, the residues involved in interaction with the AcChR seem to be spread out over one entire surface of the hand-shaped molecule (Low, 1979; Karlsson, 1979). In Fig. 6 are indicated residues implicated by the fact that chemical modification leads to partial (it is rarely complete) loss of binding activity, and the residues where evolutionary variation has been shown to affect activity (Strydom, 1979). By inference, these residues are involved in binding to AcChR (Agard *et al.*, 1983). All such side-chains are spread out on the concave surface of the hand-shaped molecule (Fig. 8). Thus the binding site between receptor and toxin is probably, as Low (1979) suggests, much more of a binding surface of about 2 × 3 nm (20 × 30 Å) in size than a binding site (Agard *et al.*, 1983). Comparison between the structures of the three toxin molecules shows that all of them would have to re-fold to achieve the congruence in AcChR interaction suggested by the high sequence homologies and conservation found among the 60 or so sequences of post-synaptic blocking neurotoxins (Strydom, 1979).

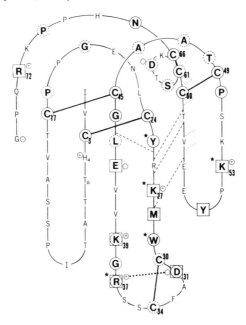

FIG. 6. A schematic view of the amino acid sequence (Mebs *et al.*, 1972) and three loop structure (Agard *et al.*, 1983) of α-bungarotoxin. Residues which are highly conserved among (i) *all* of the curaremimetic toxins, both "long" and "short" are enclosed in solid circles: (ii) all of the "long" toxins but not between the long and short are in dashed circles. Amino acids in which chemical modification of the side chain has been shown to affect toxicity (Low, 1979; Karlsson, 1979) are indicated with a star, and sites where an evolutionary change in the sequence seems to be correlated with a change in toxicity are in solid squares (Low, 1979; Karlsson, 1979).

Two residues on the neurotoxin structures are highly conserved, and it has been suggested on the basis of structure, and a loss of binding with their chemical modification (Low, 1979; Tsernoglou *et al.*, 1978) that they bind at the acetylcholine, or agonist binding site. However, these toxins lock the channel closed; they do not lead to channel opening as agonists do. While agonist binding can displace Ca^{2+} from AcChR, bound toxin can restore the Ca^{2+} sites (Chang and Neuman, 1976). Could the extended (2×3 nm; 20×30 Å) interacting surface lock the quaternary structure change in AcChR such that channel opening is no longer feasible after toxin is bound?

The neurotoxins bind to the outside crest on the synaptic side of AcChR (Klymkowsky and Stroud, 1979; Kistler *et al.*, 1982). X-ray diffraction of Fairclough and Stroud (1983) shows that the ligand binding site is located some 5 nm (50 Å) from the narrowest region of the only, single ionophoretic channel located in the monomeric AcChR (see Stroud, 1981, 1983 and Fig. 7).

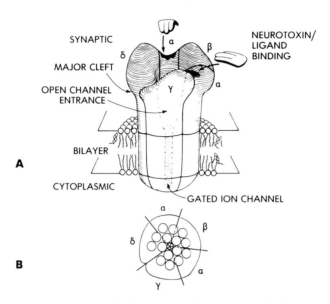

FIG. 7. A schematic drawing indicates the inferred toxin binding site (surface) located on the top crest of AcChR. Absolute definition of the toxin sites and of the individual subunit locations are still in process, and the arrangement of subunits shown is tentative at this stage (Kistler *et al.*, 1982).

The presence of neurotoxin somehow forces AcChR molecules further apart in the membrane plane, and attempts at image filtration to find the toxin site have not been successful. Zingsheim *et al.* (1982) and Holtzman *et al.* (1982) report localization of the toxin sites on unordered AcChR images. Membranes treated first with bungarotoxin are much more susceptible to digestion by trypsin than membranes not treated with neurotoxin (Klymkowsky *et al.*, 1980). The 27 kD sized fragments are generated much more rapidly suggesting that toxin, like detergents, acts to separate individual monomeric species one from another in the membrane plane, and thus increase accessibility of proteolytic enzymes to AcChR subunits. This evidence is corroborated by the finding that antitoxin antibodies bind to the toxin receptor complex (Menez *et al.*, 1979; Tamiya and Abe, 1979; Klymkowsky and Stroud, 1979). The binding sites for several anti-neurotoxin antibodies have been mapped, and essentially all are found to be associated with residues on the *back* surface of the neurotoxins rather than on the concave, active surface. This is probably because antibodies raised against neurotoxins in test animals are in fact raised against the tight complex between exogenous neurotoxin and host receptor, to which they bind extremely tightly. Nevertheless, it implies that the back surface of the hand-shaped neurotoxin molecule remains available in the receptor complex to the binding of all such antibodies. The large concave surface binds to AcChR on the outside of the AcChR molecule. This implies that the toxin and therefore ligand binding site, must be 4.5–5.5 nm (45–55 Å) away from the extracellular entrance to the most constricted region of the ionophoretic channel, the region that transposes the bilayer structure (Kistler *et al.*, 1982).

The large distance between ligand site and entrance to the ion channel ($\sim$5 nm; $\sim$50 Å) raises the question as to how the binding of acetylcholine at such a distant site could lead to opening of the channel. The effect of acetylcholine could be communicated to the channel by a detailed allosteric mechanism, or it could involve a large "global" change in conformation induced in the oligomer by ligand binding. Several factors strongly indicate the second kind of mechanism that the conformation changes involved in channel opening, channel closing, and desensitization are large. For example, the reactivity of 4 out of 5 of the subunits to photo-labelling by Bis azido ethidium bromide is altered by ligand binding (Witzemann and Raftery, 1977, 1978); i.e. four subunits are affected in a major way by ligand binding. Secondly, there are antisera which can inhibit channel opening without effect on ligand binding (Lindstrom *et al.*, 1977). Can they lock the quaternary structure and therefore prevent the necessarily large quaternary change? Most recently, Juan Korenbrot and his colleagues (Hestrin *et al.*, 1981) have shown that the entry of organic ions into the lipid regions is affected in a major way by agonist binding, suggesting that the lipid bound region of the protein undergoes a detectable structural change. This evidence strongly supports a mechanism which involves a global change between the closed channel state and the ligand-induced open channel state (Kistler *et al.*, 1982).

VI. Acknowledgements

The NIH GM-24485 and the NSF PCM80-21433 supported studies of receptor-rich membranes in my laboratory.

References

Agard, D. A. and Stroud, R. M. (1982). *Acta Cryst.* **A38**, 186–194.

Agard, D. A., Spencer, S. A. and Stroud, R. M. (1983). *Proc. Natl Acad. Sci. USA*, submitted.

Anderson, D. J. and Blobel, G. (1981). *Proc. Natl Acad. Sci. USA* **78**, 5598–5602

Anderson, C. R. and Stevens, C. F. (1973). *J. Physiol. (Lond)* **235**, 655–691.

Anholt, R., Lindstrom, J. and Montal, M. (1980). *Europ. J. Biochem.* **109**, 481–487.

Barrantes, F. J. (1979). *Ann. Rev. Biophys. Bioeng.* **8**, 287–321.

Bartfeld, D. and Fuchs, S. (1979). *Biochem. Biophys. Res. Comm.* **89**, 512–519.

Bennett, M. V. L. (1961). *Ann. N. Y. Sci.* **94**, 458–509.

Bennett, M. V. L. (1970). *Ann. Rev. Physiol.* **32**, 471–528.

Bercovici, T. and Gitler, C. (1978). *Biochemistry* **17**, 1484–1489.

Chang, H. W. and Neuman, E. (1976). *Proc. Natl Acad. Sci. USA* **73**, 3364–3368.

Chang, H. W. and Bock, E. (1977). *Biochemistry* **16**, 4513–4520.

Changeux, J.-P., Benedetti, E. L., Bourgeois, J.-P., Brisson, A., Cartaud, J., Devaux, P., Grunhagen, H., Moreau, M., Popot, J.-L., Sobel, A. and Weber, M. (1975). *Cold Spring Harbor Symp. Quant. Biol.* **40**, 211–230.

Chicheportiche, R., Vincent, J.-P., Kopeyan, C., Schweitz, H. and Lazdunski, M. (1975). *Biochemistry* **14**, 2081–2091.

Claudio, T. and Raftery, M. A. (1977). *Arch. Biochem. Biophys.* **181**, 484–489.

Claudio, T., Ballivet, M., Patrick, J. and Heinemann, S. (1983). *Proc. Natl Acad. Sci. USA* **80**, 1111–1115.

Damle, V. and Karlin, A. (1978). *Biochemistry* **17**, 2039–2045.

Damle, V., McLaughlin, M. and Karlin, A. (1978). *Biochem. Biophys. Res. Comm.* **84**, 845–851.

Dunn, S. M. J. and Raftery, M. A. (1982). *Biochemistry* **21**, 6264–6272.

Elliott, J., Blanchard, S. G., Wu, W., Miller, J., Strader, C. D., Hartig, P., Moore, H. P., Racs, J. and Raftery, M. A. (1978). *Biochem. J.* **185**, 667–677.

Epstein, M. and Racker, E. (1978). *J. Biol. Chem.* **253**, 6660–6662.

Fairclough, R. J. and Stroud, R. M. (1983). Ms in preparation.

Fambrough, D. M. (1979). *Physiol. Rev.* **59**, 165–227.

Finer-Moore, J. and Stroud, R. M. (1984). *Proc. Natl Acad. Sci. USA* **81**, 155–159.

Furukawa, T. and Furukawa, A. (1959). *Jap. J. Physiol.* **9**, 130–142.

Gonzalez-Ros, J. M., Paraschos, A. and Martinez-Carrion, M. (1980). *Proc. Natl Acad. Sci. USA* **77**, 1796–1800.

Hall, Z. W. and Kelly, R. B. (1971). *Nature New Biol.* **232**, 62–63.

Hamilton, S. L., McLaughlin, M. and Karlin, A. (1977). *Biochem. Biophys. Res. Comm.* **79**, 692–699.

Hamilton, S. L., McLaughlin, M. and Karlin, A. (1979). *Biochemistry* **18**, 155–163.

Hartig, P. R. and Raftery, M. A. (1977). *Biochem. Biophys. Res. Comm.* **78**, 16–22.

Hartig, P. R. and Raftery, M. A. (1979). *Biophys. J.* **25**, 192a.

Heidmann, T. and Changeux, J.-P. (1978). *Ann. Rev. Biochem.* **47**, 317–357.

Heidmann, T., Sobel, A., Popot, J.-L. and Changeux, J.-P. (1980). *Europ. J. Biochem.* **110**, 35–55.

Hestrin, S., Davis, C. G., Gordon, A. S., Diamond, I. and Korenbrot, J. (1981). *Soc. Neurosciences Abs.* **7**, 700.

Heuser, J. E. and Salpeter, S. R. (1979). *J. Cell Biol.* **82**, 150–173.

Holtzman, E., Wise, D., Wall, J. and Karlin, A. (1982). *Proc. Natl Acad. Sci. USA* **79**, 310–314.

Karlin, A., Weill, C. L., McNamee, M. G. and Valderrama, R. (1975). *Cold Spring Harbor Symp. Quant. Biol.* **40**, 203–213.

Karlin, A. (1980). *In* "The Cell Surface and Neuronal Function" (eds C. W. Cotman, G. Poste and G. L. Nicolson). Elsevier/North-Holland Biomedical Press, Amsterdam. pp. 191–260.

Karlsson, E. (1979). Chemistry of protein toxins in snake venoms. *In* "Handbook of Experimental Pharmacology" (ed. C.-Y. Lee), **52**, 158–212.

Katz, B. and Thesleff, S. (1957). *J. Physiol.* **138**, 63–80.

Katz, B. and Miledi, R. (1972). *J. Physiol.* **224**, 655–699.

Kistler, J. and Stroud, R. M. (1981). *Proc. Natl Acad. Sci. USA* **78**, 3678–3682.

Kistler, J., Stroud, R. M. Klymkowsky, M. W., Lalancette, R. A. and Fairclough, R. H. (1982). *Biophys. J.* **37**, 371–383.

Klymkowsky, M. W. and Stroud, R. M. (1979). *J. Mol. Biol.* **128**, 319–334.

Klymkowsky, M. W., Heuser, J. E. and Stroud, R. M. (1980). *J. Cell. Biol* **85**, 823–838.

Lester, H. (1971). *J. Gen. Physiol.* **57**, 255.

Lindstrom, J., Lennon, V., Seybold, M. E. and Whittingham, S. (1976). *Ann. N.Y. Acad. Sci.* **274**, 254–274.

Lindstrom, J., Einarson, B. and Francy, M. (1977). *In* "Cellular Neurobiology" (eds. Z. W. Hall and R. B. Kelly), pp. 119–130.

Lindstrom, J., Merlie, J. and Yogeeswaran, G. (1979) *Biochemistry* **18**, 4465–4470.

Lindstrom, J., Anholt, R., Einarson, B , Engel, A., Osame, M. and Montal, M. (1980). *J. Biol. Chem.* **255**, 8340–8350.

Lindstrom, J., Gullick, W., Conti-Tronconi, B. and Ellisman, M. (1980). *Biochemistry* **19**, 4791–4795.

Low, B. W., Preston, H. S., Sato, A., Rosen, L. S., Searl, J. E., Rudko, A. D., Richardson, J. S. (1976). *Proc. Natl Acad. Sci. USA* **73**, 2991–2994.

Low, B. W. (1979). *In* "Handbook of Experimental Pharmacology", (ed. C.-Y. Lee), **52**, 213–257.

Maeno, T., Edwards, C. and Anraku, M. (1977). *J. Neurobiol.* **8**, 173–184.

Magleby, K. L. and Stevens, C. F. (1972a). *J. Physiol.* (*Lond.*) **223**, 151–171.

Magleby, K. L. and Stevens, C. F. (1972b). *J. Physiol.* (*Lond*). **223**, 173–197.

Mebs, D., Narita, K., Iwanaga, S., Samejima, Y., and Lee, C. Y. (1972). *Hoppe-Seyler Z. Physiol. Chem.* **353**, 243–262.

Menez, A., Boulain, J. C. and Fromageot, P. (1979). *Toxicon* **17**, (suppl. 1), 123.

Moore, H. P., Hartig, P. R. and Raftery, M. A. (1979). *Proc. Natl Acad. Sci. USA* **76**, 6265–6269.

Moore, W. M., Holladay, L. A., Puett, D. and Brady, R. N. (1974). *FEBS Lett.* **45**, 145–149.

Neher, E. and Sakmann, B. (1976). *Nature* **260**, 799–802.

Neher, E. and Steinbach, J. H. (1978). *J. Physiol.* **277**, 153–176.

Neubig, R. R. and Cohen, J. B. (1980). *Biochemistry* **19**, 2770–2779.

Neubig, R. R., Krodel, E. K., Boyd, N. D. and Cohen, J. B. (1979). *Proc. Natl Acad. Sci. USA* **76**, 690–694.

Noda, M., Takahashi, H., Tanabe, T., Toyosato, M., Furutani, Y., Hirose, T., Asai, M., Inayama, S., Miyata, T. and Numa, S. (1982). *Nature* **299**, 793–797.

Noda, M., Takahashi, H., Tanabe, T., Toyosato, M., Kikyotani, S., Hirose, T., Inayama, S., Miyata, T. and Numa, S. (1983). *Nature* **301**, 251–255.

Noda, M., Takahashi, H., Tanabe, T., Toyosato, M., Kikyotani, S., Furutani, Y., Hirose, T., Takashima, H., Inayama, K. S., Miyata, T. and Numa, S. (1983). *Nature* **302**, 528–532.

Rang, H. P. (1974). *Quart Rev. Biophys.* **7**, 283–399.

Raftery, M. A., Hunkapiller, M. W., Strader, C. D. and Hood, L. E. (1980). *Science* **208**, 1454–1457.

Reader, T. A., Parisi, M. and DeRobertis, E. (1973). *Biochem. Biophys. Res. Comm.* **53**, No. 1, 10–17.

Reynolds, J. and Karlin, A. (1978). *Biochemistry* **17**, 2035–2038.

Ross, M. J., Klymkowsky, M. W., Agard, D. A. and Stroud, R. M. (1977). *J. Mol Biol.* **116**, 635–659.

Sakmann, B., Patlak, J. and Neher, E. (1980). *Nature* **286**, 71–73.

Sator, V., Gonzalez-Ros, J. M., Calvo-Fernandez, P. and Martinez-Carrion, M. (1979). *Biochemistry* **18**, 1200–1206.

Sobel, A., Heidmann, T., Cartaud, J. and Changeux, J.-P. (1980). *Europ. J. Biochem.* **110**, 13–33.

Stevens, C. F. (1977). *Nature* **270**, 391–396.

Strader, C. B. D., Revel, J.-P. and Raftery, M. A. (1979). *J. Cell. Biol.* **83**, 499–510.

Strader, C. D., Lazarides, E. and Raftery, M. A. (1980). *BBRC* **92**, No. 2., 365–373.

Strader, C. D. and Raftery, M. A. (1980). *Proc. Natl Acad. Sci. USA* **77**, 5807–5811.

Stroud, R. M. (1981). Proc. 2nd SUNYA Biomolecular Stereodynamics Meeting, Vol. II (ed. R. H. Sarma). Adenine Press, New York, pp. 55–73.

Stroud, R. M. (1983). *Neuroscience Commentaries* **1**, 124–138.

Stroud, R. M. (1983). *Neuroscience Commentaries* in press.

Stroud, R. M. and Agard, D. A. (1979). *Biophys. J.* **25**, 495–512.

Strydom, D. J. (1979). The evolution of toxins found in snake venoms. *In* "Handbook of Experimental Pharmacology" (ed. C.-Y. Lee), **52**, 258–275.

Sumikawa, K., Houghton, M., Smith, J. C., Bell, L., Richards, B. M. and Barnard, E. A. (1982). *Nucleic Acids Research* **10**, (19) 5809–5822.

Takeuchi, A. and Takeuchi, N. (1960). *J. Physiol.* **154**, 52–67.

Tamiya, N. and Abe, T. (1979). *Toxicon* **17**, (Suppl. 1), 186.

Tarrab-Hazdai, R., Geiger, B., Fuchs, S. and Amsterdam, A. (1978). *Proc. Natl Acad. Sci. USA* **75**, 2497–2501.

Tsernoglou, D., Petsko, G. A. (1976). *FEBS Lett.* **68**, 1–4.

Tsernoglou, D., Petsko, G. A. and Hudson, R. A. (1978). *Mol. Pharmacol.* **14**, 710–716.

Tzartos, S. J. and Lindstrom, J. M. (1980). *Proc Natl Acad. Sci. USA* **77**, 755–759.

Vandlen, R. L., Wu, W. C.-S., Eisenach, J. C. and Raftery, M. A. (1979). *Biochemistry* **18**, 1845–1854.

Walkinshaw, M. D., Saenger, W. and Maelicke, A. (1980). *Proc. Natl Acad. Sci. USA* **77**, 2400–2404.

Weber, M. and Changeux, J.-P. (1974a). *Mol. Pharmacol.* **10**, 1–14.

Weber, M. and Changeux, J.-P. (1974b). *Mol. Pharmacol.* **10**, 15–34.

Wennogle, L. P. and Changeux, J.-P. (1980). *Europ. J. Biochem.* **106**, 381–393.

Wise, D. S., Karlin, A. and Schoenborn, B. P. (1979). *Biophys. J.* **28**, 473–496.

Witzemann, V. and Raftery, M. A. (1977). *Biochemistry* **16**, 5862–5868.

Witzemann, V. and Raftery, M. A. (1978). *Biochemistry* **17**, 3598–3604.

Zingsheim, H. P., Barrantes, F. J., Frank, J., Hanicke, W. and Neugebauer, D. C. (1982). *Nature* **299**, 81–84.

Chapter 7

Kinetic Infrared Spectroscopy and Kinetic Light Scattering—Two New Methods for Studying Fast Trigger Processes

W. Kreutz, F. Siebert and K. P. Hofmann

Institut für Biophysik und Strahlenbiologie,
Universität Freiburg im Breisgau, Albertstrasse, 23, D-7800 Freiburg,
Federal Republic of Germany

I. Introduction

A. General Outline of the Kinetic Methods

In trigger processes taking place in signal transduction and energy conversion mechanisms as well as in the regulation of enzymatic and metabolic processes, fast biochemical and biophysical reactions play an important role.

BIOLOGICAL MEMBRANES Vol. 5
ISBN 0 12 168546 2

Triggering can proceed in light-induced processes via chromophore groups as well as by other biophysical or biochemical processes, such as potential jumps or transmitter coupling mechanisms.

Such fast processes can be conveniently detected directly at the chromophore group of a reaction partner by measuring the absorption or the fluorescence changes in the visible wavelength range. However, any optical parameter changing with the trigger processes can be used for the measurement if enough quanta per second can be detected. Examples for such measurements are changes of circular dichroism (Bayley, 1981) and of birefringence (Liebman et al., 1974).

The introduction of an infrared (IR) measuring light beam represents an essential extension in the detection of such kinetic processes. All IR-sensitive molecular changes occurring within and in the vicinity of the reaction centres become accessible. This includes changes in the chromophore groups themselves and also changes in its vicinity or possibly also in a neighbouring protein.

The introduction of kinetic light scattering enables the detection of fast structure changes, manifesting themselves by changes of polarizability, shape and volume.

The measurement of fast absorption changes in the visible wavelength range has become an almost conventional method and will not be discussed in detail. In this article we will only treat the newly developed methods of kinetic infrared spectroscopy (KIS) and the method of kinetic light scattering (KLS) especially in its realization as simultaneous kinetic light-absorption and light-scattering. The present state of development of both methods will be illustrated by examples of application. For the light sensitive systems, which were investigated up to now, a flash activation source was used as introduced by Porter (1963) and Witt (1971). It is expected, however, that also other activation sources, as for example, field jumps (Eigen and de Mayer, 1973) or pressure jumps (Attwood and Gutfreund, 1980) can be fitted to the new monitors.

B. GENERAL ASPECTS OF THE KIS METHOD

The introduction of an infrared measuring beam in kinetic measurements of absorption changes is complicated by an intrinsic property of biological systems: the requirement that biological objects have to be investigated in an aqueous milieu or at least in the hydrated state. This problem cannot be eliminated entirely, although it can be reduced considerably by dehydrating the sample prior to measurement to the minimum required where biological reactions can still be performed, i.e. by employing hydrated film samples as well as concentrated solutions in thin cuvettes. Another important point concerns the intensity of the measuring beam. In measurements of absorption changes in the visible spectral range, the intensity of the measuring beam must be kept as low as possible in order not to excite the system. The situation is different if an infrared measuring beam is applied. Due to the low energy of infrared quanta it is not possible to induce a biochemical or biophysical reaction. Therefore, the intensity of the measuring beam can be as high as technically possible, thereby considerably improving the signal-to-noise ratio. This is especially important since the detectivity of infrared detectors is lower at least by two orders of magnitude than that of detectors of visible light. For the application of the IR-method an infrared laser beam is therefore suited best. Besides the high measuring light intensity the IR-laser beam offers a further advantage: the small beam diameter allows the scanning of the cross-section area of the sample by the measuring beam. Thus, in samples with irreversible reactions which cannot be investigated repetitively, replacement of the specimen can be substituted by shifting the measuring beam across the sample area.

In the case of globar sources the warming up of the samples by the measuring beam can be neglected (provided incident energy only amounts to μWatts). Using laser beams, temperature inhomogeneities have to be taken into account and samples may have to be cooled.

Taking these general aspects into consideration, kinetic measurements in the mid-infrared spectral range (700 to 4000 cm^{-1}) can be performed without major complications. This also applies to the range of the water absorption bands (675 cm^{-1}, 1640 cm^{-1}, 3400 cm^{-1}) as well as to the absorption bands of D_2O (530 cm^{-1}, 1210 cm^{-1} and 2500 cm^{-1}). Normally, for the investigation of IR-absorption changes the following procedures may be applied:

(1) Determination of the absorption changes as a function of the wavenumbers resulting in the total amplitude difference spectrum of the system.
(2) Separation of the amplitude difference spectrum of reaction partners of the system by kinetic analysis.
(3) Setting up an amplitude–time–wavenumber diagram, i.e. determining the absorption changes of the system for all wavenumbers at any time (amplitude snap shots).
(4) Corresponding measurements in D_2O.
(5) Comparison of the kinetics with those observed in the visible spectral range.
(6) Assignment of the absorption changes by variation of chemical and physico-chemical parameters (isotopic labelling, chemical modifications, variation of pH and temperature).

These aspects will be discussed and exemplified in more detail in the case of bacteriorhodopsin.

C. GENERAL ASPECTS OF KINETIC LIGHT SCATTERING

The absolute intensity of light scattered on particles is a function of their polarizability. Intraparticle interference leads to an angular distribution of the scattered light depending on the shape and size of the particles.

If during a trigger process one of the particle parameters changes, a change of light scattering will consequently occur. The measurement of such changes of light scattering therefore allows the determination of reaction kinetics of reaction partners undergoing structural changes or of structural domains interacting with reaction partners.

"Particles" may represent proteins, reaction complexes of proteins, lipids and other components, as well as whole biomembranes, parts thereof, and biomembrane stacks. This opens a new source of information on interactions in superstructures.

Trigger processes, where possibly several structure parameters change in different time scales, spread out from local events to the dimension in the order of magnitude of the cell. Kinetic information can serve to separate the different processes which are buried in the complex total change of structure. Thereby the interpretation problem normally arising for particles $> 1/20$th of the wavelength can be partially circumvented. Principally, KLS measurements can be carried out in the following steps:

(1) Detection of light scattering changes at different scattering angles.
(2) Separation of uniform signals by kinetic analysis of the signals at different scattering angles.
(3) Detection and separation of light scattering changes by orientation of the samples (e.g. in a magnetic field).
(4) Characterization of reaction partners by variation of physical, chemical and biochemical parameters.
(5) Simultaneous measurement of absorption changes in order to assign the scattering signals to spectroscopically characterized products.

The following sections deal with the apparatus, conditions for sample preparations, and examples for the application of both methods including combined measurements with kinetic absorption changes in the visible spectral range.

II. The KIS Method

A. THE KIS APPARATUS

In Fig. 1 a schematic diagram of the apparatus used for kinetic infrared spectroscopy is shown. In the cases so far investigated the reactions were triggered by a light flash. Thus, the apparatus resembles a conventional flash photolysis instrument. The undispersed infrared beam from a globar infrared source is focused on to the sample. A filter may be inserted in between to remove the visible part of the beam. The beam is further passed through a monochromator and finally focused on to the detector. The whole optical set up, excepting the detector, is part of the normal infrared spectrophotometer from Perkin-Elmer, model PE-180. It is necessary to replace the normal thermopile detector by one more sensitive and with a shorter response time. We use an HgCdTe dectector from Santa Barbara Research Centers. The flat IR-sample is excited by two xenon flash lamps with a flash duration of 30 μs. The detector, a cooled solid state device, exhibits an $1/f$ noise characteristic dominating below 2 kHz. For time constants of 0.5 ms and longer this noise can be effectively reduced by the pulse modulation technique. The IR-beam is chopped at a frequency of 5–10 kHz, and the signal after the detector demodulated by a phase-sensitive boxcar integrator*. After the demodulation a low-pass filter of 20 kHz further reduces the noise. For the measurement of reactions with a shorter time constant the $1/f$ noise can be cut off by a high-pass filter, thereby avoiding the chopping of the infrared beam. The flash-induced signals, representing the transmission changes of the sample, are digitized in a transient recorder and transferred to a signal averager.

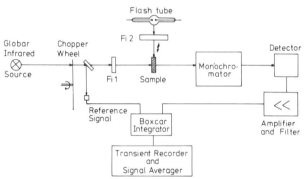

FIG. 1. Schematic diagram of the flash photolysis apparatus with infrared monitoring beam. Fi 1 and Fi 2 are optical filters for the infrared beam and the exciting flash, respectively. The boxcar integrator is synchronized by the undispersed infrared beam to compensate for inaccuracies of the chopper wheel.

The extinction coefficients of typical IR-bands are smaller at least by a factor of 10 to 100 than those of the absorption bands of typical chromophores in the visible and near UV spectral range. Due to the protein the average absorbance of a chromoprotein sample in the IR region of interest is of the same order of magnitude as the average absorbance in the visible

*In the newest version of the apparatus an elasto-optic modulator renders a chopping frequency of 140 kHz.

range. Thus, excitation of such a chromoprotein sample suitable for IR spectroscopy can effectively be performed. However, to obtain measurable transmission changes in the IR, 10 to 20 % of the sample have to be exicted by the flash.

B. PREPARATION OF SAMPLES

The presence of water is a prerequisite for most biological reactions. As already mentioned, its strong absorbance in the infrared requires very thin layers. For measurements with soluble proteins, highly concentrated solutions have to be used. The hydrated film technique (Mäntele et al., 1982) has proved very useful for membranous material. The membranes are deposited on an IR window, either by centrifugation or by simply drying the aqueous suspension. The membranes are hydrated to the degree required for the reaction via the vapour phase from a saturated salt solution. This method renders the samples sufficiently transparent over the total infrared spectral range. To identify bands involving vibrations of groups with exchangeable hydrogens, it is often necessary to perform measurements in D_2O. Hydrated film samples allow an easy exchange of H_2O for D_2O: the film may be completely dried and the aqueous solution for the hydration be substituted by the solution in D_2O or without prior drying of the film, the exchange may take place by way of the vapour phase. The latter method is more gentle though it takes longer. Many biological samples do not tolerate the drying process.

C. APPLICATIONS

1. Appropriate Systems

Kinetic infrared spectroscopy as developed by us (Siebert et al., 1980, 1981) is especially suited to measure molecular events affecting a small part of the system only. This facilitates the interpretation of the spectral changes caused by the molecular modifications. It requires, however, high sensitivity since the spectral changes are correspondingly small. It appears that these properties of the method make it especially suitable to investigate the molecular mechanism of biological trigger processes. Such processes often involve the interaction of a small molecule with a protein. For instance, the interaction of acetylcholine with the nicotinic receptor somehow leads to the opening of an ionic channel. In the visual transduction process of vertebrate rod visual cells, the molecular changes of the retinal molecule in rhodopsin, caused by the absorption of light, initiate, through a series of intramolecular reactions, changes in the protein, which, e.g. enable rhodopsin to activate the GTP-binding protein. Finally, in bacteriorhodopsin, the proton pump of *Halobacterium halobium*, light energy is first stored in the form of electronic energy of the excited state of the chromophore, *all-trans* retinal. It is transformed by the presence of the protein into an electrochemical potential, driving one to two protons per absorbed photon against its own gradient. In all cases described, the molecular changes in the protein, eventually responsible for the biological function, are difficult to detect by conventional methods. Structural methods are either not fast or not sensitive enough, while most spectroscopic methods do not respond to the molecular changes under consideration.

It is reasonable to assume that these molecular changes also cause small spectral changes in the mid-infrared. Due to the overlap of many absorption bands, a normal static infrared spectrum of such a complex system would not reveal the particular functional groups. However, if the apparatus is fast and sensitive enough to detect small spectral changes, the resulting infrared difference spectrum should unravel the functional groups and their molecular modifications during their function. Kinetic infrared spectroscopy should therefore yield valuable information on the mechanism of trigger processes. It has recently been employed successfully for studying the reactions of rhodopsin (Siebert and Mäntele, 1980), bacteriorhodopsin (Siebert *et al.*, 1981), and of the binding process of carbon monoxide to myoglobin (Siebert *et al.*, 1980).

2. KIS Applied to Bacteriorhodopsin
(a) Some general remarks on bacteriorhodopsin

Although considerable progress has recently been made in structural investigations of bacteriorhodopsin, the proton pumping mechanism is still not understood well enough. There are many models, some of which are based on general membrane transport schemes. To elucidate the situation further, more information is needed on the molecular events occurring in bacteriorhodopsin after the absorption of the photon. Spectroscopy methods, especially resonance Raman spectroscopy, have provided information on molecular processes in the chromophore. *All-trans* retinal is bound to the ε-amino-group of lysine 216 by a Schiff base linkage (Bayley *et al.*, 1981; Lemke and Oesterhelt, 1981; Mullen *et al.*, 1981). Resonance Raman spectroscopy (Terner *et al.*, 1979; Aton *et al.*, 1977, Marcus and Lewis, 1978; Stockburger *et al.*, 1979) and our method of kinetic infrared spectroscopy (Siebert *et al.*, 1981) have supplied evidence that this Schiff base is protonated. After light absorption bacteriorhodopsin undergoes a cyclic reaction involving at least the intermediates K610, L550 and M412 (the numbers refer to the absorption maxima of the respective intermediates). These intermediates and their reaction kinetics have been studied by conventional flash photolysis experiments (for a review see Stoeckenius *et al.*, 1979) and by kinetic resonance Raman spectroscopy (Terner *et al.*, 1979; Stockburger *et al.*, 1979). In these experiments it has been shown that the Schiff base in M412 is deprotonated. Chemical (Mowery and Stoeckenius, 1981; Tsuda *et al.*, 1980) and spectroscopic (Braiman and Mathies, 1980) investigations indicate that the chromophore isomerizes during the photocycle from *all-trans* to 13-*cis*. Most models assume that the Schiff base and its protonation state play a major role in the proton pumping mechanism (Schulten and Tavan, 1978; Honig *et al.*, 1979; Warshel, 1979). However, a second pathway is needed, since it has been shown recently that up to two protons can be pumped per absorbed photon (Govindjee *et al.*, 1980; Renard and Delmelle, 1980). In this context it has been suggested that the deprotonation of a tyrosine might play a role for the pumping of the second proton (Bogomolni *et al.*, 1978; Hess and Kuschmitz, 1979). Concerning the mechanism inducing the deprotonation, however, only speculations can be made as yet.

(b) KIS measurements

(i) *The chromophore* It appears appropriate to give a few general considerations as to what spectral changes can be expected in the infrared during the photocycle. IR-investigations with retinylidene Schiff base model compounds have shown (Siebert and Mäntele, 1980) that the protonated Schiff base exhibits strong absorption bands which can be assigned to the C=N vibration (*c.* 1650 cm^{-1}), to the C=C vibration (*c.* 1540 cm^{-1}), and to C–C vibrations coupled to C–H bending vibrations (region between 1300 cm^{-1} and 1150 cm^{-1}). These bands have absorption strengths comparable to or even stronger than that of the C=O vibration of carbonyl groups. The intensity of the bands is reduced by a factor of five to ten upon deprotonation. It has been further demonstrated that the position of the C–C bands is shifted if the absorption maximum in the visible is changed. The IR difference spectra between the various intermediates should be dominated by the spectral changes of the chromophore. The strongest spectral changes caused by the protein could be due to carbonyl-, carboxyl-, OH- and aromatic groups. To deduce the molecular changes in the protein from the difference spectra they have to be measured with high accuracy and criteria have to be developed, which allow the discrimination between chromophore and protein molecular changes. As an example the bacteriorhodopsin-M412 difference spectrum will be discussed in greater detail. In M412 the Schiff base is deprotonated. From the foregoing it can be concluded that the difference spectrum should exhibit mostly transmission increases.

Figure 2 shows the bacteriorhodopsin-M412 difference spectrum. It has been deduced from the kinetic signals, one of which is shown in the insert of Fig. 2. This signal has been measured with the chopped infrared beam. To resolve the fast phase of the signal it has to be measured without chopping.

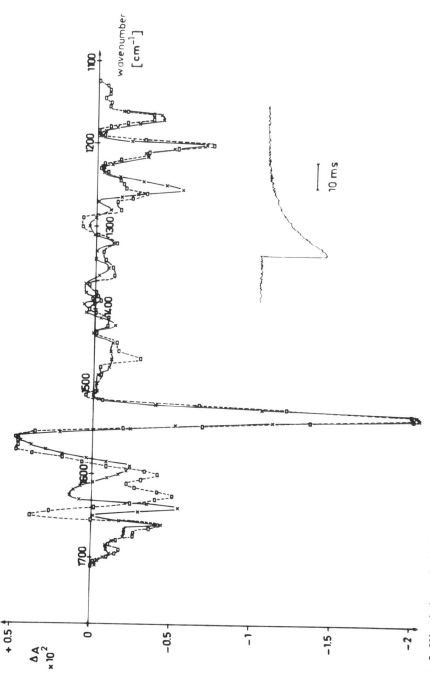

FIG. 2. Kinetic bacteriorhodopsin-M412 difference spectrum. For each data point 100 signals were averaged. Spectral resolution varied between 6 and 4 cm⁻¹, depending on the intensity at the detector. Dashed line: deuterated sample. Insert shows averaged signal at 1525 cm⁻¹ and 20°C.

The rise time of M412 from L550 is approximately 60 μs at room temperature. This is more than a factor of ten slower than the rise time of L550 from K610. M412 decays back to bacteriorhodopsin with a half time of 4 ms. The difference spectrum can therefore be deduced from the kinetic signals by taking their amplitudes at 300 μs. As expected, the spectrum exhibits mostly transmission increases. The strongest band, at 1525 cm⁻¹, can unequivocally be assigned to the C=C vibration of the retinal in bacteriorhodopsin. The band at 1200 cm⁻¹ with a shoulder at 1220 cm⁻¹ is due to the skeleton vibrations of the retinal. Both features are in agreement with resonance Raman data. To identify the C=N vibration of the protonated Schiff base measurements have been performed in D_2O. They are shown in Fig. 2 by the dashed line. It is evident that there are pronounced changes in the region between 1650 cm⁻¹ and 1600 cm⁻¹, but they are more complex than one would expect if only the protonated Schiff base vibration was shifted to lower wavenumbers. The band at 1640 cm⁻¹ for normal bacteriorhodopsin and the band at 1625 cm⁻¹ for bacteriorhodopsin in D_2O would be in agreement with resonance Raman measurements; the band at 1600 cm⁻¹, induced by D_2O, would be difficult to explain on the basis of a normal protonated Schiff base. To clarify this point we made measurements with retinal deuterated at the C15 position. The results (Fig. 3a) show that only the band at 1640 cm⁻¹, respectively 1625 cm⁻¹ in D_2O, is due to the C=N vibration of the protonated Schiff base. Since the only exchangeable proton in connection with the chromophore is the proton of the Schiff base the band at 1600 cm⁻¹ represents molecular changes of the protein (see below). Another spectral change induced by D_2O is observed at 1250 cm⁻¹. Again, to decide whether this band is caused by the chromophore or by the protein, measurements with C15-d-retinal were performed. The results are presented in Fig. 3b. It is evident that this band is also influenced by C15-d-retinal and, thus represents a molecular change of the chromophore. Since the spectral change induced by C15-d-retinal is larger than that induced by D_2O, the most likely interpretation is that the band represents the in-plane bending vibration of the hydrogen at the C15 position, which is coupled to the bending vibration of the proton at the Schiff base nitrogen. Upon deprotonation this coupling is removed, thus causing the disappearance of the band in M412.

(ii) *The protein* The results discussed so far have shown that isotopic labelling of the chromophore is an important tool to identify chromophore bands and to distinguish them from protein bands. The comparison of kinetic infrared with kinetic visible absorption change measurements provides an additional handle for this discrimination. The kinetics of chromophore molecular changes are reliably measured by flash photolysis in the visible spectral range. Spectral changes in the infrared caused by chromophore modifications must exhibit the same kinetics. If, however, clear deviations in

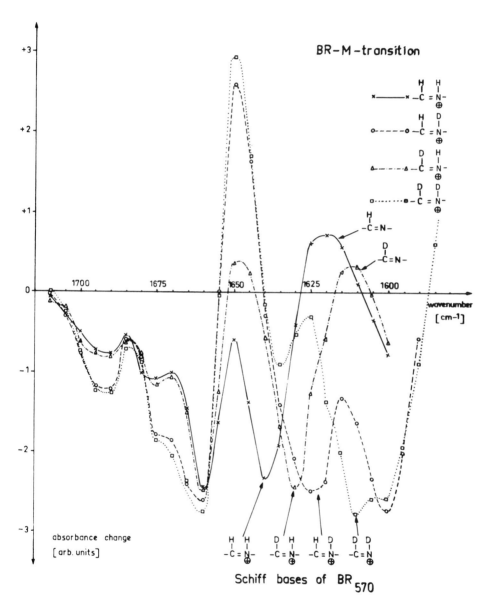

FIG. 3a. Kinetic bacteriorhodopsin-M412 difference spectrum in the region of the C=N-vibration of the Schiff base, for normal bacteriorhodopsin, bacteriorhodopsin in D_2O, bacteriorhodopsin reconstituted with C15-d-retinal, and bacteriorhodopsin reconstituted with C15-d-retinal in D_2O.

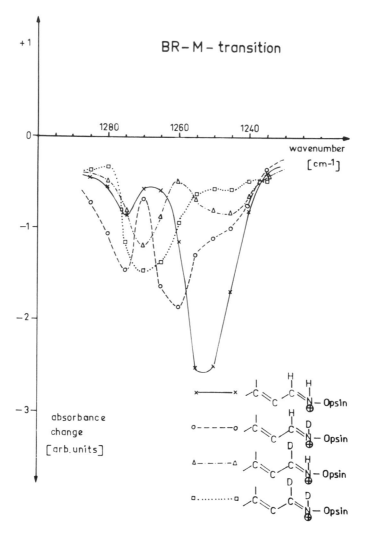

FIG. 3b. As Fig. 3a, spectral region around 1250 cm^{-1}.

the kinetics from the chromophore kinetics are observed, the corresponding spectral changes must be due to alterations in the protein. A typical signal with non-chromophore kinetics measured at 1755 cm^{-1} is shown in Fig. 4. Similar slow components (according to their slow risetime) are observed at many positions in the spectrum. It is not always easy to establish clearly the presence of the slow component, because several kinetics often overlap, especially if those belonging to the O640 intermediate are present and also

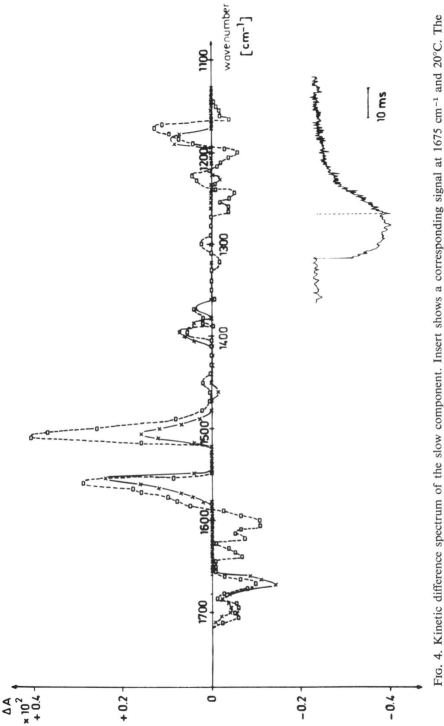

FIG. 4. Kinetic difference spectrum of the slow component. Insert shows a corresponding signal at 1675 cm^{-1} and 20°C. The time scale is expanded by a factor of 4 up to the dashed line. The difference spectrum was constructed by taking the difference between the M412 amplitude and the maximum amplitude.

show a slow risetime (Mäntele *et al.*, 1981). Nevertheless, a tentative difference spectrum is shown in Fig. 4. Herein the features around 1550 cm^{-1} and 1180 cm^{-1} must be attributed to the O640 intermediate. Deviations from chromophore kinetics are also detected at a shorter time scale. In Fig. 2 the transmission decrease at 1650 cm^{-1} observed for bacteriorhodopsin in D$_2$O may serve as an example. The risetime of the signal is between that of L550 and M412.

The interpretation of the spectral changes produced by protein molecular transformations is not easy. Apart from a few exceptions to be discussed later, unequivocal identification is not possible without additional experiments. Important tools in this connection are chemical labelling of specific amino acid side chains, modifications of the protein by proteases, and isotopic labelling of specific amino acids. Since *Halobacterium halobium* can be grown on a synthetic medium the labelled amino acids can be incorporated in the medium. This technique has already been successfully applied in resonance Raman investigations on the retinal binding site (Argade *et al.*, 1981) as well as in neutron scattering experiments (Engelman and Zaccai, 1980). Experiments with this technique in combination with kinetic infrared spectroscopy are under way.

The comparison of band positions in the infrared difference spectrum with those of the spectrum from model compounds may help to assign the bands to either protein or chromophore molecular changes. A clear discrimination, however, is only possible in exceptional cases. The band structure between 1700 cm^{-1} and 1650 cm^{-1} in the difference spectrum (Fig. 2) may serve as an example. Model compounds indicate that the C=N vibration of the Schiff base causes the band with the highest wavenumber in the spectral region up to 2000 cm^{-1}. The structure in the difference spectrum, however, is not influenced by D$_2$O or C15-d-retinal, excluding the involvement of the C=N vibration. Figure 4 also shows that slow kinetics, i.e. non-chromophoric molecular changes are present in this spectral region. From the band positions one could assume that the bands can be assigned to the deprotonation of carboxylic groups in the hydrogen-bonded form. Since D$_2$O does not influence the bands this assignment must be excluded. They are, however, influenced by high salt concentrations and moreover are sensitive to pH-changes (data not shown). Therefore, it is reasonable to assume that the groups responsible for these spectral changes are located near the surface of the purple membrane.

An example as to where the interpretation of spectral changes caused by the protein is already possible is shown in Fig. 5. The band is shifted by 10 cm^{-1} to lower frequencies in measurements in D$_2$O. From the position and from the D$_2$O-effect it is evident that the spectrum represents the protonation followed by the deuteration of carboxylic groups. The asymmetric

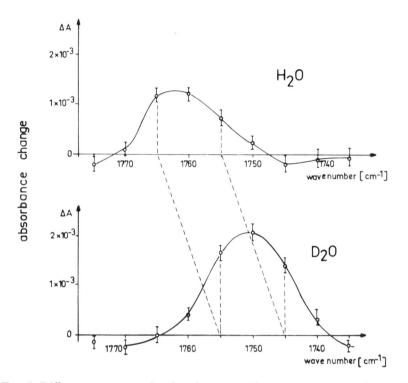

FIG. 5. Difference spectrum showing the protonation, respectively the deuteration of the two carboxylic groups.

form of the band especially in H_2O suggests that it is due to two different carboxylic groups. This is confirmed by the different kinetics observed at the two sides of the asymmetric band: the shoulder exhibits the slow kinetic, typical for protein molecular changes while the main peak at higher wavenumbers shows the same kinetics as the M412 intermediate. A detailed interpretation of the spectrum has been published recently (Siebert et al., 1982). A model for the molecular events involving the carboxylic group with the M412 kinetics is shown in Fig. 6. Since in M412 the Schiff base is deprotonated, the most simple assumption is that the carboxylic group is the direct acceptor for the proton from the Schiff base. The deprotonation of the group occurs along with the reformation of bacteriorhodopsin from M412. Since the protons ejected during the pumping process appear considerably earlier, the proton from the carboxylic group cannot be the one ejected directly. Rather, it has to be assumed that an additional group is deprotonated upon protonation of the carboxylic group and protonated by the deprotonation of the carboxylic group. The slow component is more difficult to interpret. It

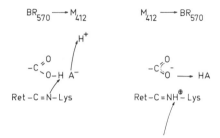

FIG. 6. Simple model for the proton migration around the "fast" carboxylic group. "A" represents the additional proton acceptor.

does not rise with a measurable delay. Therefore, M412 cannot be its precursor. Though it may play a role in the pumping of the second proton (Govindjee et al., 1980; Renard and Delmelle, 1980). As already mentioned, certain arguments indicate that the deprotonation of a tyrosine occurring prior to the formation of M412 (Bogomolni et al., 1978; Hess and Kuschmitz, 1979) is involved in the proton pumping mechanism. This might be the trigger for the protonation of the "slow" carboxylic group.

(iii) *Reaction time course* As a summary of the whole time, amplitudes, and wavenumber dependence a three-dimensional computer representation of the three parameters is given in Fig. 7. From such a diagram 'snapshots'

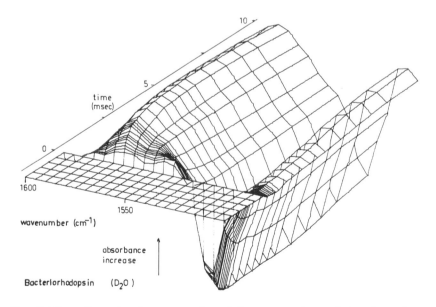

FIG. 7. Three-dimensional representation of the wavenumber and time dependence of the spectral changes between 1600 cm^{-1} and 1500 cm^{-1}.

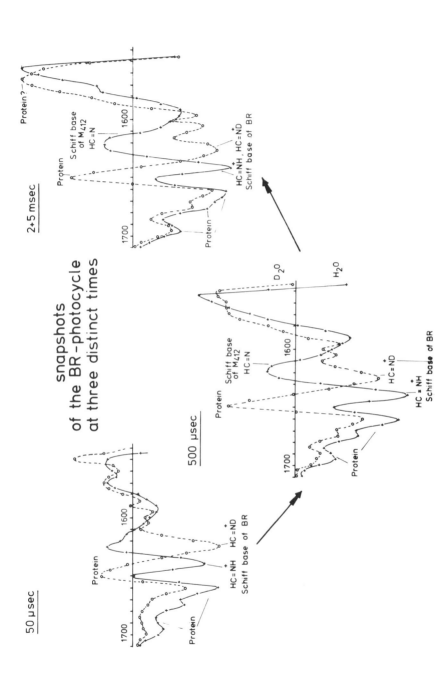

FIG. 8. Three "snapshots" taken at 50 μs, 500 μs and 2 ms, between 1700 cm^{-1} and 1550 cm^{-1}. Dashed line: measurement in D$_2$O, for which the last snapshot was taken at 5 ms.

can be taken which represent the amplitudes for the whole wavelength range at any selected time within the photo-cycle. Three of these snapshots are shown in Fig. 8.

(iv) *Concluding remarks* We have shown that with kinetic infrared spectroscopy essential information on the molecular events occurring in complex systems can be gained. The example discussed (bacteriorhodopsin) shows that information on the chromophore as well as on the protein is obtained. The method not only yields spectral information but also provides kinetic results. An important finding of the investigations is that often molecular events in the protein, exhibiting their own kinetics, decouple from processes occurring in the chromophore. The molecular changes induced by light in the chromophore act as the trigger for the processes in the protein.

D. RELATED METHODS AND PERSPECTIVES

A more general method of kinetic infrared spectroscopy would be an apparatus equipped with an additional temperature jump device. A pulse of 10 mJ from a CO_2 laser could increase the temperature of hydrous IR samples by as much as 5°C, provided that the sample only has a small heat capacity. If the sample is thermally insulated, the temperature jump could last several tens of milliseconds. This method would be especially suitable for the study of ligand binding to enzymes and their effects on the enzyme, thus serving as a tool to study trigger processes mediated by transmitter molecules.

In the past several other methods of kinetic infrared spectroscopy have been developed (Pimental, 1968) including stopped-flow equipment (Brady *et al.*, 1976). However, all of these methods are limited to investigations where large spectral changes have to be measured.

If a light-triggered reaction can be stopped by lowering temperature, static difference spectroscopy can also be employed. Several prerequisites have to be met to approach the problems under consideration. The samples must be very stable during the time course of the measurements, before and after the light-induced reaction. The spectrophotometer must be sensitive enough for the detection of the small changes. The first condition can be met by using hydrated films at low temperatures, the second fulfilled by applying Fourier transform infrared spectroscopy. Its multiplex advantage and high energy throughout shortens the measuring time considerably. Further, since it takes only a fraction of a second to record a spectrum and several hundreds of spectra can be added to attain the required accuracy, slow instabilities of the instrument as well as of the sample are averaged. First experiments have successfully been performed with bacteriorhodopsin.

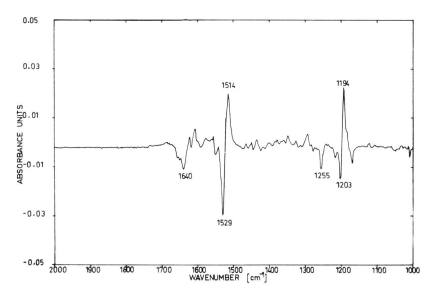

FIG. 9. Bacteriorhodopsin-K610 difference spectrum at 70 K. Measurement was performed with a Bruker Fourier transform infrared spectrophotometer, model IFS 113.

Figure 9 gives the bacteriorhodopsin–K610 difference spectrum obtained at 70 K. We have shown that the negative of this spectrum is obtained after driving back the reaction by light. The interpretation of the spectrum is beyond the scope of this article; it only serves to show that also static infrared difference spectroscopy can be employed for the study of trigger reactions in cases where the reactions can be stopped at low temperatures.

III. The KLS Method

A. THEORETICAL BACKGROUND

An interpretation of the scattering changes can be given on the basis of the classical Rayleigh-Gans-approximation. After correction of the polarization factor, the scattering intensity can be written, within the frame of this theory, by a product of two terms (for example see van de Hulst, 1957):

$$I\,(\theta) = I_\alpha \cdot I_{\vec{r}}\,(\theta) \tag{1}$$

The first term depends on the polarizability α of the particle. It is the sum of the scattering by the individual Rayleigh dipoles contained in the particle and independent on the scattering angle θ.

The second term describes the interference of the scattered light and contains the geometry, dimensions, and orientation (symbolized by $\vec{r}$) of the scattering mass within the particle.

According to these terms, one can distinguish between two principally different light scattering transformations. One of them is due to changes of the polarizability caused by changes of the individual Rayleigh dipoles. The other one, belonging to the second term, includes particle swelling, shrinkage, reorientation or any other shift of the scattering mass or parts thereof. The two scattering contributions are rather typical and can easily be distinguished from each other (Fig. 10). A convenient feature is the behaviour at small angles, where a change of shape points to zero, whereas the polarizability change remains constant. Generally, one would expect that both polarizability and shape effects contribute to real observed signals. However, as kinetic parameters are available, the separation of the effects is possible in most cases.

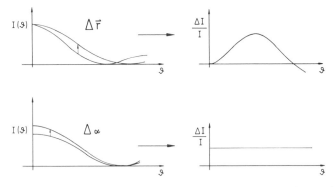

FIG. 10. Schematic difference scattering curves. The angular dependence of a given light scattering change is understood as its difference scattering curve (in analogy to the difference spectrum derived from the spectral dependence of an absorption change). The figure shows the two limiting cases for such curves as derived from the Rayleigh–Gans theory. *Above*: the interference of the light scattered on the different points of the particle is changed. *Below*: the individual scattering dipoles within the particle are changed in their scattering power. A comparison of calculated and measured difference scattering curves can help to identify the physical meaning of a signal.

B. THE KLS APPARATUS

The apparatus depicted in Fig. 11 represents a combined device for the simultaneous measurement of light scattering changes and of absorption changes. Three measuring beams, one for light scattering and two for light absorption, pass through the same sample. Both the absorption and scattering transients are evoked by the same flash, thus enabling simultaneous measurement. In systems, where, besides scattering changes absorption changes also occur, pure scattering signals can be found at spectral regions where no appreciable absorption occurs. However, absorption changes are generally superimposed by light scattering signals. The light scattering contribution can be eliminated by the dual wavelength technique, i.e. by simultaneous measurement of the absorption changes at two neighbouring wavelengths, which are very different in their absorption, e.g. at the isosbestic point and the neighbouring absorption maximum. By subtraction of the two absorption signals from each other (performed in differential detectors, see below), the scattering contribution, almost identical at the two wavelengths, is largely reduced.

So far, the arrangements in Fig. 11b were not incorporated in the combined device (Fig. 11a) but built up separately. This takes into account the fact that the absorption measurement requires sample concentrations which are too high to yield clear angular dependencies (interparticle interference, multiple scattering).

Optical compensation considerably reduces the noise caused by fluctuations of the measuring intensity because they are added to the intensity changes in the sample only as relative but not as absolute values. Zero adjustment prior to the measurement is made by automatically positioned neutral wedge filters WF (Fig. 11a). This compensation principle is also used in the devices shown in Fig. 11b.

In general, semiconductor (silicon) detectors (e.g. EG and G, UV 100 A) are used. Besides the well-known advantages (no memory and no saturation), they allow differential application (Hofmann and Emeis, 1979). Thereby, the amplifier has only to accept a small difference current between antiparallel photovoltaic detectors. The high static measuring current flows in the circuit between the detectors. The noise and signal transfer properties of this arrangement are discussed in more detail in the above cited study.

In the scattering beams, the sources are of the silicon type (Siemens LD 271, Hitachi HLP 60), matching the sensitivity curve of the detectors. The disadvantage of the scattering measurement in the far red (low scattering intensity) is thereby partly compensated. The main advantage of the semiconductor sources lies in their high luminous density; all optics are corrected Fresnel lenses (Fl and L), with apertures as high as up to 1 (Melles Griot, Arnhem, Holland). In the absorption beams, the remaining dual-wavelength scattering changes are thereby reduced, because much of the scattered light (which compensates the turbidity change in the primary beam) is collected.

In the scattering measuring beam, the scattered light of the sample is focused onto the detector D_{s1} via the large Fresnel lenses FL, the angular range being selected by diaphragms RD. Centrosymmetric ring diaphragms are used for isotropic samples, as shown in Fig. 11b1. For axially oriented samples, the measurement must select between different azimuth angles. To orientate the samples, a 12 kG magnetic field is used in the case of rod outer segments (see the measuring example below). For the orientation sensitive measurement, sector diaphragms as shown in Fig. 11b2 or vertical and horizontal detector arrays can be used. For the detection of the scattered light at higher angles (up to 60°), the self-focusing spherical arrangement in Fig. 11b3 is used. The detector arrays on the sphere can be arbitrarily switched together. In the current application, scattering changes at four angles (ring-like detector arrays on the sphere) are recorded simultaneously in four channels of the storage. At a given intensity, the source can be small enough to allow sufficiently small apertures of the incident light (1 : 30 in the present

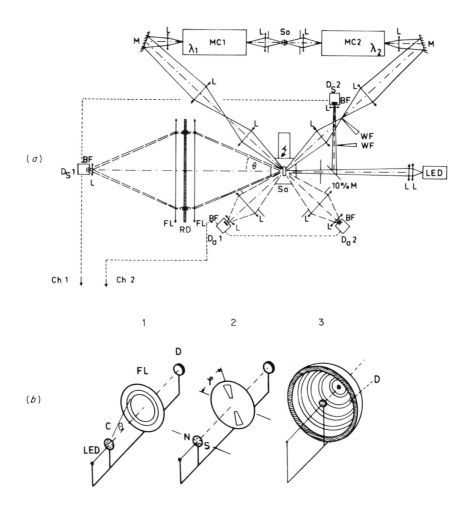

arrangement). The effective noise level is determined rather by mechanical and thermal fluctuations in the measuring beams than by the electronic detector noise. Careful mechanical insulation is essential.

To record and store the measurements, we use the signal analyzer TN 1500 (Tracor Northern). Signals from absorption and scattering, from horizontal and vertical scattering, or from several different scattering angles can be recorded simultaneously and compared to each other on continuously expanded scales.

Flat cuvettes from 0.1 to 10 mm thickness are used; good angular dependencies in the orientation device (Fig. 11b2) can only be measured with a 20 mm wide 0.1 mm thick cuvette where no reflection at the side windows occurs. A spherical cuvette (Fig. 11b3) has the advantage of self-focusing; however, the device cannot easily be placed into a magnetic field.

As to the practical performance of the measurement, the problem of different ideal sample concentrations for scattering and absorption measurements has to be considered. Absorption changes are best measured at an apparent transmittance of about 10%. For rod outer segment suspensions,

---

FIG. 11. Optics for kinetic light-scattering and -absorption measurements. (A) Optics for simultaneous measurement of pure light-absorption and -scattering changes. Symbols are: So: light-source for the absorption beams (Halogenium or Xenon high pressure); MC1: monochromator (Jobin Yvon H 20 UV); MC2: monochromator (Jobin Yvon H20 IR); L: lenses; M: mirrors; Sa: sample;   : flash (EG & G, FX 133); WF: wedge filters; D_{a1} and D_{a2}: differential detector for the absorption measurement; LED: semiconductor light source for 950 ± 20 nm (Siemens LD 271); 10% M: mirror reflecting 10% of the incident intensity; FL: Fresnel lenses (Melles-Griot, Arnhem, Holland); RD: ring diaphragms; D_{s1} and D_{s2}: differential detector for the scattering measurement; Ch1/Ch2: Output, separate storage and data manipulation; BF: Blocking filters, for example: flash: Interference broad band $450 < \lambda < 550$ nm (Hugo Anders KG, Diendorf), BG 18, OG 530 (Schott, Mainz); Scattering beam: RG 850 (Schott, Mainz); $\lambda_1 = 387$ nm: UG 5, KG 3 (Schott, Mainz); $\lambda_2 = 417$ nm: 03 SWP 011, 03 SWP 013, 03 SWP 017 (Melles-Griot, Arnhem, Holland), KG 3 (Schott, Mainz). (B) Geometry of light scattering detection. a. Isotropic sample. The parallel incident light from the LED source is scattered by the sample within the cuvette C; the scattered light is focused onto the detector D by two large Fresnel lenses FL. Selection of the scattering angle θ by ring diaphragms, centrosymmetric to the optical axis. b. Axially oriented rod outer segments. Selection of the scattering angle and an additional selection of the angle ρ relative to the axis of the outer segments oriented in the magnetic field N–S; a wide range of θ is transmitted in the above schematic picture, sharper selections are also used in some experiments. c. Isotropic samples, simultaneous measurements of several scattering angles. The cuvette is half-spherical, all parallel rays of scattered light, equal in θ, are focused on the same detector array, centrosymmetric to the optical axis and located on the larger half-sphere. The signals from several scattering angles are simultaneously measured.

which we use as an example of application in the following, for the absorption beam optics in Fig. 11a and for measuring wavelengths of c. 400 nm, an average rhodopsin concentration of 2×10^{-6} M results. For this concentration, however, considerable multiple scattering and therefore highly distorted difference scattering curves are obtained. An interpretation of scattering curves is only possible for an apparent absorption < 0.2. This means a rhodopsin concentration of $2–3 \times 10^{-7}$ M, depending on the size of the particles in the suspension (e.g. whole rod outer segments, fragments, isolated discs). A simultaneous absorption and scattering experiment, therefore, cannot be interpreted with respect to the angular dependence of the scattering signals. In practice, the difference scattering curves are best measured with oriented particles at low concentrations. The exact kinetics of the scattering signals are then measured at higher concentrations, simultaneously with the absorption signals.

C. KLS APPLIED TO VISUAL TRANSDUCTION

1. Rod Outer Segment System

In photoreceptors, trigger processes are found on two levels of organization: firstly, the light quantum triggers photochemical pigment reactions, and secondly, at least one of these reactions must trigger the ensuing cellular events. The KLS-method was especially developed for research on the second trigger step. It is probably more generally applicable but so far it was mainly tested on isolated vertebrate rod organelles. To introduce the subject, the morphology and function of such systems will be briefly outlined.

The rod outer segment consists of a regular stack of biological membranes. The membranes close pairwise to form a very flat vesicle, the so-called disc; the aqueous lumen between the membranes is only about 3.0 nm (30 Å) thick (Funk et al., 1981). This means that every membrane is in the contact zone of another, i.e. within the range of the Gouy-Chapman potential and the van der Waals forces. Between the discs, the distance amounts to about 15 nm (150 Å) and stabilizing material is present in this cytoplasmic space.

One half of the disc membrane consists of the visual pigment rhodopsin and about 10% of other functional proteins (Daemen, 1973; Siebert et al., 1977; Godchaux and Zimmermann, 1979). The remainder of the mass are lipids in a composition which is very accurately controlled by the organism (Dudley et al., 1975). The disc membrane is highly fluid; as is well-known, rhodopsin and presumably also the other proteins are free to rotate and to laterally diffuse within the membrane (Cone, 1972; Liebman and Entine, 1974; Poo and Cone, 1973). While rhodopsin is a so-called core protein (Dewrey et al., 1969; Funk et al., 1981), the other functional proteins are peripherally bound to the cytoplasmic surface of the disc membrane (Kühn,

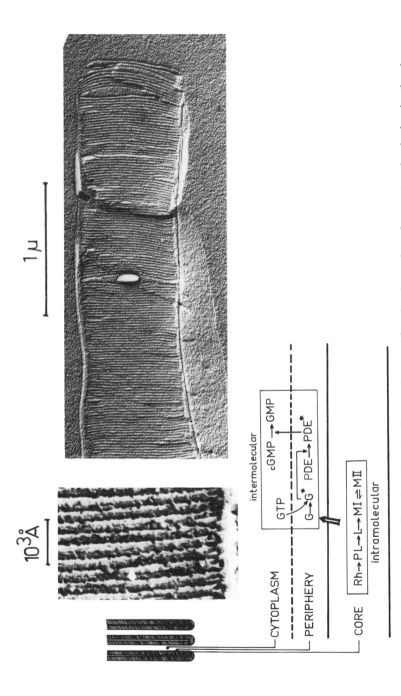

Fig. 12. Simplified scheme of reactions in and at the disc membrane. After the absorption of a quantum by rhodopsin, the chromoprotein runs through a cascade of intramolecular reactions in the time scale of ps (rhodopsin → prelumirhodopsin) to ms (metarhodopsin I → metarhodopsin II). Many G-molecules can be activated by one rhodopsin and many phosphodiesterases (PDE's) by G*. The arrow indicates the trigger problem: which of the photoproducts induces the enzymatic reactions? Freeze fracture electron micrographs by W. Welte.

1980). The transduction of the signal starts with intermolecular reactions of the rhodopsin molecule, similar to the bacteriorhodopsin cycle (see above). Prelumi-, lumi-, metarhodopsin I and metarhodopsin II follow each other in the ps to ms time scale (see, e.g. Abrahamson, 1972).

At least one of these rhodopsin products must act as the trigger of transduction on the cellular level, but it cannot be excluded that more than one species is functionally active. The metarhodopsin transition is known to be sensitive to the molecular surroundings of the rhodopsin molecule. It is the first which only takes place in the presence of water (Kühne, 1878) and which is markedly influenced by detergents (Applebury et al., 1974), pH (Emrich, 1971) and outer pressure (Attwood and Gutfreund, 1980). Furthermore, it is the last which is sufficiently fast to be involved in visual transduction. Special attention has been paid to the equilibrium between metarhodopsin I and metarhodopsin II. This equilibrium depends on pH and temperature (Matthews et al., 1963) as well as the outer pressure (Lamola et al., 1974) and other parameters, such as the membrane surface potential (Schnetkamp et al., 1981).

All these intramolecular rhodopsin reactions are presumably more or less restricted to the core of the membrane. At the membrane periphery, membrane-bound or membrane-associated biochemical reactions take place (Fig. 12). The so-called GTP-binding protein is activated by rhodopsin and itself activates the phosphodiesterase which then is able to change the cytoplasmic concentration of cyclic GMP (see, e.g. Fung et al., 1981). cGMP is currently discussed as a cytoplasmic transmitter. The main question of interest in the study discussed below is: which rhodopsin species acts as trigger of the enzymatic cascade (arrow in Fig. 12)?

Optical probes are available for the intramolecular as well as for the enzymatic reactions. The intramolecular reactions of rhodopsin can be studied by measuring the accompanying absorption changes; this is a well-known application of the flash-photometric technique. The species metarhodopsin II absorbs maximally at 380 nm, the isosbestic point with metarhodopsin I is at 417 nm. The application of light scattering to monitoring the enzymatic reactions at the disc membrane has been demonstrated by Kühn et al. (1981). They obtained a stoichiometric relation between the GTP-binding protein and light-induced near infrared transmission changes. Light scattering measurements are best made in this system at 800–900 nm where no absorption occurs in the rhodopsin reaction cascade.

2. KLS Measurements on the ROS System

The aim of the measurements treated in the following is first to give evidence on the trigger process by kinetic analysis and second to obtain a physical interpretation of the scattering changes in ordinarily stacked disc membranes in order to come closer to the physics of the trigger mechanism.

(a) *Unoriented ROS-preparations* For such measurements, a good compromise between signal, noise and stability was achieved at a concentration of $c_{Rhod} = 3 \times 10^{-6}$ M and a scattering angular range of 2–6°. Under these conditions, the intensity change from the primary beam (transferred by multiple scattering) predominates.

Figure 13a shows recordings for different flash intensities of the so-called scattering signal P (Hofmann *et al.*, 1976; Uhl *et al.*, 1977). It is a positive change of the scattered intensity at all angles. The signal displays a typical delay (sigmoidal time course) and is not proportional to the flash-induced rhodopsin turnover but saturates at a flash intensity corresponding to about 10% rhodopsin bleaching (Fig. 13b). The effect recovers spontaneously within a quarter of an hour (Schnetkamp and Hofmann, unpublished). Computer fits of the saturation function with different model functions were performed. The best fit was obtained by using the exponential function

$$P(\rho) = P_0(1 - e^{-\beta \alpha}) \tag{2}$$

where ρ is the fraction of rhodopsin transformed by the flash and P_0 the maximal light-induced signal amplitude. β was found $23 \leq \beta \leq 30$ in all measurements, this means a half-saturation at about 3.5%.

We calculated a model for the time course where a rhodopsin state (arising in ms and causing the observed delay) induces two consecutive events which are both observed in light scattering. Transition times of 5, 12 and 50–150 ms (20°C, pH 6) were obtained for the three assumed concatenated first order processes (Reichert and Hofmann, 1981).

As shown by the solid line drawn into the measured signals (Fig. 13a), the fit is fairly good. The following is to show that this fit is not only of formal significance, but that the above assumed three processes must be physically real.

For the time constants k_2 and k_3 this was achieved by axial orientation of the rod outer segments in a magnetic field (Hofmann *et al.*, 1981).

(b) *Oriented ROS-preparations* The P-signal of oriented samples splits into two kinetically different signals (Fig. 14a). One of them is observed in the axial scattering of the disc stacks, the other in the radial scattering. Both signals are positive changes of the scattering intensity. The kinetics are comparable to those which were to be introduced for the consecutive processes of the P-signal from the isotropic sample. To interpret the signals, the angular dependence was measured. The resulting difference scattering curve is shown for the fast axial signal in Fig. 14b, left side. The measuring points are obtained from four different preparations, all of them being from the first flash on a fresh aliquot.

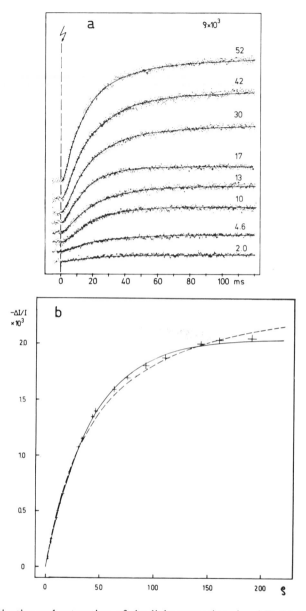

FIG. 13. Kinetics and saturation of the light scattering signal P. (a) P-signals for various flash intensities. Parameter of the signals is the relative rhodopsin turnover induced by the flash applied at $t = 0$. Measuring parameters: $c_{rhod} = 3 \times 10^{-6}$ M, scattering angle; $2° \leq \theta \leq 6°$, pH = 6.0, T = 22°C. It is seen that the signals become apparently faster with increasing turnover. (b) Amplitude of the P-signal as a function of the relative rhodopsin turnover ρ induced by the flash. Every measuring point corresponds to a new sample from the same preparation. The fit for exponential (solid line) and hyperbolic functions (dotted line) are shown.

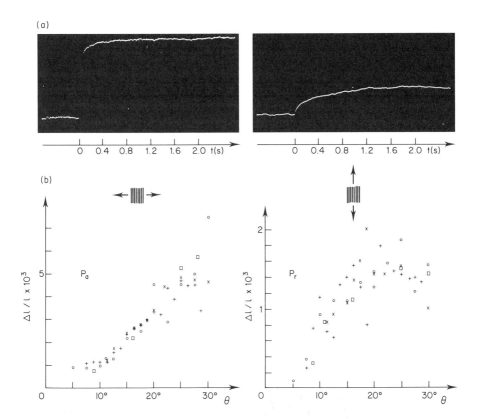

FIG. 14. Original recordings and angular dependence of the axial and radial scattering signals P_a and P_r. (a) Fast signal P of a suspension of axially oriented ROS; simultaneous measurement in vertical (P_r) and horizontal (P_a) direction. Horizontal orientation of the ROS axes; signals are the normalized intensity changes $\Delta I/I$; transmitted angular range $15° \leq \theta \leq 25°$. It is seen that the signal P (see Fig. 13a) splits into two kinetically different signals. (b) Experimental difference scattering curves—angular dependence of the signals P_a and P_r. The different symbols are obviously not constant and points to zero for $\theta \rightarrow 0$; for the P_r-curve, the angular dependence is not so clear since only a few reliable measuring points for $\theta < 10°$ are available.

The difference scattering curve is of the type expected for a shift of the scattering mass and appears not to contain an essential contribution of a polarizability change (no intercept on the ordinate). An axial shrinkage in coherent disc packets of a mean length of 2–3 μm fits well to the data (Hofmann *et al.*, 1981).

The radial signal is more difficult to measure and the values scatter considerably (Fig. 14b, right side). However, the isotropic curve, which is essentially the sum of the curves in Fig. 14b, points to zero (Hofmann *et al.*,

1981). We therefore concluded that light-activated rhodopsin induces a fast axial and a slower radial shrinkage of the disc stack. It should be mentioned that it cannot be decided whether the whole particle is contracted or if only parts of the scattering mass contribute to the effects. The reaction times in any case fit into those obtained by calculation from the isotropic measurement.

Recent experiments show that osmotic disc compression mimics pre-illumination with respect to the axial shrinkage. This behaviour suggests that this effect is due to a pairwise attraction of the adjacent membranes of the single discs, summing up to the observed stack effect. The mechanism of the radial effect is still unclear.

Light-induced changes of the polarizability are also found in rod outer segments. However, they are much slower, in the order of seconds (Hofmann et al., 1981).

3. Isolated Discs

On isolated disc vesicles, obtained by osmotic dissolution of the disc stack (Smith et al., 1975), the fast signals are not observed. A slower polarizability change, the so-called signal P_D, is found in this system. It will be discussed below (see Fig. 18). Obviously, the intact disc stack is a prerequisite for the fast signals.

4. Simultaneous Light Scattering and Absorption Change Measurements

In order to identify the inducing process manifested in the delay of the scattering signal P (Fig. 13a), simultaneous measurements of the meta-rhodopsin transition and of the near infrared scattering in the low bleaching range were performed.

Scattering and absorption signals are shown for a temperature of 7°C, for different pH values and for different stages of bleaching, in Fig. 15. Under the conditions of these measurements, the three kinetic components of the time course of the signal P (see above) degenerate to only one rate limiting reaction.

It turns out that, under all pH conditions, the scattering and absorption signals from the first flash are kinetically congruent. It may therefore be concluded that the inducing state for the shrinkage effect is metarhodopsin II. This means that the metarhodopsin transition acts as a trigger in the sense that it closes the gap between rhodopsin and the postponed membrane processes observed in light scattering.

One might ask how an equilibrium reaction like the metarhodopsin transition can act as a reliable trigger. In an earlier study we proposed that the equilibrium could be postponed to the actual irreversible trigger step (Uhl et al., 1978). A simpler solution of the problem is apparently now given by recent simultaneous measurements. They show that the equilibrium between metarhodopsin I (MI) and metarhodopsin II (MII) is shifted to MII by the trigger process.

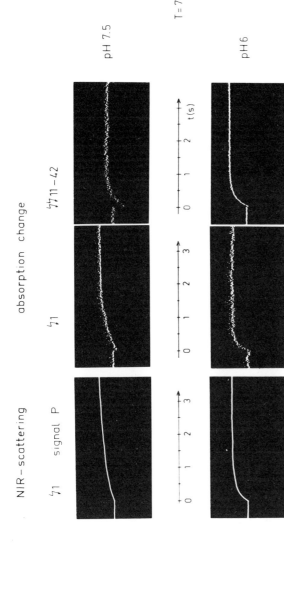

NIR-scattering   absorption change

signal P $\sqrt{1}$ $\sqrt{1}$11-42

pH 7.5

T = 7°

pH6

FIG. 15. Simultaneous measurement of flash-induced absorption and scattering changes in rod outer segments. Flash applied at $t = 0$, amplitudes are normalized to the same value (amplitudes are plotted in Fig. 16). Scattering signals are the relative intensity changes at the measuring wavelength $\lambda = 800$ nm and in an angular range $6.5° < \theta < 23°$; absorption signals are the absorption changes, i.e. the negative difference between the relative intensity changes at 387 nm and 417 nm. The signals in the first and in the second line belong to 2 measurements on 2 different samples; the scattering signal P and the left absorption signal are a simultaneous measurement from the first flash, the absorption signal at the right is an average of flashes no. 11–42. Measuring parameters as indicated in the figure; relative rhodopsin turnover by one flash $\rho = 0.02$.

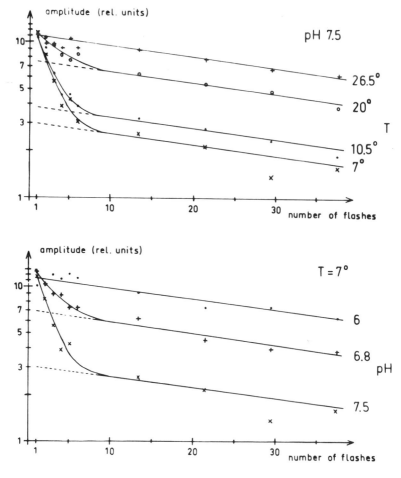

FIG. 16. Successive photolyse of rod outer segment (ROS) suspensions in series of flashes (exhaustion curves). Every flash bleaches the system a further step; the amplitude of the flash-induced signals are plotted as single recordings (flashes no. 1–7) respectively as the average of groups of 8 flashes (higher bleaching). Signals are defined as in Fig. 1, measuring parameters as indicated in the figure; relative rhodopsin turnover by one flash $\rho = 0.02$. For a bleaching of $< 10\%$, each flash produces more than the normal equilibrium amount of metarhodopsin II (given by the straight line).

An indication of this result is already found by more careful consideration of Fig. 15. Comparing the MI/MII-signals of the first and of the later flashes, it becomes evident that they are equal to each other at pH 6 but different at pH 7.5. The kinetics of the signal at higher bleachings corresponds to the "normal" conditions, i.e. all rhodopsin molecules run into MI/II equilibrium.

Because the equilibration goes with the sum of the time constants for forward and back reaction, it is apparently the faster the more MI it produces. It is therefore suggested that the slower kinetics at flash No. 1 is due to an equilibrium shift to MII in the dark adapted membrane.

This becomes evident by plotting the signal amplitudes in series of flashes (exhaustion curves) as shown in Fig. 16. For bleachings higher than about 10%, the successive illumination yields a logarithmic decrease of the metarhodopsin amplitude, as is expected because of the successive decrease of rhodopsin available for activation. The more temperature and pH favour MI, the smaller is the MI/MII-signal amplitude and the whole exhaustion curve is consequently shifted to lower values. For the first flash, however, the bulk conditions are virtually irrelevant, since it essentially only produces MII. A comparison of the saturation curve of the P-signal (Fig. 13b) and of the non-equilibrium amount of MII ("extra-MII") with increasing photolysis suggests that both signals reflect the same process. It appears that the induction of the membrane processes seen in the scattering signals reacts on the trigger and withdraws it from the equilibrium (Emeis and Hofmann, 1981).

In recent experiments, the extra MII-effect was explained by a complex formation between MII and the GTP-binding protein (G-protein), one of the above mentioned peripheral proteins of the disc membrane. First evidence for this is offered by the stoichiometric relation between the G-protein and the saturation of the near infrared scattering signal found by Kühn et al. (1981), which is comparable to that of the P-signals. The 'binding signal' observed by these authors is most probably a superposition of the signals P_a and P_r. Furthermore, our experiments on the extra-MII showed that the fraction of peripheral proteins must contain the determining factor also for this effect because, in the case of isolated discs, it was only observed after recombination of the proteins (Emeis and Hofmann, 1981). Most recently it was shown that recombination with purified G-protein has the same effect, revealing that the factor is given by this protein. As is seen in Fig. 17, there is a qualitative stoichiometry between the extra-MII and the amount of G-protein added (Emeis et al., 1982).

The simplest scheme explaining data is:

$$MI \rightleftharpoons MII \rightleftharpoons MII\text{-}G$$

This scheme was supported by quantitative evaluation of the amplitude and kinetics relations between the MI/MII-signals in different stages of bleaching (Emeis et al., 1982).

Kinetic analysis shows that the scattering indicator of the complex formation cannot be the polarizability change P_D observed in isolated disc vesicles. The kinetic contradiction is easily seen in Fig. 18. A comparison of the signal P_D with the MI/MII-signal shows that the shift of the equilibrium

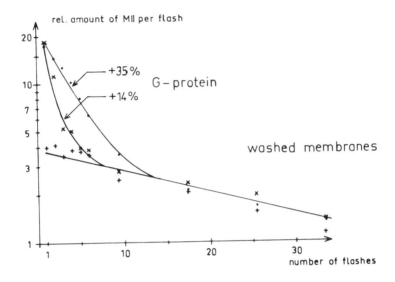

FIG. 17. Exhaustion curves in dependence on the GTP-binding protein G. A qualitative stoichiometry is found between the amount of recombined G and the "extra MII".

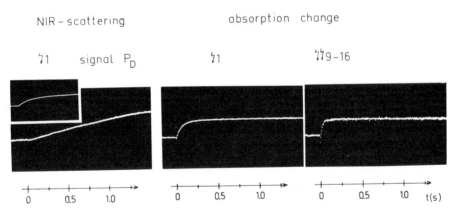

FIG. 18. Simultaneous measurement of flash-induced absorption and scattering changes in isolated disc vesicles. Flash applied at $t = 0$, all amplitudes are normalized to the same value (amplitudes are plotted in Fig. 17). Scattering and absorption signals are defined as in Fig. 15. In analogy to Fig. 15, the scattering signal P_D and the first absorption signal are a simultaneous measurement from the first flash of a series, the absorption signal at the right is an average of the flashes no. 9–16. Measuring parameters as in Fig. 1. It is seen that, in contrast to the finding in ROS (Fig. 6), the scattering signal is much slower in this system (full time scale of P_D in the inset: 8 s).

is much faster than the scattering signal. Thus, the influence on the equilibrium must be established before the polarizability change seen in P_D occurs. By its kinetic congruence with the MI/MII-equilibrium shift, the fast axial shrinkage signal P_a is recognized as the actual indicator of the MII-G complex. This is corroborated by the finding shown in Fig. 19: if the G-protein is removed from the intact ROS-structure, the fast scattering signal is no longer observed.

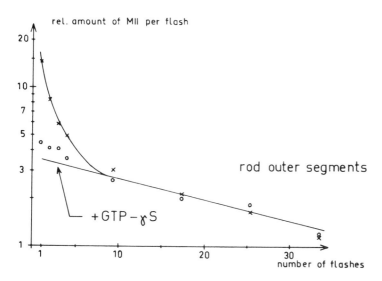

FIG. 19. Exhaustion in rod outer segments in dependence on the GTP-binding protein G. As in the isolated washed membranes (Fig. 17), the extra MII is only found in the presence of G; if G is removed by GTP-γS, the effect is no longer observed.

For sufficiently high GTP-concentrations (c. 10 μM), the "extra-MII" and the scattering signals are considerably reduced. The most probable explanation, suggested by the work of Kühn et al., (1981), is a rapid re-dissociation of the MII-G complex induced by the binding of GTP.

The dominant physical concomitant of the MII-G complex formation, as far as is observable in light scattering, is the axial shrinkage of the ROS. The mechanism and induction of the underlying disc membrane attraction is currently under investigation.

IV. Acknowledgements

The authors wish to thank D. Emeis, W. Mäntele, J. Reichert and A. Schleicher for essential contributions to this work; they further wish to express their appreciation to Mrs W. Herbst and Miss S. Filter for their assistance in preparing the manuscript. The work was supported by the Deutsche Forschungsgemeinschaft (SFB 70).

References

Abrahamson, E. W. (1972). *In* "Biochemistry and Physiology of Visual Pigments" (ed. H. Langer), Springer, Berlin, Heidelberg, New York.

Applebury, M. L., Zuckermann, D. M., Lamola, A. A. and Jovin, T. M. (1974). *Biochemistry* **13**, 3448–3458.

Argade, P. V., Rothschild, K. J., Kawamoto, A. H., Herzfeld, J. and Herlihy, W. C. (1981). *Proc. Natl Acad. Sci. USA* **78**, 1643–1646.

Aton, B., Doukas, A. G., Callender, R. H., Becher, B. and Ebrey, T. G. (1977). *Biochemistry* **16**, 2995–2999.

Attwood, P. V. and Gutfreund, H. (1980). *FEBS Lett.* **119**, 323–326.

Bayley, H., Huang, K.-S., Radhakrishnan, R., Ross, A. H., Takagaki, Y. and Khoran, H. G. (1981). *Proc. Natl Acad. Sci. USA* **78**, 2225–2229.

Bayley, P. M. (1981). *Progr. Biophys. Molec. Biol.* **37**, 149–180.

Bogomolni, R. A., Stubbs, L. and Lanyi, J. K. (1978). *Biochemistry* **17**, 1037–1041.

Brady, S. E., Maher, J. P., Bromfield, J., Stewart, K. and Ford, M. (1976). *J. Phys. E. Sci. Instrum.* **9**, 19–24.

Braiman, M. and Mathies, R. (1980). *Biochemistry* **19**, 5421–5428.

Cone, R. A. (1972). *Nature New Biol.* **236**, 39–42.

Daemen, F. J. M. (1973). *Biochim. Biophys. Acta* **300**, 255–288.

Dewrey, M. M., Davis, P. K., Blasie, J. K. and Barr, L. (1969). *J. Mol. Biol.* **39**, 395–405.

Dudley, P. A., Landis, D. J. and Anderson, R. E. (1975). *Exp. Eye Res.* **21**, 549.

Eigen, M. and de Mayer, L. (1973). Theoretical basis of relaxation spectroscopy. *In* "Techniques of Chemistry" Vol. 6, Part 2. (ed. A. Weisberger). Wiley, New York, London, Toronto.

Emeis, D. and Hofmann, K. P. (1981). *FEBS Lett.* **136**, 201–207.

Emeis, D., Kühn, H., Reichert, J. and Hofmann, K. P. (1982). *FEBS Lett.* **143**, 29–34.

Emrich, H. M. (1971). *Z. Naturforsch.* **26b**, 352–356.

Engelman, D. M. and Zaccai, G. (1980). *Proc. Natl Acad. Sci. USA* **77**, 5894–5898.

Fung, B. K.-K., Hurley, J. B., Stryer, L. B. (1981). *Proc. Natl Acad. Sci. USA* **78**, 152–156.

Funk, J., Welte, W., Hodapp, Wutschel, I. and Kreutz, W. (1981). *Biochim. Biophys. Acta* **640**, 142–158.

Godchaux, W. and Zimmermann, W. F. (1979). *Exp. Eye Res.* **28**, 483–500.

Govindjee, R., Ebrey, T. G. and Crofts, R. (1980). *Biophys. J.* **30**, 231–242.

Hess, B. and Kuschmitz, D. (1979). *FEBS Lett.* **100**, 334–340.

Hofmann, K. P., Uhl, R., Hoffmann, W., Kreutz, W. (1976). *Biophys. Struct. Mech.* **2**, 61–77.

Hofmann, K. P. and Emeis, D. (1979). *Rev. Sci. Instr.* **50**, 249–252.

Hofmann, K. P., Schleicher, A., Emeis, D., Reichert, J. (1981). *Biophys. Struct. Mech.* **8**, 67–93.

Honig, B., Ebrey, T. G., Callender, R. H., Dinur, U. and Ottolenghi, M. (1979). *Proc. Natl Acad. Sci. USA* **76**, 2503–2507.

Hulst, H. C. van de (1957). *In* "Light Scattering by Small Particles". J. Wiley and Sons, New York; Chapman and Hall, London.

Kühn, H. (1980). *Nature* **283**, 587–589.

Kühn, H., Bennett, N., Michel-Villaz, M., Chabre, M. (1981). *Proc. Natl Acad. Sci. USA* **78**, 6873–6877.

Kühne, W. (1878) *In* "Untersuchungen aus dem Physiol. Institut der Universität Heidelberg", Carl Winters Universitäts-Buchhandlung, Heidelberg.

Lamola, A. A., Yamane, T. and Zipp, A. (1974). *Biochemistry* **13**, 738–745.

Lemke, H.-D. and Oesterhelt, D. (1981). *FEBS Lett.* **128**, 255–260.

Liebman, P. A. and Entine, G. (1974). *Science* **185**, 457–459.

Liebman, P. A., Jagger, W. S., Kaplan, M. W. and Bargot, F. G. (1974). *Nature* **251**, 31–36.

Mäntele, W., Siebert, F. and Kreutz, W. (1981). *FEBS Lett.* **128**, 249–254.

Mäntele, W., Siebert, F. and Kreutz, W. (1982). *Methods Enzymol.* **88**, 729–740.

Marcus, M. A. and Lewis, A. (1978). *Biochemistry* **17**, 4722–4735.

Matthews, R. G., Hubbard, R., Brown, P.-K. and Wald, G. (1963). *J. Gen. Physiol.* **47**, 215–240.

Mowery, P. C. and Stoeckenius, W. (1981). *Biochemistry* **20**, 2302–2306.

Mullen, E., Johnson, A. H. and Akhtar, M. (1981). *FEBS Lett.* **130**, 187–193.

Pimental, G. C. (1968). *Appl. Opt.* **7**, 2155–2159.

Poo, M. M. and Cone, R. A. (1973). *Exp. Eye Res.* **17**, 503–510.

Porter, G. (1963). *In* "Techniques of Organic Chemistry, 2nd ed., Vol. 8, Part 2". Interscience Publishers, New York, pp. 1055–1066.

Reichert, J. and Hofmann, K. P. (1981). *Biophys. Struct. Mech.* **8**, 95–105.

Renard, M. and Delmelle, M. (1980). *Biophys. J.* **32**, 993–1006.

Schnetkamp, P. C. M., Kaupp, U. B. and June, W. (1981). *Biochim. Biophys. Acta* **642**, 213–230.

Schulten, K., and Tavan, P. (1978). *Nature* **272**, 85–86.

Siebert, F., Schmid, H. and Mull, R. H. (1977). *Biochem. Biophys. Res. Comm.* **75**, 1071–1077.

Siebert, F., Mäntele, W. and Kreutz, W. (1980). *Biophys. Struct. Mech.* **6**, 139–146.

Siebert, F. and Mäntele W. (1980). *Biophys. Struct. Mech.* **6**, 147–164.

Siebert, F., Mäntele, W. and Kreutz, W. (1981). *Can. J. Spectrosc.* **26**, 119–125.

Siebert, F., Mäntele, W. and Kreutz, W. (1982). *FEBS Lett.* **143**, 82–87.

Smith, H. G., Stubbs, G. W. and Litman, B. J. (1975). *Exp. Eye Res.* **20**, 211–217.

Stockburger, M., Klusmann, W., Gattermann, H., Massig, G., and Peters, R. (1979). *Biochemistry* **18**, 4886–4900.

Stoeckenius, W., Lozier, R. H. and Bogomolni, R. A. (1979). *Biochim. Biophys. Acta* **505**, 215–278.

Terner, J., Hsieh, C.-L., Burns, A. R. and El-Sayed, M. A. (1979). *Proc. Natl Acad. Sci. USA* **76**, 3046–3050.

Tsuda, M., Glaccum, M., Nelson, B. and Ebrey, T. G. (1980). *Nature* **287**, 351–353.

Uhl, R., Hofmann, K. P. and Kreutz, W. (1977). *Biochim. Biophys. Acta* **469**, 113–122.

Uhl, R., Hofmann, K. P. and Kreutz, W. (1978). *Biochemistry* **17**, 5347–5352.

Warshel, A. (1979). *Photochem. Photobiol.* **30**, 285–290.

Witt, H. T. (1971). *Quart. Rev. Biophys.* **4**, 365–477.

Chapter 8

Optical Probes and the Detection of Conformational Changes in Membrane Proteins

PETER B. GARLAND

*Department of Biochemistry, University of Dundee,
Dundee DD1 4HN, Scotland*

BIOLOGICAL MEMBRANES Vol. 5
ISBN 0 12 168546 2

I. Introduction. Possible Consequences of Ligand Binding at the Cell Surface

Many of the physiological signals received at the outer surface of a cell in a multicellular organism involve signal molecules which do not cross the cell membrane. Yet such molecules can exert a profound effect on the cell: for example, the stimulation of glucose transport by insulin. In other cases, such as in the inward transport of epidermal growth factor into a cell, it is clear that the possibility exists for more direct effects of regulatory signals on aspects of cell function that are well removed from the cell surface. For the most part a detailed description of the molecular events whereby signal molecules act at the cell membrane is currently unavailable. I use the term "signal molecules" in the widest possible sense to include all molecules which, by arriving at a cell-surface of appropriate receptivity, provoke some physiological response. Thus signal molecules can refer to neurotransmitters, hormones, lymphokines, interferons, viruses and even other cell surfaces. Not surprisingly, in view of this list, the study of cell-surface events at the molecular level is one of intense activity.

The purpose of this chapter is to review briefly some recent advances in the methods available for studying early molecular events in the binding of signal molecules to their receptors at a cell surface. The methods to be described are all optical, and some are applicable to small numbers of cells, even single cells, by use of a fluorescence microscope. They therefore represent a move away from cuvette spectroscopy towards microscopy. In all instances the purpose of the method is to determine in what way a membrane receptor may be physically modified by binding to itself some specific ligand. Implicit in this approach is the underlying assumption that if the receptor is to do something on binding its ligand, then it must do it as a consequence of first changing its conformation to one that enables different properties to be expressed—for example, formation of an ion channel or binding to some other effector protein.

II. Diffusional Movements of Membrane Proteins

A. LATERAL DIFFUSION

The movement of a single molecule undergoing two-dimensional diffusion in the plane of a membrane cannot be predicted. But for a set of such molecules moving over a relatively short distance in time t, their mean squared displacement $(\bar{x})^2$ is given by

$$(\bar{x})^2 = 4D_L t \tag{1}$$

where D_L is the lateral diffusion coefficient. D_L is usually given in CGS units of $cm^2 s^{-1}$. Saffman and Delbrück (1975) developed a theoretical treatment for predicting D_L of a membrane protein modelled as a cylinder of radius r spanning a bilayer of width h. If the viscosities of the bilayer interior and of the aqueous phase are η and η' respectively, then D_L is given by

$$D_L = \frac{KT}{4\pi\eta h}\left[\ln\left(\frac{\eta h}{\eta' r} - \gamma\right)\right] \tag{2}$$

where K is Boltzmann's constant (1.38×10^{-6} erg K^{-1}), T is absolute temperature and γ is Euler's constant (0.5772). The viscosity of water is approximately 0.01 poise, whereas that of the hydrophobic interior of a phospholipid bilayer in the liquid crystalline state is some hundred-fold greater. Hence $\eta/\eta' > 1$ and, from equation (2), there will be only a very weak dependence of the lateral diffusion coefficient on the molecular radius.

B. ROTATIONAL DIFFUSION

There is an abundance of structural evidence demonstrating that integral membrane proteins are asymmetrically arranged across the bilayer and that this asymmetry is preserved throughout the life of the protein. This means that rotation of an integral membrane protein about axes lying in the membrane plane does not occur at a significant rate. However, rotation around an axis perpendicular (normal) to the membrane plane is not restricted in a fluid membrane model. As with lateral diffusion, the rotational displacements of a given molecule about an axis cannot be predicted. But for a set of such molecules undergoing uniaxial rotation the mean squared angular displacement $(\bar{\theta})^2$ occurring in time t is given by

$$(\bar{\theta})^2 = 2D_R t \tag{3}$$

where D_R is the rotational diffusional coefficient. The value of D_R predicted by the model of Saffman and Delbrück (1975) is given by

$$D_R = \frac{KT}{4\pi\eta h r^2} \tag{4}$$

The viscosity of the aqueous phase bordering the membrane is unimportant in this model provided that $\eta/\eta' > 1$. It is clear from equation (4) that D_R has a very strong dependence on molecular radius. It is convenient to define a rotational relaxation time as D_R^{-1}. Rotational relaxation times of membrane proteins can be as low as a few µs, or be immeasurably long (Cherry, 1979).

C. OTHER MOVEMENTS

Uniaxial rotation of the type just described regards the protein molecule as a rigid body. Proteins are not rigid bodies, but neither are they so devoid of structure that movement of amino acid residues, of the polypeptide chain or of protein domains is unrestricted in amplitude. In the present context two broad classes of movement are of interest. Firstly, the movements of amino acid residues, such as the side-chain of lysine or cysteine, will be reported upon by rotationally sensitive probes that are covalently attached at these residues. Such rotation is fast, in the nanosecond region, but limited in extent by structural and steric constraints (Yguerabide, 1972; Wahl, 1975). Secondly, the movement of whole segments or domains of the protein will also be limited in extent by structural and steric constraints. For example, the Fab arms of immunoglobulin G have a rotational relaxation time of about 15 ns, but of limited physical range: the hinge region of immunoglobulin G is not a universal joint (Yguerabide *et al.*, 1970; Hanson *et al.*, 1981). It is not possible to put an upper time limit on these so-called segmental motions of proteins.

III. The Measurement of Lateral Diffusion

The first quantitative measurements of the lateral diffusion of a membrane protein were made on rhodopsin in the membrane discs of retinal rod outer segments (Poo and Cone, 1974; Liebman and Entine, 1974). The experiments involved the use of a microscopic spectrophotometer, and made use of the fact that rhodopsin can be photobleached by a brief flash of intense light. Only part of the rhodopsin-pigmented disc was photobleached. Subsequent return of pigment to the bleached area occurred by lateral diffusion of unbleached rhodopsin in exchange for bleached rhodopsin. This type of photobleaching recovery experiment is limited in applicability by the infrequency with which light sensitive proteins occur in membranes. To overcome this drawback Peters *et al.* (1974) and Edidin and Farmbrough (1973) introduced fluorescent labelling of cell surface components. The labelled components were detected by their fluorescence, and irreversible photobleaching of part of the cell surface labelled was achieved by excessive exposure to high light intensity. The subsequent recovery of fluorescence in the bleached area occurred through lateral diffusion from the unbleached area. The technique is accordingly known as fluorescence photobleaching recovery.

Considerable improvements in the fluorescence photobleaching recovery method were made by Webb and colleagues (Axelrod *et al.*, 1976; Koppel *et al.*, 1976). A continuous wave argon or krypton laser was used in conjunction with a fluorescence microscope equipped with an epi-illuminator and photon-counting photomultiplier detection of fluorescence. The laser beam has a Gaussian transverse energy profile, and the beam is greatly attenuated to a power of a microwatt or so. The beam is focused by the microscope objective to a small spot of radius 0.5 to 5μm, according to the application. The attenuated laser beam is sufficiently powerful to excite easily detectable fluorescence from as little as a few thousand fluorescently labelled membrane molecules within the illuminated area, but not so powerful as to cause photobleaching. However, irreversible photobleaching can be caused by increasing the beam power to a few milliwatts for several hundred milliseconds, so that on returning the laser beam to its measuring intensity the fluorescence signal is 30–70% diminished. Subsequent recovery of fluorescence occurs by lateral diffusion of unbleached molecules into the illuminated area.

It is important to note that photobleaching recovery methods measure lateral diffusion over distance of a micron or more, according to the laser spot size. These are vast distances on a molecular scale, and it is therefore possible that a molecule could have unhindered lateral diffusion over short or local distances and yet show hindered diffusion over longer distances. There are no satisfactory methods for measuring short range lateral diffusion of proteins.

In view of the overall dependence of D_L on molecular radius, measurements of D_L would appear at first sight to be of limited value in exploring early effects of ligand binding to cell surface receptors. However, this is not the case. The Saffman and Delbrück model applies to a simplified arrangement wherein the membrane is free from underlying cytoskeleton. When cytoskeleton is present, it can hinder lateral diffusion in at least two ways. Firstly, by direct binding of cytoskeletal proteins to membrane proteins. Secondly, by providing physical barriers to the diffusion of membrane proteins by obstructing the passage of their extramembranous hydrophilic parts that protrude beyond the membrane surface. Both of these types of cytoskeletal constraint have been invoked to explain the frequent experimental observation that in most eukaryotic cells there is usually a large fraction of protein that shows no lateral mobility (i.e. $D_L < 10^{-13} \text{ cm}^2 \text{ s}^{-1}$) and that the remaining fraction that does diffuse ($D_L = 10^{-10} \text{ cm}^2 \text{ s}^{-1}$) does so with a diffusion coefficient that is one to two orders of magnitude slower than that expected from measurements either with lipid probes or from reconstituted systems (Webb, 1978; Cherry, 1979; Webb *et al.*, 1981).

Thus although there may not as yet be a good molecular model to explain changes of D_L or of the relative proportions of mobile and immobile cell surface receptors in response to ligand binding, such changes provide compelling evidence for alterations in the relationship between the membrane protein and the underlying cytoskeleton. Interesting examples of biological situations where such alterations occur are the now-classical patching and capping phenomenon in lymphocytes (Taylor *et al.*, 1971), the effects of platelet binding on concanavalin A receptors in 3T3 fibroblasts (Schlessinger *et al.*, 1977), the effects of fertilization on surface antigen mobility in mouse eggs (Johnson and Edidin, 1978), and in general internalization of occupied receptors in clathrin-coated pits (Schlessinger *et al.*, 1978).

IV. The Measurement of Rotational Diffusion

A. PRINCIPLES OF PHOTOSELECTION

In the fluorescence photobleaching recovery method for measuring lateral diffusion, a flash of light is used to set up an asymmetric lateral distribution of molecules, and a weaker measuring beam is used to detect the diffusional decay of the asymmetry. Analogous methods are used for measuring rotational diffusion. A flash of light is used to set up an asymmetric angular distribution of molecules, and a weaker measuring beam is used to detect the diffusional decay of the asymmetry. Plane (linearly) polarized light is used both to establish and to detect the asymmetric angular distribution of molecules. The underlying principle is that absorption of plane polarized light by a molecule is proportional to the square of the cosine of the angle between a particular axis in the molecule (the transition dipole moment) and the plane of polarization of the incident light. If we start with a random set of light-sensitive molecules and expose them to a brief but intense flash of plane polarized light, then those molecules with dipole moments most nearly parallel to that plane will have the highest probability of absorbing light and undergoing some light-driven change. The resulting population of molecules will be anisotropic, but the anisotropy will decay if rotational diffusion can occur. Anisotropy can be detected as linear dichroism, the phenomenon wherein greater absorption occurs for light polarized in one plane than for light polarized in the orthogonal plane (i.e. at 90° to the first). Alternatively, emission from the light activated molecules may itself be polarized and the polarization measured through appropriately orientated polarizing filters.

The initial establishment of an anisotropic distribution of molecules is known as photoselection. The underlying photochemical event must have a lifetime that is similar to or greater than the rotational relaxation time under study, otherwise no significant decay of anistropy will occur before the photochemistry is extinguished. Accordingly, prompt fluorescence with lifetimes in the range of a few nanoseconds or even up to 100 ns is of no use for measuring rotations with relaxation times of several μs or more.

B. LINEAR DICHROISM

The first measurements of rotational diffusion of a membrane protein were made by Cone (1972) using a microspectrophotometer and polarized flash photolysis of rhodopsin in membrane discs of retinal rod outer segments. The flash induced linear dichroism decayed over a few μsec. Subsequently flash photolysis and time-resolved measurement of linear dichroism was made on cytochrome *c* oxidase of mitochondria, using the CO-complex of cytochrome a_3 as the light-sensitive probe (Junge, 1972; Junge and De Vault, 1975).

A more general approach was provided by the development of so-called triplet probes such as eosin (Cherry and Schneider, 1976), dibromofluorescein (Lavalette *et al.*, 1977) or erythrosin (Garland and Moore, 1979), synthesized with reactive groups (e.g. isothiocyanate, maleimide, iodoacetamide) whereby they could be covalently linked to proteins. In the absence of oxygen these molecules can be converted by light absorption to the triplet state which has a lifetime of several hundred μs or more. A polarized laser flash provides photoselection, and a much weaker measuring beam monitors anisotropy as linear dichroism (Cherry, 1978; Garland and Moore, 1979; Austin *et al.*, 1979).

C. PHOSPHORESCENCE

The triplet state of eosin and of erythrosin decays to the ground state with a low but measurable red phosphorescence, several times brighter for erythrosin than eosin (Garland and Moore, 1979; Austin *et al.*, 1979; Moore *et al.*, 1979; Jovin *et al.*, 1981). If the excitation flash is polarized then the phosphorescence is also polarized. Rotation of the flash excited eosin- or erythrosin-labelled membrane protein is measured as a time-resolved decay of phosphorescence polarization. The phosphorescence method is one to two orders of magnitude more sensitive than linear dichroism. It suffers from unpredictable variations, dependent on the site of labelling, in the extent to which the phosphorescent emission is polarized. This is because the dipole moments for light absorption and light emission may not be parallel to each other.

D. FLUORESCENCE DEPLETION

This method combines the long lifetime of the triplet state with the high sensitivity of prompt fluorescence (Johnson and Garland, 1981, 1982, 1983). It resembles the linear dichroism method in principle, but in practice it does not measure absorption of light but rather the fluorescence that such absorption excites. Fuller descriptions appear elsewhere. In its present form our apparatus has high sensitivity ($< 10^5$ molecules) but low time resolution (10–20 μs). As with the linear dichroism method, the dipole moments for photoselection and for measurement are parallel and the anisotropy of the measuring signals is maximized.

V. Examples of Conformational Change Detected by Rotational Diffusion Measurements

A. SARCOPLASMIC RETICULUM CA²⁺-DEPENDENT ATPASE

The high concentration of the Ca^{2+}-dependent ATPase in sarcoplasmic reticulum vesicles from rabbit skeletal muscle facilitates quite specific labelling of this protein with chemically reactive probes such as eosin- or erythrosin isothiocyanate. It could be inferred from linear dichroism measurements that the protein underwent some segmental motion in the first few μs or less after the photoselection flash, and before rotation of the whole molecule occurred (Bürkli and Cherry, 1981; Hoffman et al., 1979). This segmental motion was directly observed using the better sensitivity and time resolution of the phosphorescence method (Spiers et al., 1983). Furthermore, large changes in the extent and rate of this segmental motion occurred as the temperature of the sarcoplasmic reticulum preparation was passed through 11–13°C. It was already known from laser Raman spectroscopy that the enzyme undergoes a conformational change at around 13°C (Lippert et al., 1981).

B. CHLOROPLAST H⁺-TRANSLOCATING ATPASE

Junge and colleagues labelled the F_1 component of the H⁺-translocating ATPase of spinach chloroplasts thylakoid membranes under a variety of conditions with eosin isothiocyanate and measured the time-resolved decays of laser-flash induced linear dichroism. They concluded that "energization" of the membrane loosened the eosin-labelled site(s) on the F_1 and allowed greater access of oxygen to the eosin probe. It was also concluded that the diameter of the intramembranous part of the ATPase (i.e. the F_0 component) was contracted when carrying out the light-driven synthesis of ATP (Wagner and Junge, 1980; Wagner et al., 1981).

C. CHLOROPLAST FERREDOXIN: NADP⁺ OXIDOREDUCTASE

Wagner et al. (1982) isolated ferredoxin : NADP⁺ oxidoreductase reductase from spinach chloroplast thylakoid membranes, labelled it with eosin iso-thiocyanate, and used the labelled enzyme to reconstitute depleted membranes. In the absence of ferredoxin the re-bound oxidoreductase had a rotational relaxation time of less than 1 μs at 10°C. This was slowed to 40 μs by the addition of ferredoxin and further altered by illumination. Changes of the triplet lifetime in response to NADP$^+$ or transmembrane pH gradient were interpreted as changes in polypeptide chain flexibility.

D. EPIDERMAL GROWTH FACTOR RECEPTOR

Zidovetski et al. (1981) applied erythrosin-labelled epidermal growth factor to human epidermis carcinoma cells (A-431) at 4°C and measured the time resolved decay of phosphorescence polarization as a function of temperature. At 4°C the rotational relaxation time was 25–90 s. Prolonged incubation and higher temperature (23°–37°C) increased the rotational relaxation time to 350 μs. It was concluded that the epidermal growth factor receptors formed microclusters. By contrast, lateral diffusion rates increased with temperature.

VI. Future Directions

The application of optical methods to the analysis of early molecular events following the binding of ligands to their receptors is in its infancy. Indeed, of the four examples given above only one was for a receptor. This state of affairs reflects the rather recent development of the experimental methods and the inevitability of the need to work at first with systems that offer particular advantages of high occupancy of protein (sarcoplasmic reticulum) or ease of triggering (light-driven events on thylakoid membranes). The value of developing methods of high sensitivity is clearly shown by the use of erythrosin phosphorescence with epidermal growth-factor receptor. Further technical advances can be anticipated, in sensitivity and time resolution of rotational segmental motions of membrane proteins, and of the time resolution with which changes in these properties can be detected in response to physiological stimuli (Garland and Johnson, 1983).

References

Axelrod, D., Koppel, D. E., Schlessinger, J., Elson, E. and Webb, W. W. (1976). *Biophys. J.* **16**, 1055–1069.

Austin, R. H., Chan, S. S. and Jovin, T. M. (1979). *Proc. Natl Acad. Sci. USA* **76**, 5650–5654.

Bürkli, A. and Cherry, R. J. (1981). *Biochemistry* **20**, 138–145.

Cherry, R. J. (1978). *Methods Enzymol.* **54**, 47–61.

Cherry, R. J. (1979). *Biochim. Biophys. Acta* **559**, 289–327.

Cherry, R. J. and Schneider, G. (1976). *Biochemistry* **15**, 3657–3661.

Cone, R. A. (1972). *Nature New Biol.* **236**, 39–43.

Edidin, M. and Fambrough, D. (1973). *J. Cell Biol.* **57**, 27–37.

Garland, P. B. and Johnson, P. (1983). *In* "Spectroscopy and the Dynamics of Biological Systems" (eds P. Bayley and R. E. Dale). Academic Press, London, Orlando and New York.

Garland, P. B. and Moore, C. H. (1979). *Biochem. J.* **183**, 561–572.

Hanson, P. C., Yguerabide, J. and Schumaker, V. N. (1981). *Biochemistry* **20**, 6842–6852.

Hoffman, W., Sarzala, M. G. and Chapman, D. (1979). *Proc. Natl Acad. Sci. USA* **76**, 3860–3864.

Johnson, M. and Edidin, M. (1978). *Nature* **272**, 448–450.

Johnson, P. and Garland, P. B. (1981). *FEBS Lett.* **132**, 252–256.

Johnson, P. and Garland, P. B. (1982). *Biochem. J.* **203**, 313–321.

Johnson, P. and Garland, P. B. (1983). *FEBS Lett.* **153**, 391–394.

Jovin, T. M., Bartholdi, M., Vaz, W. L. C. and Austin, R. M. (1981). *Ann. N. Y. Acad. Sci. USA* **78**, 389–397.

Junge, W. (1972). *FEBS Lett.* **25**, 109–112.

Junge, W. and DeVault, D. (1975). *Biochim. Biophys. Acta* **408**, 200–214.

Koppel, D. E., Axelrod, D., Schlessinger, J., Elson, E. L. and Webb, W. W. (1976). *Biophys. J.* **16**, 1313–1329.

Lavalette, D., Amand, B. and Pochon, F. (1977). *Proc. Natl Acad. Sci. USA* **74**, 1407–1411.

Liebman, P. A. and Entine, G. (1974). *Science* **185**, 457–459.

Lippert, J. L., Lindsay, R. M. and Schultz, R. (1981). *J. Biol. Chem.* **256**, 12411–12416.

Moore, C. H., Boxer, D. H. and Garland, P. B. (1979). *FEBS Lett.* **108**, 161–166.

Peters, R., Peters, J., Tews, K. M. and Bähr, W. (1974). *Biochim. Biophys. Acta* **367**, 282–294.

Poo, M.-M. and Cone, R. A. (1974). *Nature* **247**, 438–441.

Saffman, P. G. and Delbrück, M. (1975). *Proc. Natl Acad. Sci. USA* **72**, 3111–3113.

Schlessinger, J., Elson, E. L., Webb, W. W., Yahara, I., Rutishauer, U. and Edelman, G. M. (1977). *Proc. Natl Acad. Sci. USA* **74**, 1110–1114.

Schlessinger, J., Schechter, Y., Cuatrecacas, P., Willingham, M. C. and Pastan, I. (1978). *Proc. Natl Acad. Sci. USA* **75**, 5353–5357.

Spiers, A., Moore, C. H., Boxer, D. H. and Garland, P. B. (1983). *Biochem. J.* (in press).

Taylor, R. B., Duffus, P. H., Raff, M. C. and de Petris, S. (1971). *Nature New Biol.* **233**, 225–229.

Wagner, R., Carrillo, N., Junge, W. and Vallejos, R. H. (1981). *FEBS Lett.* **136**, 208–212.

Wagner, R., Carrillo, N., Junge, W. and Vallejos, R. H. (1982). *Biochim. Biophys. Acta* **680**, 317–330.

Wagner, R. and Junge, W. (1980). *FEBS Lett.* **114**, 327–333.

Wahl, P. (1975). *New Tech. Biophys. Cell Biol.* **2**, 233–285.

Webb, W. W. (1978). *Front. Biol. Energ.* **2**, 133–139.

Webb, W. W., Barak, L. S., Tank, D. W. and Wu, E.-S. (1981). *Biochem. Soc. Symp.* **46**, 191–205.

Yguerabide, J. (1972). *Methods Enzymol.* **26**, 498–577.

Yguerabide, J., Epstein, H. F. and Stryer, L. (1970). *J. Mol. Biol.* **51**, 573–590.

Zidovetski, R., Yarden, Y., Schlessinger, J. and Jovin, T. M. (1981). *Proc. Natl Acad. Sci. USA* **78**, 6981–6985.

Chapter 9

Calcium and Cellular Activation

B. D. GOMPERTS

*Department of Experimental Pathology,
University College, London, England*

BIOLOGICAL MEMBRANES Vol. 5
ISBN 0 12 168546 2

I. Introduction

The presence of extracellular Ca^{2+} is a common requirement for cells undergoing activation to undertake their specialized functions. The first description of a Ca^{2+} requiring process was surely due to Ringer (1884) in his study of frog cardiac muscle contractility and since that time the list of Ca^{2+} requiring processes has grown to include such disparate activities as contraction of all muscle types, some synthetic processes including prostaglandins, leukotrienes and steroids, cell mediated cytotoxicity, exocytotic and fluid secretion, cell communication through gap junctions, parthenogenic reactions and many metabolic responses.

It is the purpose of this chapter to consider some of the developing interests in the regulation by intracellular Ca^{2+} of cell function. The field is an enormous one; as partial justification for the foundation of the new journal *Cell Calcium*, it was pointed out that over 2000 papers in this field are published annually (Case, 1980). To restrict the scope of this chapter, the role and regulation of intracellular Ca^{2+} will be considered with respect to a limited range of cell types, mainly mammalian in origin. Among these, two, the rat peritoneal mast cell and the rabbit neutrophil will be presented as particular case studies. The justification for this choice is straightforward: treated as secretory cells, the extent of activation is simply assessed as the proportion of the contained secretory products which becomes externalized following stimulation. Furthermore, these two cell types are the main objects of study in my own laboratory, and therefore form the basis of my own experience. The questions for consideration are as follows:

(1) What is the effect of extracellular Ca^{2+} on cell function?
(2) What concentrations of cytosol Ca^{2+} are required to cause cellular activation?
(3) What are the sources whence cytosol Ca^{2+} is derived?
(4) What are the kinetics of cytosol Ca^{2+}?
(5) What does elevated cytosol Ca^{2+} do?
(6) What is the molecular mechanism of receptor-induced Ca^{2+} mobilization?

II. The Requirement for Extracellular Ca²⁺

A. THE MEANING OF Ca²⁺-DEPENDENCE

It was already clear to Ringer that the site of action of the Ca^{2+} was obscured from the fluid bathing his experimental tissue, and for many years now, it has been accepted that Ca^{2+} acts within cells to initiate the subsequent steps of the activation process.

Confirmation of this axiom for most of the microscopic mammalian cells mentioned in this chapter had to await the advent of the ionophores for Ca^{2+}. These are low molecular weight compounds capable of solubilizing inorganic cations in low dielectric organic solvents, and by virtue of this property they can move cations across the hydrocarbon phase of biological membranes (Pressman, 1968; Gomperts, 1977). For example, the *Streptomyces*-derived Ca^{2+}-ionophores A23187 (Reed and Lardy, 1972) and ionomycin (Liu and Hermann, 1978) and the synthetic compound 1,1,1,2,2,3,3-heptafluoro-7,7-dimethyl-4,6-octanedione (familiar in NMR spectroscopy as the shift reagent, fod) can all release Ca^{2+} from liposomes into the bulk aqueous phase in exchange for protons (Gomperts *et al.*, 1981a). The two antibiotic compounds have also been shown to release accumulated Ca^{2+} from sarcoplasmic reticulum vesicles (Caswell and Pressman, 1972; Beeler *et al.*, 1979). Although there have been occasional reports of effects of A23187 quite separate from its ionophorous properties (Hovi *et al.*, 1975; Bottenstein and de Vellis, 1976; Klausner *et al.*, 1979), the finding that the chemically dissimilar antibiotic ionomycin and the synthetic compound fod provoke similar effects on both model systems and on cells argues strongly in favour of the idea that these molecules are biologically active solely by virtue of their ability to transport Ca^{2+} across membranes.

The number of cell types and tissues now accepted as being activated by an elevation in intracellular Ca^{2+} is actually much greater than those which merely have a dependence on extracellular Ca^{2+}. These cells contain intracellular Ca^{2+} stores which can be mobilized when the cells are suitably stimulated, but in many cases a dependence on extracellular Ca^{2+} can be established when steps are taken to deplete these internal sources. If we bear in mind the widespread requirement for extracellular Ca^{2+}, then it is somewhat surprising to find that only a few systems have been subjected to any systematic scrutiny so as to establish in detail what this dependence on Ca^{2+} actually entails. Most authors have been content to demonstrate dependence on Ca^{2+} merely by the presentation of histograms indicating the extent of cellular activity which occurs in the presence of a defined concentration of Ca^{2+} and in its absence. If we are interested in defining the means by which Ca^{2+} controls cellular activity, then surely it is worth considering the question

TABLE I

Dependence on extracellular Ca^{2+} of various stimulated secretory processes

	Stimulus	Product	Range of Ca to stimulate	Mg(mM)	Comment	Reference
Frog neuromuscular junction	Electrode	End plate potential	0.2–3 mM	1.0	Stimulation $\alpha(Ca^{2+})^4$	Dodge and Rahamimoff (1967)
Cat neurohypophysis	K⁺ depolarization	Vasopressin	0.2–2 mM	?	Ca^{2+} inhibits 4–8 mM	Douglas and Poisner (1964)
Cat parotid	Carbachol	Amylase	0.1–1 mM	0	Ca^{2+} inhibits 1–3 mM	Putney (1976)
Rat endocrine pancreas	Pancreozymin	Amylase	still rising at 2.5 mM	0	Estimated $K_M(Ca^{2+})$ 4.8 mM no effect of Mg^{2+} at 10 mM	Kanno (1972)
Perfused cat adrenal medulla	Acetylcholine	Catecholamine	incomplete at 18.8 mM	0	Apparently biphasic	Douglas and Rubin (1961)
Adrenal chromaffin cells	Acetylcholine	Catecholamine	half max. 1.5 mM	1.2		Hochman and Perlman (1976)
Adrenal chromaffin cells	K⁺ depolarization	Catecholamine	rises linearly between 2–8 mM	1.2	Non-saturable in range tested	Hochman and Perlman (1976)
Adrenal chromaffin cells	Acetylcholine	Catecholamine	half max. 1 mM	1.2	pH 6.7–6.9	Fenwick et al. (1978)
Adrenal chromaffin cells	K⁺ depolarization	Catecholamine	saturates at 1 mM	0.8	Long-term cultures	Kilpatrick et al. (1980)
Transplantable rat phaeochromocytoma	K⁺ depolarization	Catecholamine	half max. 1 mM	1.2	Saturable	Chalfie and Perlman (1976)
Rabbit neutrophils	f-MetLeuPhe	β-glucuronidase	half max. 0.1 mM	1	Cells treated to deplete internal stores	Cockcroft et al. (1981)
Islets of Langerhans	Glucose	Insulin	0.3–4 mM	2	Sustained insulin release from islets derived from fed rats	Malaisse et al. (1978)
Islets of Langerhans	Glucose	Insulin	1–4 mM	2	Sustained insulin release from islets derived from starved ob/ob mice	Hellman (1975)
Islets of Langerhans	Glucose	Insulin	1–10 mM	2	As above: islets incubated in absence of phosphate, sulphate, bicarbonate	Hellman (1975)
Mast cells	Antigen	Histamine	half max. 0.3 mM	1	Sr^{2+} (half max. ~ 3 mM) substitutes for Ca^{2+}	Foreman and Mongar (1972)

of the dependence on extracellular Ca^{2+} concentration rather more critically. There is the possibility that the affinity of the transport system for Ca^{2+} and the factors that modulate it could reveal information about its identity, although when we come to examine the concentrations of Ca^{2+} required in the activation of different cell types, it becomes clear that these do not fall into any typical or discrete range.

B. THE DEPENDENCE ON Ca^{2+} OF DIFFERENT SYSTEMS

Even among the few examples presented in Table I, the published estimates for half-maximal stimulation cover a 50-fold range. Note that not all the systems show evidence of saturation with Ca^{2+}, and in others there is inhibition at higher concentrations of Ca^{2+} so that it is not possible to determine directly the extent of maximal activation. Even so, some authors have attempted to develop their data according to the equations for the standard adsorption isotherm so as to characterize notional Ca^{2+}-binding sites in terms of specificity, affinity and cooperativity (Foreman and Mongar, 1972; Dodge and Rahamimoff, 1967). It is doubtful whether such cell activation experiments, even in the most straightforward of situations, can really support such a simplifying analysis. It entirely neglects the effects of the electrostatic potential due to the surface charge at the cell surface which must cause considerable accretion of divalent cations relative to the bulk solution (McLaughlin et al., 1971). The availability of the kind of information which would be needed to make a full analysis is fragmentary to say the least. In one such study however, it was found that the effect of polyvalent metal ions (on the threshold potential of crayfish axons) under physiological conditions is due to screening effects (i.e. in which there is no direct contact between the ion and oppositely charged groups on the membrane surface); only at low pH, when most surface ionizable groups are protonated so reducing the surface charge density, is true point-to-point ion-binding accountable (d'Arrigo, 1974, 1978). In many of these systems Mg^{2+} acts as a competitive inhibitor for Ca^{2+} (e.g. Jenkinson, 1957; Foreman and Mongar, 1972), so that in comparing the concentrations of Ca^{2+} required to stimulate different cell types, an eye must be kept open to check whether Mg^{2+} was present, and if so, at what concentration.

C. CASE STUDY 1: RAT PERITONEAL MAST CELLS

The normal agonist for mast cells is an antigen (Mota, 1957) which binds to and cross-links immunoglobulin (IgE) which is present on the surface of the mast cells of suitably immunized animals (Ishizaka and Ishizaka, 1967). In addition to the specific antigen, any ligand which can bind to and cross-link either the IgE or the cell surface receptor for the IgE will cause stimulation of these cells provided there is Ca^{2+} in the extracellular fluid. These and other agents which induce Ca^{2+}-dependent histamine secretion from rat mast cells are listed in Table II. There are a large number of other agonistic ligands (not listed), mostly cationic polypeptides which can activate mast cells in a manner independent of extracellular Ca^{2+}. These agonists become Ca^{2+}-dependent when the cells have been pretreated to deplete intracellular Ca^{2+} stores.

TABLE II

Agents for Ca^{2+}-dependent histamine secretion

(A) IgE-directed ligands:	
(1) Specific antigen	Mota (1957)
(2) Anti-immunoglobulin	Humphrey et al. (1963)
(3) Concanavalin A	Keller (1973)
(B) IgE receptor-directed ligands:	
(1) Dimerised IgE	Segal et al. (1977)
(2) Anti-receptor antibody	Ishizaka et al. (1978)
(C) Ca^{2+}-ionophores:	
(1) A23187	Foreman et al. (1973)
(2) Ionomycin	Bennett et al. (1979)
(3) Fod	Gomperts et al. (1981a)
(D) Membrane permeabilizing agents:	
(1) ATP^{4-}	Bennett et al. (1981)
(2) Sendai virus	Gomperts et al. (1983)

Stimulation of mast cells causes exocytotic degranulation (Lawson et al., 1977) with release of histamine and a number of other substances which are collectively referred to as mediators (Lewis and Austen, 1981). Under standard experimental conditions (of pH, ionic strength, temperature etc.) histamine secretion is half-maximal with Ca^{2+} at 0.3 mM (range 0.1–1.0 mM) and Mg^{2+} at 1 mM (Foreman and Mongar, 1972). Sr^{2+} can substitute for Ca^{2+} but the concentration required is about ten times greater. For this reason, one may have some confidence that these measurements are probing a binding location for the divalent cations rather than the characteristics of electrostatic screening, but even then its identity remains obscure. If the detailed examination of the Ca^{2+} dependence of agonist stimulated cell activation is less revealing than might have been hoped, this is not necessarily the case when other methods for stimulating mast cells are applied.

D. Ca²⁺-DEPENDENCE OF IONOPHORE-INDUCED HISTAMINE SECRETION

1. *Ionomycin*

Figure 1 illustrates the dependence on Ca^{2+} of secretion of histamine due to different concentrations of an antigen (Fig. 1a) and the Ca^{2+}-carrying ionophore ionomycin (Fig. 1b) (Bennett *et al.*, 1979). As the concentration of the ionophore is increased, the required concentration of Ca^{2+} declines,

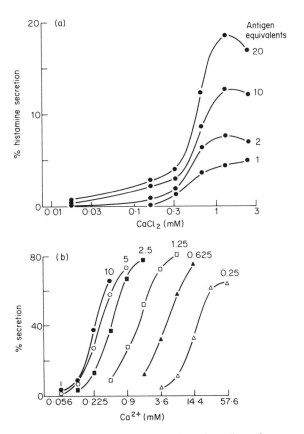

FIG. 1. (a) Dependence on extracellular Ca^{2+} of histamine release from rat peritoneal mast cells stimulated with a series of antigen concentrations. In this experiment the animals had been sensitized to the antigens of the helminth *Nippostrongylus braziliensis* by previous injection of larvae; the cells isolated after approx. 20 days were then sensitive to antigens present in fluid conditioned by contact with the worms. (Unpublished experiment of Dr. John R. White.) (b) Dependence on Ca^{2+} of histamine secretion due to a series of ionomycin concentrations. In both these experiments, the cells were quenched after 10 min with iced saline, and histamine in the supernatant was estimated fluorimetrically after formation of the *o*-phthalaldehyde adduct. (Bennett *et al.*, 1979; *Nature*.)

whereas although the maximal extent of secretion due to the receptor-directed agonist varies, its dependence on Ca^{2+} remains more or less invariant. It was also shown (Fig. 2) that the time course of secretion was not sensitive to ionophore concentration and that addition of further ionophore after secretion had terminated (even at low levels) was without much effect. Thus, all the separate determinations presented in Fig. 1b represent completed secretory events; the cells possess a mechanism for terminating secretion even though the increase in membrane permeability due to the ionophore is likely to be of long duration.

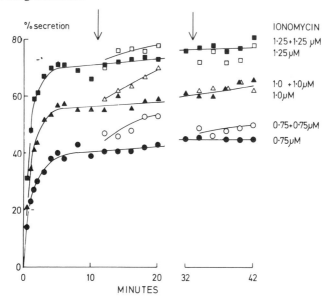

FIG. 2. Time course of histamine secretion from mast cells treated with ionomycin at 0.75 (●———●), 1.0 (▲———▲) and 1.25 μM (■———■). At the times indicated by the vertical arrows, portions of the cells were transferred so as to double the ionomycin concentration. (Bennett et al., 1979; Nature.)

These results confirm that the active agent in ionomycin-induced secretion is the [Ca.ionomycin] complex, the species which would be capable of transporting Ca^{2+} across the lipid phase of the cell membrane. The size of this fraction under all separate conditions represented by the individual points in Fig. 1b was determined by the well-known two-phase extraction experiment (Pressman et al., 1967) which measures the equilibrium

$$H_2I_{org} + Ca^{2+}_{aq} \rightleftharpoons Ca^{2+}.I_{org} + 2H^+_{aq}$$

H_2I and I^{2-} are the protonated and the fully dissociated forms of ionomycin (which possesses two ionizable acid groups). In this experiment, the partition of different concentrations of $^{45}Ca^{2+}$ was measured in the presence of the ionophore, between a buffered salt solution (similar to that used to incubate the cells) and an organic phase (toluene : butanol 7 : 3 having dielectric properties approximating those of the organic phase of cell membranes). The extraction equilibrium is composed of the partial equilibria

$$H_2I_{org} \rightleftharpoons H_2I_{aq} \tag{1}$$

$$H_2I_{aq} \rightleftharpoons HI^-_{aq} + H^+_{aq} = I^{2-}_{aq} + 2H^+_{aq} \tag{2}$$

$$I^{2-}_{aq} + Ca^{2+}_{aq} \rightleftharpoons Ca.I_{aq} \tag{3}$$

$$Ca.I_{aq} \rightleftharpoons Ca.I_{org} \tag{4}$$

Equations 1 and 4 represent the partition coefficients of the ionophore and its calcium complex and are thereby insensitive to the concentration of Ca^{2+}. Also as the aqueous phase was buffered at fixed pH, the proton dissociation equilibrium (equation 2) was similarly insensitive. Because of the vanishingly small solubility of metal ions in organic phases we have assumed that the ion complexation reaction takes place in the aqueous phase and that it is the position of this equilibrium which therefore determines the concentration of the transportable complex [Ca.ionomycin]. As further evidence for this we were able to show by UV absorption measurements, that the dependence on Ca^{2+} for the extraction equilibrium and for the complexation reaction in solution (equation 3) are indeed the same. In this way we determined the concentration of [Ca.ionomycin] for all the conditions represented by the separate data points of the secretion experiment of Fig. 1b. In Fig. 3 histamine secretion is now presented as a function of [Ca.ionomycin]. The original concentration–effect curves, which encompassed a 70-fold range of Ca^{2+} concentrations are now represented by a unique relationship indicating cell stimulation over a concentration range for [Ca.ionomycin] of 0.1–0.3 μM. This concentration dependence is unlikely to be distorted by the electrostatic potential at the cell–water interface in the manner described above because the [Ca.ionomycin] complex is neutral.

The knowledge that [Ca.ionomycin] is the catalytic entity (equivalent to an active site in enzymology) points to the conclusion that in these cells the extent (but not the time course) of secretion is regulated by the rate of entry of Ca^{2+}. Other things being equal, the enhanced rate of influx would have the effect of elevating the steady state concentration of Ca^{2+} which might be the controlling factor. Alternatively, as the ionophore-induced Ca^{2+} influx occurs at the plasma membrane and the sites of Ca^{2+} sequestration are located elsewhere (e.g. mitochondria, secretory granules) it could be that it is the Ca^{2+} concentration gradient necessarily set up by the opposing effects of these

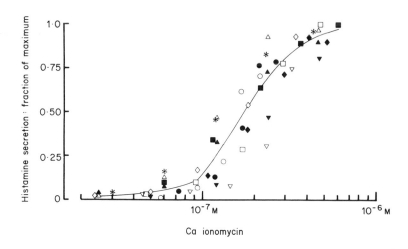

FIG. 3. The dependence of histamine secretion on the calculated concentration of [Ca.ionomycin] complex. The fraction of ionomycin which was present as the complex for each of the data points of Fig. 1b (and another similar experiment (open symbols)) was calculated as described in the text. (Bennett *et al.*, 1979; *Nature.*)

two processes which acts as the second messenger in the exocytotic reaction.

The importance of working with the ionophores is not what they can tell about themselves but what they can tell about the more complex physiological processes, regulated by the receptors. We attempt to extrapolate from the easily understood and well established towards the hard-to-understand and the unknown. In the present case we can now try to rationalize the finding that the Ca^{2+}-dependence of receptor mediated secretion is invariant with changes in agonist concentration (Fig. 1a). Here, it is likely that activation of a receptor leads to the generation of a limited number of Ca^{2+}-channels (possibly only one, though an amplifying process leading to more than one channel per receptor cannot be discounted). Thus, the greater the extent of receptor occupation, the more channels will be generated, the faster will be the rate of Ca^{2+} influx and the greater will be the extent of secretion. The channels being alike in respect of their affinity for Ca^{2+}, the effect of increasing the agonist concentration is solely to enhance Ca^{2+} flux and hence the extent of secretion. Clearly, the results of the experiment presented in Fig. 1 would argue against the idea of the generation of an endogenous ionophore for Ca^{2+} resulting from activation of the receptor: if this were the case, we would expect to see an inverse relationship between receptor occupancy and the concentration dependence of Ca^{2+}.

2. *A23187 and Fod*

The evidence from other Ca^{2+}-carrying ionophores is similar though the data have not been analysed in detail. Both the *Streptomyces*-derived antibiotic A23187 (Johansen, 1978), and the synthetic compound fod (Gomperts *et al.*, 1981a) stimulate secretion in a manner showing inverse concentration dependence on the concentration of Ca^{2+} resembling that of ionomycin presented in Fig. 1b.

A similar study of cytochalasin-B-treated rabbit neutrophils did not yield such a systematic result, nor did it point to a simple solution. In this case we found that although there exists an inverse relationship between the concentrations of ionomycin and Ca^{2+}, the maximal extent of secretion (of the lysosomal enzyme β-glucuronidase) declined as the concentration of ionomycin was reduced (unpublished observation). Thus, here (and presumably in other cell types as well) the extent of secretion must be controlled by factors other than the rate of Ca^{2+} entry into the cytosol.

III. Extracellular Control of Cytosol Ca^{2+}

Quite clearly, the closer the approach to the final site of action the simpler will be the interpretation of experimental data relating to the dependence of secretion on Ca^{2+} concentration. We are now asking: what is the concentration of intracellular Ca^{2+} required to stimulate the secretory process (or different expressions of activation in other cell types)? For this purpose we need to gain direct access to the cytosol and to introduce this Ca^{2+} at appropriate concentrations.

A. BUFFERS FOR Ca^{2+}

In order to control Ca^{2+} precisely at micromolar concentrations, especially in the presence of millimolar concentrations of Mg^{2+} it is necessary to make use of Ca^{2+} buffers. Generally these are anionic chelates having pK_As in the physiological pH range. There exists a wide range of such substances (e.g. EDTA, EGTA, HEDTA, NTA, citrate) so that buffers having practical affinity for Ca^{2+} requisite to the experimental situation can normally be chosen. The very widespread use of EGTA relies on its great selectivity between Ca^{2+} and Mg^{2+} (about 10^5) though for many purposes its buffering range centred around 10^{-7} M at physiological pH is on the low side. It is normal to prepare a series of buffers having a fixed concentration of the chelate and various concentrations of Ca^{2+} lying between 0.1 and 0.9 of the chelate concentration. The concentrations of the free $[Ca^{2+}]$ are commonly calculated by reference to tables of affinity constants (Sillen and Martell, 1971) or more simply but restrictively to calibration curves which relate to

Ca^{2+} titration curves of the buffer under selected conditions of pH, Mg^{2+}, ionic strength and temperature (Portzehl *et al.*, 1964). There is some danger in this reliance on published data relating even to the most commonly used Ca-buffers (EGTA) and the values are far from being universally accepted (Scharff, 1979; Harafuji and Ogawa, 1980), so in our work we have preferred to make individual measurements of $[Ca^{2+}]$ after dilution of concentrated buffered solutions into media directly relevant to the situations encountered in our experiments. Recent reports of Ca-buffering chelates having pK_As well below the physiological pH range give hope of some simplification in the near future (Tsien, 1980); thus not only will the concentration of Ca^{2+} be invariant with change in pH but the pH too will be invariant with change in Ca^{2+}.

There are several ways by which Ca^{2+} and Ca^{2+}-buffers can be introduced into the cytosol. Some of these are more or less destructive of cell integrity, though there is no doubting their value in selected situations. Thus, much of the earlier work in this area was done with muscle fibres in which the sarcolemma had been irreversibly damaged either mechanically ("skinned" fibres) (Ford and Podolsky, 1972) or by treatment with glycerol (to cause osmotic damage) (Heinl *et al.*, 1974). Contraction of the exposed actomyosin filaments occurs when the Ca^{2+} is applied in the range 1–10 μM in the presence of Mg.ATP.

B. BUFFERED Ca^{2+} IN LARGE SINGLE CELLS

Microinjection is another method for introducing Ca^{2+} and Ca-buffers directly into the cytosol and so long as the expression of cellular activity can be assessed at the level of the single cell, this approach yields important information. Thus microinjection of neurones from Helix aspersa with Ca^{2+} buffered at 1 μM (with EGTA) causes a drop in the membrane resistance due to increased K^+ permeability with consequent hyperpolarization (Meech, 1974). This result caused some surprise at first, as it was known that the presence of extracellular Ca^{2+} can have the effect of increasing membrane resistance and in some quarters it was even understood that neuronal exitability could arise from the effects of K^+/Ca^{2+} competition (Singer and Tasaki, 1968). The phenomenon of Ca^{2+}-induced K^+ efflux has since been observed in a wide range of tissues. The inclusion of Ca-buffers to regulate Ca^{2+} in the artificial cytosol of resealed osmotically lysed red cell ghosts makes them readily permeable to K^+ (Simons, 1976a,b), and K^+ efflux from cells is now sometimes accepted as an indication of the presence of micromolar concentrations of Ca^{2+} within the cytosol. It has been suggested that Ca^{2+}-induced hyperpolarisation might be a mechanism for maintaining elevated cytosol Ca^{2+} levels in the period following stimulation (Rasmussen *et al.*, 1979).

C. BUFFERED Ca^{2+} IN POPULATIONS OF MICROSCOPIC CELLS

Recently we have seen developments which allow the non-destructive probing of the cytosol of large populations of microscopic mammalian cells. The principle involved here is to render the cell membranes selectively leaky to low molecular weight aqueous solutes while maintaining integrity towards the larger components of the cytosol. In cells so treated it is then generally possible to introduce controlled amounts of Ca^{2+} (buffered as described above) and measure the cellular response. A number of practical methods for probing the cytosol of otherwise intact cells have been described.

1. Cell Permeabilization by High-voltage Discharge

A method which should prove almost universal in terms of susceptible cell types has been pioneered principally by Knight and Baker (1982). It involves discharging a capacitor at stainless steel electrodes placed in a suspension of the cells. Field strengths of about 2 kV cm^{-1} have been used to generate lesions on the plasma membranes of bovine adrenal chromaffin cells without causing damage to the membranes of the internal organelles. In principle, two holes appear at the extremities of the cells facing the electrodes at each discharge of the capacitor, and so after exposing the cells to several discharges they become freely accessible to solutes present in the extracellular medium. Since surfaces having greater curvature are less prone to damage, the membranes of the intracellular organelles remain intact. What is more, the lesions formed in the plasma membrane do not reseal on subsequent incubation and so the cells can be washed and transferred into new environments as demanded by experimental protocols.

When chromaffin cells are made leaky in this way, and then transferred to solutions containing Mg.ATP and Ca^{2+} buffered in the range 1–10 μM, they release catecholamines and the enzyme dopamine-β-hydroxylase, both of which are components of the secretory granules. By a number of criteria the release of the granule components appear to be due to the normal mechanisms of exocytotic secretion. Thus, the cytosol enzyme lactate dehydrogenase is retained by the cells and the requirement for Mg.ATP demonstrates the dependence of the secretion on a metabolic process which would otherwise be depleted by leakage of low molecular weight intermediates. Hill-plot analysis of the dependence of the secretory process on Ca^{2+} in the permeabilized cells indicates the involvement of 2[Ca^{2+}] at the exocytotic site but gives no further indication of its identity (Baker and Knight, 1981; Knight and Baker, 1982).

Similar studies with platelets (secretion of 5-hydroxytryptamine), pancreatic β-cells (secretion of insulin) and with sea urchin eggs (cortical granule release) also point to the central role of Ca^{2+} (1–10 μM) and the need for metabolic supplementation in this kind of preparation (Knight and Scrutton, 1980, 1982; Baker and Whitaker, 1978; Baker *et al.*, 1980; Pace *et al.*, 1980).

2. *Permeabilization of Mast Cells by Extracellular ATP^4-*

Working with mast cells we have developed another method for gaining access to the cytosol. While this method seems to be far more restricted in terms of susceptible cell types, it offers the compelling advantage of being reversible: the cells reseal about their artificially manipulated cytosol. We found that treatment of mast cells with ATP^{4-} in the range 1–5 μM causes leakage of intermediary metabolites including nucleotides and sugar phosphates (Cockcroft and Gomperts, 1979a,b). By measuring the leakage of ^{32}P-labelled metabolites we found that the effective filtration size of the membrane lesions increases with the concentration of ATP^{4-}. At low concentrations the cells lost inorganic phosphate only but became leaky to nucleotides and sugar phosphates as the ATP^{4-} was raised above 5 μM.

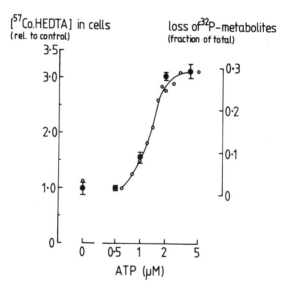

FIG. 4. Dependence on ATP^{4-} concentration of ^{32}P-metabolite release (○——○) and of [^{57}Co.HEDTA] uptake (●) by mast cells. (Bennett *et al.*, 1981; *J. Physiol.*)

Figure 4 illustrates the dependence on [ATP^{4-}] of the leakage of (total) ^{32}P-metabolites and the uptake of a low molecular weight normally impermeant aqueous solute, ^{57}Co.HEDTA (MW 400) (Bennett *et al.*, 1981). Other normally impermeant water soluble markers are able to penetrate ATP^{4-}-treated mast cells; these include ^{45}Ca.EDTA (MW 337), 6-carboxyfluorescein (MW 376) and lissamine rhodamine B sulphonate (MW 557), and all these can occupy up to about 90% of the water accessible space. Inulin (MW 5200) remains impermeant. Figure 5 illustrates the timecourse of uptake of ^{57}Co.HEDTA by mast cells treated with different concentrations of ATP^{4-}. Note that even the lowest concentration of ATP^{4-} used in this experiment allows ultimate equilibration (see also Fig. 4) but that the onset of permeability is delayed.

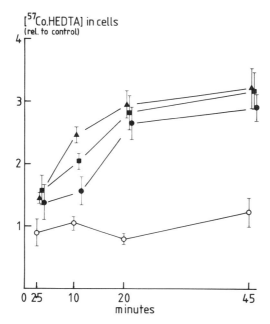

FIG. 5. Time course of [^{57}Co.HEDTA] uptake by mast cells treated with ATP^{4-} at 2.5 (●——●), 4 (■——■) and 10 μM (▲——▲). Open symbols indicate zero ATP. (Bennett *et al.*, 1981; *J. Physiol.*)

The chelates used were high affinity complexes of Co^{2+} and Ca^{2+} prepared by the addition of carrier free isotopes (i.e. about 10^{-9} M) to a great excess of the chelate (3 × 10^{-3} M) so as to ensure the absence of any uncomplexed divalent cation. Thus uptake of the isotope registered uptake of the complex salts. The complex of Ca^{2+} with HEDTA is of a lower affinity ($k' \sim 1.4$ μM at I = 0.1 and 37°C) and with the knowledge that salts of similar molecular weight can penetrate the ATP^{4-}-permeabilized cells, this chelate could be used to buffer intracellular Ca^{2+}. As in other systems secretion occurs as the concentration of buffered Ca^{2+} is presented to the mast cell cytosol in the range 1–10 μM (Fig. 6) (Bennett *et al.*, 1981). Lactate dehydrogenase is retained.

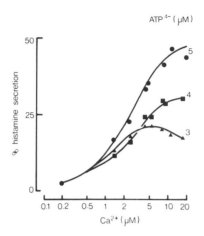

FIG. 6. Dependence on extracellular Ca^{2+} for ATP^{4-}-induced histamine secretion. Ca^{2+} was maintained in the micromolar range with HEDTA buffers. (Bennett *et al.*, 1981; *J. Physiol.*)

In Fig. 6 it will be noted that the extent of secretion increases as the concentration of ATP^{4-} is increased from 3 to 5 μM, although the data of Fig. 4 indicate that the maximal permeabilization is achieved at concentrations below this level. We think that this paradox is well explained by the earlier experience of ionophores: here it was shown that the extent of secretion from mast cells is controlled by the rate of Ca^{2+} entry into the cytosol. In the present case we find that the rate of uptake of high stability extracellular markers having molecular weight similar to the Ca-buffer continues to increase with increase in ATP^{4-} concentration over this range.

(a) *Duration of permeability lesions* There is one important difference between cells rendered permeable by dielectric damage and by treatment with ATP^{4-}. The permeability lesions generated by high voltage discharge through suspensions of cells are (with the exception of red blood cells (Zimmerman *et al.*, 1975)) sustained for periods of one hour or more (Baker and Knight, 1981) and this offers some manipulative advantages. Because the permeability of the lesions is a constant factor, it is possible to subject the cells to operations such as the sequential addition and removal of Ca^{2+}, metabolites, inhibitors, in order to test their effect on secretion and the steps leading up to it. Mast cells treated with ATP^{4-} being only transiently permeable offer the prospect of enclosing extraneous solutes in the cytosol and then treating the cells with their normal agonists. We have found that mast cells treated with ATP^{4-} at 2 μM recover within 2 min so that they can be stimulated with ionophores and millimolar Ca^{2+} in the normal way. On addition of Mg^{2+} to remove ATP^{4-}, recovery is accelerated. The future obviously holds out the possibility of enclosing indicators (for Ca^{2+}) and selective inhibitors of individual enzymes in the investigation of the normal agonist triggered activation processes.

(b) *Other cell types susceptible to ATP* Mast cells are not unique in their susceptibility to permeabilization by ATP, though this does appear to be a property shared by rather a restricted range of cell types. Mostly these are transformed cell lines (Trams, 1974; Rosengurt *et al.*, 1977). The particular ionic form of ATP which is agonistic for permeabilization has not been determined, as most experiments have been carried out using cells maintained in tissue culture media. Glycolysis in 3T6 cells treated with ATP becomes dependent on extracellular provision of ADP and NAD^+ and also the normal intracellular electrolytes K^+, Mg^{2+} and phosphate (Makan and Heppel, 1978). Clearly this is a preparation which should offer scope for investigating the role of intracellular Ca^{2+} on the regulation of the cell cycle.

Dog red blood cells become permeable to Na^+ and K^+ within seconds of applying ATP, and as this effect is rapidly reversed on adding divalent cations it is possible that the tetrabasic acid ATP^{4-} is the agonist (Parker *et al.*, 1977). ATP also allows exchange of monovalent cations when applied to renal tubules (Rorive and Kleinzeller, 1972) and there are descriptions of extracellular ATP on other cell types. These include the ^{32}P-induced phosphorylation of intracellular proteins in squid giant axon (Pant *et al.*, 1979), and the generation of intracellular ^{32}P-cyclicAMP from externally applied ^{32}P-ATP in mouse LM fibroblasts (Westcott *et al.*, 1979). Although different explanations for these actions of ATP have been proposed, some hindsight would suggest permeabilization as a possible mechanism.

3. *Other Methods for Generating Hyperpermeable Cells.*

There are other ways in which permeability lesions can be generated in cells to allow the introduction of normally impermeant aqueous solutes such as the Ca^{2+}-buffers. We have used Sendai virus to permeabilize mast cells, and once again release of histamine occurs when this is done in the presence of Ca^{2+} buffered in the micromolar range (Gomperts *et al.*, 1983). There are a large number of bacterial haemolytic toxins which cause discrete permeability lesions in model and mammalian cell membranes (Benz and Hancock, 1981; Thelestam and Mollby, 1975), but their application in the manner described here has not been reported. Prolonged soaking of bundles of cardiac muscle in solutions of EGTA causes permeabilization and access of low molecular weight solutes so that contractility requires the provision of Ca^{2+}, buffered in the micromolar range, together with a supply of ATP (or creatine phosphate plus ADP) (Mope *et al.*, 1980).

All the evidence of the Ca-ionophores, and latterly with the permeabilized cells, points to the conclusion that the key common step in cellular activation for a wide variety of cells, is the elevation of cytosol Ca^{2+} into the range 1–10 μM.

IV. Sources of Ca^{2+}

A. USE OF $^{45}Ca^{2+}$: FLUXES AND ARTEFACTS

Although it is easy to demonstrate movements of Ca^{2+} by the use of $^{45}Ca^{2+}$ in stimulated cells, it has not proved a simple matter even to assign the directionality of Ca^{2+}-flux, or to determine whether tracer movements are due to initial increases in the permeability of the plasma membrane or due to alteration in the subcellular distribution of stored Ca. For cells incubated in the presence of labelled $^{45}Ca^{2+}$ at or around 1 mM, any manipulation which has the effect of increasing the concentration of cytosol Ca^{2+} will increase the amount of label associated with the cells after they have been isolated from the extracellular medium. This is because the cytosol Ca^{2+} constitutes a large proportion of the rapidly exchangeable Ca^{2+} within the cells. Similarly, in the measurement of $^{45}Ca^{2+}$-release from pulse labelled cells, stimulation can only increase the rate of efflux. In secretory cells there is the additional problem of distinguishing the artefactual movements of $^{45}Ca^{2+}$ which are consequential on cellular degranulation (eg binding of $^{45}Ca^{2+}$ to newly revealed sites, and ejection of $^{45}Ca^{2+}$ bound to the secretory granules) from those movements of $^{45}Ca^{2+}$ which might precede and hence indicate regulation of the exocytotic mechanism.

B. MEASUREMENTS OF $^{45}Ca^{2+}$ FLUX UNDER STEADY STATE CONDITIONS

The most diagnostic experiments are those designed to perturb the steady state (i.e. when the specific activity of the intracellular exchangeable $^{45}Ca^{2+}$ is the same as that of the extracellular medium) (Sha'afi and Naccache, 1981). One can predict the changes in cell associated ^{45}Ca which would result either from an increase in membrane permeability or from redistribution from internal stores for cells treated in the presence of high (mM) or low (μM) concentrations of extracellular Ca^{2+}. These are illustrated in Fig. 7a,b. In the case of receptor activation exclusively resulting in an increase in membrane permeability, for cells incubated in the presence of low Ca^{2+} there should be no perturbation of the steady state and no increase in the amount of cell associated $^{45}Ca^{2+}$; for similarly responsive cells treated in the presence of extracellular Ca^{2+} there would be a net influx of Ca^{2+} leading to an increase in the amount of cell-associated radioactivity (curves A). If the increase in cytosol Ca^{2+} arises exclusively from redistribution from previously non-exchangeable sources, this will result in a decrease in the specific activity of the $^{45}Ca^{2+}$ in the cytosol and a decrease in the cell-associated radioactivity as Ca^{2+} is transferred to the exterior by operation of the Ca^{2+}-pumps acting to restore homeostasis. Since the redistribution causes an alteration in the specific activity of the exchangeable Ca^{2+} there must follow a period of readjustment as this falls back towards a new steady state. This situation is represented by the curves B in Fig. 7.

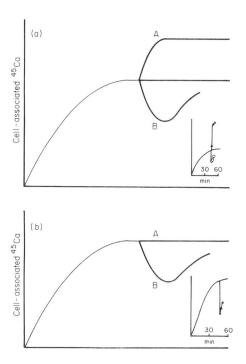

FIG. 7. Effect of stimulation on the steady state level of cell-associated ^{45}Ca in the presence of (a) millimolar concentrations of Ca^{2+} or (b) micromolar concentrations of Ca^{2+}. In these experiments, cells are incubated with the radioactive $^{45}Ca^{2+}$ for a time sufficient to reach the steady state and the stimulus is added without intervening washes. In the schematic presentations, the curves A indicate the changes in cell-associated $^{45}Ca^{2+}$ resulting from an increase in membrane permeability, and the curves B indicate the effect of an increase in the level of intracellular exchangeable $^{45}Ca^{2+}$ without alteration in membrane permeability. Insets illustrate results of experiments carried out on rabbit neutrophils stimulated with f-MetLeuPhe at 10^{-11}M ($\bigcirc$——$\bigcirc$) and 10^{-8}M ($\blacktriangle$——$\blacktriangle$). (Adapted from Sha'afi and Naccache, 1981; Raven Press, NY).

C. CASE STUDY 2: RABBIT NEUTROPHILS AS SECRETORY CELLS

The normal response of neutrophils to soluble agonists is chemotaxis up the gradient of agonist concentration. This can be suppressed, and the cells can be switched into a secreting mode by application of cytochalasin B (Goldstein *et al.*, 1973; Becker *et al.*, 1974, 1979). In both these modes the agonists operate at the same receptors (Showell *et al.*, 1976), so the cytochalasin B can be used to distinguish between events directly related to receptor activation (early events) and the subsequent processes such as secretion of lysosomal enzymes and O_2^- production (Bennett *et al.*, 1980a). The commonly applied agonists used in experimental situations are *N*-formylmethionyl peptides (Schiffmann *et al.*, 1975), analagous to the signal sequences of bacterial proteins (Bennett *et al.*, 1980b) of which the archetype, f-MetLeuPhe has $k_d' \sim 3 \times 10^{-9}$ M (Sha'afi *et al.*, 1978). The secretory function of neutrophils can be elicited with Ca^{2+}-ionophores in the presence of Ca^{2+}, and it is accepted that activation involves an increase in the concentration of cytosol Ca^{2+}. When stimulated with receptor-directed agonists, the dependence on extracellular Ca^{2+} is not absolute, though in its absence the concentration requirement for agonists increases about tenfold and the maximal extent of secretion declines (Cockcroft *et al.*, 1980b). There is thus a real possibility that in these cells, the stimulation of receptors mobilizes Ca^{2+} from both intracellular and extracellular sources (Naccache *et al.*, 1977; Sha'afi and Naccache, 1981).

D. $^{45}Ca^{2+}$ FLUXES IN STIMULATED NEUTROPHILS

On addition of f-MetLeuPhe to neutrophils incubated in steady state conditions with $^{45}Ca^{2+}$ in the presence of low concentrations of Ca^{2+} (Fig. 7a), it was found that the cell-associated radioactivity declined transiently by more than 50% (Naccache *et al.*, 1977). As this indicates a decline in specific activity, it was understood that under these conditions the source of Ca^{2+} which enlarges the exchangeable pool is derived from previously non-exchangeable sources. (There might be a paradox here: on preincubating neutrophils in the presence of EGTA, the f-MetLeuPhe induced stimulation becomes fully dependent on the presence of extracellular Ca^{2+} (Cockcroft *et al.*, 1981). Does this not imply that the intracellular sources of Ca^{2+} are indeed exchangeable?) In the presence of extracellular Ca^{2+}, two patterns were observed depending on the concentration of the agonist. At the higher concentrations commonly used to maximise cell stimulation *in vitro*, the cell-associated radioactivity rapidly increased, but at the lower concentrations which might be more relevant to *in vivo* situations cell-associated radioactivity declined transiently. These findings are consistent with the conclusion that at least two mechanisms can operate to increase the amount of exchangeable Ca^{2+} and, by inference, the concentration of Ca^{2+} in the cytosol. The internal sources are the more accessible to low agonist concentrations, but at higher concentrations there is also an increase in the permeability of the plasma membrane to Ca^{2+} which allows movement of extracellular Ca^{2+} down its concentration gradient into the cytosol.

Further evidence for internal sources of Ca^{2+} accessible to stimulated receptors comes from the finding of a net efflux of Ca^{2+} from neutrophils (bovine cells stimulated with a phorbol diester) into a medium containing 100 μM Ca^{2+} (Mottola and Romeo, 1982). The efflux (detected with a Ca^{2+}-electrode), amounts to 30% of the total cell Ca^{2+}, and clearly precedes the exocytotic release of the secretory granules. As its extent declines in parallel with loss of ATP on incubation of the cells with a glycolytic inhibitor, it most probably represents activation of the plasma membrane extrusion pump (Ca^{2+}-ATPase) due to the elevated level of cytosol Ca^{2+}.

Neither these experiments nor the work with $^{45}Ca^{2+}$ have given any clues as to the actual sources of intracellular Ca^{2+} whence cytosol Ca^{2+} is derived upon stimulation of surface receptors.

V. Measurement of Cytosol Ca^{2+}

A. Ca^{2+}-SENSITIVE ELECTRODES AND PHOTOPROTEINS

1. *Measurements on Large Single Cells*

Whilst it has been possible for some years to introduce Ca^{2+}-sensing probes into the cytosol of the large cells of some lower organisms and so measure stimulus induced Ca^{2+}-transients, practical techniques for doing this in microscopic mammalian cells are only just emerging. For individual large cells, measurements can be made either with Ca^{2+}-specific microelectrodes, or by microinjection of the Ca^{2+}-sensitive photoproteins aequorin (Baker, 1972; Ashley and Caldwell, 1974) or obelin (Campbell, 1974) and by subsequent monitoring of light emission. The methods have different advantages and disadvantages and can be regarded in some senses as complementary. The problems of working with either become magnified when they are applied to microscopic mammalian cells. Thus the tip resistance and hence the time constant for the response of microelectrodes increases as the diameter is reduced down to the scale needed to penetrate these cells; on the other hand, the composition of the sensor materials which confer Ca^{2+}-sensitivity on these electrodes now permits measurements at the level of Ca^{2+} in resting cardiac muscle cells with negligible interference from competing ions such as H^+ and Mg^{2+} (Marban *et al.*, 1980; O'Doherty *et al.*, 1980; Lee *et al.*, 1980). While there are a number of uncertainties in the determination of free Ca^{2+} when using the photoproteins for intracellular recording (e.g. Mg^{2+} is a competitive inhibitor) the response kinetics are independent of Ca^{2+} concentration and allow resolution of transients down to the 10 ms level (Hastings *et al.*, 1969). Applied to cardiac muscle, it has been found necessary to inject a large number of cells and also to use signal averaging techniques to accumulate data from many contractures. The traces obtained give indications of two temporally distinct processes that lead to the elevation in cytosol Ca^{2+} (probably indicating two sources) and show that the peak of Ca^{2+} concentration coincides with the maximum rate of tension development (Allen and Blinks, 1978; Wier, 1980).

2. *Populations of Microscopic Cells*

It has long been evident that one way to introduce the Ca^{2+}-sensitive photoproteins into the cytosol of populations of cells would be by fusion after enclosing the indicators in suitable lipid carriers. It was also evident that there would be many practical problems to be overcome, but a start has now been made (Hallett and Campbell, 1982). Red cell ghosts, loaded with obelin, were fused to rat neutrophils using inactivated Sendai virus, and under favourable conditions up to 80% of the contained material could be transferred into the cytosol of the resulting hybrids. These remained physiologically responsive to phagocytic (particulate) stimuli as monitored by the generation of active oxygen (luminol chemiluminescence measurements), but there was no detectable change in obelin fluorescence. Only when the cells were subjected to gross over-stimulation such as saturating concentrations of f-MetLeuPhe, or the application of lytic complement could changes in intracellular Ca^{2+} be detected. Whilst these changes certainly preceded the generation of active oxygen, the raised Ca^{2+} levels (approximately 8 μM for 10^{-6} M f-MetLeuPhe) were thereafter maintained for long periods, suggesting defective operation of the homeostatic mechanisms which might have been expected to return cytosol Ca^{2+} towards resting levels (Naccache *et al.*, 1979a). These experiments give some indications of future possibilities, but the methods will need much refinement before they can provide useful data upon which conclusions regarding mechanisms can be based.

1. *Water Soluble Absorption Indicators*

There exist a number of low molecular weight-soluble indicators having sensitivity for Ca^{2+} in the range relevant to the cytosol. These include arsenazo III (sensitive to changes of less than 10^{-7} M Ca^{2+}), murexide (sensitive to changes down to about 10^{-6} M Ca^{2+}) and antipyrylazo III (of intermediate sensitivity) (Scarpa *et al.*, 1978). Apart from the practical difficulties of introducing such indicators into the cytosol of populations of microscopic cells, there are also considerable problems in the interpretation of the data, obtained as changes in absorption, which these indicators provide. These arise from the sensitivity of the dyes to small changes in pH, and also to the variability depending on conditions, of the stoichiometry of the Ca^{2+}–dye complexes. Thus, as has been pointed out by Brown and Rydqvist (1981), a change of Ca^{2+} of approx. 3.5×10^{-6} M would be calculated on the basis of $1:1$ stoichiometry for an optical density change of 0.01 in a cell injected with arsenazo III at 0.75×10^{-3} M (assuming $k_d \sim 30 \times 10^{-6}$ M (Brinley *et al.*, 1977)). With stoichiometry of $1:2$ (and $k_d = k_1 \times k_2 = 1.8 \times 10^{-9}$ M^2 (Thomas, 1979)) then the change in cytosol Ca^{2+} would be $\sim 2.8 \times 10^{-7}$ M. The non-linearity of the optical response of arsenazo III to Ca^{2+} is a property shared by antipyrylazo III (Ogawa *et al.*, 1980) and both dyes furthermore, are sensitive to changes in Mg^{2+}. Murexide provides a linear response in Ca^{2+} concentration but even when the problems of introducing it into the cytosol have been overcome, its use will be limited by its rather low sensitivity. Other metallochromic dyes sensitive to Ca^{2+} in the range of interest include chlorophosphonazo III (Brown *et al.*, 1975) and tetramethylmurexide (Ohnishi, 1978; Ogawa *et al.*, 1980) but so far there have been few reports recording their use as intracellular indicators.

2. *Chlorotetracycline*

The antibiotic chlorotetracycline forms organic soluble complexes with divalent cations which fluoresce with 50- to 200-fold greater intensity than the uncomplexed ligand in aqueous solution. Because it penetrates membranes readily and because the emission spectra of the Ca^{2+} and the Mg^{2+} salts are easily distinguished, this compound has been used as a probe for monitoring changes in intracellular Ca^{2+} (Caswell, 1972). The main problems in the use of chlorotetracycline concern its location within the cell; since the fluorescent complex is soluble in organic phases, it has been assumed that the cytosol is not the locality which is probed by this indicator. The real situation is likely to be more complicated than this, because the complex of chlorotetracycline with divalent cations forms ternary complexes with divalent cation binding sites such as ADP and ATP normally present in the cytosol (Caswell, 1972;

Gains, 1980). Nor has it proved feasible to calibrate the fluorescence signals in terms of Ca^{2+} concentration in any meaningful way. At concentrations in excess of 10^{-5} M, chlorotetracycline modulates the intracellular Ca^{2+} of pancreatic islet cells in a manner reminiscent of the ionophores (Sehlin and Taljedal, 1979).

Chlorotetracycline has most usefully been applied to the study of Ca^{2+} mobilization and redistribution from intracellular sources in tissues such as pancreatic islets (Sehlin and Taljedal, 1979; Gagerman et al., 1980), pancreatic acinar exocrine cells (Chandler and Williams, 1978a,b), platelets (LeBreton et al., 1976; Feinstein, 1980) and neutrophils (Naccache et al., 1979a). As generally used, the dye is allowed to equilibrate at low concentration ($\sim 10^{-5}$ M) with the cells present in suspension in cuvettes in the fluorimeter, and the changes consequent on adding agonistic ligands or other agents are monitored directly. It is assumed that the fluorescence emission is quite blind to events taking place in the aqueous phases both inside and outside the cells. The normal effect of stimulating the cells is to cause a decrease in the fluorescence emission, and this is taken to imply that there is a decrease in the amount of Ca^{2+} associated with the organic (i.e. membrane) phase of the cells. In all of the above-mentioned tissues, the agonist-induced changes in chlorotetracycline fluorescence have been shown to occur before the appearance of secreted material, and by this criterion the probe would appear to be monitoring early events involving Ca^{2+} initiated by the occupation of relevant receptors. In neutrophils (treated in the absence of cytochalasin B so as to prevent artefactual signals arising from the exocytotic degranulation) it was shown that the pool of interior Ca^{2+} could be depleted by addition of f-MetLeuPhe as subsequent addition of another soluble agonist (complement derived c5a) was then without effect on the fluorescence emission.

3. Quin2

Quin2 is a water soluble indicator having some structural resemblance to EGTA (Tsien, 1980), and it shares with this compound a considerable specificity between Ca^{2+} and Mg^{2+}. In the presence of the normal cytosol electrolytes it has an apparent affinity for Ca^{2+} of 115×10^{-9} M, and shows a sixfold increase in fluorescence (339–492 nm) on saturation with Ca^{2+} (Tsien et al., 1982a,b). The problem of transferring the dye into the cytosol of intact cells was neatly solved by the use of the acetomethoxy ester of quin2 (Tsien, 1981). This has considerable organic solubility, and traverses the membrane into the cytosol where it is rapidly and completely saponified to the tetracarboxylic acid which is then trapped. Cells are thus provided with a low molecular weight Ca^{2+}-sensitive fluorescent indicator having affinity for Ca^{2+} in the biologically interesting range, and one furthermore which can be monitored with rather simple instrumentation. Unlike the Ca^{2+}-sensitive photoproteins which are effectively consumed at rates which depend on the concentration of Ca^{2+}, and for which photon counting instrumentation must be used, the Ca^{2+}-sensing dye can be detected and monitored by steady state measurements using a normal laboratory fluorimeter. The dye response to Ca^{2+} can be simply calibrated by point interpolation within the range of fluorescence signals resulting from lysis of the cells after the experiments into solutions containing zero or saturating concentrations of Ca^{2+}. This new approach to the problem of cytosol Ca^{2+} measurement has seen few applications so far, but it certainly offers the method of greatest scope and promise.

4. Ca^{2+} Kinetics in Lymphocytes

Quin2 has been applied to lymphocytes—cells for which cytosol Ca^{2+} had not previously been measurable (Tsien et al., 1982a,b). After a delay of about 30 s after applying T-cell mitogens (con A or PHA) the population average cytosol Ca^{2+} was found to double during 2 min from a resting level of 120 nM. In terms of the changes which take place in the responding (T) cells this probably represents an underestimate. The lectin induced changes in cytosol Ca^{2+} preceded membrane hyperpolarization (from −50 to −70 mV), probably the result of Ca^{2+}-induced K^+ efflux referred to earlier. Early changes in cytosol Ca^{2+} monitored by quin2 could be prevented by agents which have the effect of elevating cyclicAMP (such as theophylline or cholera toxin) and which also prevent the sequence of events which lead to eventual DNA synthesis after a period of about 48 hours.

It is a fundamental truism that the act of measurement must perturb the subject system, and in consideration of introduction of an intracellular chelator, one might expect the levels of Ca^{2+} to be depressed. However, while the Ca^{2+}-transients due to A23187 or ionomycin are indeed blunted at higher quin2 concentrations, the fluorescence due to the Ca.quin2 complex under resting conditions remains rather invariant as quin2 is varied over the range 0.5–5 mM. It appears that the plasma membrane compensates for the presence of the chelator by allowing the entry of Ca^{2+} during loading so as to maintain the homeostatic condition (Tsien et al., 1982b). It will be apparent that quin2 offers many advantages over other methods for the measurement of intracellular Ca^{2+}, and its application to other cell systems may be expected in the near future. For the present the main disadvantage of quin2 derives from the fluorescence of the free acid as this restricts the signal range to a sixfold increase at full saturation; by comparison with this, the sensitivity of the Ca^{2+}-sensitive photoproteins cover ranges encompassing many orders of magnitude.

VI. What Does Cytosol Ca^{2+} Do?

A. Ca^{2+} AND CYCLIC AMP

Given a tissue whose functions are dependent on Ca^{2+}, and which can be activated by the Ca^{2+}-ionophores, then the simplest conclusion has to be that Ca^{2+} acts as a second messenger in the activation process. The question of how Ca^{2+} actually causes cell activation generally remains a thorny problem and furthermore one complicated by its interplay with systems known also to be controlled by cyclicAMP.

A proper consideration of this field (recently reviewed by Rasmussen *et al.*, 1979) is well beyond the scope of the present account. Suffice to say here, that the enzymes through which the concentration of cyclicAMP is controlled, and the enzymes through which the effects of cyclicAMP are expressed (protein kinases) are subject to modulation by Ca^{2+} (see Table III). Similarly, the concentration of Ca^{2+} and the processes which it regulates are subject to modulation by cyclicAMP. However one attempts to control either of the second messengers unilaterally, it is almost impossible to be sure that the other remains unchanged, or that the sensitivity (i.e. K_M) of systems subject to them remain unaltered.

The investigation of single processes in isolation obviates the above conundra. The problem now becomes that of identifying which of the many Ca^{2+}-dependent reactions are essential to the expression of activity in intact cells, and which should be counted as also-rans (possibly of importance as subsidiary activities in prolonging or terminating activation, or in post-activation repair and homeostasis etc.).

B. CALMODULIN

1. *Enzymes Regulated by Calmodulin*

Many but not all of the intracellular functions of Ca^{2+} are mediated through the ubiquitous Ca^{2+}-binding protein now called calmodulin. This has become an integral word in the vocabulary of all those who concern themselves with the role and the functions of intracellular Ca^{2+}. Some enzymes which are regulated by Ca^{2+} through the mediation of calmodulin are listed in Table III. Beyond indicating that calmodulin (or troponin C) has been linked with each of the categories of cell function listed in the first paragraph of this chapter, it is not the purpose of this section to attempt any general survey of the field (see reviews by Klee *et al.*, (1980), Means and Dedman (1980), Means *et al.* (1982)). Rather the aim of this section is to suggest that the criteria commonly used for diagnosing a role for calmodulin in the activities of intact cells and tissues are far from ideal, and that there also exist a number of intracellular systems which, though sensitive to Ca^{2+} at micromolar concentrations, may not be regulated by calmodulin or any other Ca^{2+}-regulatory proteins.

TABLE III

Enzymes activated by Ca.calmodulin

Adenylate cyclase	Brostrom *et al.* (1975)
Phosphodiesterase	Brostrom *et al.* (1975) Cheung (1970) Kakiuchi *et al.* (1970)
Phosphorylase-*b* kinase	Cohen *et al.* (1978)
Myosin light chain kinase skeletal muscle smooth muscle non-muscle	Yagi *et al.* (1978) Dabrowska *et al.* (1978) Yerna *et al.* (1979) Dabrowska and Hartshorne (1978) Hathaway and Adelstein (1979)
Ca^{2+}-ATPase red cells cardiac muscle	 Gopinath and Vincenzi (1977) Jarrett and Penniston (1977) Sobue *et al.* (1979) Katz and Remtulla (1978)
NAD^+ kinase	Anderson and Cormier (1978)
Tryptophan mono-oxygenase	Yamauchi and Fujisawa (1979)
Phospholipase A_2	Walenga *et al.* (1981)

2. *Criteria for a Role in Activated Cell Functions*

Calmodulin is present in most tissues of eukaryotic origin. The test most frequently applied for identifying its role in cellular processes is to measure the inhibitory effects of drugs of the phenothiazine or butyrophenone series. These bind to calmodulin in the presence of Ca^{2+} and prevent the sequence of Ca^{2+} activation of those enzymes which are normally sensitive to it (Levin and Weiss, 1978, 1979; Vincenzi, 1981). By such experiments, a role for calmodulin has been assigned in the secretory processes of mast cells (Douglas and Nemeth, 1982), neutrophils (Elferink, 1978; Naccache *et al.*, 1980) and platelets (White and Raynor, 1980). However, these drugs have other effects.

Among these may be cited the inhibition of phosphatidate phosphohydrolase which results in the redirection of glyceride and phospholipid synthesis, and leads to accumulation of phosphatidylinositol (Allan and Michell, 1975; Brindley *et al.*, 1975). Further, these drugs are also known to interfere with the binding of some hormones (in particular, α-adrenergic agonists) to their receptors (Blackmore *et al.*, 1981). In lymphocytes, some of the phenothiazines actually reduce the concentration of ATP and depress glycolytic flux when applied at concentrations at which calmodulin regulated functions are known to be sensitive (Corps *et al.*, 1982). The finding (Landry *et al.*, 1981) of a close correlation between the inhibition of six different neuroleptic drugs of a cyclic nucleotide phosphodiesterase (a well recognised Ca.calmodulin-dependent system (Levin and Weiss, 1978, 1979)) and the protection of red blood cells against osmotic lysis (so-called membrane stabilization (Seeman and Kwant, 1969; Alhanaty and Livne, 1974), for which a role for Ca.calmodulin seems most implausible) and the knowledge that inhibition by these compounds also correlates with their partition coefficients in organic phases (Norman *et al.*, 1979) should act as a salutary warning. There is a real possibility that these drugs could alter cell functions through non-stereospecific effects at unknown locations within the cell and that inhibition does not unequivocally implicate a role for calmodulin.

The inhibition of cortical granule release following injection of antibodies to calmodulin into sea urchin eggs can certainly be regarded as strong evidence that it is calmodulin which is responsible for the sensitivity of this exocytotic reaction to Ca^{2+} (Steinhardt and Alderton, 1982). (Such a role had been previously suggested, but not proven, by the demonstration of inhibition with trifluoperazine (Baker and Whitaker, 1979).) A study of the effects of trifluoperazine and pimozide on insulin secretion from islet β-cells in which care was taken to account for non-specific effects and to avoid the pitfalls listed above (Henquin, 1981) is also strongly suggestive of a role for calmodulin at some point in this process. What this might be remains a matter for speculation.

C. Ca^{2+} IN THE CONTROL OF ACTIN FUNCTION

Calmodulin shares with the troponins much sequence and functional similarity (Amphlett *et al.*, 1976; Dedman *et al.*, 1977; Cohen *et al.*, 1979; Yerna *et al.*, 1979), and on this basis its role in conferring sensitivity to Ca^{2+} on the contractile activity of the actomyosin filamentous network of non-muscle cells can be understood. Localized movements in non-muscle cells (e.g. in the mechanism of chemotaxis and phagocytosis in neutrophils and macrophages, and shape change in platelets) are also mediated through

TABLE IV

Ca^{2+} dependent actin-binding proteins

Name	Source	Mol wt. ($\times 10^{-3}$)	Subunits	Ca^{2+} sites	$k'd$ (μM)	Reference
(A) *F-actin severing proteins*						
Gelsolin	Macrophages Neutrophils	91	1	~2	1.1	Yin and Stossel (1979) Yin et al. (1980)
Villin	Intestinal epithelial microvilli	95	1	1	2.5	Glenney et al. (1980)
Fimbrin	Intestinal epithelial microvilli	68	1	1	9 (Mg^{2+} ~ 1 mM)	Glenney et al. (1981)
	Platelets	95				Wang and Bryan (1980)
(B) *Dimeric extended rod-like crosslinkers*						
Actinogelin	Ehrlich ascites cells	115				Mimura and Asano (1979)
α-actinin	Platelets	105				Rosenberg et al. (1982)
α-actinin	Hela cells	100	2			Burridge and Feramisco (1981) Feramisco and Burridge (in prep. 1982)

cycles of gel–sol transformation in the cortical cytoplasm which are controlled through the reversible polymerization of actin into three dimensional lattices. The actin binding proteins which control the gel–sol transition are of two main types: there are those which act by regulating the length of the actin filaments and there are those which function as crosslinkers analagous to α-actinin of smooth muscle. Some properties of the Ca^{2+}-sensitive actin binding proteins are summarized in Table IV. The proteins which act to control filament length are characteristically monomeric globular molecules, and a number of these are certainly Ca^{2+}-binding proteins. The best described protein of this class is gelsolin (Yin and Stossel, 1979; Yin et al., 1980). The purified protein binds 1.7 moles of Ca^{2+} per mole at a single class of binding site having high affinity ($K' = 1.09 \times 10^6$ M^{-1} in the presence of K^+ (0.1 M) and Mg^{2+} (2 mM) (Yin and Stossel, 1980)). The ability of gelsolin to bind Ca^{2+} and to depolymerize actin filaments in the presence of Ca^{2+} is destroyed by prior heating, and this property alone sets it quite apart from calmodulin, which is characteristically heat stable (Beal et al., 1977). The actin crosslinkers, dimeric extended flexible molecules (Hartwig and Stossel, 1981), are mainly insensitive to Ca^{2+} (Brotschi et al., 1978; Yin and Stossel, 1980) and so it is unlikely that the Ca^{2+}-sensitive crosslinkers (listed in Table IV) play any important role in the gel–sol transformation. They might, however, be involved in anchoring organelles, such as the secretory granules of the chromaffin cells which are known to contain actin (Wilkins and Lin, 1981) to the F-actin matrix in the cytoplasm of resting cells (Fowler and Pollard, 1982).

D. Ca^{2+} AND MEMBRANE FUSION

It is widely believed that the membrane fusion which underlies the mechanism of exocytosis and endocytosis is controlled by intracellular Ca^{2+}, and there have been many attempts to identify the mechanism of this process.

1. Synexin: Adrenal Medulla Fusion Protein

The secretory granules of adrenal chromaffin cells will aggregate and fuse (giving the characteristic pentalaminar cross-sectional image) when treated with high (2–20 mM) concentrations of Ca^{2+} or Mg^{2+} (Edwards et al., 1974; Schober et al., 1977). Addition of a cytoplasmic protein, synexin, reduces the threshold concentration for Ca^{2+} to 6 μM (Creutz et al., 1978), (Mg^{2+} at these concentrations is without effect). It is suggested that in the presence of Ca^{2+} the synexin self-associates to form hydrophobic rods. These might interdigitate into the membranes of the secretory granules and the plasma membrane and form nucleation sites about which the reorganization of phospholipids which determines membrane fusion could commence (Creutz et al.,

1979). Synexin is probably specific in its interaction with appropriate membranes (it aggregates and fuses the secretory granules and plasma membranes of chromaffin cells while having no effect on adrenal cortical mitochondria and red cell membranes (Dabrow et al., 1980)). However, one should distinguish carefully between what synexin can do and what it actually does do. Certainly, optimism that synexin might really be the endogenous fusion protein responsible for causing Ca^{2+}-dependent exocytosis should be tempered by the knowledge that even in its presence the half-maximal concentration of Ca^{2+} for fusion of chromaffin granules is 200 μM.

2. Ca^{2+}-dependent Membrane Fusion without Intervention of Synexin

It is not known whether this is a general mechanism for fusion of intracellular membranes or even whether the fusion protein is widely distributed among the tissues. Indeed Ca^{2+}-dependent fusion of secretory vesicles from other sources appears to occur without the intervention of any cytosol protein. Conversion of proalbumin to mature albumin is catalysed by cleavage of the N-terminal pentapeptide by a cathepsin B-like proteinase (Judah and Quinn, 1978; Quinn and Judah, 1978). As isolated, the substrate and the cleaving enzyme are located in different populations of golgi-derived vesicles, and conversion of the proalbumin then proceeds on addition of Ca^{2+} in the range 1–10 μM. Electronmicroscopy observations support the inference that this conversion occurs upon fusion of the vesicles (Gratzl and Dahl, 1976, 1978), and it is this which allows access of enzyme to substrate. Similarly, Ca^{2+} in the micromolar range induces fusion of the secretory granules of islet β-cells (Dahl and Gratzl, 1976) neurohypophysis (Gratzl et al., 1977), and adrenal chromaffin cells (Ekerdt et al., 1981). The latter study points to the existence of membrane proteins having affinity for Ca^{2+} in the micromolar range and which appear to be involved in the fusion process.

3. Ca^{2+}-dependent Fusion of Model Membranes

Alternatively, Ca^{2+} could cause fusion by direct interaction with the phospholipids, but while divalent cations can undoubtedly cause fusion and non-lytic intermixing of the contents of liposomes formed predominantly of acidic phospholipids, the concentration requirements are generally up in the millimolar range (Papahadjopoulos et al., 1979). While synexin, the adrenal medulla fusion protein, can enhance the rate of fusion of liposomes by up to 100-fold, the concentration of Ca^{2+} required still remains high (Hong et al., 1981) until phosphatidate is included in the mixture of lipids used to generate the liposomes. In such systems, threshold concentration for Ca^{2+} can drop to around 10 μM, a level which is above, but tantalizingly close to those pertaining in the cytosol of activated cells (Sundler and

Papahadjopoulos, 1981; Hong *et al.*, 1982a,b). The presence of phosphatidylinositol opposes this. Since the amount of phosphatidylinositol decreases and the amount of phosphatidate increases following stimulation of cells having Ca^{2+}-mobilizing receptors (see below), it is possible that a function of the phosphatidylinositol cycle is to control the propensity of membranes to fuse.

The action of synexin in these model systems appears to be exclusively catalytic: it accelerates fusion only of those vesicles having phospholipid composition that would undergo fusion in its absence (albeit at high and biologically irrelevant Ca^{2+} concentrations). Thus, synexin accelerates the fusion process to a rate such that it can be detected at lower levels of Ca^{2+}; even in the presence of Mg^{2+} (millimolar concentrations) synexin enhanced fusion retains its sensitivity to Ca^{2+} and also its lipid specificity (Hong *et al.*, 1982a). Other Ca^{2+}-binding proteins which have been tested (these include calmodulin from various sources, muscle parvalbumin and prothrombin) not only fail to enhance the rate of fusion, but actually retard the rate of synexin-induced fusion (Hong *et al.*, 1982b). Clearly these bind to the phospholipid vesicles, and so the inhibition has been rationalized on the supposition that the exposed areas of these proteins, after binding to one lipid surface are unavailable to bind another: they are effectively monovalent with respect to the lipid vesicles.

In all likelihood synexin is but one of a family of related fusogenic proteins having different tissue and membrane specificity. Calelectrin, recently identified in a number of excitable tissues (Walker, 1982), also undergoes self aggregation and accelerates the fusion of biological membranes and acidic liposomes. It is reported to be sensitive to Ca^{2+} in the micromolar range (Sudhof *et al.*, 1982).

E. Ca^{2+} IN THE CONTROL OF SOME METABOLIC PROCESSES

There are many roles for intracellular Ca^{2+} in the control of metabolic processes, and while some of these are unquestionably mediated through calmodulin (see Table III), others appear to be due to direct effects of Ca^{2+} on target enzymes. In particular, since rat liver mitochondria are devoid of calmodulin (Smoake *et al.*, 1974), effects of Ca^{2+} on the enzymes of the citric acid cycle and related processes of the inner mitochondrial matrix are likely to be exerted either directly or through the mediation of other, as yet unidentified Ca^{2+}-sensitizing proteins. Three dehydrogenases exclusively located in the mitochondria of mammalian cells have been shown to be activated by Ca^{2+} (Denton and McCormack, 1980, 1981). These are pyruvate dehydrogenase, isocitrate dehydrogenase and 2-oxoglutarate dehydrogenase (Hansford and Chappell, 1967). When purified, these can be activated by

Ca^{2+} in the range 0.1–10 μM. While the effect of Ca^{2+} on the two citric acid cycle enzymes is most probably direct, due to the enhancement of the affinity of the enzymes for their substrates (Denton *et al.*, 1978; McCormack and Denton, 1979) the effect on pyruvate dehydrogenase is more complex. Here, Ca^{2+} activates the phosphatase which controls the proportion of the inactive (phosphorylated) and active (non-phosphorylated) form of the dehydrogenase (Denton *et al.*, 1972). Measurements with intact mitochondria indicate that alteration of extramitochondrial Ca^{2+} in the range 0.1–10 μM leads to parallel activation of all three enzymes, which suggests that the concentration of Ca^{2+} in the mitochondrial sol is not very different from that of the cytosol (McCormack and Denton, 1980; Denton *et al.*, 1980). If this is indeed the case, then there exists a line of communication between the cell surface Ca^{2+}-mobilizing receptors and oxidative metabolism not requiring the imposition of metabolic loads (i.e. altered ATP/ADP or NAD+/NADH ratios). Some situations in which this may occur have been identified.

Citric acid cycle activity increases in heart muscle on perfusion with adrenaline and other inotropic agents (McCormack and Denton, 1981) and in liver on stimulation with vasopressin or α-adrenergic agonists (Babcock *et al.*, 1979; Hems, 1979). In both these systems, this correlates with increases in the proportion of pyruvate dehydrogenase in its active (non-phosphorylated) form (Hems *et al.*, 1978; Hiraoka *et al.*, 1980; McCormack and Denton, 1981).

Is it also possible that the early increase in lymphocyte metabolism following stimulation with mitogenic lectins could be controlled by a similar mechanism? A Ca^{2+}-imposed enhancement in citric acid cycle activity could provide metabolic intermediates for protein and nucleic acid synthesis which might limit the synthesis of macromolecules in the quiescent cells and an increase in the rate of pyruvate oxidation has been shown to occur within minutes of applying mitogenic lectins (Beachy *et al.*, 1981). Unfortunately for this idea, the only data relating to the question of Ca^{2+} and oxidative metabolism in lymphocytes derive from experiments in which Mg^{2+} was excluded from the media used for mitochondrial incubations (Vijayakumar and Weidemann, 1976). The extraordinarily low concentrations of Ca^{2+} required both to activate (20 nM) and to inhibit (50 nM) pyruvate dehydrogenase under these conditions (Hume *et al.*, 1978) should therefore be taken merely as an indicator of Ca^{2+}-dependent events at this level.

VII. The Search for Biochemical Mechanisms of Ca^{2+} Mobilization

While we can state the concentrations of Ca^{2+} which are commensurate with cellular activation and are now in sight of practical methods for determining the sources and the kinetics of Ca^{2+} for many cell types, knowledge of the mechanism whereby Ca^{2+} passes into the cytosol upon stimulation of receptors eludes us almost completely. This contrasts with the considerable knowledge we possess concerning the other main class of receptors—those that control adenylate cyclase. The reason for our ignorance about the process of receptor induced Ca^{2+} mobilization is fairly obvious. We possess no enzymic handle comparable to adenylate cyclase which is an intrinsic part of the transduction function of its associated receptors. The molecular details of this system are fairly well understood: it is known for example that there are at least three different proteins involved and that these interact within the membrane. These are the receptor protein, a regulatory protein having a binding site for GTP, and adenylate cyclase itself. Such a system seems to operate in all cells studied and has also been highly conserved in evolution so that it is possible to reconstitute *in vitro* the receptor from one cell with the catalytic protein from a different cell (or even organism) (Schulster *et al.*, 1978).

In the case of the Ca^{2+}-mobilizing receptors we do not know whether these operate according to a single or even to a few common mechanisms, though this is an assumption which has formed the basis of much biochemical investigation. The aim has been to identify intermediate enzyme-catalysed steps, initiated by occupation of receptors by agonistic ligands, which then determine the mobilization of Ca^{2+}.

A. PHOSPHATIDYLINOSITOL HYDROLYSIS: CAUSE OR CONSEQUENCE OF Ca^{2+} MOBILIZATION?

1. *Universal Occurrence*

The primary impetus to the idea of enzyme-catalysed intermediate steps was the realization that there is an increase in the rate of metabolic turnover of phosphatidylinositol, when receptors known to mobilize Ca^{2+} are activated by specific ligands (Michell, 1975). Many tissues have been examined in this respect and it is clear (Table V) that there is a high degree of coincidence between those tissues exhibiting agonist-mediated phosphatidylinositol responses and those which can be activated by the Ca-ionophore A23187. Receptors mediating such responses include the widely distributed muscarinic cholinergic and α-adrenergic receptors as well as the more sparsely distributed receptors such as those for IgE and f-MetLeuPhe which have been considered in this chapter. On the other hand, there is no change in the metabolism of phosphatidylinositol when β-adrenergic receptors linked to adenylate cyclase are activated (Michell, 1975). The initiating step of the phosphatidylinositol cycle (Fig. 8) is probably the hydrolysis of phosphatidylinositol by a phospholipase C specific for this lipid. The diacylglycerol so formed is rapidly phosphorylated to form phosphatidate, and

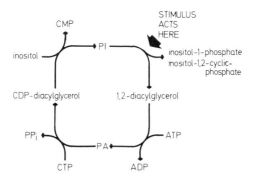

FIG. 8. The reactions of the phosphatidylinositol cycle.

phosphatidylinositol is then regenerated via CDP-diacylglycerol. Until recently the common experimental approach has been to measure the incorporation of ^{32}P (or less frequently ^{3}H-inositol) into phosphatidylinositol on the assumption that labelling of phosphatidylinositol is indicative of earlier events involving its breakdown. This presumption is not really tenable since any reaction leading to the production of diacylglycerol or phosphatidate would provide substrates capable of generating phosphatidylinositol. Clearly, direct measurement of phosphatidylinositol hydrolysis is to be preferred and this can be done by measuring the loss of phosphatidylinositol, previously pulse labelled with ^{3}H-glycerol, or by direct chemical estimation of the separated lipids. An alternative and probably better approach is to measure the appearance of anionic labelled water-soluble products of phosphatidylinositol hydrolysis (i.e. inositol phosphates) from cells previously labelled with ^{3}H-inositol.

2. Criteria for a Role of Phosphatidylinositol in Ca^{2+} Mobilization

In seeking to establish whether such a step is universal (or even widespread) in the mobilization by receptors of Ca^{2+}, the following criteria should ideally be fulfilled:

(a) *It should not be dependent on a rise in intracellular Ca^{2+}* Because until recently it has not been possible either to measure intracellular Ca^{2+} nor to manipulate cytosol Ca^{2+} with precision, it has been common to test: (1) the effect of extracellular Ca^{2+}; (2) the effect of Ca^{2+}-ionophores; and (3) the effect of Ca^{2+}-antagonists. Thus, in the case of parotid acinar tissue stimulated with cholinergic muscarinic or with α-adrenergic agonists, there is no effect of extracellular Ca^{2+} (Jones and Michell, 1975; Oron *et al.*, 1975) on the phosphatidylinositol response although Ca^{2+} deprivation is known to prevent fluid secretion. In this same tissue, the ionophore A23187 is without effect on

TABLE V

Coincidence of ionophore stimulation and phosphatidylinositol responsiveness

Tissue functions	Tissue activation by A23187 (references)	Agonist stimulation of phosphatidylinositol turnover (references)
Exocytotic secretion		
Synaptosomes : acetylcholine	Michaelson and Sokolovsky (1978)	Hawthorne and Bleasdale (1975)
Synaptosomes : dopamine	Holz (1975)	Hawthorne and Pickard (1979)
CNS : noradrenaline and GABA	Vargas et al. (1976)	
	Cotman et al. (1976)	
	Foreman et al. (1973)	
Mast cells : histamine	Lewis et al. (1975)	Cockcroft and Gomperts (1979c)
Basophilic leukaemia cells : mediators	Smith and Ignarro (1975)	Kennerley et al. (1979)
(SRS, ECF-A, PAF, histamine)	Schneider et al. (1978)	
Neutrophils : lysosomal enzymes	Ashby and Speke (1975)	Bennett et al. (1980a), Cockcroft et al. (1981)
Macrophages : lysosomal enzymes	Bicknell and Schofield (1976)	Ogmundsdottir and Weir (1979)
Islets of Langerhans : insulin and glucagon	Eimerl et al. (1974)	Clements and Rhoten (1976)
Pituitary : growth hormone	Feinman and Detwiler (1974)	Bicknell et al. (1979)
Pancreas : amylase	Cochrane et al. (1975)	Hokin (1968)
Platelets : ATP		Lapetina and Cuatrecasas (1979)
Adrenal medulla : catecholamines		Trifaro (1969)
Fluid secretion		
Fly salivary gland	Prince et al. (1973)	Fain and Berridge (1979a)
Rabbit ileal mucosa	Bolton and Field (1977)	Eggman and Hokin (1960)
Rabbit colon	Frizzell (1977)	
Rabbit lacrimal gland	Pholramool and Tangkrisanavinont (1976)	Jones et al. (1979)
Frog cornea	Candia et al. (1977)	
Rat parotid	Selinger et al. (1974)	Oron et al. (1975)
Amphiuma red cells	Gardos et al. (1976)	

Contractile activity		
Smooth muscle	Triggle et al. (1975)	Jafferji and Michell (1976a,b)
Cardiac muscle	Holland et al. (1975)	
Skeletal muscle	Mobley (1977)	
Synthesis of inflammatory mediators		
Slow reacting substance of anaphylaxis in		
basophilic leukaemia cells	Jakschik et al. (1977)	
Prostaglandins and thromboxane in platelets		Lloyd and Mustard (1974)
and renal medulla	Knapp et al. (1977)	
Prostaglandin E in macrophages	Gemsa et al. (1979)	
Motility		
Leucocyte locomotion	Wilkinson (1975)	Bennett et al. (1980a)
Bacterial locomotion	Ordal (1977)	
Control of flagella wave motion	Holwill and McGregor (1975)	
Proliferative responses		
Lymphocyte transformation	Maino et al. (1974)	Maino et al. (1975)
DNA synthesis in bone marrow stem cells	Gallien-Lartigue (1976)	
Parthenogenesis of *Xenopus oocyte*	Belle et al. (1977)	
Activation of sea urchin eggs	Steinhardt and Epel (1974)	
Metabolic responses		
Glycogenolysis	Keppens et al. (1977)	Kirk et al. (1977)
Changes in renal gluconeogenesis	Mennes et al. (1978)	Bennett et al. (1980b)
Oxidative metabolism in neutrophils	Schell-Frederick (1974)	

phosphatidylinositol hydrolysis (Jones and Michell, 1975). Similarly, there is no effect of Ca^{2+}-antagonists (D600, cinnarizine, lidoflazine etc.) on ^{32}P-incorporation in phosphatidylinositol in guinea pig ileal smooth muscle although these compounds prevent contraction (Jafferji and Michell, 1976c).

(b) *There should be a relevant timecourse* Phosphatidylinositol responses are certainly rapid: in platelets (Bell and Majerus, 1981) and rabbit neutrophils (Cockcroft *et al.*, 1981) the rate of breakdown is at least as fast as other measured tissue responses but the question of the precursor–product relationship has not been resolved.

(c) *The responses should be triggered only by relevant receptors* As indicated earlier, this was the observation that provided the initial impetus for experimental investigation in this field. Further, the extent of phosphatidylinositol labelling in ileal smooth muscle and of breakdown in pancreas correlates very closely with the extent of occupancy of the muscarinic cholinergic receptors (Michell *et al.*, 1976). Since activity in these tissues maximizes with less than 5% of receptors occupied it would appear that for these systems, phosphatidylinositol hydrolysis is tied to activities at the receptor rather than events downstream of this. Similarly, phosphatidylinositol hydrolysis correlates closely (but hydrolysis of phosphatidylinositol-4,5-bisphosphate correlates even more closely) with the extent of vasopressin binding to hepatocytes, whereas phosphorylase-a activity maximizes when only a small proportion of the receptors are occupied (Michell *et al.*, 1981).

(d) *Phosphatidylinositol hydrolysis should occur in the plasma membrane* There are few studies relating to the subcellular location of phosphatidylinositol responses (Table VI), and there are only three which address the question of phosphatidylinositol hydrolysis directly. These have produced conflicting answers. While our work with f-MetLeuPhe stimulated neutrophils points to the plasma membrane being the site of both phosphatidylinositol hydrolysis and phosphatidate labelling (Bennett *et al.*, 1982), no single site for phosphatidylinositol hydrolysis could be discerned in carbachol-stimulated adrenal medulla (Azila and Hawthorne, 1982) and vasopressin stimulated hepatocytes (Kirk *et al.*, 1981). (But contrast Lin and Fain (1981) who identify the plasma membrane as the site of phosphatidylinositol hydrolysis.) The failure to identify a unique site for phosphatidylinositol hydrolysis in these cells is possibly related to the operation of phospholipid exchange enzymes which could cause rapid redistribution among the subcellular organelles. While the findings with lymphocytes (Fisher and Mueller, 1971) and avian salt gland (Hokin-Neaverson, 1977) are in agreement with our own work with rabbit neutrophils, other locations for phosphatidylinositol hydrolysis were identified in islets of Langerhans (Clements *et al.*, 1977) and in synaptosomes (Pickard and Hawthorne, 1978). In these last two tissues it was not demonstrated that the cell fractionation procedure

TABLE VI

Subcellular location of phosphatidylinositol responses in different tissues

Tissue	Stimulus	Measurement	Location	Comments	References
Lymphocytes	PHA	$[^{32}P]$-PA labelling	Plasma membrane	3 min stimulation	Fisher and Mueller (1971)
Avian salt gland	Acetylcholine	PI hydrolysis	Plasma membrane	Chemical determination	Hokin-Neaverson (1977)
Hepatocytes	Vasopressin	$[^3H$-inositol]-PI hydrolysis	No specific location	5 min stimulation	Kirk et al. (1981)
Islets of Langerhans	Glucose	$[^3H$-inositol]-PI hydrolysis	Secretory vesicles	30 min stimulation; failure to account for possible membrane fusion artefacts	Clements et al. (1977)
Exocrine pancreas	Acetylcholine	PI hydrolysis	Rough endoplasmic reticulum	Chemical determinations 80% of cell PI localized in e.r.	Hokin-Neaverson (1977)
Synaptosomes	Depolarization	$[^{32}P]$-PI hydrolysis	Secretory vesicles	Failure to account for possible membrane	Pickard and Hawthorne (1978)
Neutrophils	f-MetLeuPhe	$[^3H$-glycerol]-PI hydrolysis and $[^{32}P]$-PA labelling	Plasma membrane	Cytochalasin not added in order to prevent artefactual redistribution due to membrane fusion	Bennett et al. (1982)
Adrenal medulla	Carbachol	Specific activity of $[^{32}P]$-PI and PA	No specific location	Similar results obtained when Ca^{2+} withheld to prevent secretion	Azila and Hawthorne (1982)

(in both cases defined for unstimulated cells) remained valid after secretion had occurred and so in my view, it is still possible the initial location of phosphatidylinositol hydrolysis could have been the plasma membrane.

3. *Phosphatidylinositol-dependent Ca^{2+} Mobilization*

A particularly strong case for a role of phosphatidylinositol hydrolysis in the mobilization of Ca^{2+} has been made for blowfly salivary glands (Berridge and Fain, 1979; Fain and Berridge, 1979a,b). These secrete iso-osmotic solutions of KCl in response to 5-hydroxytryptamine. On stimulation, Ca^{2+} enters the cells through the serosal surface and is extruded into the saliva through the apical membrane, thus generating a transepithelial Ca^{2+} current which can be measured as a second indicator of receptor activation. In contrast to mammalian systems, the free inositol generated from hydrolysis of phosphatidylinositol during prolonged stimulation leaks out of these cells, which thus become deprived of an essential metabolite for phosphatidyl-inositol resynthesis. Under these circumstances the transepithelial flux of Ca^{2+} declines to basal levels and the tissue remains refractory to hormonal stimulation until the receptor accessible pool of phosphatidylinositol is restored. The ionophore A23187 is without effect on phosphatidylinositol hydrolysis in this tissue (Fain and Berridge, 1979a). Another tissue for which there exist serious reasons for considering that phosphatidylinositol hydro-lysis plays a mediating role in the action of Ca^{2+}-mobilizing receptors is the mammalian parotid gland, which satisfactorily fulfils the criteria listed above (Jones and Michell, 1975). Indeed, a mechanism for this has been proposed.

4. *Is Phosphatidate a Ca^{2+} Ionophore?*

Phosphatidate, an intermediate in the phosphatidylinositol cycle, forms organic soluble complexes with Ca^{2+} (Tyson *et al.*, 1976), and it has therefore been suggested that it might be capable of acting as an ionophore for Ca^{2+} in those membranes in which it is generated. Indeed, phosphatidate appears to cause some slight enhancement of Ca^{2+} permeability when it is incor-porated in phospholipid liposomes (Serhan *et al.*, 1981). An ionophoric mechanism for cell stimulation has been given credence by reports showing activation on application of exogenous phosphatidate to parotid (but not lacrimal) gland (Putney *et al.*, 1980; 1981) and some other tissues (Salmon and Honeyman, 1980; Harris *et al.*, 1981; Ohsako and Deguchi, 1981). A fully saturated synthetic phosphatidate applied at high concentrations (1–3 mM) to rat mast cells causes histamine secretion. This occurs in a manner independent of extracellular Ca^{2+} and the mechanism is thought to involve release of Ca^{2+} from intracellular stores into the cytosol (Pearce and Messis, 1982).

However, doubts must be expressed about whether phosphatidate really acts as an ionophore in these systems (i.e. as a diffusible ion bearer). Experience of the antibiotic ionophores has shown that these mediate ion movements and consequential cellular processes in all systems to which they have been applied. This cannot be said for phosphatidate which is without discernible effect when generated *in situ* in red blood cells (Allan and Michell, 1977a). In reconstituted sarcoplasmic reticulum, which is another relevant system for testing the action of Ca^{2+}-ionophores, it has been shown that the vesicles continue to accumulate Ca^{2+} when supplied with ATP even if the pump protein is in an environment consisting almost entirely of phosphatidate (Bennett *et al.*, 1978), whereas the effect of the ionophores A23187 and ionomycin is to uncouple the pump completely (Caswell and Pressman, 1972; Beeler *et al.*, 1979).

In consideration of the effects of exogenous phosphatidate in biological and model systems, it is well worth pointing out that diacyl (i.e. intact) phospholipids are unable to interdigitate into preformed phospholipid membranes (Uemura and Kinsky, 1972) though their lyso derivatives do so readily. As lysophosphatidate is a very common contaminant of phosphatidate preparations, some of the reported effects of phosphatidate on cells might better be ascribed to its lyso derivative. When applied to platelets, lysophosphatidate certainly causes cell activation in a manner quite similar to A23187 (Gerrard *et al.*, 1979; Lapetina *et al.*, 1981) but with the important difference that the response to the lyso lipid exhibits desensitization (Schumacher *et al.*, 1979; Tokomura *et al.*, 1981). The propensity to undergo such a tachyphylactic response is widely accepted as a criterion in the definition of a receptor mechanism, and this observation surely rules out the possibility that lysophosphatidate stimulates platelets by acting as an ionophore.

In concluding this argument, it is worth recalling (see p. 295) that ionophore-mediated cellular processes would be expected to show an inverse concentration dependence on Ca^{2+}. This is certainly not the case for antigen stimulated mast cells (see Fig. 1a).

5. Ca^{2+}-dependent Phosphatidylinositol Hydrolysis

If there are doubts about phosphatidate being an endogenous ionophore for Ca^{2+} and thereby being the active instrument of phosphatidylinositol hydrolysis in mediating Ca^{2+} fluxes, then it should also be said that for some tissues, phosphatidylinositol metabolism may be quite irrelevant to the operation of Ca^{2+}-mobilizing receptors (Cockcroft, 1982). There are a number of tissues for which receptor-mediated phosphatidylinositol responses are dependent on or certainly enhanced by the presence of extracellular Ca^{2+}. These include neutrophils (Cockcroft *et al.*, 1980a,b, 1981), lymphocytes (Maino *et al.*, 1974; Hui and Harmony, 1980), exocrine pancreas (Farese *et al.*, 1980) and mast cells (ATP⁴⁻ receptor (Cockcroft and Gomperts, 1980), but not the IgE-system (Cockcroft and Gomperts, 1979c)). Furthermore, phosphatidylinositol responses in all these tissues can be initiated by Ca^{2+} ionophores (Allan and Michell, 1977b; Kennerly *et al.*, 1979; Farese *et al.*, 1980; Cockcroft *et al.*, 1981). Phosphatidylinositol hydrolysis, although an early event, and one located at the plasma membrane of rabbit neutrophils, has a dependence on Ca^{2+} which is, for both agonist (Fig. 9a,b) and iono-phore (Fig. 10a,b) stimulated cells, identical to that of lysosomal enzyme secretion. Furthermore, the extent of phosphatidylinositol hydrolysis correlates with the extent of secretion and maximizes when less than 10% of the receptors are occupied (Cockcroft *et al.*, 1981). Clearly, in these cells, phosphatidylinositol hydrolysis is a consequence, not a cause of Ca^{2+} mobilization.

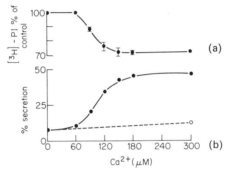

FIG. 9. Dependence on extracellular Ca^{2+} of (a) phosphatidylinositol hydrolysis and (b) β-glucuronidase secretion from cytochalasin-B treated rabbit neutrophils stimulated with f-MetLeuPhe (10^{-8} M). (Cockcroft *et al.*, 1981; *Biochem. J.*)

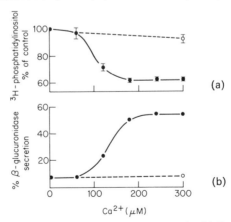

FIG. 10. Dependence on extracellular Ca^{2+} of (a) phosphatidylinositol hydrolysis and (b) β-glucuronidase secretion from cytochalasin-B treated rabbit neutrophils stimulated with ionomycin (5×10^{-6} M). (Cockcroft *et al.*, 1981; *Biochem. J.*).

The realization that there are some tissues in which phosphatidylinositol hydrolysis is probably causal and others in which it appears to be a consequence of Ca^{2+} mobilization suggests that a re-examination of the earlier data, culled from at least 30 tissues would be appropriate. With recent developments we are now in a position to answer the main question by the direct approach: thus we can measure the relevant reaction (phosphatidylinositol hydrolysis instead of phosphatidylinositol labelling) and we could do this under the relevant conditions (control of cytosol Ca^{2+} in the micromolar range by the use of permeabilized cells).

6. Other Roles for Phosphatidylinositol Hydrolysis

The correlation between events involving intracellular Ca^{2+} and the hydrolysis of phosphatidylinositol (first discerned by Michell (1975)) is now more compelling than ever. However, it has become apparent that this broad and possibly universal correlation may have blinded us to complexities involving the function of phosphatidylinositol hydrolysis in different cell types and under different circumstances. Instead of searching for a unique role for this metabolic cycle, it would appear that phosphatidylinositol hydrolysis (and hence the generation of its immediate products, diacylglycerol, phosphatidate and the water soluble inositol phosphates) may perform different functions in different tissues. The possibility that activation of the phosphatidylinositol cycle may be permissive in membrane fusion processes has already been discussed. Other suggested roles for the involvement of the phosphatidylinositol cycle in cell activation include:

(1) Turnover of phosphatidylinositol may act as a transmembrane signal for protein phosphorylation during platelet activation (Kawahara *et al.*, 1980).
(2) Also in platelets, phosphatidylinositol or one of its products may control the release of arachidonate from the major phospholipids (i.e. phosphatidylcholine and phosphatidylethanolamine) (Bell *et al.*, 1979; Billah *et al.*, 1981).
(3) Phosphatidylinositol might act as an anchor for proteins on the inner surface of the plasma membrane as has already been demonstrated for acetylcholine esterase and 5'-nucleotidase on the outer surface of several cell types (Low and Finean, 1978).
(4) Hydrolysis of phosphatidylinositol could be a regulator in the termination of Ca^{2+} signals (Cockcroft, 1982).

B. LIPID METHYLATION

Although the hydrolysis of phosphatidylinositol appears to be the only metabolic process universally and expressly linked to the mechanism of Ca^{2+}

mobilization, other pathways of lipid metabolism, notably that involving methylation of phosphatidylethanolamine are also under consideration as partial reactions in the transduction of signals from activated receptors (Hirata and Axelrod, 1980). When mast cells (Hirata *et al.*, 1979a) or lymphocytes (Hirata *et al.*, 1980) are stimulated after preincubation with [^{3}H-methyl]-methionine the label is transferred initially to phosphatidylethanolamine (present on the cytosol face of the plasma membrane) to generate phosphatidylcholine (at the exterior) (Hirata and Axelrod, 1978, 1980). Two methylating enzymes are involved: the first generates monomethylphosphatidylethanolamine, and it is thought that this is transferred from the inner to the outer leaflet of the lipid bilayer, where it becomes trapped following further methylation at the external surface to form phosphatidylcholine. The methylation and the translation of the phospholipid are thought to have the effect of reducing the membrane viscosity, as monitored by measurements of fluorescence depolarization of an extrinsic probe, diphenylhexatriene. All these processes are independent of the presence of extracellular Ca^{2+} and they do not occur when the ionophore A23187 is used to stimulate the cells. Since these events occur early in the train of processes which lead to cellular activation, and since inhibitors of the transmethylation reactions also prevent histamine secretion and lymphocyte transformation, it has been proposed that methylation of phosphatidylethanolamine is an early determinant of receptor induced Ca^{2+}-mobilization.

1. Problems of interpretation

The identification of the methylated lipids has been challenged: due to the very low transfer of radioactivity to the phospholipid, the possibility has been raised that the radioactivity could be derived, not from the ^{3}H-methyl donor itself, but from impurities present in the commercial radiochemical (Vance and De Kruijff, 1980). There is also some doubt as to whether the disposition of the lipids has been correctly assigned; of the two phospholipases-c used in this exercise (Hirata and Axelrod, 1978, 1980), that derived from *C. welchii* is commonly lytic (Mauco *et al.*, 1978), while that from *B. cereus* is normally inactive when applied to intact cells (Roelofsen *et al.*, 1971). In addition, it seems very unlikely that the changes in microviscosity could be determined by the very low proportion of phospholipids involved in the methylation pathway (it has been pointed out (Vance and De Kruijff, 1980) that in a typical experiment, as little as 0.0012% of the phosphatidylethanolamine was converted to the monomethyl derivative).

There is also the question of whether lipid methylation arises from activation of relevant receptors. While there has been a failure to detect methylation in thrombin-stimulated platelets (Randon *et al.*, 1981; Hotchkiss *et al.*, 1981) (where it might have been expected) it does occur in rat reticulocytes treated with isopreterenol (Hirata *et al.*, 1979b), which is a well defined example of a β-adrenergic system linked to adenylate cyclase. There is, however, no link between adenylate cyclase activation and lipid methylation in hepatocytes (Colard and Breton, 1981; Schanche *et al.*, 1982).

2. *Mast Cells: Evidence of Inhibitors*

The finding that secretion and methylation can both be inhibited by reagents known to be inhibitors of the methyltransferase enzymes has also been taken as an indication that these processes are obligatorily coupled under conditions of cell stimulation. Thus both secretion and methylation are inhibited in mast cells preincubated with *S*-isobutyryl-3-deazaadenosine in the range 1–100 μM (Ishizaka *et al.*, 1980). However, the concentration of the inhibitor required to cause 50% inhibition of ³H-methyl incorporation into phospholipid is one tenth of that needed to cause a similar degree of inhibition of secretion. While it is thought unlikely that other methyl trans-ferases are affected by the inhibitor at the concentrations used, the difference in sensitivity of secretion and methylation does suggest the possibility that the mechanism of inhibition of secretion by this compound might not be related to its effects on lipid transmethylation.

3. *Temporal Relationship: Mast Cells*

If we turn to other criteria discussed above with reference to phosphatidyl-inositol hydrolysis, we find that the question of relevancy in the time course has not yet been properly resolved. Methylation of phosphatidylethanolamine in mast cells stimulated with anti-receptor antibody is certainly rapid, and it is transient: it maximizes at 15 s and declines to zero by 1 min (Ishizaka *et al.*, 1980). As histamine secretion is only half-completed at this time, methylation certainly has a precursor relationship to secretion. But does it precede Ca^{2+} mobilization? While the published conclusions based on measurements of $^{45}Ca^{2+}$ uptake certainly support this idea, we have seen how this method is unlikely to discern directionality or the sources of Ca^{2+}, let alone the question of kinetics. The resolution of precursor relationships will have to await results of more rigorous experiments using internal Ca^{2+} indicators.

4. *Temporal Relationship: Lymphocytes*

Mouse T lymphocytes appear to be the one example for which there exists information concerning the rate of both phospholipid methylation (Hirata *et al.*, 1980) and for Ca^{2+} mobilization (Tsien *et al.*, 1982a,b). As with mast cells, when T cells are stimulated with appropriate ligands in the presence of ³H-methyl donors, there is an increased rate of radioactive incorporation into phospholipids, and application of methyltransferase inhibitors has the effect of preventing both this early event and later processes including DNA synthesis (Hirata *et al.*, 1980). Following treatment with con A (or other T cell mitogens) methylation maximizes at 10 min, while as we have noted earlier, cytosol Ca^{2+} has already achieved a new plateau level by 3 min (Tsien *et al.*, 1982a,b). Although it must be acknowledged that the tissue origin of the lymphocytes (spleen and thymus respectively) differed in the two investigations, here is a situation in which the inference at the present time has to be that lipid methylation, rather than being a precursor of Ca^{2+} mobilization, may yet turn out to be a product. Certainly more critical experiments concerning the temporal relationships of the two processes are required, and in addition, it will be of some importance to examine the effects of the inhibitors of lipid methylation enzymes directly on Ca^{2+}-mobilization rather than as in previous work, on more distal events such as DNA synthesis (in lymphocytes) or secretion (in mast cells).

In a recent re-examination of phospholipid metabolism in lymphocytes derived from three different sources, no alterations could be detected in the methylation pathway leading to phosphatidylcholine synthesis following stimulation with mitogenic lectins (Moore *et al.*, 1982). In this work, especial care was taken to maximize and quantitate the recovery of the methylated lipids, and then to separate and identify them positively. The inference has to be that the organic extractable $[^3H]$-methionine-derived material earlier considered to be methylated homologues of phosphatidylethanolamine may have been wrongly assigned.

5. Evidence from Drug-resistant Mutants

Further evidence for a precursor role for the transmethylation pathway in control of Ca^{2+}-dependent events comes from experiments involving formation of hybrid cells from transmethylase-deficient mutants. These were derived from separate clones of thioguanine and chloramphenicol resistant, ethane methylsulphonate generated mutants of rat basophilic leukaemia cells which were shown to be deficient in the first and second methyl transferase enzymes respectively. Although the mutant cells have receptors for IgE, and respond normally to the Ca^{2+}-ionophore A23187, neither is able to respond to IgE-directed ligands. Hybrids, generated by fusion with polyethylene glycol recover the capacity both to carry out the complete sequence of reactions catalysed by the transmethylase pathway and to secrete histamine on treatment with specific antigen (McGivney *et al.*, 1981). This surely provides the strongest argument so far in favour of the idea that phospholipid methylation precedes and possibly controls Ca^{2+}-mobilization in mast cells and related cell types. But how strong is this argument?

What is far from clear is the extent of the genetic alteration in the clones of non-secreting and methyltransferase deficient mutants from which the active recombinants were generated. Unless these can be shown to present with unique genetic mutations relating to the processes, both recognized and unrecognized, which are involved in the stimulus–response sequence, then it must be said that the argument remains ambiguous. Indeed, the finding that hybridization of the mutant lines led to eight independent hybrids, and that the rate of growth of the hybrid cells was slowed (this too possibly genetically determined) suggests that the mutations are not unique and that the relationship between phospholipid methylation and Ca^{2+}-mobilization, far from being causal, may yet be consequential.

In spite of this, the experimental strategy is certainly a powerful one and its introduction into this field will surely extend knowledge and understanding. An early extension of this work into the reactions of the phosphatidylinositol cycle, and their relation with stimulus–response coupling would be greeted with the greatest possible interest.

C. PHOSPHOLIPASE A₂ ACTIVATION: NEUTROPHILS

Neutrophils do not respond to stimulation by an increased rate of phospholipid methylation. Instead there is an increase in the rate of degradation of phosphatidylcholine (which is the main methylated phospholipid) due to the action of phospholipase A_2 (Hirata et al., 1979c) and this results in the generation of fatty acids, including arachidonate and its metabolites of the lipoxygenase pathway which are themselves agonists for neutrophil chemotaxis and secretion (Naccache et al., 1979b). (Interestingly, only that very minor fraction of phosphatidylcholine which is synthesized by the methylation pathway is subject to f-MetLeuPhe activation of phospholipase A_2; phosphatidylcholine synthesized by the major CDP-choline pathway is unaffected (Hirata et al., 1979c).) While arachidonate metabolites (HETES and leukotrienes) can mobilize Ca^{2+} from both internal sources and from the cell exterior in ways similar to f-MetLeuPhe (Volpi et al., 1980) it must be noted that the phospholipase A_2 which results in the generation of arachidonate is itself dependent on Ca^{2+} (Walenga et al., 1981). In vivo, there is the possibility that arachidonate and its metabolites might be released as chemotactic signals so as to recruit and hasten the arrival of more neutrophils at sites of infection. The finding that neutrophil activation can be modulated by inhibitors that act on the lipoxygenase pathway of arachidonate metabolism suggests that even among cells sharing a common reticuloendothelial origin, quite separate mechanisms for the control of Ca^{2+} exist.

VIII. The Receptors

From the discussion earlier in this chapter, it should be apparent that Ca^{2+} plays many roles even within a single cell. Thus, in hepatocytes, there is a requirement for Ca^{2+} (in the range 1–10 μM) in the fusion of golgi vesicles for the secretion of serum albumin. This is a continuous process, not subject to hormonal regulation, so the Ca^{2+} must be ever present in the secreting cells. On the other hand, the same cells when stimulated with vasopressin or α-adrenergic agonists respond by enhanced mitochondrial oxidative activity, and this is also under control of Ca^{2+} in a rather similar range of concentrations. Obviously, the control of Ca^{2+} at the point of entry, and later, the systems which are controlled by Ca^{2+} are complex. By contrast, the systems which control the formation of cyclicAMP are universal, and when generated, cyclicAMP appears to have a single function, that of coenzyme in the phosphorylation of proteins. While the role of calmodulin in mediating Ca^{2+}-dependent processes is certainly widespread among tissues and classes of living organisms, we have seen that it is not universal. It should therefore come as no surprise that there may exist multiple and diverse mechanisms for the mobilization of Ca^{2+}. It is likely that future developments in this field will require the isolation and the reconstitution of the Ca^{2+}-mobilizing receptors, together with other proteins (possibly analogous to the GTP-binding protein in the adenylate cyclase system) associated with them.

For the two cells which have been the main objects of consideration in this chapter, a start in this direction has been made. For the receptor for IgE (isolated from a rat basophilic leukaemia cell line) this certainly goes well beyond the initial stages of solubilization and fractionation (Kanellopoulos et al., 1980; Metzger et al., 1981; Hempstead et al., 1981). It comprises two subunits which are firmly but non-covalently bound in 1 : 1 ratio. The IgE binds to the exposed α-subunit which is extensively glycosylated. The β-subunit is in contact with the hydrophobic region of the lipid bilayer and is also exposed at the cytoplasmic surface of the membrane. While much has now been learned about the structural interactions of the IgE receptor system, there is little understanding of how it might function as a signal transducer. Attempts to isolate and characterize the receptor for f-MetLeuPhe from neutrophils are also in progress, though much less advanced (Niedel, 1981; Goetzl et al., 1981; Baldwin et al., 1983).

These two receptors, so strikingly different in their interactions with the ligands that stimulate them are both clearly of the class of Ca^{2+}-mobilizing receptors. The receptor for IgE can be activated by any divalent ligand capable of binding and crosslinking it (see Table II): in this respect it is similar to the insulin receptor which can also be activated by anti-receptor antibodies (Kahn et al., 1978; Belsham et al., 1980). The receptor for f-MetLeuPhe, like those for the catecholamines and acetylcholine, has a strict requirement for occupancy of a specific site (Freer *et al.*, 1982). In view of this, and bearing in mind the highly conserved receptor mechanisms for the control of adenylate cyclase on the one hand, and the apparent profusion of interactions of the Ca^{2+}-mobilizing receptors with the pathways of phospholipid metabolism on the other, it is now important to find out to what extent the receptors which control cytoplasmic Ca^{2+} share common structural and mechanistic features.

References

Alhanaty, E. and Livne, A. (1974). *Biochim. Biophys. Acta* **229**, 146–155.
Allan, D. and Michell, R. H. (1975). *Biochem. J.* **148**, 471–478.
Allan, D. and Michell, R. H. (1977a). *Biochem. J.* **166**, 495–499.
Allan, D. and Michell, R. H. (1977b). *Biochem. J.* **164**, 389–397.
Allen, D. G. and Blinks, J. R. (1978). *Nature* **273**, 509–513.
Amphlett, G., Vanaman, T. and Perry, S. (1976). *FEBS Lett.* **72**, 163–168.
Anderson, J. M. and Cormier, M. J. (1978). *Biochem. Biophys. Res. Commun.* **84**, 595–602.
Ashby, J. P. and Speke, R. N. (1975). *Biochem. J.* **150**, 89–96.
Ashley, C. C. and Caldwell, P. C. (1974). *Biochem. Soc. Symp.* **39**, 29–50.
Azila, N. and Hawthorne, J. N. (1982). *Biochem. J.* **204**, 291–299.
Babcock, D. F., Chen, J.-L. J., Yip, B. P. and Lardy, H. A. (1979). *J. Biol. Chem.* **254**, 8117–8120.
Baker, P. F. (1972). *Prog. Biophys.* **24**, 177–223.
Baker, P. F. and Knight, D. E. (1981). *Phil Trans. R. Soc. Lond. B* **296**, 83–103.
Baker, P. F. and Whitaker, M. J. (1978). *Nature* **276**, 513–515.
Baker, P. F. and Whitaker, M. J. (1979). *J. Physiol.* **298**, 55.
Baker, P. F., Knight, D. E. and Whitaker, M. J. (1980). *Proc. R. Soc. Lond. B* **207**, 149–161.
Baldwin, J. M., Bennett, J. P. and Gomperts, B. D. (1983). *Eur. J. Biochem.* **135**, 515–518.
Beachy, J. C., Goldman, D. and Czech, M. P. (1981). *Proc. Natl Acad. Sci. USA* **78**, 6256–6260.
Beal, E. G., Dedman, J. R. and Means, A. R. (1977). *Endocrinology* **101**, 1621–1634.
Becker, E. L., Showell, H. J., Henson, P. M. and Hsu, L. S. (1974). *J. Immunol.* **112**, 2047–2054.
Becker, E. L., Sigman, M. and Oliver, J. M. (1979). *Am. J. Pathol.* **95**, 81–97.
Beeler, T. J., Jona, I. and Martonosi, A. (1979). *J. Biol. Chem.* **254**, 6229–6231.
Bell, R. L. and Majerus, P. W. (1981). *J. Biol. Chem.* **255**, 1790–1792.
Bell, R. L., Kennerly, D. A., Stanford, N. and Majerus, P. W. (1979). *Proc. Natl Acad. Sci. USA* **76**, 3238–3241.
Belle, R., Oron, R. and Stinnakre, J. (1977). *Molec. Cell. Endocrinol.* **8**, 65–72.
Belsham, G. J., Brownsey, R. W., Hughes, W. A. and Denton, R. M. (1980). *Diabetologia* **18**, 307–312.
Bennett, J. P., Smith, G. A., Houslay, M. D., Hesketh, T. R., Metcalfe, J. C. and Warren, G. B. (1978). *Biochim. Biophys. Acta* **513**, 310–320.
Bennett, J. P., Cockcroft, S. and Gomperts, B. D. (1979). *Nature* **282**, 851–853.

Bennett, J. P., Cockcroft, S. and Gomperts, B. D. (1980a). *Biochim. Biophys. Acta* **601**, 584–591.

Bennett, J. P., Hirth, K. P., Fuchs, E., Sarvas, M. and Warren, G. B. (1980b). *FEBS Lett.* **116**, 57–61.

Bennett, J. P., Cockcroft, S. and Gomperts, B. D. (1981). *J. Physiol.* **317**, 335–345.

Bennett, J. P., Cockcroft, S., Caswell, A. H. and Gomperts, B. D. (1982). *Biochem. J.* **208**, 801–808.

Benz, R. H. and Hancock, R. E. W. (1981). *Biochim. Biophys. Acta* **646**, 298–308.

Berridge, M. J. and Fain, J. N. (1979). *Biochem. J.* **178**, 59–69.

Bicknell, R. J. and Schofield, J. G. (1976). *FEBS Lett.* **68**, 23–26.

Bicknell, R. J., Young, P. W. and Schofield, J. G. (1979). *Molec. Cell. Endocrinol.* **13**, 167–180.

Billah, M. M., Lapetina, E. G. and Cuatrecasas, P. (1981). *J. Biol. Chem.* **256**, 5399–5403.

Blackmore, P. F., El-Refai, M. F., Dehaye, J.-P., Strickland, W. G., Hughes, B. P. and Exton, J. H. (1981). *FEBS Lett.* **123**, 245–248.

Bolton, J. E. and Field, M. (1977). *J. Membrane Biol.* **35**, 159–173.

Bottenstein, J. and de Vellis, J. (1976). *Biochem. Biophys. Res. Commun.* **73**, 486–493.

Brindley, D. N., Allan, D. and Michell, R. H. (1975). *J. Pharm. Pharmacol.* **27**, 462–464.

Brinley, F. J., Tiffert, T., Scarpa, A. and Mullins, L. J. (1977). *J. Gen. Physiol.* **70**, 355–384.

Brostrom, C. O., Huang, Y.-C., Breckenridge, B. McL. and Wolff, D. J. (1975). *Proc. Natl Acad. Sci. USA* **72**, 64–68.

Brotschi, E. A., Hartwig, J. G. and Stossel, T. P. (1978). *J. Biol. Chem.* **253**, 8988–8993.

Brown, H. M. and Rydqvist, B. (1981). *Biophys. J.* **36**, 117–137.

Brown, J. E., Cohen, L. B., De Weer, P., Pinto, L. H., Ross, W. N. and Salzberg, B. N. (1975). *Biophys. J.* **15**, 1155–1160.

Burridge, K. and Feramisco, J. R. (1981). *Nature* **294**, 565–567.

Campbell, A. K. (1974). *Biochem. J.* **143**, 411–418.

Candia, D. A., Montoreano, R. and Podos, S. M. (1977). *Am. J. Physiol.* **233**, F94–101.

Case, R. M. (1980). *Cell Calcium* **1**, 1–5.

Caswell, A. H. (1972). *J. Membrane Biol.* **7**, 345–364.

Caswell, A. H. and Pressman, B. C. (1972). *Biochem. Biophys. Res. Commun.* **49**, 292–298.

Chalfie, M. and Perlman, R. L. (1976). *J. Pharm. Exp. Ther.* **197**, 615–622.

Chandler, D. E. and Williams, J. A. (1978a). *J. Cell. Biol.* **76**, 371–385.

Chandler, D. E. and Williams, J. A. (1978b). *J. Cell Biol.* **76**, 386–399.

Cheung, W. Y. (1970). *Biochem. Biophys Res. Commun.* **33**, 533–538.

Clements, R. S. and Rhoten, W. B. (1976). *J. Clin. Invest.* **57**, 684–691.

Clements, R. S., Rhoten, W. B. and Starnes, W. R. (1977). *Diabetes* **26**, 1109–1116

Cockcroft, S. (1982). *Trends in Pharmac. Sci.* **12**, 340–342.

Cockcroft, S. (1982). *Cell Calcium* (in press).

Cockcroft, S. and Gomperts, B. D. (1979a). *J. Physiol.* **296**, 229–243.

Cockcroft, S. and Gomperts, B. D. (1979b). *Nature* **279**, 541–542.

Cockcroft, S. and Gomperts, B. D. (1979c). *Biochem. J.* **178**, 681–687.

Cockcroft, S. and Gomperts, B. D. (1980). *Biochem. J.* **188**, 789–798.

Cockcroft, S., Bennett, J. P. and Gomperts, B. D. (1980a). *FEBS Lett.* **110**, 115–118.

Cockcroft, S., Bennett, J. P. and Gomperts, B. D. (1980b). *Nature* **288**, 275–277.
Cockcroft, S., Bennett, J. P. and Gomperts, B. D, (1981). *Biochem. J.* **200**, 501–508.
Cochrane, D. E., Douglas, W. W., Moori, T. and Nakazatu, Y. (1975). *J. Physiol.* **252**, 363–378.
Cohen, P., Burchall, A., Foulkes, G., Cohen, P. T. W., Vanaman, T. C. and Nairn, A. C. (1978). *FEBS Lett.* **92**, 278–293.
Cohen, P., Picton, C. and Klee, C. B. (1979). *FEBS Lett.* **104**, 25–30.
Colard, O. and Breton, M. (1981). *Biochem. Biophys. Res. Commun.* **101**, 727–733.
Corps, A. N., Hesketh, T. R. and Metcalfe, J. C. (1982). *FEBS Lett.* **128**, 280–284.
Cotman, C. W., Haycock, J. W. and White, W. F. (1976). *J. Physiol.* **254**, 475–506.
Creutz, C. E., Pazoles, C. J. and Pollard, H. B. (1978). *J. Biol. Chem.* **253**, 2858–2866.
Creutz, C. E. Pazoles, C. J. and Pollard, H. B. (1979). *J. Biol. Chem.* **254**, 553–558.
Dabrow, M., Zaremba, S. and Hogue-Angeletti, R. A. (1980). *Biochem. Biophys. Res. Commun.* **96**, 1164–1171.
Dabrowska, R. and Hartshorne, D. J. (1978). *Biochem. Biophys. Res. Commun.* **85**, 1352–1359.
Dabrowska, R., Sherry, J. M. F., Aromatorio, D. K. and Hartshorne, D. J. (1978). *Biochemistry* **17**, 253–258.
Dahl, G. and Gratzl, M. (1976). *Cytobiologie* **12**, 334–355.
D'Arrigo, J. S. (1974). *J. Physiol.* **243**, 757–764.
D'Arrigo, J. S. (1978). *Am. J. Physiol.* **253**, C109–117.
Dedman, J. R., Potter, J. D. and Means, A. R. (1977). *J. Biol. Chem.* **252**, 2537–2440.
Denton, R. M. and McCormack, J. G. (1980). *FEBS Lett.* **119**, 1–8.
Denton, R. M. and McCormack, J. G. (1981). *Clin. Sci.* **61**, 135–140.
Denton, R. M., Randle, P. J. and Martin, B. R. (1972). *Biochem. J.* **128**, 161–163.
Denton, R. M., Richards, D. A. and Chin, J. G. (1978). *Biochem. J.* **176**, 899–906.
Denton, R. M., McCormack, J. G. and Edgell, N. J. (1980). *Biochem. J.* **190**, 107–117.
Dodge, F. and Rahamimoff, R. (1967). *J. Physiol.* **193**, 419–432.
Douglas, W. W. and Nemeth, E. F. (1982). *J. Physiol.* **323**, 229–244.
Douglas, W. W. and Poisner, A. M. (1964). *J. Physiol.* **172**, 1–18.
Douglas, W. W. and Rubin, R. P. (1961). *J. Physiol.* **159**, 40–57.
Edwards, W., Phillips, J. H. and Morris, S. J. (1974). *Biochim. Biophys. Acta* **356**, 164–173.
Eggman, L. D. and Hokin, L. E. (1960). *J. Biol. Chem.* **235**, 2569–2571.
Eimerl, S., Savion, N., Heichal, O. and Selinger, Z. (1974). *J. Biol. Chem.* **249**, 3991–3993.
Ekerdt, R., Dahl, G. and Gratzl, M. (1981). *Biochim. Biophys. Acta* **646**, 10–22.
Elferink, J. G. R. (1978). *Biochem. Pharmacol.* **28**, 965–968.
Fain, J. N. and Berridge, M. F. (1979a). *Biochem. J.* **178**, 45–58.
Fain, J. N. and Berridge, M. F. (1979b). *Biochem. J.* **180**, 655–661.
Farese, R. V., Larson, R. E. and Sabir, M. A. (1980). *Biochim. Biophys. Acta* **633**, 479–484.
Feinman, R. D. and Detwiler, P. C. (1974). *Nature* **249**, 172–173.
Feinstein, M. B. (1980). *Biochem. Biophys. Res. Commun.* **93**, 593–600.
Fenwick, E. M., Fajdiga, P. B., Howe, N. B. S. and Livett, B. G. (1978). *J. Cell Biol.* **76**, 12–30.
Feramisco, J. R. and Burridge, K. (1982). in prep. ref. B and F, 1981.
Fisher, D. B. and Mueller, G. C. (1971). *Biochim. Biophys. Acta* **248**, 434–448.
Ford, L. E. and Podolsky, R. J. (1972). *J. Physiol.* **223**, 1–19.
Foreman, J. C. and Mongar, J. L. (1972). *J. Physiol.* **224**, 753–769

Foreman, J. C., Mongar, J. L. and Gomperts, B. D. (1973). *Nature* 245, 249–251.

Fowler, V. M. and Pollard, H. B. (1982). *Nature* 295, 336–339.

Freer, R. J., Day, A. R., Muthukumaraswamy, N., Pinon, D., Wu, A., Showell, H. J. and Becker, E. L. (1982). *Biochemistry* 21, 257–263.

Frizzell, R. A. (1977). *J. Membrane Biol.* 35, 175–187.

Gagerman, E., Sehlin, J. and Taljedal, I.-B. (1980). *J. Physiol.* 300, 505–513.

Gains, N. (1980). *Eur. J. Biochem.* 111, 199–202.

Gallien-Lartigue, O. (1976). *Cell Tissue Kinet.* 9, 533–540.

Gardos, G., Lassen, U. V. and Pape, L. (1976). *Biochim. Biophys. Acta* 448, 599–606.

Gemsa, D., Seitz, M., Kramer, W., Grimm, W., Trill, G., and Resch, K. (1979). *Exp. Cell Res.* 188, 55–62.

Gerrard, J. M., Kindon, S. E., Peterson, D. A., Peller, J., Krantz, K. E. and White, J. G. (1979). *Am. J. Pathol.* 96, 423–438.

Glenney, J. R., Bretscher, A. and Weber, K. (1980). *Proc. Natl Acad. Sci. USA* 77, 6458–6462.

Glenney, J. R., Kaulfus, P., Matsudaira, P. and Weber, K. (1981). *J. Biol. Chem.* 9283–9288.

Goetzl, E. J., Foster, D. W. and Goldman, D. W. (1981). *Biochemistry* 20, 5717–5722.

Goldstein, I., Hoffstein, S., Gallin, J. and Weissmann, G. (1973). *Proc. Natl Acad. Sci. USA* 70, 2916–2920.

Gomperts, B. D. (1977) "The Plasma Membrane: Models for Structure and Function". Academic Press, London.

Gomperts, B. D., Bennett, J. P. and Allan, D. (1981a). *Eur. J. Biochem.* 117, 559–562.

Gomperts, B. D., Baldwin, J. M. and Micklem, K. J. (1983). *Biochem. J.* 210, 737–743.

Gopinath, R. M. and Vincenzi, F. F. (1977). *Biochem. Biophys. Res. Commun,* 77, 1203–1209.

Gratzl, M. and Dahl, G. (1976). *FEBS Lett.* 62, 142–145.

Gratzl, M. and Dahl, G. (1978). *J. Membrane Biol.* 40, 343–364.

Gratzl, M., Dahl, G., Russell, J. T. and Thorn, N. A. (1977). *Biochim. Biophys. Acta* 470, 45–57.

Hallett, M. B. and Campbell, A. K. (1982). *Nature* 295, 155–158.

Hansford, R. G. and Chappell, J. B. (1967). *Biochem. Biophys. Res. Commun.* 27, 686–692.

Harafuji, H. and Ogawa, Y. (1980). *J. Biochem (Tokyo)* 87, 1305–1312.

Harris, R. A., Schmidt, J., Hitzeman, B. A. and Hitzeman, R. J. (1981). *Science* 212, 1290–1291.

Hartwig, J. H. and Stossel, T. P. (1981). *J. Mol. Biol.* 145, 563–581.

Hastings, J. W., Mitchell, G., Mattingly, P. H., Blinks, J. R. and van Leeuwen, M. (1969). *Nature* 222, 1047–1050.

Hathaway, D. R. and Adelstein, R. S. (1979). *Proc. Natl Acad. Sci. USA* 76, 1653–1657.

Hawthorne, J. N. and Bleasdale, J. E. (1975). *Mol. Cell Biochem.* 8, 83–87.

Hawthorne, J. N. and Pickard, M. R. (1979). *J. Neurochem.* 32, 5–14.

Heinl, P., Kuhn, H. J. and Ruegg, J. C. (1974). *J. Physiol.* 237, 243–258.

Hellman, B. (1975). *Endocrinology* 97, 392–398.

Hempstead, B. L., Parker, C. W. and Kulczycki, A. (1981). *J. Biol. Chem.* 256, 10717–10723.

Hems, D. A. (1979). *Clin. Sci.* **56**, 197–202.
Hems, D. A., McCormack, J. G. and Denton, R. M. (1978). *Biochem. J.* **176**, 627–629.
Henquin, J. C. (1981). *Biochim. J.* **196**, 771–780.
Hiraoka, T., De Buysere, M. and Olsen, M. S. (1980). *J. Biol. Chem.* **255**, 7604–7609.
Hirata, F. and Axelrod, J. (1978). *Proc. Natl Acad. Sci USA* **75**, 2348–2352.
Hirata, F. and Axelrod, J. (1980). *Science* **209**, 1082–1090.
Hirata, F., Axelrod, J. and Crews, F. T. (1979a). *Proc. Natl Acad. Sci. USA* **76**, 4813–4816.
Hirata, F., Strittmatter, W. J. and Axelrod, J. (1979b). *Proc. Natl Acad. Sci. USA* **76**, 368–372.
Hirata, F., Corcoran, B. A., Venkatasubramanian, K., Schiffman, E. and Axelrod, J. (1979c). *Proc. Natl Acad. Sci. USA* **76**, 2640–2643.
Hirata, F., Toyoshima, S., Axelrod, J. and Waxdal, M. J. (1980). *Proc. Natl Acad. Sci. USA* **77**, 862–865.
Hochman, J. and Perlman, R. L. (1976). *Biochim. Biophys. Acta* **421**, 168–175.
Hokin, M. R. (1968). *Arch. Biochem. Biophys.* **124**, 280–284.
Hokin-Neaverson, M. (1977). *Adv. Exp. Med. Biol.* **83**, 429–446.
Holland, D. R., Steinberg, M. I. and Armstrong, W. McD. (1975). *Proc. Soc. Exp. Biol. Med.* **148**, 1141–1145.
Holwill, M. J. and McGregor, J. L. (1975). *Nature* **255**, 157–158.
Holz, R. W. (1975). *Biochim. Biophys. Acta* **375**, 138–152.
Hong, K., Duzgunes, N. and Papahadjopoulos, D. (1981). *J. Biol. Chem.* **256**, 3461–3644.
Hong, K., Duzgunes, N., Ekerdt, R. and Papahadjopoulos, D. (1982a). *Proc. Natl Acad. Sci. USA* **79**, 4642–4644.
Hong, K., Duzgunes, N. and Papahadjopoulos, D. (1982b). *Biophys. J.* **37**, 297–305.
Hotchkiss, A., Jordan, J. V., Hirata, F., Schulman, N. R. and Axelrod, J. (1981). *Biochem. Pharmacol.* **30**, 2089–2095.
Hovi, T., Williams, S. C. and Allison, A. C. (1975). *Nature* **256**, 70–72.
Hui, D. Y. and Harmony, J. A. K. (1980). *Biochem. J.* **192**, 91–98.
Hume, D. A., Vijayakumar, E. K., Schweinberger, F., Russell, L. M. and Weidemann, M. J. (1978). *Biochem. J.* **174**, 711–716.
Humphrey, J. H., Austen, K. F. and Rapp, H. J. (1963). *Immunology* **6**, 226–245.
Ishizaka, K. and Ishizaka, T. (1967). *J. Immunol.* **99**, 1187–1198.
Ishizaka, T., Ishizaka, K., Conrad, D. H. and Froese, A. (1978). *J. Allergy Clin. Immunol.* **61**, 320–330.
Ishizaka, T., Hirata, F., Ishizaka, K. and Axelrod. J. (1980). *Proc. Natl Acad. Sci. USA* **77**, 1903–1906.
Jafferji, S. S. and Michell, R. H. (1976a). *Biochem. J.* **154**, 653–657.
Jafferji, S. S. and Michell, R. H. (1976b). *Biochem. Pharmacol.* **25**, 1429–1430.
Jafferji, S. S. and Michell, R. H. (1976c). *Biochem. J.* **160**, 163–169.
Jakschik, B. A., Kulczycki, A., MacDonald, H. H. and Parker, C. W. (1977). *J. Immunol.* **119**, 618–622.
Jarrett, H. W. and Penniston, J. T. (1977). *Biochem. Biophys. Res. Commun.* **77**, 1210–1216.
Jenkinson, D. H. (1957). *J. Physiol.* **138**, 434–444.
Johansen, T. (1978). *Brit. J. Pharmacol.* **63**, 643–649.
Jones, L. M. and Michell, R. H. (1975). *Biochem. J.* **148**, 479–485.
Jones, L. M., Cockcroft, S. and Michell, R. H. (1979). *Biochem. J.* **182**, 669–676.
Judah, J. D. and Quinn, P. S. (1978). *Nature* **271**, 384–385.

Kahn, C. R., Baird, K. L., Jarrett, D. B. and Flier, J. S. (1978). *Proc. Natl Acad. Sci. USA* **75**, 4209–4213.

Kakiuchi, S., Yamazaki, R. and Nakajima, H. (1970). *Proc. Japan Acad.* **46**, 587–591.

Kanellopoulos, J. M., Liu, T-Y., Poy, G. and Metzger, H. (1980). *J. Biol. Chem.* **255**, 9060–9066.

Kanno, T. (1972). *J. Physiol.* **226**, 353–371.

Katz, S. and Remtulla, M. A. (1978). *Biochem. Biophys. Res. Commun.* **83**, 1373–1379.

Kawahara, Y., Takai, Y., Minakuchi, R., Sano, K. and Nishizuka, Y. (1980). *Biochem. Biophys. Res. Commun.* **97**, 309–317.

Keller, R. (1973). *Clin. Exp. Immunol.* **13**, 139–147.

Kennerly, D. A., Sullivan, T. J. and Parker, C. W. (1979). *J. Immunol.* **122**, 152–159.

Keppens, S., van den Heede, J. R. and de Wulf, H. (1977). *Biochim. Biophys. Acta* **496**, 448–457.

Kilpatrick, D. L., Ledbetter, F. H., Carson, K. A., Kirshner, A. G., Slepetis, R. and Kirshner, N. (1980). *J. Neurochem.* **35**, 679–692.

Kirk, C. J., Verrinder, T. R. and Hems, D. A. (1977). *FEBS Lett.* **83**, 267–271.

Kirk, C. J., Michell, R. H. and Hems, D. A. (1981). *Biochem. J.* **194**, 155–165.

Klausner, R. D., Fishman, M. C. and Karnovsky, M. J. (1979). *Nature* **281**, 82–83.

Klee, C. B., Crouch, T. H. and Richman, P. G. (1980). *Ann. Rev. Biochem.* **49**, 489–515.

Knapp, H. R., Oelz, O., Roberts, B. J., Sweetman, B. J., Oates, J. A. and Reed, P. W. (1977). *Proc. Natl Acad. Sci. USA* **74**, 4251–4255.

Knight, D. E. and Baker, P. F. (1982). *J. Membrane Biol.* **68**, 107–140.

Knight, D. E. and Scrutton, M. C. (1980). *Thrombosis Res.* **20**, 437–446.

Knight, D. E. and Scrutton, M. C. (1982). *Nature* **296**, 256–257.

Landry, Y., Amallal, M. and Ruckstuhl, M. (1981). *Biochem. Pharmacol.* **30**, 2031–2032.

Lapetina, E. G. and Cuatrecasas, P. (1979). *Biochim. Biophys. Acta* **573**, 394–402.

Lapetina, E. G., Billah, M. and Cuatrecasas, P. (1981). *J. Biol. Chem.* **256**, 11 984–11 987.

Lawson, D., Raff, M. C., Gomperts, B. D., Fewtrell, C. and Gilula, N. B. (1977). *J. Cell. Biol.* **72**, 242–259.

LeBreton, G. C., Dinerstein, R. J., Roth, L. J. and Feinberg, H. (1976). *Biochem. Biophys. Res. Commun.* **71**, 362–370.

Lee, C. O., Uhm, D. Y. and Dresdner, K. (1980). *Science* **209**, 699–701.

Levin, R. M. and Weiss, B. (1978). *Biochim. Biophys. Acta* **540**, 197–204.

Levin, R. M. and Weiss, B. (1979). *J. Pharm. Exp. Ther.* **208**, 454–459.

Lewis, R. A. and Austen, K. F. (1981). *Nature* **293**, 103–108.

Lewis, R. A., Goetzl, E., Wasserman, S. I., Valone, F. H., Rubin, R. H. and Austen, K. F. (1975). *J. Immunol.* **114**, 87–92.

Lin, S. H. and Fain, J. N. (1981). *Life Sci.* **29**, 1905–1912.

Liu, C. M. and Hermann, T. E. (1978). *J. Biol. Chem.* **253**, 5892–5894.

Lloyd, J. V. and Mustard, J. F. (1974). *Brit. J. Haematol.* **26**, 243–253.

Low, M. S. and Finean, J. B. (1978). *Biochim. Biophys. Acta* **508**, 565–570.

McCormack, J. G. and Denton, R. M. (1979). *Biochem. J.* **180**, 533–544.

McCormack, J. G. and Denton, R. M. (1980). *Biochem. J.* **190**, 95–105.

McCormack, J. G. and Denton, R. M. (1981). *Biochem. J.* **194**, 639–643.

McGivney, A., Crews, F. T., Hirata, F., Axelrod, J. and Siraganian, R. (1981). *Proc. Natl Acad. Sci. USA* **78**, 6176–6180.

McLaughlin, S. G. A., Szabo, G. and Eisenmann, G. (1971). *J. Gen. Physiol.* **58**, 667–687.

Maino, V. C., Green, N. M. and Crumpton, M. J. (1974). *Nature* **251**, 324–327.

Maino, V. C., Hayman, M. J. and Crumpton, M. J. (1975). *Biochem. J.* **146**, 247–252.

Makan, N. R. and Heppel, L. A. (1978). *J. Cell Physiol.* **96**, 87–93.

Malaisse, W. J., Hutton, J. C., Sener, A., Levy, J., Herchuelz, A., Devis, G. and Somers, G. (1978). *J. Membrane Biol.* **38**, 193–208.

Marban, E., Rink, T., Tsien, R. Y. (1980). *Nature* **286**, 845–850.

Mauco, G., Chap, H., Simon, M-F. and Douste-Blazy, L. (1978). *Biochimie* **60**, 653–661.

Means, A. R. and Dedman, J. R. (1980). *Nature* **285**, 73–77.

Means, A. R., Tash, J. S. and Chafouleas, J. G. (1982). *Physiol. Rev.* **62**, 1–39.

Meech, R. W. (1974). *J. Physiol.* **237**, 259–277.

Mennes, P. A., Yates, J. and Klahr, S. (1978). *Proc. Soc. Exp. Biol. Med.* **157**, 168–174.

Metzger, H., Kanellopoulos, J., Holowka, D., Goetze, A. and Fewtrell, C. (1981). *In* "Biochemistry of the Acute Allergic Reactions" (E. L. Becker, A. S. Simon and K. F. Austen, eds), 183–194. Alan Liss, New York.

Michaelson, D. M. and Sokolovsky, M. (1978). *J. Neurochem.* **30**, 217–230.

Michell, R. H. (1975). *Biochim. Biophys. Acta* **415**, 81–147.

Michell, R. H., Jafferji, S. S. and Jones, L. M. (1976). *FEBS Lett.* **69**, 1–5.

Michell, R. H., Kirk, C. J., Jones, L. M., Downes, C. P. and Creba, J. A. (1981). *Phil. Trans R. Soc. Lond. B.* **296**, 123–137.

Mimura, N. and Asano, A. (1979). *Nature* **282**, 44–48.

Mobley, B. A. (1977). *Eur. J. Pharmacol.* **45**, 101–104.

Moore, J. P., Smith, G. A., Hesketh, T. R. and Metcalfe, J. C. (1982). *J. Biol. Chem.* **257**, 8183–8189.

Mope, L., McClellan, G. B. and Winegrad, S. (1980). *J. Gen. Physiol.* **75**, 271–282.

Mota, I. (1957). *Brit. J. Pharmacol.* **12**, 453–456.

Mottola, C. and Romeo, D. (1982). *J. Cell Biol.* **93** 129–134.

Naccache, P. H., Showell, H. J., Becker, E. L. and Sha'afi, R. I. (1977). *J. Cell. Biol.* **73**, 428–444.

Naccache, P. H., Showell, H. J., Becker, E. L. and Sha'afi, R. I. (1979a). *J. Cell. Biol.* **83**, 179–186.

Naccache, P. H., Showell, H. J., Becker, E. L. and Sha'afi, R. I. (1979b). *Biochem. Biophys. Res. Commun.* **87**, 292–299.

Naccache, P. H., Molski, T. F. P., Alobaidi, T., Becker, E. L., Showell, H. J. and Sha'afi, R. I. (1980). *Biochem. Biophys. Res. Commun.* **97**, 62–68.

Niedel, J. (1981). *J. Biol. Chem.* **256**, 9295–9299.

Norman, J. A., Drummond, A. H. and Moser, P. (1979). *Mol. Pharmacol.* **16**, 1089–1094.

O'Doherty, J. O., Youmans, S. J., Armstrong, W. McD. and Stark, R. J. (1980). *Science* **209**, 510–513.

Ogawa, Y., Harafuji, H. and Kurebayashi, N. (1980). *J. Biochem. (Tokyo)* **87**, 1293–1303.

Ogmundsdottir, H. M. and Weir, D. M. (1979). *Immunology* **37**, 689–696.

Ohnishi, S. T. (1978). *Analyt. Biochem.* **85**, 165–179.

Ohsako, S. and Deguchi, T. (1981). *J. Biol. Chem.* **256**, 10 945–10 948.

Ordal, G. W. (1977). *Nature* **270**, 66–67.

Oron, Y., Lowe, M. and Selinger, Z. (1975). *Mol. Pharmacol.* **11**, 79–86.

Pace, C. S., Tarvin, J. T., Neighbors, A. S., Pirkle, J. A. and Greider, M. H. (1980). *Diabetes* **29**, 911–918.

Pant, H. C., Terakawa, S., Yoshioka, T., Tasaki, I. and Garner, H. (1979). *Biochim. Biophys. Acta* **582**, 107–114.

Papahadjopoulos, D., Poste, G. and Vail, W. J. (1979). *Meth. Membrane Biol.* **10**, 1–121.

Parker, J. C., Castranova, V. and Goldinger, J. M. (1977). *J. Gen. Physiol.* **69**, 417–430.

Pearce, F. L. and Messis, P. D. (1982). *Int. Archs. Allerg. Appl. Immunol.* **68**, 93–95.

Pholramool, C. and Tangkrisanavinont, V. (1976). *Life Sci.* **19**, 381–388.

Pickard, M. R. and Hawthorne, J. N. (1978). *J. Neurochem.* **30**, 145–155.

Portzehl, H., Caldwell, P. C. and Ruegg, J. C. (1964). *Biochim. Biophys, Acta* **79**, 581–591.

Pressman, B. C. (1968). *Fed. Proc.* **27**, 1283–1288.

Pressman, B. C., Harris, E. J., Jagger, W. S. and Johnson, J. H. (1967). *Proc. Natl Acad. Sci. USA* **58**, 1949–1956.

Prince, W. T., Rasmussen, H. and Berridge, M. J. (1973). *Biochim. Biophys. Acta* **329**, 98–107.

Putney, J. W. (1976). *J. Pharm. Exp. Ther.* **199**, 526–537.

Putney, J. W., Weiss, S. J., van de Walle, C. M. and Haddas, R. A. (1980). *Nature* **284**, 345–347.

Putney, J. W., Poggioli, J. and Weiss, S. J. (1981). *Phil. Trans. R. Soc. B.* **296**, 37–45.

Quinn, P. S. and Judah, J. D. (1978). *Biochem. J.* **172**, 301–309.

Randon, J., Lecompte, T., Chignard, M., Siess, W., Marlas, G., Dray, F. and Vargaftig, B. B. (1981). *Nature* **293**, 660–662.

Rasmussen, H., Clayberger, C. and Gustin, M. C. (1979). *Symp. Soc. Exp. Biol.* **33**, 161–197.

Reed, P. W. and Lardy, H. A. (1972). *J. Biol. Chem.* **247**, 6970–6977.

Ringer, S. (1884). *J. Physiol.* **4**, 29–42.

Roelofsen, B., Zwaal, R. F. A., Comfurius, P., Woodward, C. B. and van Deenen, L. L. M. (1971). *Biochim. Biophys. Acta* **241**, 925–929.

Rorive, G. and Kleinzeller, A. (1972). *Biochim. Biophys. Acta* **274**, 226–239.

Rosenberg, S., Stracher, A. and Burridge, K. (1982). *J. Biol. Chem.* (in press).

Rosengurt, E., Heppel, L. A. and Frieberg, I. (1977). *J. Biol. Chem.* **252**, 4584–4590.

Salmon, D. M. and Honeyman, T. W. (1980). *Nature* **284**, 344–345.

Scarpa, A., Brinley, F. and Dubyak, G. (1978). *Biochemistry* **17**, 1378–1386.

Schanche, J. S., Ogreid, D., Doskeland, S. O., Refsnes, M., Sand, T. E., Veland, P. M. and Christoffersen, T. (1982). *FEBS Lett.* **138**, 167–172.

Scharff, O. (1979). *Analyt. Clin. Acta* **109**, 291–305.

Schell-Frederick, E. (1974). *FEBS Lett.* **48**, 37–40.

Schiffmann, E., Showell, H. J., Corcoran, B. A., Ward, P. A., Smith, E. and Becker, E. L. (1975). *J. Immunol.* **114**, 1831–1837.

Schneider, C., Gennaro, R., de Nicola, G. and Romeo, D. (1978). *Exp. Cell Res.* **112**, 249–256.

Schober, R., Nitsch, C., Rinne, V. and Morris, S. J. (1977). *Science* **195**, 495–497.

Schulster, D., Orly, J., Seidel, G. and Schramm, M. (1978). *J. Biol. Chem.* **253**, 1201–1206.

Schumacher, K. A., Classen, H. G. and Spath, M. (1979). *Thrombos. Hemostas.* **42**, 631–640.

Seeman, P. and Kwant, W. O. (1969). *Biochim. Biophys. Acta* **183**, 512–519.
Segal, D. M., Taurog, J. D. and Metzger, H. (1977). *Proc. Natl Acad. Sci. USA* **74**, 2993–2997.
Sehlin, J. and Taljedal, I-B. (1979). *Eur. J. Physiol.* **381**, 281–285.
Selinger, Z., Eimerl, S. and Schramm, M. (1974). *Proc. Natl Acad. Sci. USA* **71**, 128–131.
Serhan, C., Anderson, P., Goodman, E., Dunham, P. and Weissmann, G. (1981). *J. Biol. Chem.* **256**, 2736–2741.
Sha'afi, R. I. and Naccache, P. (1981). *In* "Advances in Inflammation Research" (ed. G. Weissmann), Vol. 2, 115–148. Raven Press, New York.
Sha'afi, R. I., Williams, K., Wacholtz, M. C. and Becker, E. L. (1978). *FEBS Lett.* **91**, 305–309.
Showell, H. J., Freer, R. J., Zigmond, S. H., Schiffman, E., Aswanikumar, S., Corcoran, B. and Becker, E. L. (1976). *J. Exp. Med.* **143**, 1154–1169.
Sillen, L. G. and Martell, A. E. (1971). "Stability Constants of Metal–Ion Complexes." The Chemical Society, London.
Simons, T. J. B. (1976a). *J. Physiol.* **256**, 209–225.
Simons, T. J. B. (1976b). *J. Physiol.* **256**, 227–244.
Singer, I. and Tasaki, I. (1968). *In* "Biological Membranes: Physical Fact and Function" (ed. D. Chapman), 347–410. Academic Press, London and New York.
Smith, R. J. and Ignarro, L. J. (1975). *Proc. Natl Acad. Sci. USA* **72**, 108–112.
Smoake, J. A., Song, S-Y. and Cheung, W. Y. (1974). *Biochim. Biophys. Acta* **341**, 402–411.
Sobue, K., Ichida, S., Yoshida, H., Yamazaki, R. and Kakoicjo, S. (1979). *FEBS Lett.* **99**, 199–202.
Steinhardt, R. A. and Alderton, J. M. (1982). *Nature* **295**, 154–155.
Steinhardt, R. A. and Epel, D. D. (1974). *Proc. Natl Acad. Sci. USA* **71**, 1915–1919.
Sudhoff, T. C., Walker, J. H. and Obrocki, J. (1982). *EMBO J.* (in press).
Sundler, R. and Papahadjopoulos, D. (1981). *Biochim. Biophys. Acta* **649**, 743–750.
Thelestam, M. and Mollby, R. (1975). *Infect. Immun.* **12**, 225–232.
Thomas, M. V. (1979). *Biophys. J.* **25**, 541–548.
Tokomura, A., Fukuzawa, K., Isobe, J. and Tsukatani, H. (1981). *Biochem. Biophys. Res. Commun.* **99**, 391–398.
Trams, E. G. (1974). *Nature* **252**, 480–482.
Trifaro, J. M. (1969). *Mol. Pharmacol.* **5**, 382–393.
Triggle, C. R., Grant, W. F. and Triggle, D. J. (1975). *J. Pharmacol. Exp. Ther.* **194**, 182–190.
Tsien, R. Y. (1980). *Biochemistry* **19**, 2396–2404.
Tsien, R. Y. (1981). *Nature* **290**, 527–528.
Tsien, R. Y., Pozzan, T. and Rink, T. J. (1982a). *Nature* **295**, 68–71.
Tsien, R. Y., Pozzan, T. and Rink, T. J. (1982b). *J. Cell Biol.* **94**, 325–334.
Tyson, C. A., Zande, H. V. and Green, D. E. (1976). *J. Biol. Chem.* **251**, 1326–1332.
Uemura, K. and Kinsky, S. C. (1972). *Biochemistry* **11**, 4085–4093.
Vance, D. E. and De Kruijff, B. (1980). *Nature* **288**, 277–278.
Vargas, O., Miranda, R. and Orrego, F. (1976). *Neuroscience* **1**, 137–145.
Vijayakumar, E. K. and Weidemann, M. J. (1976). *Biochem. J.* **160**, 383–393.
Vincenzi, F. (1981). *Cell Calcium* **2**, 387–409.
Volpi, M., Naccache, P. H. and Sha'afi, R. I. (1980). *Biochem. Biophys. Res. Commun.* **92**, 1231–1237.
Walenga, R. W., Opas, E. E. and Feinstein, M. B. (1981). *J. Biol. Chem.* **256**, 12 523–12 528.

Walker, J. H. (1982). *J. Neurochem.* **39**, 815–823.
Wang, L-L. and Bryan, J. (1980). *Eur. J. Cell Biol.* **22**, 329a.
Westcott, K. R., Engelhard, V. H. and Storm, D. R. (1979). *Biochim. Biophys. Acta* **583**, 47–54.
White, G. C. and Raynor, S. T. (1980). *Thrombosis. Res.* **18**, 279–284.
Wier, W. G. (1980). *Science* **207**, 1085–1087.
Wilkins, J. A. and Lin, S. (1981). *Biochim. Biophys. Acta* **642**, 55–66.
Wilkinson, P. C. (1975). *Exp. Cell Res.* **93**, 420–426.
Yagi, K., Yazawa, M., Kakiuchi, S., Ohshima, M. and Uenishi, K. (1978). *J. Biol. Chem.* **253**, 1338–1340.
Yamauchi, T. and Fugisawa, H. (1979). *Biochem. Biophys. Res. Commun.* **90**, 28–35.
Yerna, M. J., Hartshorne, D. J. and Goldman, R. D. (1979). *Biochemistry* **18**, 673–678.
Yin, H. L. and Stossel, T. P. (1979). *Nature* **281**, 583–586.
Yin, H. L. and Stossel, T. P. (1980). *J. Biol. Chem.* **255**, 9490–9493.
Yin, H. L., Zaner, K. S. and Stossel, T. P. (1980). *J. Biol. Chem.* **255**, 9494–9500.
Zimmerman, U., Reimann, F. and Pilwat, G. (1975). *Biochim. Biophys. Acta* **375**, 209–219.

Note Added in Proof

While most of the fundamentals of this paper remain much as they were when it was written during the early months of 1982, there are some aspects which have altered almost out of recognition. In particular, there have been considerable developments in the field of receptor mechanisms, which formed a substantial aspect of this paper, although I should stress that confusion remains rife and that understanding appears as elusive as ever. That these developments have come about is largely due to the availability of the intracellular Ca^{2+} indicator quin2. This has now been applied to a wide range of cells and under a wide range of circumstances.

Most studies have led to predictable results: the level of cytosol Ca^{2+} does indeed generally increase in response to receptor stimulation. However, there are also a number of instances in which this has been found not to occur. In particular, this is true of the platelets, which when stimulated with thrombin, or with the phorbol ester PMA, undergo typical shape changes and secrete ATP and serotonin without any detectable elevation in the level of cytosol Ca^{2+} (Rink *et al.*, 1982, 1983). This indicates that there may be an alternative pathway for activation, even in those cells in which a role for Ca^{2+} has already been well defined. Quin2 experiments in neutrophils reveal that although secretion due to ionomycin is dependent on the presence of extracellular Ca^{2+}, the ionophore is none the less able to release Ca^{2+} from intracellular stores (Pozzan *et al.*, 1983). In contrast, the Na^+ ionophore monensin causes neutrophils to secrete in a manner dependent on extracellular Na^+, and it does so by releasing intracellular Ca^{2+} stores: there is no requirement for extracellular Ca^{2+} (Di Virgilio and Gomperts, 1983). Hints that the neutrophils were not as straightforward as some other systems have already been alluded to (e.g. the simple inverse requirement for secretion of Ca^{2+} and ionomycin which applies so well in the mast cells, does not apply with the neutrophils (p. 299); also phagocytic stimulation of active oxygen radical formation in obelin-loaded neutrophils can proceed without detectable elevation of cytosol Ca^{2+} (p. 310). As with the platelets, phorbol activation occurs without elevation of cytosol Ca^{2+}, and indeed it is possible use the quin2 both as indicator and to clamp cytosol Ca^{2+} at 10^{-8} M and yet secretion (of vitamin B binding protein) still occurs (Di Virgilio and Pozzan, pers. commun).

To explain these findings, a role for the lipid-dependent protein kinase C, and for diglyceride (the phospholipase C derived product of inositol lipid breakdown) has been invoked (Michell, 1983). The phorbol ester which appears uniquely to affect the protein kinase C pathway of activation, is understood to substitute for diglyceride, the normal protein kinase C activator, and it does so in a manner which is quite independent of Ca^{2+}. The kinase causes phosphorylation of only some of those proteins which are normally labelled following receptor stimulation. Likewise (in hepatocytes at least), application of specific Ca^{2+} stimuli (i.e. the ionophores) results in an alternative pattern of labelling. The combination of the phorbol and the ionophore labelling looks more or less like that which is normally achieved by stimulation of receptors (Garrison et al., 1984).

Thus, in searching for a mechanism of exocytosis, the fundamental question "what does cytosol Ca^{2+} do?" should perhaps now be shifted to the rhetorical "what does protein phosphorylation do?". If we accept that exocytosis can be triggered by two independent pathways, one dependent on Ca–calmodulin, and the other on protein kinase C, both acting through separate families of protein phosphorylations, then it will be seen that there is scope, not for a few remarks of update, but a new paper.

The biochemistry underlying the mobilization of Ca^{2+} has developed too. While less has been heard of lipid methylation (indeed, a role for a methylation pathway in the mast cells has now been as good as ruled out (Moore et al., 1983)) there are indications that one of the water-soluble products of inositol lipid hydrolysis, inositol 1,4,5-trisphosphate (IP_3) might act to release Ca^{2+} from intracellular stores. IP_3 is the product of phospholipase C induced degradation of phosphatidylinositol 4,5-bisphosphate. There are a number of systems in which it now appears that this, rather than phosphatidylinositol, is the lipid whose breakdown is under the direct control of the receptors. In these situations, phosphatidylinositol acts as a buffer or source for synthesis (by phosphorylation) of the polyphosphoinositides. Introduced at 5 μM into the cytosol of pancreatic acinar cells permeabilized by washing in a low-Ca^{2+} solution, or hepatocyctes permeabilized with saponin, IP_3 causes release of Ca^{2+} from non-mitochondrial intracellular stores into the external medium (Streb et al., 1983; Joseph et al., 1984). If the compound can be shown to be generated at or around such a concentration as a result of inositol lipid breakdown due to receptor activation, then it becomes a serious candidate for the mediator of communication between the cell surface receptors and the intracellular Ca^{2+} stores which they control.

How do these roles for the breakdown products of the polyphosphoinositides square with our knowledge of the neutrophils? Here, in addition to previous scepticism regarding the role of inositol lipid metabolism in the mechanism of Ca^{2+} mobilization (due to Ca^{2+} dependence and the finding that PI breakdown is controlled not through occupation of receptors, but by events far downstream of this (p. 329)), we now find that (1) diglyceride and the inositol phosphates, the products of phospholipase C catalysed breakdown of the inositol lipids are not liberated on stimulation with f-MetLeuPhe, and (2) that the products of stimulated inositol lipid metabolism in these cells are those due to the action of a phospholipase D acting on PI, namely phosphatidate and inositol. In cells labelled with ^{23}P, the phosphatidate liberated is of low specific activity compared with ATP, the nominal donor

(Cockcroft, 1984). Thus, although roles for both the water and the organic soluble products of inositol lipid hydrolysis (in particular, the products of phosphatidylinositol 4,5-bisphosphate) have been defined in the mechanism of cellular activation, in the neutrophils at least, the mechanism leading to their generation just does not operate. Lest confusion remains about a possible ionophoric role for the phosphatidate generated as a result of phospholipase D action on PI, a recent reinvestigation has revealed that it is the presence of oxidation derived impurities, not the phosphatidate itself, which causes Ca^{2+} in model membrane preparations (Holmes and Yoss, 1983).

As to the mechanism of receptor-induced Ca^{2+} movement across the plasma membrane, there is now the suggestion that the mechanism for control may have a closer relationship to that which operates in the adenylate cyclase system than was formerly thought. Here it is known that the receptor–transducer system is composed of at least three separate entities. These are the receptor itself, the catalytic unit, and a guanine nucleotide regulatory protein which communicates between the two. Adenylate cyclase in isolated membranes can be activated in the absence of receptor-directed ligands by application of GTP or its so-called "non-hydrolysable" analogues such as GppNHp. When introduced into the cytosol of the mast cells through permeabilization (application of ATP^{4-}: see pp. 302–305) and then trapped by addition of excess Mg^{2+} (converts ATP^{4-} to the non-agonistic Mg^{2+} salt) to reseal the ATP^{4-}-induced lesions, GppNHp causes the cells to become spontaneously and selectively permeable to extracellular Ca^{2+}. They undergo exocytotic secretion when Ca^{2+} is added to the external medium (Gomperts, 1983).

If some generality for this mechanism can be discerned among other cell types, then understanding of the Ca^{2+} mobilizing receptors had best be sought at the level of protein–protein interactions and the lessons of the well-known and understood adenylate cyclase systems taken closer to heart. There has been a tendency for most investigators to regard the two main classes of receptors as something apart.

References

Cockcroft, S. (1984). *Biochim. Biophys. Acta* (in press).
Di Virgilio, F. and Gomperts, B. D. (1983). *Biochim. Biophys. Acta* **763**, 292–298.
Garrison, J. C., Johnsen, D. E. and Campanile, C. P. (1984). *J. Biol. Chem.* **259**, 3283–3292.
Gomperts, B. D. (1983). *Nature* **306**, 64–66.
Holmes, R. P. and Yoss, N. L. (1983). *Nature* **305**, 637–638.
Joseph, S. K., Thomas, A. P., Williams, R. J., Irvine, R. F. and Williamson, J. R. (1984). *J. Biol. Chem.* **259**, 3077–3081.
Michell, R. H. (1983). *Trends Biochem. Sci.* **8**, 263–265.
Moore, J. P., Johannsson, A., Hesketh, T. R., Smith, G. A. and Metcalfe, J. C. (1984). *Biochem. J.* (in press).
Pozzan, T., Lew, D. P., Wollheim, C. B. and Tsien, R. Y. (1983). *Science* **221**, 1413–1415.
Rink, T. J., Smith, S. W. and Tsien, R. Y. (1982). *FEBS Lett.* **148**, 21–26.
Rink, T. J., Sanchez, A. and Hallam, T. J. (1983). *Nature* **305**, 317–319.
Streb, H., Irvine, R. F., Berridge, M. J. and Schulz, I. (1983). *Nature* **306**, 67–69.

Chapter 10

The Ordered Water Model of Membrane Ion Channels

D. T. EDMONDS

*Department of Physics, The Clarendon Laboratory,
University of Oxford, Oxford, England*

BIOLOGICAL MEMBRANES Vol. 5
ISBN 0 12 168546 2

I. Introduction

The purpose of this chapter is to discuss some general features of ion channel models and then to describe in some detail the ordered water channel model. Many of the predictions of this model are common to a whole class of channel models in which the flow of ions is controlled by electric fields within the channel rather than blocked or opened by the movement of mechanical gating particles, as in some more traditional models.

A model is useful if it suggests further experiments. The process of model building, followed by experiment aimed at verifying or discarding particular features of the model, is a well-known technique. The ultimate aim is to arrive at an accurate simulation of a real system based firmly on experiment. In this sense the Hodgkin and Huxley (1952) equations, which so accurately describe some of the properties of the sodium and potassium channels of the squid giant axon, do not constitute a model of the mechanism of the ion channels. To deduce mechanism purely by the simulation of output and without further physical input is notoriously unreliable, as was clearly recognized by Hodgkin and Huxley (1952) who wrote:

> It was pointed out in Part II of this paper that certain features of our equations were capable of a physical interpretation, but the success of the equations is no evidence in favour of the mechanism of permeability change that we tentatively had in mind when formulating them.

A good example of output modelling is a tape recorder which can accurately simulate the acoustical output of a flute while having nothing in common with its mechanism of sound production.

No attempt can be made in this short chapter to review the previous work on models of membrane channels. However, mention will be made of some of the more interesting suggestions that have been made directly relating to aqueous channels and most particularly to those that rely upon an ordered water structure. The known permeability of real cell membranes to small molecules and to ions has lead to many models of membranes incorporating pores. The large negative hydration energy of small ions discussed below has suggested to many authors that some of the pores must be more or less "aqueous".

Models of the cell membrane based upon the lipid bilayer (Gorter and Grendel, 1925) may be thought to date from the paper of Danielli and Davson (1935) in which they account for the protein content of the membrane by sandwiching the bilayer between layers of protein. An early example of such a model incorporating pores is that of Stein and Danielli (1956) in which the bilayer is penetrated by cylindrical pores, lined with the same protein layer that coats the membrane surface, to ensure that they are hydrophilic. Some of the models from this period, including those of Fernandez-Moran

(1959) and Hechter (1965), incorporated the concept of water that was more or less ordered. Hechter's model included ordered water structures between the protein plates that coated the lipid bilayer. The thin water layers were ordered by their proximity to the protein surface which presented an hexagonal array of hydrogen bonding sites that matched the ice structure in the manner suggested by Warner (1961).

With the postulation (Singer and Nicolson, 1972) and acceptance of the fluid mosaic model of the membrane, attention has been increasingly focused on the amphipathic protein units that span the membrane for the sites of ion channels. In some cases, such as that of the acetylcholine activated post-synaptic ion channel, it has proved possible to remove the protein from its native membrane, incorporate it in an entirely artificial lipid bilayer and still to retain its function (Schindler and Quast, 1980; Nelson *et al.*, 1980).

When modelling aqueous ion channels a conflict exists between the need to hydrate the ions in order to lower their energy and the necessity to provide the highly discriminating ion selectivity found with some real ion channels. An early and ingenious solution to this conflict was suggested by Mullins (1959) who supposed the ions to traverse the channel with either one or two hydration "shells" intact. Selectivity was provided by the match between the diameter of the partially hydrated ion and the inner diameter of the channel. The walls of the channel were lined with charged or polarized groups capable of completing the effective hydration of the ion to ensure an acceptably low energy of the ion within the channel. This model is often criticized on the grounds that the hydration shell of an ion is at best a statistical concept but modern low-angle X-ray and neutron diffraction experiments (Enderby and Neilson, 1979) do reveal a remarkable stability and a definite structure for at least the first hydration layer of many ions.

Another influential attempt to account for the selectivity of ion channels is that of Eisenman (1962) who supposed there to be charged binding sites within the channel. Using simplifying assumptions he calculated the change in electrostatic energy of the ion between a fully hydrated state in solution and that bound at the site within the channel. For example, this energy difference for a spherical non-polarizable ion binding to a non-polarizable channel site of known charge and fixed effective radius may be shown to depend on the radius of the binding site thus yielding sequences of ion selectivity for ions of different unhydrated radius and charge. Attempts were also made to incorporate the influence that some water molecules close to the binding site would have on the energy difference. It is often said that such binding energy models are irrelevant to the selectivity of channels because it can be shown by Eyring rate theory (Eyring *et al.*, 1949) that the equilibrium transfer rate of a channel represented as a series of energy barriers depends only on the heights of the energy peaks and not on the depths of the energy

D. T. EDMONDS

wells which would represent the binding sites. This conclusion is true (Oosting, 1979) when applied to models in which the number of ions occupying any internal location is not limited, so that they each move independently through the channel. In more realistic models which, because of electrostatic repulsion, limit to a small number the ions that may occupy a channel at any time, the depths of the potential wells do have a profound effect upon the dwell time of an ion within a channel and hence upon the ion transfer rate. Even assuming independent ion motion it is doubtful if the equilibrium transfer rate is relevant in describing the transient transfer rate of rapidly switched channels.

For a review of more modern models that attempt to account for the selectivity of ion channels, including his own, see Hille (1975).

II. Ionic Hydration Energy

The electrostatic self energy in joules of an isolated spherical ion of radius R metres and charge Q coulombs is given by

$$U_{ION} = Q^2/(8\pi\varepsilon_0 R) = (1/2)QV_S, \tag{1}$$

where V_S is the voltage at the surface of the ion and $\varepsilon_0 = 8.85 \times 10^{-12}$ Fm. This may be calculated either as the energy necessary to assemble the ionic charge from infinitesimal charges brought from infinity or as the energy stored in the electric field that surrounds the ion so that

$$U_{ION} = (1/2)\int \varepsilon_0 E^2 d\tau, \tag{2}$$

where E is the electric field in volts/metre and the volume integral is taken over all space outside the ion. For a small ion such as Na^+ with a radius of 0.095 nm the value of U_{ION} is 12×10^{-19} J which is about $290 k_B T_R$, where k_B is the Boltzmann constant and T_R represents room temperature.

If the ion is now immersed in a fictitious continuous fluid with relative dielectric constant ε then the expressions for U_{ION} in equations (1) and (2) are multiplied by a factor $1/\varepsilon$. If ε is large this is a substantial reduction in the energy of the ion. The physical reason for this is that the time-averaged electric dipole moments P per unit volume in the dielectric fluid point radially away from a cation with their negative ends toward the ion. In terms of equation (1) the induced dipoles lower the potential at the surface of the ion from V_S to V_S/ε and in terms of equation (2) the volume integral of E^2 is reduced by a factor $1/\varepsilon$ because the electric field originating at the cation terminates at the net negative charge —div P per unit volume induced in the fluid, rather than at infinity.

When dealing with small ions it is a poor approximation to replace water by a continuous fluid, but more realistic calculation and experiment agree

(Buckingham, 1957) that when a bare Na^+ ion is immersed in water its energy ΔH is lowered by 6.6×10^{-19} J or about $160 \, k_B T_R$. Thus for a sodium ion to pass from an aqueous environment with a high relative dielectric constant and to enter as a bare ion the lipid bilayer with a low dielectric constant would require it to surmount a very large energy barrier of the order of $100 \, k_B T_R$. At normal temperatures the probability of such a transit is quite negligible. Thus the barrier presented by a lipid bilayer to the passage of unhydrated ions is not steric but electrostatic.

III. Types of Channel Model

A. PAIRED ION MODELS

At first sight it may be thought that the electrostatic barrier presented by a membrane with a low dielectric constant could be bypassed by the transfer simultaneously of pairs of ions with cancelling charges. That this is not so is easily seen by writing an expression for the electrostatic self energy of an isolated pair of spherical ions in contact, the first with charge Q and radius A and the second with charge $-Q$ and radius B.

$$U_{PAIR} = \frac{Q^2}{4\pi\varepsilon_0}\left(\frac{1}{2A} + \frac{1}{2B} - \frac{1}{A+B}\right) \qquad (3)$$

The first two terms represent the electrostatic self energies of the two ions and the third their attractive interaction energy. As has been previously pointed by Parsegian (1969), if $A = B$ the energy of the pair becomes exactly the same as that of the original single ion so

$$U_{PAIR} = U_{ION}, \quad \text{if } A = B$$

Even taking the optimum value of $B = 2.4 \, A$ results in a reduction of less than 20% from the single ion value so that ion pairing offers no solution provided there is no charge transfer or polarization of the charges. Of course large and complex ion carriers such as valinomycin do exist but are not capable of the very high ion transfer rates of 10^6 to 10^7 ions/channel/second encountered in excitable tissue.

B. CHARGED SITE MODELS

Most models of selective ion channels to date have been charged site models in which the ion traverses the channel by transit between these sites. Because of the low dielectric constant of the bulk lipid with $\varepsilon = 2.5$ and of most protein with $\varepsilon = 3$ (Pethig, 1979) the electrostatic energy of such models may be very high, even in the absence of an ion, unless the channel also contains electrically polarizable material such as water.

To illustrate this we consider a simple model system of a line of 7 negative charges each half buried in a plane slab of ideal dielectric of relative dielectric constant $\varepsilon = 5$. The charges, equally spaced along a distance of 5 nm represent a line of charges embedded in the lining of an otherwise empty pore penetrating a membrane. If each wall charge has a single electron charge the electrostatic potential experienced by an ion travelling along the line of charges a distance 0.2 nm from the dielectric is displayed in Fig. 1(a). One aspect of the model is satisfactory in that for much of the traverse the ion experiences an average electrostatic potential of around -4 V which lowers its energy by 6.4×10^{-19} J which is comparable to the energy reduction of a bare sodium ion when fully hydrated of 6.6×10^{-19} J. Other aspects are much less satisfactory in that the maximum barriers to motion are about 1 V corresponding to energy barriers of $39 \, k_B T_R$ so that ion mobility would be very low indeed. Further, the electrostatic energy of the wall charges including self and interaction energies is 3.7×10^{-18} J ($897 \, k_B T_R$) assuming a charge radius of 0.1 nm. Doubling the radius reduces the self energy by a factor of 2 but leaves the interaction energy unaltered leading to a total energy of $580 \, k_B T_R$, still so large as to make any such configuration highly improbable. One way that would lower the barrier height is to include more charges along the line but this raises even higher the total energy of the configuration of wall charges. In Fig. 1(b) is plotted the maximum barrier height and total electrostatic energy for a linear array of 3, 5, 7 and 9 single electron charges of radius 0.1 nm disposed along a 5 nm length. Were the charges to be embedded in the lining of a pore it is easily shown that the approach of high dielectric constant fluid to both ends of the channel would not reduce the barrier height appreciably. It would reduce the total energy somewhat but as shown by Parsegian (1969) the effect is much less than that which would result in allowing electrically polarizable material within the pore.

To sum up, exposed electrical charges within a pore can lower the energy of a bare ion to the energy of a fully hydrated ion. It is also possible that selectivity of ion channels could be modelled by suitably configured charged binding sites. However, in the exceptionally low dielectric constant environment within the membrane the similarly charged sites may not be close without incurring very high energy penalties. Well-separated charged sites lead to high barriers to motion and to low mobilities. The energy of the charged wall sites may be substantially reduced by introducing highly polarizable material in the pore but they remain high even in the absence of an ion. As shown below, water, the most common highly polarizable material, is capable of providing both the selectivity and the low energy sites within the pore without the necessity of wall charges.

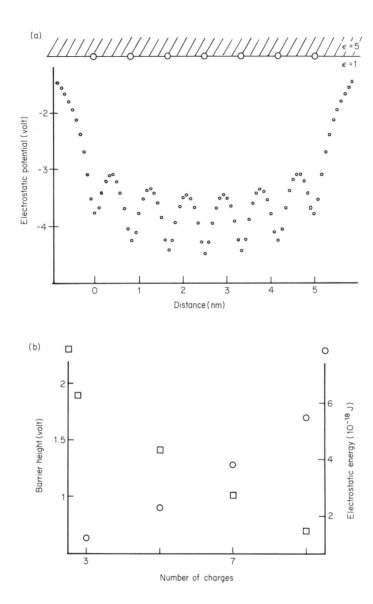

FIG. 1. (a) The calculated electrostatic potential along a line parallel to and distant 0.2 nm from, a linear array of 7 electron charges half buried in a plane slab of ideal dielectric of relative dielectric constant 5. (b) The height of the maximum voltage barrier encountered by an ion travelling as in (a) and the total electrostatic energy of the embedded charges as a function of their number in a linear array 5 nm long.

C. DIPOLE SITE MODELS

Models of ion channels based upon electrical dipoles as opposed to charges have as their most attractive feature that in the absence of the ion a closely spaced dipole array is capable of adopting a low-energy configuration. The array can then create closely spaced, selective, low-energy binding sites for ions by local rotation of a few dipoles while retaining elsewhere the low-energy resting dipole configuration. Some of these effects are illustrated in Fig. 2. In Fig. 2(a) is shown a linear array of dipoles fixed in position but free to rotate such that each dipole is aligned in the electric field of its neighbours. The electrostatic interaction energy of the array is negative thus contributing to the stability of the whole molecular configuration. In Fig. 2(b) is shown how, by rotation of a few dipoles to present their negative ends to a cation, an attractive binding site may be created. If the attractive interaction between the cation and the dipoles lowers the electrostatic energy of the system by more than the rise in the dipole–dipole interaction energy due to the disruption of the low energy state depicted in Fig. 2(a), then the dipole rotation will occur spontaneously and the new configuration remains stable. Comparison between Figs 2(b) and 2(c) shows that, neglecting end effects, the ion is free to move along the whole dipole chain whilst the energy at each resting

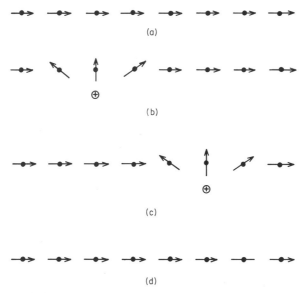

FIG. 2. (a) An idealized linear array of electric dipoles free to rotate but not to translate, in their electrostatic energy ground state. (b), (c) The creation of a low energy binding site for a cation by local rotation of the dipoles. (d) The restored ground state of the dipole array when the cation has moved away.

site remains constant. The mobility would depend on the energy barrier between two adjacent rest sites along the dipole chain, which can be low. Figure 2(d) shows the return to the lowest energy resting dipole configuration when the ion has left the channel.

While Fig. 2 illustrates some of the features of dipole channel models it is unrealistic in one important respect. Because the field of a dipole falls off with distance as $1/R^3$, dipoles have to be numerous and close to effectively lower the energy of an unhydrated ion. Even with dipole moments as large as those of water, calculations show (Edmonds, 1980a) that the ion must be in contact with at least 4 or 5 water molecules to have an energy as low as the hydrated ion energy. An attempt to ensure continuous contact with this number of water molecules leads to models incorporating water ring or net structures. Intermittent contact with a few electric dipoles such as are found along a protein α-helix is, for example, inadequate to produce the low energy environment needed for an unhydrated ion to enter a channel from an aqueous region.

IV. A Brief Description of
the Ordered Water Channel Model

The channel is supposed to consist of a ring or net structure of water molecules which lines a pore about 1 nm in diameter defined by parallel helical protein rods which span the membrane perpendicularly as sketched in Fig. 3(a). The helical protein rods support the water array by providing hydrogen bonding sites that match some of the unsaturated hydrogen bonds of the array so that the water array and the protein rods together form the channel protein or group of proteins. The water array is thought to be composed largely of planar pentagon and puckered hexagon rings which are two small ring structures that preserve the tetrahedral bonding angles required by a water molecule in its ground state. The ions find low energy sites at the centres of the water rings which are selective for unhydrated radius, valence state and the algebraic sign of the charge. The ions move over the water structure by passing from water ring to contiguous water ring.

Due to the peculiarly low dielectric constant and mechanically shielded environment found within the membrane and also to the elongated shape of the water channel, the water molecule electric dipoles are ordered. Any water molecule in a static tetrahedrally bonded array may point its dipole moment in one of six directions while maintaining the directions of its four hydrogen bonds to its neighbours. In the two equally low-energy ground states of the channel water array the water dipoles have a predominantly positive or negative projection on the outward normal to the membrane.

An ordered array leads to the presence of very large electric fields within the channel so that ion transfer may be controlled by these electric fields

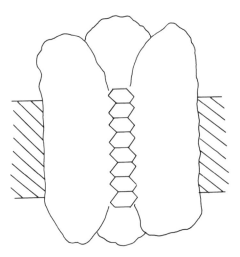

FIG. 3(a). A sketch of a section through the postulated ordered water array supported by hydrogen bonding to parallel helical protein rods that span the membrane perpendicularly.

alone and there is no need of mechanical gating particles. A switch between the two oppositely polarized states with the lowest energy involves no translation but only rotation of the water molecules. It can be brought about by the electric field across the membrane (electrical gating) or by the proximity of a suitably charged group (chemical gating). Thus each ordered water channel has two distinct ground states each with a different current versus voltage relationship depending on the direction of the polarization within the channel. An individual channel in a given polarized configuration would not conduct continuously although an assembly of such channels would display the current versus voltage characteristics of that polarization. The conduction of an individual channel would depend, for example, on freedom from the electrically charged hydrogen bonding defects which would block the channel and so conduct only in randomly timed bursts. A distinction needs to be made between activation, which corresponds to the dipole polarization switch and subsequent conduction of ions.

The degree to which a channel is selective to particular ions will depend on the extent to which the protein rods provide hydrogen bonding sites that discriminate in favour of one particular type of water molecule ring structure that provides particularly low energy sites for one particular ion. No configurational change is required of the protein support rods when a channel switches as the hydrogen bonds between the water array and the protein do not change. The protein rods do, however, have a vital role in providing specific binding sites for chemical messengers at both ends of the channel.

For example, the specific tetrodotoxin and amiloride binding sites on the outer ends of sodium channels in nerve and tight epithelia respectively would be provided by particular protein rod structures.

The action of internal pronase (Armstrong *et al.*, 1973) suggests that inactivation may be the result of a conformational change of the protein structure protruding from the inner surface of the membrane. A physically plausible model of inactivation may be constructed based upon the modulation of the binding properties of charged groups at the channel ends that necessarily result from the changed electric fields caused by the charge transfer or dipole reversal measured on activation (Edmonds, 1983). The model simulates many of the equilibrium and kinetic properties of sodium channels in the squid giant axon.

V. Calculation of the Properties of an Ordered Water Channel

A. THE SELECTIVE RING SITES

The planar pentagon and puckered hexagon water rings which preserve the tetrahedral bonding angles of the water molecules, and which are commonly found in the water cage structures of crystalline clathrate hydrates (Davidson, 1973), have been investigated as low-energy binding sites for ions. The electric field of a cation situated within the ring is sufficient to tip the dipoles of the ring water molecules to point radially away from the ion thus rupturing two of the four hydrogen bonds binding each ring molecule in the ordered water array. In this configuration the electrostatic energy of the array may be calculated (Edmonds, 1980a) including ion–water dipole, ion–water quadrupole, water dipole–water dipole and water dipole–water quadrupole interactions. The important effects of induced water dipole moments must be included as they are substantial for the water molecule in such large fields (Coulson and Eisenberg, 1966). The attractive ion–water molecule interactions lead to a large reduction in the energy of the unhydrated ion which is partly compensated by the rise in energy due to the disruption of the low-energy water structure present in the absence of the ion. The electrostatic energies are so large that the entropic term $T\Delta S$ in

$$\Delta G = \Delta H - T\Delta S$$

contributes less than 2%, and the lowering of the energy of the ion–water complex due to dispersion forces (Buckingham, 1957) is estimated to be less than 5% of the calculated electrostatic contribution.

The net reduction in energy of the ion–water molecule assembly from an initial state of isolated unhydrated ion and undisturbed ordered water array is found to be comparable with experimentally determined reductions due to full hydration of the ion. The ring sites are selective for unhydrated ion size and particularly low energies are found for the Na^+ ion which fits almost exactly within a planar pentagon water ring and for K^+ which similarly fits the puckered hexagon ring. Thus a channel composed largely of planar pentagons would discriminate in favour of Na^+ as opposed to K^+ and the inverse would be true if the puckered hexagon ring predominates in the water array.

To treat a divalent ion such as Ca^{2+} as a simple unpolarizable charged sphere is a much more questionable assumption than the same assumption for simpler ions as Li^+, Na^+ or even K^+. However, if such an assumption is made the ring sites do discriminate between divalent and univalent ions of the same unhydrated size such as Ca^{2+} and Na^+.

For cations the assumption that the water molecules in the ring reorientate with their dipole moments pointing radially away from the ion is supported by the known structures of the first hydration shells of ions in aqueous solution (Enderby and Neilson, 1979). For anions the situation is much less clear as the first hydration layer of many anions consists of water molecules with an O–H bond rather than the dipole moment directed towards the ion. In simple cage structures (Edmonds, 1979, 1980a) the outward radial dipole moment can be accomodated by rupture of only two of the four hydrogen bonds supporting the ring water molecules. To accommodate the radially inward pointing O–H bond requires the rupture of at least three hydrogen bonds per molecule which would result in considerably increased energies. Were this to prove the required orientation of the water molecules around an anion then the ordered water channels would discriminate strongly in favour of cations over anions. Even assuming the radial dipole moment the calculations do predict higher ion–water ring energies for anions compared to cations due to the ion–water quadrupole interaction.

B. THE ELECTRIC ORDER OF THE WATER MOLECULE ARRAY

The postulated water array, forming an essentially one- or two-molecule thick lining of the channel defined by the protein rods, is in a highly anomalous environment. Except at its ends it is surrounded by low dielectric constant material. In these circumstances a given array of charges or dipoles can easily give rise to electric fields as much as ten times larger than that of an identical array in an aqueous environment. In addition, the array is protected except for its ends from mechanical collision and hence mechanical exchange with water molecules of the bulk aqueous phase. In these protected circumstances, calculations predict (Edmonds, 1979, 1980a) that the dipoles of the array will be electrically ordered in the lowest energy configurations.

FIG. 3(b). Two dodecahedral sections of one possible columnar water array represented by CPK atomic models. Each dodecahedron has 5 donor and 5 acceptor hydrogen bonding sites directed radially outward from its centre.

Normal room temperature ferromagnetic substances order under the influence of the essentially electrostatic exchange interaction. However, paramagnetic crystals will order as ferromagnets or antiferromagnets under the influence of purely magnetic dipolar forces (Cooke et al., 1962) at sufficiently low temperatures, usually of the order of 0.1 K. Even assemblies of nuclei with their very much smaller magnetic dipole moments order due to the nuclear magnetic dipolar interaction at temperatures of order 10^{-6} K (Jacquinot et al., 1974). The interaction in a vacuum of two electric dipoles each of strength 1 D is 11 644 times as strong as that between two magnetic dipoles of strength 1 Bohr magneton separated by the same distance, so that ferroelectric order in suitable arrays of electric dipoles may be expected at temperatures of hundreds of Kelvin. Of course, there do exist many examples of ferroelectrically ordered solids but in a solid the electrostatic energy inevitably is coupled to the elastic energy of the lattice through the piezoelectric effect so that the analogy between the water array and magnetism is probably much closer.

Calculations including dipole and quadrupole effects over particular isolated columnar water arrays, such as that shown in Fig. 3(b) composed entirely of planar pentagon rings which can be visualized as the stacking of regular dodecahedra one on top of the other (Edmonds, 1979, 1980a), do show that the lowest energy state has electrical order. In the two equally low-energy ground states the projections of the dipole moments on the axis of the channel are all positive or all negative. The particular directions of the dipole moments are constrained by the need to preserve intact hydrogen bonding throughout the fixed array with no defects and to conserve as invariant the structural hydrogen bonds between the array and the protein support helices.

Several other properties of the array help to reinforce the conclusion that electrical order exists. Besides the long-range electric dipole interaction, there is for tetrahedrally bonded water molecules, a short range force as has been shown for ice by Coulson and Eisenberg (1966). In the absence of hydrogen bonding defects, so that each oxygen to oxygen bond contains exactly one proton, there is a correlation in the alignment of the electric dipole moments to second nearest neighbours. This may be thought of as a short-range contribution to the Onsager (1936) reaction field although Onsager only considered long-range forces in his calculation. Following the arguments of Onsager this correlation cannot by itself cause electric order but can aid its stability once established. A second property of the array that aids order is the large polarizability of the water dipole (Coulson and Eisenberg, 1966). The magnitude of the dipoles that interact grow as the order is established because the order leads to bigger electric fields at a molecular site with a positive projection along the dipole direction. It is this property which makes realistic calculations most difficult. Again, the presence of high dielectric constant liquid at the ends of the array is an aid in establishing a polarization parallel or antiparallel to its axis as can be verified approximately by the method of electrical images (Smythe, 1939).

Finally in the resting membrane of a cell there often exists a voltage of 50 mV or more across the membrane. For a membrane of thickness 5 nm the electric field across the membrane is 10^7 V/m. In such a field, the energy of 100 dipoles each of strength 1.84 D and making angles θ with the field direction such that the average value of cos θ is $1/\sqrt{2}$, would have an energy about 1 $k_B T_R$ less than the same array randomly oriented. Thus the membrane field itself has a substantial ordering influence and promotes axial order.

Such calculations cannot definitely establish the presence of electrical order in an elongated electrically shielded water array at 300K, but do suggest that it is likely. Strong experimental evidence does suggest that the equilibrium ground state of normal bulk hexagonal ice is electrically ordered below 72K (Tajima *et al.*, 1982).

C. ELECTRIC FIELDS WITHIN AN ORDERED WATER ARRAY

A horizontal planar array of dipoles each pointing vertically upward gives rise to fields that point downward at all points in the plane. The consequence is that in a columnar array of dipoles pointing upward the electric fields point predominantly downward. To illustrate this using realistic water molecule arrays is much complicated by the need to preserve tetrahedral bonding between the molecules. A simplified structure consisting of five planar, upward pointing hexagonal arrays of dipoles stacked one on top of the other which illustrates many of the features of more complicated arrays is shown in Fig. 4. For purposes of calculation each non-polarizable dipole has a moment of 1 D, each dipole is a distance of 0.3 nm from the Z axis and the separation of the planes of dipoles along the Z axis is also taken as 0.3 nm.

The first important property of the array is that the Z component of the electric field at the site of each dipole is positive, helping the dipole to point upward. The size of this Z component is 8.3×10^8, 3.2×10^9, 3.6×10^9, 3.2×10^9, and 8.3×10^8 V/m for the five layers. This means that the energy of the array is 1.2×10^{-19} J, or about 28 $k_B T_R$ lower than would be a randomly oriented identical array of dipoles.

The electric field along the axis of the array is always parallel or antiparallel to the Z axis by symmetry. In Fig. 5(a) is plotted the size of the field as a function of position along the axis. Far from the array and approaching it from above the field is seen to be positive (upward). This is to be expected as at large distances the individual dipoles may be thought of as combining to form a giant upward pointing dipole. However, with closer approach to the

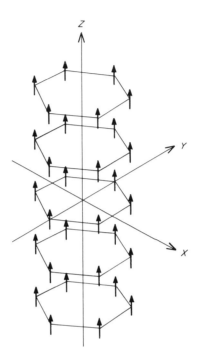

FIG. 4. An array of dipoles which simulates in simplified form some of the electrical properties of real polarized water dipole arrays.

array the field changes sign and remains negative (downward) until after the lowest dipole layer has been penetrated. Also in Fig. 5(a) is plotted the Z component of the field as a function of Z along a parallel line outside the array defined by $X = 0.6$ nm, $Y = 0$. Here again the field near the array is downward. Finally in Fig. 5(b) is plotted the Z component of the field as a function of distance along the X axis which is again seen to be everywhere downward.

It is thus demonstrated that such arrays can provide fields at the dipole sites which support the upward directed dipoles but that in the neighbourhood of the arrays the fields are predominantly downward. The presence of low dielectric constant material around the array and high dielectric constant material at its ends materially affects some of the magnitudes of the calculated fields but does not change the conclusions arrived at above. These conclusions are confirmed by detailed calculations on particular tetrahedrally bonded water arrays (Edmonds, 1979, 1980a).

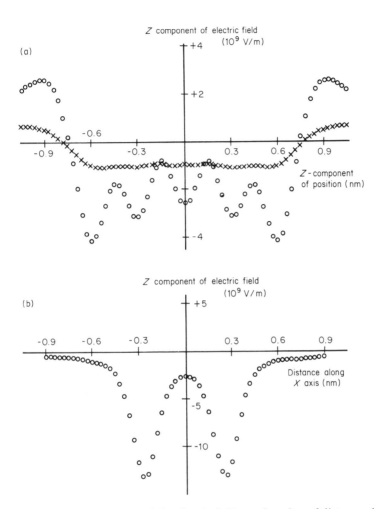

FIG. 5. (a) The Z component of the electric field as a function of distance along the axis ($\bigcirc$) and along a line parallel to the axis ($\times$) defined by $X = 0.6$ nm, $Y = 0$ for the dipole array in Fig. 4. (b) The Z component of the electric field as a function of distance along the X axis for the dipole array in Fig. 4.

D. THE CURRENT VERSUS VOLTAGE CHARACTERISTIC OF THE CHANNEL

A calculation of the electrostatic potential that would be experienced by a cation at the centre of each pentagon water ring in an isolated dodecahedral channel with its dipole moment pointing to the right is shown in Fig. 6(a) (Edmonds, 1980a). The calculation includes the dipolar and quadrupolar

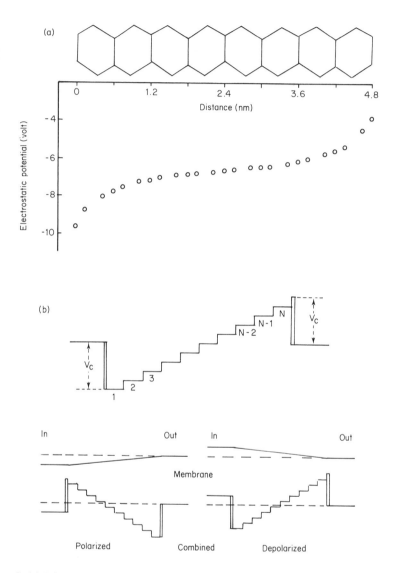

FIG. 6. (a) The calculated electrostatic potential that would be experienced by a sodium ion at the centre of each water pentagon in the dodecahedral channel shown in Fig. 3(b) with its total composite dipole moment pointing to the right. (b) An idealized potential distribution in the form of a linear stepped ramp and the idealized potentials when combined with the membrane potentials in the polarized and the depolarized configurations.

fields of the water molecules and assumes that the dipole moments of the water molecules in the ring occupied by the ion are tipped to point radially away from it. The more remote dipoles are assumed to remain in their low-energy ordered state undisturbed by the presence of the ion. In practice these water molecules would tip to some extent causing a further lowering of the electrostatic energy partially compensated by a rise in the strain energy, represented by slight distortion of their hydrogen bonds. The presence of high dielectric fluid at each end of the channel and the presence of slightly polarizable protein surrounding the array reduces considerably the change in potential along the channel but has much less effect on the average potential experienced, which is largely the very short range effect of the tipped water dipoles of the particular ring occupied by the ion. There would be potential barriers between adjacent sites impeding the passage between contiguous rings. It is difficult to estimate the size of these barriers but they may not be large as the presence of the ion has already destroyed the hydrogen bonds around the ring, so that ion tunnelling between adjacent rings is possible.

Viewed mathematically, the ions perform a biased one-dimensional random walk as they move between the resting sites and experience a semi-reflecting boundary at one end. The bias is provided by the electric field within the channel represented by the electrostatic potential gradient in Fig. 6. A particularly simple calculation of the current versus voltage characteristic of a channel with such a structure is possible (Edmonds, 1981b) if the calculated slightly curved potential distribution is replaced by a linear stepped potential ramp. It is easy to include the effects of potential barriers between sites in the calculation but for simplicity we show in Fig. 6(b) a simple stepped ramp and also the combined potential for the channel in its two lowest energy polarized states in equilibrium with a polarized (P) and a depolarized (D) membrane.

A channel will tend to switch when the electric field it experiences changes sign. This need not be at a membrane voltage ($V_M = V_{IN} - V_{OUT}$) of zero because of the additional local fields due to the dipole moments of the helical protein rods (Hol *et al.*, 1978), and any charged groups on the proteins near the channel. If the protein rods are inserted in their lowest energy state with the dipole moments parallel to the membrane field, the channel switching centres on a membrane voltage $V_M = V_S$, with V_S negative. Due to statistical fluctuations the channel will not switch abruptly when the field it experiences changes sign but there will exist a Boltzmann distribution of the two ground state configurations leading to the usual sigmoid distribution function over a finite range of V_M centred on $V_M = V_S$ (Edmonds, 1981b).

The calculated shape of the current versus voltage characteristics are not very sensitive to the number of steps in the ramp N or to the peak to peak voltage difference $2V_C$, although the absolute value of the transfer rate does vary with these parameters. An estimate of the number of steps can vary from about 30 for a channel spanning the entire membrane width of about 5 nm to perhaps 10 steps for one that forms a short selective gate within a longer length of pore. A calculation of the peak to peak voltage is complicated by the presence of polarizable material near the channel as discussed above. Estimates based on the method of images (Smythe, 1939) lead to values of a few hundred millivolts. For definiteness a channel of 30 steps with a peak to peak voltage of 300 mV is used in all the calculations in this chapter.

In Fig. 7(a) is shown a diagrammatic sketch of a channel with its axial dipole moment components represented by short arrows in its two polarized ground states in equilibrium with a sufficiently polarized (P) or a sufficiently depolarized (D) membrane. The long arrows represent the direction of the dipole moments of the protein helices which do not switch on depolarization of the membrane. Assuming that only one ion may occupy the channel at any time and that concentration of free ions externally and internally are given by $C_O = 440$ mM and $C_I = 50$ mM like those of sodium in the squid axon, the two current versus voltage curves shown in Fig. 7(b) are obtained corresponding to the P and D polarizations (Edmonds, 1981a,b). The dotted line is the equilibrium characteristic centred on a switching voltage $V_S = -26$ mV and the broken line is the normalized prediction of the Hodgkin and Huxley (1952) empirical equations discussed later. There are no adjustable parameters in the calculations and it is gratifying to see that realistic transfer rates of order 10^7 ions/channel/second are obtained with the physically plausible assumptions that have been made and with only one ion allowed in the channel at a time. It is clear that the rise in the predicted inward Na^+ transfer rate that would result from the depolarization of a membrane from a resting voltage $V_M < V_S$ to a depolarized voltage $V_M > V_S$ would be sufficient to initiate and propagate an action potential in the normal manner.

Two features of the predicted current versus voltage characteristic will be discussed further here. The first is illustrated in Fig. 8(a) which displays the inward transfer rate predicted at $V_M = 0$ for a D configuration channel with the channel parameters chosen as above as a function of C_O and C_I the external and internal free ion concentrations. At constant C_I the transfer rate is linearly dependent on $C_O - C_I$ as for simple diffusion but the slope of the dependence varies inversely with C_I. At constant C_O the transfer rate rises as some power of $C_O - C_I$ as C_I is reduced. In some sense the inward transfer rate is controlled by C_I much more than by C_O. The physical reason for this anisotropy is seen by examining the assumed stepped potential. In the D polarization the inner end has a potential well permitting free entry but the outer end is protected by a potential barrier. In the competition to occupy a newly vacated channel the ions at the inner end in the D configuration have the advantage. In the P configuration the advantage is reversed. This

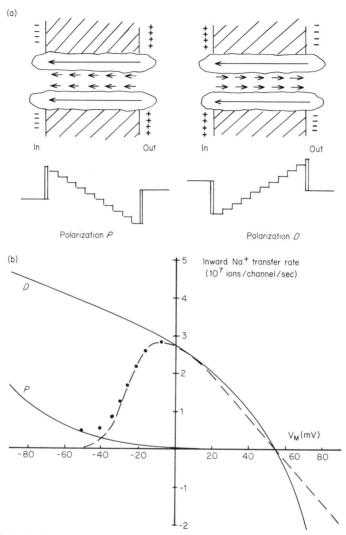

FIG. 7. (a) A sketch showing the directions of the water channel axial dipole components (short arrows) and the helical protein rod dipole moments (long arrows) in the P and D polarizations. Below each sketch is shown the simplest electrostatic potential distribution (the linear stepped ramp) that can represent the interstitial water ring ion sites within the water array in that polarization state. (b) The model predicted sodium ion transfer rates for the P and D channel polarizations as a function of the membrane voltage. The external, $C_O = 440$ mM, and internal, $C_I = 50$ mM, ion concentrations were chosen to mimic those for sodium in the squid axon. The broken line gives the prediction of the Hodgkin and Huxley equations normalized to equal the model prediction at a membrane voltage of $V_M = 0$. The dotted line gives the predicted equilibrium transfer rate showing the switching between the P and D polarizations of the channel centred on a membrane voltage given by $V_S = -26$ mV.

is a direct result of the assumption, supported by electrostatic calculation (Edmonds, 1980a), that the mean energy of a Na^+ ion in a pentagon ring channel or a K^+ ion in a hexagon ring channel is approximately equal to the appropriate fully hydrated energy. In Fig. 8(b) the mean ionic energy in the channel is assumed to be higher than that in solution by 2.8×10^{-20} J (0.175 V for a monovalent ion) in order to remove the potential well. Now the calculation shows the transfer rate linearly proportional to $C_O - C_I$ but the slope is almost independent of C_I as is the case with simple diffusion. The absolute rates for inward transfer are much less, due to the high potential barrier at the outer end. Model voltage controlled channels with and without potential wells may well simulate different types of real channels.

The second feature of the predicted current versus voltage curves is their sensitivity to the presence of divalent ions in solution capable of passing through the same channel as the more abundant monovalent ions. Ca^{2+} which has approximately the same unhydrated radius as Na^+ is known to pass through the sodium channel in the squid axon (Baker et al., 1971; Meves and Vogel, 1973). To a divalent ion all electrostatic potential barriers are twice as high in energy and the dwell time of a divalent ion in a voltage controlled channel may be very long, thus denying access to monovalent ions and drastically reducing the average transfer rate of the monovalent ions (Edmonds, 1982). This situation for voltage controlled channels is in marked contrast to that for mechanically blocked channels.

The anisotropy of the channels with electrostatic potential distributions like those of Fig. 7 or Fig. 8(a) is important here in that a channel is only very sensitive to divalent ion entry at the end with the potential well, the other end being protected by a potential barrier twice as high in energy to divalent ions as to monovalent ions. As an example, the current versus voltage curves shown in Fig. 7(b) are recalculated on the assumption of simple competition between divalent and univalent ions in the access volumes at each end of the channel (Edmonds, 1981b) with the sodium ion concentrations $C_O = 440$ mM, $C_I = 50$ mM as before but with in addition an external free Ca^{2+} ion concentration of 10 mM as in seawater or in the blood of the squid. The potential barriers experienced by the two types of ion are assumed the same. The result is that the total current, due to both the Na^+ and Ca^+ ions in a channel with the P polarization is everywhere reduced by a factor of 10 but a channel with the D polarization has its current reduced by less than 1 part in 1000. A free ion concentration of Ca^{2+} inside the axon would conversely lead to reduction of the current in the D polarization with no effect on the P polarization. In this respect it is interesting to note that the free Ca^{2+} ion concentration in the axoplasm is strongly buffered at a very low value of about 20–50 nM (Brinley, 1978). The sensitivity of voltage controlled channels to divalent

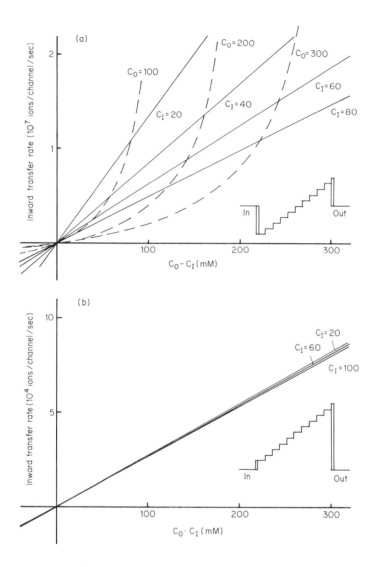

Fig. 8. (a) The model predicted inward transfer rate for the D polarization of the channel at a membrane voltage of zero as a function of $C_O - C_I$ where C_O and C_I are the external and internal free ion concentrations. The full lines connect points corresponding to a constant value of C_I and the broken lines connect points corresponding to a constant value of C_O. The sketched potential distribution assumed is as in Fig. 6(b). (b) The model predictions of the same inward transfer rates as a function of $C_O - C_I$ using the modified potential distribution shown with no potential well at the inner end.

ions capable of passing through them may be much enhanced by the inclusion of potential barriers between the ion resting sites which has a marked effect on the divalent ion resting time, with much less effect on monovalent ions.

VI. Evidence for the Ordered Water Model

A. STRUCTURAL EVIDENCE

Real but non-biological systems which have much in common with the model are the hydrated Zeolites. These have alumino-silicate cage structures that contain ordered water arrays and mobile ions. Taking as an example hydrated Zeolite Na A, the alumino-silicate structures may be thought to comprise large, approximately spherical, α-cages and smaller, approximately spherical β-cages connected by tube-like structures (Gamlich and Meier, 1971). X-ray diffraction reveals that the β-cages contain 4 water molecules forming a distorted tetrahedron and the α-cages contain 20 water molecules forming the dodecahedron composed of planar pentagons. The water structures are stabilized by hydrogen bonds to the negatively charged oxygen atoms of the cage structure and to the Na$^+$ ions. A major difference between the Zeolites and the model is that the alumino-silicate lattice bears a large negative charge whereas the helical protein rods do not. As a consequence the Na$^+$ ion resting sites in the Zeolites tend to be close to the cage.

One feature of the model is that only unhydrated ions move through the channel but that while they do so they are continuously hydrated by an array of water molecules that rotate but do not translate. That a similar situation applies to the Zeolites is illustrated in Fig. 9. In Fig. 9(a) is plotted the Arrhenius activation energy barrier to ion diffusion for various cations in hydrated chabazite (Barrer et al., 1963). The diameter of the smallest cage aperture limiting motion is 0.39 nm which is bigger than the unhydrated diameter of any of the ions. The divalent ions all experience roughly twice the barrier experienced by the monovalent ions irrespective of diameter, demonstrating an electrostatic but not a steric barrier. In contrast in Fig. 9(b) is plotted the Arrhenius barrier to diffusion in hydrated analcite (Barrer and Rees, 1960) in which the diameter of the limiting aperture is 0.23 nm comparable to the ionic diameters. Now evidence of steric hindrance is clearly seen which argues for motion of the unhydrated ion, a process known as ion sieving.

To obtain direct structural evidence for small ordered water arrays within membranes in aqueous surroundings is difficult. However, structural data relating to channel proteins is accumulating rapidly and it may be possible (Edmonds, 1980b) to deduce water array structure from protein structure.

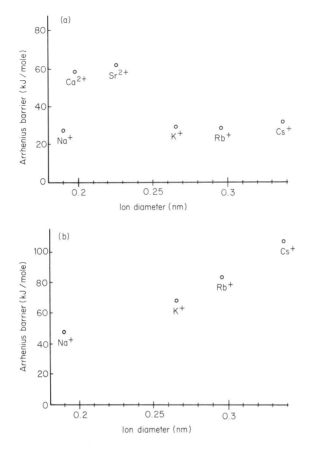

FIG. 9. (a) The measured Arrhenius barrier to self diffusion in hydrated chabazite (Barrer *et al.*, 1963). The smallest aperture limiting diffusion has a diameter of 0.39 nm, larger than the unhydrated diameter of any of the ions. The plot gives evidence of an electrostatic but not of a steric barrier. (b) The same Arrhenius barrier measured (Barrer and Rees, 1960) in hydrated analcite in which the limiting aperture has a diameter of 0.23 nm, comparable with the unhydrated diameters of the diffusing ions. The effects of steric hindrance are seen.

For example, the postulated dodecahedral array shown in Fig. 3(b) based on pentagon rings would require 5 protein rods to support it. It is approximately 1 nm in diameter and has a repetition length of 0.6 nm. Other water columns based on hexagon rings would require 6 rods and have a repetition length of 0.45 nm. The best characterized membrane protein is that of the post-synaptic acetylcholine receptor. This is roughly rod shaped, spans the membrane perpendicularly and protrudes at each side. When viewed end on by electron microscopy it resembles a rosette (Meunier *et al.*, 1974; Eldefrawi *et al.*, 1975) which is now known to consist of 5 parallel protein rods (Raftery *et al.*, 1980) spanning the membrane. Preliminary X-ray diffraction data (Ross *et al.*, 1977) reveal that the rods are helical with repetition lengths of 0.52 nm and 0.63 nm. The central pit in the rosette is variously described (Famborough, 1979) as having a diameter of 0.65, 0.8 or 1.5 nm. All these observations match well with the presence of an ordered water structure lining the protein defined pore.

Another much studied channel protein is "porin" extracted from the outer membrane of *Escherichia coli*. When incorporated into artificially prepared lipid bilayers it displays voltage sensitive but not ion selective ionic conductivity with negative resistance regions, indicative of gating (Schindler and Rosenbusch, 1978; Benz *et al.*, 1978). Besides conducting ions the channels pass uncharged hydrophilic small molecules like sugars. Model building from the known protein sequences suggests (Inouye, 1974) 6 protein rods defining a pore 1.25 nm in diameter. Computer enhanced, negatively stained electron microscopy reveals (Steven *et al.*, 1977) a hexagonal lattice but shows a triplet structure surrounding a pit 2 nm in diameter. Both conductivity studies agree on an aqueous pore diameter close to 1 nm. Circular dichroism studies confirm (Braun, 1975) that 80% of the protein is α-helical and that in this configuration most of the hydrophobic residues line up on one side of the helix, again suggesting the parallel rod barrel-stave model with an outer hydrophobic face to match the lipid and a hydrophilic inner pore for conduction.

Finally, alamethicin and suzukacilin are two rod-shaped proteins which promote rapid ion conduction in real and artificial membranes. When the conductivity of single channels is monitored it is observed (Baumann and Mueller, 1974; Boheim and Kolb, 1978) to jump, always sequentially, between well-defined quantized levels. This is widely interpreted as the result of forming conducting pores defined by sequential numbers of parallel protein rods that span the membrane. Fluctuation measurements (Kolb and Boheim, 1978 measure the probabilities of each of these structures and if the first conducting pore is taken as that incorporating 3 protein rods, the probabilities of those incorporating 5 and 6 rods are 29.4% and 51.7%, showing these to be the most stable. Another interesting feature of these proteins is the large number of 2-aminoisobutyrate residues that they incorporate. X-ray diffraction of model compounds (Venkataram *et al.*, 1980) and alamethicin fragments (Smith *et al.*, 1981) show that they form 3_{10} helices. These are particularly suitable for supporting columnar water structures as the hydrogen bonding sites lie on a line parallel to the helical axis rather than spiralling about it as in the α-helix. The repetition length in the 3_{10} helix is 0.6 nm exactly like that of the dodecahedral water array.

B. CURRENT VERSUS VOLTAGE CHARACTERISTICS

In Fig. 7(b) is displayed, as full lines connected by the dotted line, the predicted equilibrium characteristic of the water channel with the external $C_O = 440$ mM and internal $C_I = 50$ mM ion concentrations chosen to match those found for sodium in the squid axon. The characteristic predicted by the Hodgkin and Huxley (1952) equations is shown as the broken line and has been scaled to equal the model prediction at $V_M = 0$. The switching voltage $V_S = -26$ mV and the proportion switched as a function of V_M (Edmonds, 1981b) have been chosen to agree with those measured. The prediction is seen to be in reasonable agreement with experiment except for the P characteristic which is the equilibrium state at very negative V_M. However, as described in Section VI.D, the presence of divalent ions such as Ca^{2+} in the external fluid which are capable of passing through the pore would suppress this current.

Turning briefly to another system, the flattening of the D characteristic at negative V_M, which is often called outward rectification, is found for the current in the light-controlled Na^+ channel in the outer segment membrane of isolated retinal rods of the Salamander (Bader et al., 1979). Choosing a value of V_S about -70 mV would ensure that at all physiological values of V_M the model channel would exist in the D polarization and would not switch. In these circumstances the model prediction is like that marked D in Fig. 7(b) and is close to that found for the Salamander rod. Values of V_S much more negative than -70 mV are known to exist in the sodium channel of the apical membrane of the granular cells of tight epithelia (Lindemann, 1968) as discussed below in Section VII.C.

In Fig. 10 is displayed (Edmonds, 1981a,b) as a full line the model-prediction for the equilibrium transfer rate assuming the same channel potential distribution as before but with $C_O = 20$ mM and $C_2 = 400$ mM which pertain to potassium in the squid axon. V_S is chosen as -50 mV. The Hodgkin and Huxley equations predict the broken curve which has again been scaled to equal the prediction at $V_M = 0$ in order to compare shapes. It is seen that once again reasonable agreement is obtained and that the switching characteristic between the two curves predicted for the P and D polarizations is largely suppressed by the proximity of V_S to V_K, the Nernst potential (V_M for no transfer) of -75 mV for potassium. Using the same assumptions as before but with $C_O = C_I = 400$ mM, which ensures a Nernst potential of zero, restores the typical S-shaped switching characteristic as shown by the dotted curve. To obtain such a curve requires both the P and D polarizations to contribute to the transfer rate and this could not be predicted by a mechanical gating particle model unless one set of channels

happened to open just as another set of channels started to close. A characteristic just like this is in fact measured with $C_O = C_I = 400$ mM as potassium concentrations for the squid axon (Moore, 1959) although at the time
the current was attributed partly to K$^+$ and partly to Cl$^-$ ions. That this is
not so may be inferred from the experiments of Segal (1958) who obtained
results consistent with Moore using a constant current technique but with
400 mM potassium acetate. An unstable (negative resistance) portion of the
current versus voltage characteristic has also been observed by Werblin
(1979) in the voltage- but not light-controlled membrane channels of the
outer segment of the Salamander retinal rod with artificially high external
K$^+$ concentrations. Using $V_S = -60$ mV the shape of the measured characteristic is simulated by the model when the experimental values of C_O and
C_I are used.

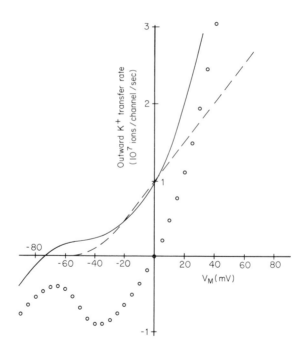

FIG. 10. The full line gives the model predicted equilibrium potassium ion outward
transfer rate as a function of membrane voltage V_M using $C_O = 20$ mM and
$C_I = 400$ mM as in the squid axon, and with a switching membrane voltage of $V_S =$
-50 mV. The broken line is the prediction of the Hodgkin and Huxley equations
normalized to equal the model prediction at $V_M = 0$. The dotted curve is the
model prediction using $C_O = C_I = 400$ mM.

Exact agreement is not to be expected with such simplified models but whether or not they reflect physical reality, models based on linear stepped potential ramps do seem effective in predicting the current versus voltage characteristics found both at physiological and non-physiological ionic concentrations. The temperature dependence of the model transfer rates may be determined by comparing the predictions at 290 K and 280 K, a parameter known as the Q_{10}. This depends on V_M slightly but at $V_M = 0$ it is 1.53 for the potential distribution used here for both P and D polarizations. Experimentally, the Q_{10} of the peak conductivity is found to be 1.5 for both Na^+ and K^+ channels of the squid axon (Moore, 1958).

C. GATING CHARGE TRANSFER

One of the most sensitive methods of determining the mode of operation of ion channels is to eliminate the transfer of ions by the use of suitable bathing fluids and selective poisons and then to measure the charge transfer that occurs within the membrane connected with the voltage-sensitive activation of the channels (Armstrong and Bezanilla, 1974; Keynes and Rojas, 1974). Some of the properties of the gating charge are (1) the equality of the charge transfer on opening and shutting for short activation times, (2) the saturation of charge transferred with membrane voltage applied and (3) the lack of temperature dependence of the saturated value of the charge transferred. Clearly these are all intrinsic features of a reversing dipole array. If the charged gating structure experiences a change in energy given by $Zq(V_M - V_S)$ when it moves, where q is the proton charge, then it is possible to predict the percentage of gating charge movement that will be achieved at any membrane voltage V_M using Boltzmann statistics. By comparison with experiment the groups quoted above deduced a value of Z about 1.3 for the sodium channel of the squid axon. In an array of N dipoles each of strength p Debye and making angles θ with the membrane electric field E_M such that the average value of $\cos \theta$ is given by $<\cos \theta>$, the change in energy on reversing the direction of the dipoles is given by ΔU in

$$\Delta U = 2\,N p < \cos \theta > E_M$$
$$= 2\,(N/t)\,p < \cos \theta > (V_M - V_S)$$

where t is the effective membrane thickness such that $E_M = (V_M - V_S)/t$. For the dodecahedral channel $2\,(N/t)\,p < \cos \theta >$ is equal to $1.6\,q$ and for another possible water array (Edmonds, 1979) it is equal to $1.26\,q$ so that there is good agreement between the model predictions and experiment.

The fact that the gating charge transfer is not prevented when a sodium channel is poisoned by an agent such as tetrodotoxin is often quoted as an argument that the selective filter and the gate must be separated within the channel. For the ordered water channel model this separation is not necessary as the poisons can only bind selectively to the protein rods at the outer face of the channel and will only impede channel switching if the poison has a sufficiently large electric charge to interfere with the channel dipoles.

VII. Predictions of the Ordered Water Model

A. EACH CHANNEL CAN EXIST IN TWO POLARIZATIONS EACH WITH A DISTINCT CURRENT VERSUS VOLTAGE CHARACTERISTIC

For all channels, however gated, as opposed to pumps which require a direct energy input, the sign of the cation transfer rate is necessarily positive (outward) for membrane voltages more positive than the Nernst potential and negative (inward) for membrane voltages less positive than the Nernst potential for that particular ion. A major difference exists, however, between the predictions of mechanically blocked (gating particle) models and electrically controlled channels in the amplitudes of these currents. The mechanically gated channel is activated by membrane voltage excursions of one sign and blocked by those of the opposite sign so that each channel has one current versus voltage characteristic representing its open configuration. With a voltage-controlled channel of the type discussed above the situation is more complex with two channel configurations P and D each of which conducts over appreciable ranges of membrane voltage. A channel switch results in a transition between the two characteristics.

Some properties of the current versus voltage characteristics are common to all models controlled by an electrostatic potential gradient (an electric field) within the channel. V_S is defined above as the value of membrane voltage at which the external electric field experienced by the channel, due both to the membrane field and any other charged structures such as the protein rods that define the channel, is zero. Then at sufficiently positive values of $V_M - V_S$ the membrane voltage will dominate the voltages created internal to the channel and a rapidly rising outward transfer rate with increasing V_M will be predicted for the D channel polarization. Were it possible to maintain a P polarization at these values of V_M, the outward transfer rate of such channels would tend to a constant value independent of V_M and dependent only on C_I and the potential barrier to ion entry at the inner end of the channel. At sufficiently negative values of $V_M - V_S$ a rapidly rising inward transfer rate would be predicted as V_M decreased for the P configuration whilst a constant inward current dependent on C_O would be predicted for the D polarization. At values of $V_M - V_S$ comparable to or smaller than the internally generated voltages very non-linear current characteristics would be predicted. In the current physiological notation this means that P polarization channels, stable at sufficiently negative values of V_M, are necessarily inward rectifying while D polarization channels, stable at sufficiently positive V_M are necessarily outward rectifying.

To take a particular example it has been found that channels in the outer segment of isolated retinal rods of the Salamander have the following properties (Werblin, 1979; Attwell and Wilson, 1980). In normal Ringer solution for excursions of V_M positive of the resting potential of about -35 mV the current shows marked outward rectification but no time dependence. For hyperpolarizing pulses of sufficient size a new inward rectifying channel is activated with a time delay. The activation obeys first order kinetics, with the typical curve of relaxation time against V_M centred on $V_M = V_S$. Both tetraethylammonium and caesium ions block the channel suggesting that much of the current is carried by potassium ions.

For comparison the P and D current versus voltage curves predicted by the model are shown in Fig. 11(a) using $V_S = -60$ mV with $C_O = 2.5$ mM and $C_I = 50$ mM which are the measured K$^+$ ion concentrations. At a resting potential given by $V_M = -35$ mV, well positive of V_S, the D polarization will predominate. For positive excursions of V_M this is a non-time-dependent (instantaneous) outward rectifying current characteristic. For negative excursions of V_M approaching V_S the channel would progressively switch to the P polarization which is inward rectifying with a delay time representative of channel switching. Voltage clamped hyperpolarizing pulses from the resting potential would result in a predicted instantaneous positive change in inward current for voltage excursions to $V_M > -75$ mV and an instantaneous increase followed by further delayed increases in inward current for larger hyperpolarising pulses. For small inward current pulses the model predicts an instantaneous decrease in V_M and, for larger inward current pulses, an instantaneous large decrease in V_M followed by a delayed change to a less negative V_M. Both such effects are commonly measured and the model does seem to simulate the experiments well in broad outline. The prediction of the unstable (negative resistance) region of the current characteristics of these channels in artificially high external potassium measured by Werblin (1979) has been mentioned above in Section VI.B. Not yet measured but predicted by the characteristics of Fig. 11(a) is that this rod membrane held at values of V_M more negative than V_S and subsequently pulsed to values of V_M much more positive than V_S would first follow the P characteristic and then switch with a delay to the D polarization with the consequent time-delayed increase in outward current.

Another inward rectifying, delayed potassium channel that is activated by hyperpolarizing pulses is the so-called anomalous rectifier found in

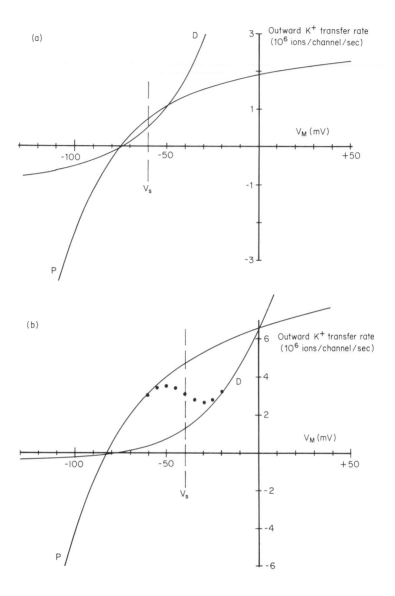

FIG. 11. (a) The model predicted potassium ion outward transfer rate with the channel in the **P** and **D** polarizations for $C_O = 2.5$ mM and $C_I = 50$ mM approximately as found for the outer segment membrane of the retinal rod of the Salamander. The switching membrane voltage assumed is $V_S = -60$ mV. (b) A similar prediction by the model for $C_O = 5$ mM and $C_I = 140$ mM as found for potassium in some cardiac muscle fibres. The switching membrane voltage assumed here is
$$V_S = -40 \text{ mV.}$$

mammalian muscle (Adrian, 1969) and starfish eggs (Hagiwara *et al.*, 1976). These characteristics can be stimulated by the model in a very similar way by the switching between a resting D and an activated P polarization of the channels.

A second type of behaviour is shown in Fig. 11(b) which is the model prediction for $V_S = -40$ mV with $C_O = 5$ mM and $C_I = 140$ mM as found for potassium ions in mammalian heart muscle. Assuming a typical resting membrane potential of $V_M = -80$ mV well negative of V_S, the P curve now represents the dominant resting channel polarization. Thus the strongly inward rectifying P curve represents a non-time-dependent (instantaneous) channel for both hyperpolarizing and small depolarizing excursions of V_M. The D channel polarization becomes a delayed outward rectifying channel progressively activated by positive excursions of V_M such that $V_M \geqslant V_S$. Just two such channels were found necessary by Noble (1962) in a detailed description of the experimentally determined spontaneous pacemaker activity of Purkinje fibres. Many advances have been made since that time in the study of such fibres (Noble, 1979), most notably the understanding of the role of calcium channels. However, the instantaneous inward rectifying potassium channel and the delayed outward rectifying channel that carries mostly potassium ions are still needed to explain the behaviour of Purkinje fibres. One particularly interesting feature measured for the instantaneous inward rectifying channels is that the outward current at $V_M = 0$ is greater the greater is C_O for potassium. To explain this requires a model that allows occupation of the channel by more than one ion at a time (Hille and Schwarz, 1978). Such effects can be incorporated into an ordered water model by a Monte Carlo type of calculation in which the interaction of a few ions, allowed simultaneously to occupy the channel, is taken account of during their biased random walk through the channel. The electrostatic considerations of Section III and V above would preclude simultaneous occupancy by more than a few ions.

B. THE ENDS OF THE WATER CHANNEL ARE CHARGED AND THE SIGN OF THE CHARGE REVERSES ON SWITCHING

An ordered water channel resting in a membrane hyperpolarized to a membrane voltage V_M more negative than the switching voltage V_S has its outer end negatively charged and its inner end positively charged as shown in Fig. 6(a). The electric field at its outer end is the cumulative effect of all the water dipoles but the largest contribution is from the local dominance of negatively charged lone-pair oxygen orbitals over positively charged hydrogen atoms at that end. The attractiveness of the protein rod structures at each end as binding sites for charged chemical messengers will change when the channel switches. These definite predictions should be readily verified by noting the effects of externally and internally applied charged molecules that act as channel poisons, blockers or even activators.

In fact there is already evidence of such effects. Experiments involving both pH variation and external cation binding (Hille *et al.*, 1975) demonstrate the presence of a high density of negative surface charge near the outer mouth of resting sodium channels in myelinated nerve fibres. Externally applied Zn^{2+} is known to delay the onset of both the Na^+ current and the gating charge transfer on activation but not to effect the decay of the Na^+ current or the back transfer of gating charge on repolarization (Armstrong and Gilly, 1979). This evidence led these authors to postulate

> A possible explanation of the action of Zn is that it binds to a negatively charged group that is part of the gating apparatus. The negative group is at the outer surface of the membrane and accessible to external Zn only when the channel is fully closed. On depolarization the extra step required as Zn dissociates slows activation. Once the channel is activated the Zn binding site has migrated inward and becomes inaccessible to Zn, which has no further effect on the kinetics.

This explanation would fit very well an ordered water channel.

Turning to the inner end of the channel experiments on the sodium channel with local anaesthetics (Courtney, 1975) such as procaine, procaine amide, lidocaine and lidocaine derivatives give evidence that:

(1) repetitive opening (activation) enhances inhibition by these drugs;
(2) if activation is preceded with large hyperpolarizing pulses the inhibition is reversed;
(3) the drugs act at the inner end of the channel most probably in cationic form;

Once again this is consistent with channel charge reversal as predicted by the model in that a negatively charged inner binding site is exposed only in an activated channel.

Very similar evidence is found for K^+ channels in the squid axon (Armstrong, 1971) and at the node of Ranvier in frog nerve (Armstrong and Hille, 1972). Experiments with positively charged internally applied tetraethyl-ammonium (TEA) ions, and derivatives with hydrophobic tails attached, demonstrate that:

(1) The ions only block K^+ channels that have been activated.
(2) Hyperpolarization helps to clear the occluded channels.
(3) Raising external K^+ concentration helps to clear the channels.

Yet again this is evidence for the negatively charged binding site near the inner mouth of a K^+ channel accessible only when it is activated. Interesting additional evidence is that although the TEA ion itself is easily dislodged by repolarization, the derivatives with hydrophobic tails longer than a critical length are much more tightly bound. If the TEA ion blocks by entering the inner mouth of the channel defined by the protein rods it is tempting to view it as being further anchored by the hydrophobic tail locked between protein rods or in the lipid beyond.

When considering the charge at the channel ends predicted by the ordered water model it is important to distinguish the switching charge due to the electrically ordered water array and the effective charges at the ends of the protein rods due to their dipole moments which do not change on channel switching as sketched in Fig. 6(a). The possible contribution to the mechanism of inactivation of the switching charges at the channel ends (Edmonds, 1983) has already been mentioned in Section IV.

C. MOST CHANNELS SHOULD SWITCH POLARITY IF SUITABLE ELECTRIC FIELDS ARE APPLIED

An electrically ordered water molecule array possesses a large electric dipole moment and therefore interacts with applied electric fields. Even if such a channel is not required to switch in its normal physiological role, it should usually be possible to force it to do so by applying suitable electric fields across it.

An example of such behaviour is found in the sodium selective channels that penetrate the apical or mucosal face of the bipolar granular cells of tight epithelia (Cuthbert, 1980). One important physiological role of such channels is to extract sodium from the mucosal fluid even if the concentration of Na^+ is already much higher in the cell than in the fluid. The cell achieves this by arranging a large negative voltage across the apical membrane maintained by a Na^+/K^+ exchange pump and a K^+ channel in the basal or serosal face of the cell. For a mucosal fluid to cell Na ion concentration ratio of $1:100$, corresponding to a sodium Nernst potential of $-115\,mV$, the voltage across the apical membrane must be maintained at more negative voltages than $-115\,mV$ in order to ensure a net inward flow of Na^+ ions. Despite these large negative membrane voltages there is no evidence that these sodium channels switch under normal physiological conditions. However, if the membrane is made artificially even more negative, the channels do display the well known S-shaped current versus voltage curve typical of switching channels (Lindemann, 1968).

If other channels that need not switch to fulfil their physiological role can be shown to switch by applying electric fields outside the physiological range, this would be good evidence that an overall electric dipole moment is an intrinsic property of channel structure and argue for an electrically ordered array within the channel. On the other hand, an electrically ordered array will switch only if the total external electric field it experiences changes sign. The membrane field contributes to this total field but so also do the dipole moments of the proteins and any local charged groups. It is therefore possible that even if particular types of channel do possess dipole moments that it will not be possible to cause them to switch by applying membrane fields less than those required to rupture the membranes.

D. CHANNEL STABILITY MAY BE ALTERED BY THE PRESENCE OF SMALL UNCHARGED MOLECULES

The clathrate hydrates possess (Davidson, 1973) water cage structures some of which are very like those postulated for the ordered water channel. These are stabilized by the inclusion within the water cages of small uncharged molecules of particular sizes rather than by hydrogen bonds to an external structure as in the Zeolites. It is therefore possible that an ordered water array forming a channel and supported by hydrogen bonds to the protein rods may be further stabilized by inclusion within the water cages of suitably sized uncharged molecules. As the predicted motion of the ions within a channel is over the surface of the water cages rather than through them, the inclusion of uncharged molecules need not block the ion transfer except perhaps by sterically hindering the rotation of water molecules necessary to provide the low energy ion sites.

The effects of such small molecules might best be studied when the channels are formed by barrel stave oligomers of single rod shaped proteins such as alamethicin or suzukacilin mentioned above in Section VII.A. If particular sized small molecules were to be shown to stabilize or perhaps to block either pentameric or hexameric protein structures selectively, this would be evidence of particular water structures within the pores. Unfortunately these channels have been shown to be not very selective for particular ions which argues against the predominance of one particular water ring in the structure. The dodecahedral cages of the pentagonal ring structure shown in Fig. 3(b) have a free inner diameter of 0.5 nm and could include molecules such as N_2O, CF_4, CO_2, H_2S, Xe, N_2 or O_2.

VIII. Conclusions

The ordered water model of the membrane ion channel may be considered either as one representative of a wider class of models in which the transfer of ions is controlled by electric fields rather than by mechanical gating particles, or it may be seen as an attempt at a detailed molecular model of at least one type of channel in excitable tissue.

From the first standpoint it seems to my doubtless partisan view that the evidence, such as that in Sections VI and VII above, is much more strongly supportive of electrically controlled ion flow with the consequent switching between two different configurations of the channel, than it is of the open or shut mechanically gated channel. If the results of Moore (1959) and Segal (1958) could be repeated on the squid axon with the much advanced techniques of today and if in particular the S-shaped switching characteristic were measured in an internally perfused axon such that the only permeant ion was potassium, the case would be made. The probability that there exist two sets of mechanically blocked potassium channels such that a hyperpolarizing excursion of the membrane voltage which shuts one set simultaneously opens the other set must surely be remote. Were voltage control to be established it would be a considerable step in the understanding of channels such as the anomalous rectifier that are activated by hyperpolarising membrane voltage transients in that they would be seen as the inevitable consequence of the switching of the more usual potassium channels activated by depolarization.

From the second viewpoint of the ordered water channel as a detailed molecular model of certain ion channels, the evidence is harder to gather. There is at least evidence for an electrical dipolar structure which switches and is exposed at both channel ends, in Section VII.B on selective blocking by charged molecules and in Section VII.C on gating charge transfer.

Strong evidence for the presence of a multiply connected, hydrogen bonded water molecule array would be gated proton conduction. Such a water structure is perhaps unique in allowing efficient proton conduction by the propagation of ionic and hydrogen bonding defects as in ice while simultaneously providing a highly polarizable, low energy environment for metal cations in the manner I have discussed. The first experimental evidence for gated proton currents in heavy ion channels has recently been obtained (Thomas and Meech 1982) in moluscan neurons.

IX. Acknowledgements

I am most grateful to the Royal Society for the award of a Senior Research Fellowship which has enabled me to pursue these interests in parallel with physics research.

References

Adrian, R. H. (1969). *Prog. Biophys. Mol. Biol.* **19**, 339–369.

Armstrong, C. M. (1971). *J. Gen. Physiol.* **58**, 413–437.

Armstrong, C. M. and Bezanilla, F. (1974). *J. Gen. Physiol.* **63**, 533–552.

Armstrong, C. M. and Gilly, W. F. (1979). *J. Gen. Physiol.* **74**, 691–711.

Armstrong, C. M. and Hille, B. (1972). *J. Gen. Physiol.* **59**, 388–400.

Armstrong, C. M., Bezanilla, F. and Rojas, E. (1973). *J. Gen. Physiol.* **62**, 375–391.

Attwell, D. and Wilson, M. (1980). *J. Physiol., Lond.* **309**, 287–315.

Bader, C. R., Macleish, P. R. and Schwartz, E. A. (1979). *J. Physiol., Lond.* **296**, 1–26.

Baker, P. F., Hodgkin, A. L. and Ridgway, E. B. (1971). *J. Physiol., Lond.* **218**, 709–755.

Barrer, R. M. and Rees, L. V. C. (1960). *Trans. Faraday Soc.* **56**, 709–721.

Barrer, R. M., Bartholomew, R. F. and Rees, L. V. C. (1963). *J. Phys. Chem. Solids* **24**, 51–62.

Baumann, G. and Mueller, P. (1974). *J. Supramol. Struct.* **2**, 538–557.

Benz, R., Janko, K., Boos, W. and Lauger, P. (1978). *Biochim. Biophys. Acta.* **511**, 305–319.

Boheim, G. and Kolb, H. A. (1978). *J. Membr. Biol.* **38**, 99–150.

Braun, V. (1975). *Biochim. Biophys. Acta.* **415**, 335–377.

Brinley, F. F. (1978). *Ann. Rev. Biophys. Bioeng.* **7**, 363–392.

Buckingham, A. D. (1957). *Discuss. Faraday Soc.* **24**, 151–157.

Cooke, A. H., Edmonds, D. T. and Finn, C. B. P. (1962). *J. Phys. Soc. Jpn.* **17**(B1), 481–486.

Coulson, C. A. and Eisenberg, D. (1966). *Proc. Roy. Soc. Lond.* **A291**, 445–453.

Courtney, K. R. (1975). *J. Pharm. Exp. Ther.* **195**, 225–236.

Cuthbert, A. W. (1980). *In* "Membrane Structure and Function" (ed. E. E. Bittar), pp. 171–206. Wiley, New York.

Danielli, J. F. and Davson, H. (1935). *J. Cell. Comp. Physiol.* **5**, 495–508.

Davidson, D. W. (1973). *In* "Water: a Comprehensive Treatise", Vol. 2, Chap. 4. Plenum Press, London.

Edmonds, D. T. (1979). *Chem. Phys. Lett.* **65**, 429–433.

Edmonds, D. T. (1980a). *Proc. Roy. Soc. Lond.* **B211**, 51–62.

Edmonds, D. T. (1980b). *Biochem. Soc. Symp.* **46**, 91–101.

Edmonds, D. T. (1981a). *Trends Biochem. Sci.* **6**, 92–93.

Edmonds, D. T. (1981b). *Proc. Roy. Soc. Lond.* **B214**, 125–136.

Edmonds, D. T. (1982). *Proc. Roy. Soc. Lond.* **B217**, 111–115.

Edmonds, D. T. (1983). *Proc. Roy. Soc. Lond.* **B219**, 423–438.

Eisenman, G. (1962). *Biophys. J.* **2**, 259–323.

Eldefrawi, M. E., Eldefrawi, A. T. and Shamóó, A. E. (1975). *Ann. N.Y. Acad. Sci.* **264**, 183–202.

Enderby, J. E. and Neilson, G. W. (1979). *In* "Water: a Comprehensive Treatise". Vol. 6, Chap. 1. Plenum Press, London.

Eyring, H., Lumry, R. and Woodbury, J. W. (1949). *Rec. Chem. Prog.* **10**, 100.

Famborough, D. M. (1979). *Physiol. Rev.* **59**, 165–227.

Fernandez-Moran, H. (1959). *In* "Biophysical Science: a Study Program" (ed. J. L. Oncley), pp. 319–330. Wiley, New York.

Gamlich, V. and Meier, W. M. (1971). *Z. Kristallog.* **133**, 134–149.

Gorter, E. and Grendel, F. (1925). *J. Exp. Med.* **41**, 439–443.

Hagiwara, S., Miyazaki, S. and Rosenthal, N. P. (1976). *J. Gen. Physiol.* **67**, 621–638.

Hechter, O. (1965). *Fedn. Proc. Fedn. Amer. Soc. Exp. Biol.* **24**(2), S91–S102.

Hille, B. (1975). *In* "Membranes" (ed. G. Eisenman), Vol. 3, pp. 255–323.

Hille, B. and Schwarz, W. (1978). *J. Gen. Physiol.* **72**, 409–442.
Hille, B., Woodhull, A. M. and Shapiro, B. I. (1975). *Phil. Trans. Roy. Soc. Lond.* **B270**, 301–318.
Hodgkin, A. L. and Huxley, A. F. (1952). *J. Physiol.* **117**, 500–544.
Hol, W. G. J., van Duijnen, P. T. and Berendsen, H. J. C. (1978). *Nature* **273**, 443–446.
Inouye, M. (1974). *Proc. N.Y. Acad. Sci.* **71**, 2396–2400.
Jacquinot, F., Wenckenbach, W. T., Chapellier, M., Goldman, M. and Abragam, A. (1974). *C.R. Acad. Sci. Paris* **B278**, 93–96.
Keynes, R. D. and Rojas, E. (1974). *J. Physiol.* (*London*) **239**, 393–434.
Kolb, H. A. and Boheim, G. (1978). *J. Membr. Biol.* **38**, 151–191.
Lindemann, B. (1968). *Biochim. Biophys. Acta.* **163**, 424–426.
Meunier, J. C., Sealock, R., Olsen, R. and Changeux, J. P. (1974). *Eur. J. Biochem.* **45**, 371–394.
Meves, H. and Vogel, W. (1973). *J. Physiol.* **235**, 225–265.
Moore, J. W. (1958). *Fedn. Proc. Fedn. Am. Soc. Exp. Biol.* **17**, 113–114.
Moore, J. W. (1959). *Nature* **183**, 265–266.
Mullins, L. J. (1959). *J. Gen. Physiol.* **42**, 817–829.
Nelson, N., Anholt, R., Lundstrum, J. and Montal, M. (1980). *Proc. Natl Acad. Sci. USA* **77**, 3057–3061.
Noble, D. (1962). *J. Physiol.* (*London*) **160**, 317–352.
Noble, D. (1979). *In* "The Initiation of the Heartbeat" (2nd edn) Oxford University Press, Oxford.
Onsager, L. (1936). *J. Am. Chem. Soc.* **58**, 1486–1493.
Oosting, P. H. (1979). *Rep. Prog. Phys.* **42**, 1479–1532.
Parsegian, A. (1969). *Nature* **221**, 844–846.
Pethig, R. (1979). *In* "Dielectric and Electronic Properties of Biological Materials," Chap. 2. Wiley, Chichester.
Raftery, M. A., Hunkapiller, M. W., Strader, C. D. and Hood, L. E. (1980). *Science* **208**, 1454–1457.
Ross, M. J., Klymkowsky, M. W., Agard, D. A. and Stroud, R. M. (1977). *J. Mol. Biol.* **116**, 635–659.
Schindler, H. and Rosenbusch, J. P. (1978). *Proc. Natl Acad. Sci. USA* **75**, 3751–3755.
Schindler, H. and Quast, U. (1980). *Proc. Natl Acad. Sci. USA* **77**, 3052–3056.
Segal, J. R. (1958). *Nature* **182**, 1370.
Singer, S. J. and Nicolson, G. L. (1972). *Science* **175**, 720–731.
Smith, G. D., Pletnev, V. Z., Duax, W. L., Balasubramanian, T. M., Bosshard, H. E., Czerwinski, E. W., Kendrick, N. E., Mathews, F. S. and Marshall, G. R. (1981). *J. Am. Chem. Soc.* **103**, 1493–1501.
Smythe, W. R. (1939). *In* "Static and Dynamic Electricity," Chap. 4. McGraw-Hill, New York.
Stein, W. D. and Danielli, J. F. (1956). *Discuss. Faraday Soc.* **21**, 238–251.
Steven, A. C., Heggeler, B., Muller, R., Kistler, J. and Rosenbusch, J. P. (1977). *J. Cell. Biol.* **72**, 292–301.
Tajima, Y., Matsuo, T. and Suga, H. (1982). *Nature* **299**, 810–812.
Thomas, R. C. and Meech, R. W. (1982). *Nature* **299**, 826.
Venkataram Prasad, B. V., Shamala, N., Nagaraj, R. and Balaram, P. (1980). *Acta. Cryst.* **B36**, 107–110.
Warner, D. T. (1961). *J. Theor. Biol.* **1**, 514–000
Werblin, F. S. (1979). *J. Physiol.* **294**, 613–626.

Chapter 11

Electric Field-induced Fusion and Rotation of Cells

W. M. Arnold and U. Zimmermann

Lehrstuhl für Biotechnologie, Universität Würzburg,
Röntgenring 11, D8700 Würzburg, West Germany

I. Introduction

The influence and interactions of electricity on biological systems have held the interest of scientists for more than two centuries (reviewed by Krauss, 1932). It was well known from this early work that direct, alternating or pulsed electric fields can disrupt biological reactions and irreversibly influence

BIOLOGICAL MEMBRANES Vol. 5
ISBN 0 12 168546 2

cellular systems. The techniques employed in these studies were neither precise nor specific enough to distinguish between those effects which were directly attributable to the electric field and those due to thermal effects or those resulting from electrolysis. The applied nature of most of the research work (which was directed, for example, towards the sterilization of food, disinfection of rooms or extraction of fat from cellular systems) did not always require the detailed investigation of the fundamental cause of cell death (e.g. Anderson and Finkelstein, 1919; Beatlle and Lewis, 1925; Doevenspeck, 1961, 1962; Ingram and Page, 1953). The use of electric fields to kill cells has recently become of interest again (Allen and Soike, 1966; Barraneo et al., 1974; Ciezynski, 1967; Gutknecht et al., 1981; Hamilton and Sale, 1967; Hülsheger et al., 1981; Jakob et al., 1981; Pareilleux and Sicard, 1970; Rowley, 1972; Sale and Hamilton, 1967, 1968; Schulz et al., 1982).

Conversely, completely reversible non-thermal effects of electric fields on living cells were also known 40 years ago. Pearl chain formation in an inhomogeneous alternating field belongs to this reversible class (Krasny-Ergen, 1936; Liebesny, 1939; Ludloff, 1895; Mast, 1931; Muth, 1927; Verworn, 1896).

Later, alternating-field-induced rotation of cells was also reported (Teixeira-Pinto et al., 1960). Even though these discoveries are now considered milestones in the investigation of the passive electrical properties of membranes, they did not stimulate much research in other laboratories. Perhaps this was due to the concentration of interest on excitable membranes (reviewed by Almers, 1978). Research on passive properties was mainly concerned with the interactions of radio and microwaves with bulk tissue (e.g. Cole, 1968; Schwan, 1957, 1977; Schwan and Forster, 1977; Tell and Harlen, 1979).

The discovery that reversible electrical breakdown of cell membranes occurs in response to short pulses of high field strength (Zimmermann et al., 1973) led to more interest in the effects of strong fields on the membranes of non-excitable cells. The impetus behind this was the possibility of using reversible breakdown to carry substances into the cell without permanent damage to membrane function or integrity (reviewed by Zimmermann et al., 1981b). The ability of this method to manipulate the contents of the cell gives it great potential in membrane research, medicine, and technology (genetic engineering). The most dramatic development has been the combination of this technique with those of orientation and pearl chain formation to give in vitro cell fusion mediated by electric fields (Zimmermann et al., 1980c,d).

This technique can be applied to all living cells because the underlying mechanism depends only on the general properties common to cells and their membranes. In contrast to earlier fusion techniques using chemical agents or viruses (Ahkong *et al.*, 1975a,b, 1980; Kao and Michayluk, 1974; Keller and Melchers, 1973; Knutton, 1977; Knutton and Pasternak, 1979; Lucy, 1970, 1978, 1982, Peretz *et al.*, 1974; Poste, 1972; Poste and Pasternak, 1978; Power *et al.*, 1978; Ringertz and Savage, 1976) the electric field method allows the fusion of particular species of cells to be predicted, and then guided under the microscope on the basis of individual or larger numbers of cells as necessary.

The procedure is very mild and leads to a high yield of viable and reproductive hybrids. Knowledge of the characteristics of pearl chain formation has enabled control of the number of cells to be fused. In this way, anything from two-cell hybrids to giant cells (containing thousands of like cells) may be produced at will (Zimmermann and Vienken, 1982, 1984). The size of the giant cells permits direct investigation of the electrical properties of their membranes by use of intracellular microelectrodes or patch clamp techniques (Neher *et al.*, 1978) even when the natural cells are too small to allow this. The study of the fusion process and the distinct steps involved in it has shed new light on the effects on which it is based, namely cell orientation, pearl-chain formation and electrical breakdown. The close relationship between fusion and these other topics has made their inclusion natural here. In addition, emphasis has been placed on cell rotation. Although a phenomenon which can interfere with cell fusion, this is itself becoming a new method for revealing the electrical properties of membranes and other parts of the cell (Arnold and Zimmermann, 1982a,b; Zimmermann and Arnold, 1983).

Electric field-induced fusion and rotation have only recently become of wide interest. We have not, therefore, restricted ourselves only to a summary of the data so far obtained but have also included new concepts to spur further research in these directions. We have a strong belief that studies of the interaction of electric fields with biological materials will give deeper insight into the structure and function of cell membranes and stimulate advances in medicine and technology.

II. Context of the Electric Field Pulse Method

If two cells can be brought together they can be made to fuse if the membrane structure in the area of contact is disturbed. Historically the only methods of doing this were the use of chemicals, viruses, or by freezing and thawing (Poste and Pasternak, 1978; Peberdy, 1980; Hui *et al.*, 1981; Lucy, 1982). The recently introduced electric field pulse technique (Zimmermann *et al.*, 1980c,d; Zimmermann and Scheurich, 1981a) has the distinct advantage that the extent of the disturbance can be restricted to the zone of membrane contact or widened to include a larger area if necessary. The earlier methods were bound to non-specifically affect the entire surface of the cells.

The field pulse technique is, therefore, much more likely to produce fusion products with largely undisturbed membranes, leading to better survival (viability).

Orientation, pearl-chain formation and rotation of cells can be readily produced using weak electric fields which do not cause the membrane voltage to approach breakdown. By contrast, fusion requires strong fields to give the necessary structural changes in the membrane. Various membrane perturbations can occur depending on the length of time the membrane is subjected to voltage: these are membrane deformation (Allan and Mason, 1962; Helfrich, 1973a,b, 1974; Friend et al., 1975; Scheurich et al., 1980b), punch-through effect (Coster, 1965; Zimmermann, 1977a), mechanical breakdown (Crowley, 1973; Tien, 1974; Benz et al., 1979; Zimmermann et al., 1980c) and reversible electrical breakdown (Zimmermann et al., 1973). Electric field induced fusion is only possible when the field causes reversible electrical breakdown. On the other hand, inappropriate field conditions can produce mechanical breakdown or punch-through effects which either prevent fusion or kill the cells.

A full description of all these effects is available, and reversible electrical breakdown has itself been reviewed (Zimmermann et al., 1976a, 1980c, 1981b). We will now consider those aspects of electrical breakdown (which we shall use in the restricted sense of reversible electrical breakdown) which are required to understand electro-fusion.

A. ELECTRICAL BREAKDOWN IN MEMBRANES OF SINGLE CELLS

Breakdown is to be expected if the membrane is considered as a layer of dielectric (the hydrocarbon region and intrinsic proteins) with the aqueous solutions on either side serving as capacitor plates. Figure 1a shows the equivalent circuit of the membrane, consisting of a capacitor filled with a dielectric with a resistor in parallel. At high frequencies, or when using short pulse lengths, the membrane resistance is normally high enough to be considered infinite. Application of too high a voltage to the analogous electronic component causes a dramatic increase in the conductance which is termed breakdown (Franz, 1956). The short circuit developed is usually permanent and irreversible so the component is useless, although "self-healing" capacitors are increasingly available. In the case of cell membranes, breakdown produced by very short field pulses (some microseconds or shorter) has been shown to allow the membrane to regain its high resistance and impermeability after breakdown. This is the (reversible) electrical breakdown referred to previously. Longer pulses or unreasonable voltages cause irreversible damage (referred to as irreversible or mechanical breakdown). Benz et al. (1979) and Zimmermann et al. (1981b) describe these effects and terms in more detail.

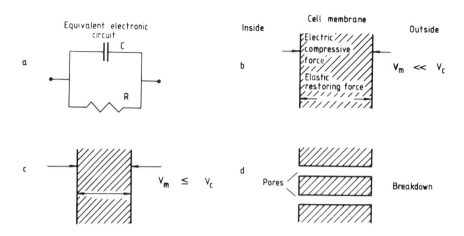

FIG. 1. The electro-mechanical model of membrane breakdown. (a) The membrane can be represented as a capacitance C in parallel with a resistance R. (b) A given membrane voltage V_m is associated with opposite capacitive charges on the two sides of the membrane which exert compression on the membrane. (c) Increasing the voltage causes the membrane to become thinner (the compressive force is thereby increased) until the membrane voltage reaches the breakdown value, V_c, and an instability occurs. The membrane breaks down by formation of low resistance pores, which quickly discharge the voltage, (d).

The following macroscopic "electro-mechanical" model provides insight into electrical breakdown (Fig. 1). The opposite electrical charges on the capacitor "plates" exert a mutal attractive force (Fig. 1a,b), which causes membrane thinning (assuming a non-zero compressibility of the membrane) (Fig. 1c). Further thinning leads to an increase in the mutual attraction, so that at a critical membrane potential localized instability occurs, leading to the formation of pores filled with medium or cytoplasm (Fig. 1d) (Coster and Zimmermann, 1975b, 1976; Coster et al., 1977; Zimmermann, 1977a, 1978, 1980). The conductance of these pores gives a drastic decrease in membrane resistance which is observed as breakdown.

There is a body of evidence that this macroscopic picture and the key assumptions are basically sound (Péqueux et al., 1980; Zimmermann et al., 1977a,b; 1980b; Zimmermann, 1980, 1982). Theories for the production of pores in membranes by molecular fluctuations have also been put forward (Abidor et al., 1979; Petrov et al., 1980; Chizmadzhev and Abidor, 1980; Chernomordik and Abidor, 1980), although these theories are valid only for the time domain of irreversible mechanical breakdown. Sugár (1979) discussed field-induced phase transitions of phospholipid bilayers which may also be involved in mechanical breakdown.

The membrane voltage reached during the short pulse (microseconds or less) giving reversible breakdown is approximately 1 V in amplitude for most cell membranes at room termperature. The breakdown is seen to vary according to temperature and duration of the field pulse (Coster and Zimmermann, 1975b; Zimmermann et al., 1977b; Benz et al., 1979, Zimmermann et al., 1981b). Zimmermann et al. (1973) were the first to understand this principle and exploit it by showing that rapid polarization of membranes to 1 V or more need not cause permanent membrane perturbations (see also Turnbull, 1973). This finding was supported by several other laboratories (Akeson and Mel, 1981; Benz and Conti, 1981; Berg et al., 1978; Coster and Smith, 1980; Kinosita and Tsong, 1977a, 1979; Teissie and Tsong, 1981; Mishra et al., 1981; Gauger and Bentrup, 1979; Steinbiss, 1978; Jausel-Hüsken and Deuticke, 1981; Puchkova et al., 1981) even though Tsong and Kingsley (1975) and Tsong et al. (1976) initially believed thermal effects were responsible. There are several methods available to produce reversible electrical breakdown. In individual giant cells, intracellular and extracellular electrodes can be used (Coster and Zimmermann, 1975a,c; Zimmermann and Benz, 1980; Benz and Zimmermann, 1980b, 1981b). Conversely, when working with smaller cells or with suspensions it is more practical to use the fact that an induced cellular dipole (see below) and a corresponding membrane voltage are created by setting up a field in the bulk medium.

The necessary short pulse may be achieved in two ways. In the first method cells are made to transit a stationary field as is found in the orifice of a Coulter Counter (Zimmermann et al., 1973, 1974a,b, 1980a, 1981b; Pilwat et al., 1980b). The second method is to apply a voltage pulse to a pair of relatively large parallel electrodes which has a cell suspension between them. This may be accomplished by discharging a capacitor into the arrangement (Neumann and Rosenheck, 1972, 1973; Zimmermann et al., 1974a, 1975; Riemann et al., 1975; Saleemuddin et al., 1977; Schneeweiss et al., 1977; Jausel-Hüsken and Deuticke, 1981). Under these conditions the field is approximately uniform and the membrane voltage produced can be derived by integrating the Laplace equation with boundary conditions appropriate to the cell shape (Schwan, 1977; Jeltsch and Zimmermann, 1979). For a spherical cell the result is:

$$V_m = 1.5 \cdot E \cdot a \cdot \cos \vartheta \tag{1}$$

where V_m is the membrane voltage at any point on the cell surface defined by the angle ϑ in Fig. 2, E is the field strength in V cm^{-1}, and a is the cell radius in cm. The equation is only valid for the "stationary" case, which means that the membrane voltage must have reached its final value in response to the field. In practice the pulse length must be about five times longer than the electrical time constant, τ, of the cell membrane. In the case of a sinusoidal alternating field of frequency f, one can say that the peak amplitude is given by (Schwan, 1977; Bernhardt and Pauly, 1973; Holzapfel *et al.*, 1982):

$$V_m^0 = \frac{1.5 \cdot E^0 \cdot a \cdot \cos \vartheta}{\sqrt{(1 + (2\pi f \cdot \tau)^2)}} \tag{2}$$

where V_m^0 and E^0 are peak values of V_m and E, and τ is the time constant of the charging process of the membrane (see below).

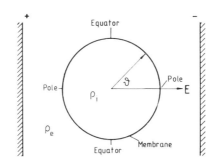

FIG. 2. A cell subjected to a uniform field, such as exists between large parallel electrodes. The size of the cell relative to the electrode spacing is much exaggerated. E defines the direction and magnitude of the field; the angle ϑ describes the position of a given area of the membrane; ρ_i and ρ_e are the resistivities of the cell interior and the medium respectively.

At higher frequencies the membrane voltage decreases because of the frequency term, f, in the denominator. This is of relevance to oriéntation, pearl-chain formation, and rotation which are discussed in the following sections. For a short rectangular field pulse (of length t seconds) applied to the cell suspension, the membrane voltage at the end of the pulse is given by (Jeltsch and Zimmermann, 1979):

$$V_m = 1.5 \cdot E \cdot a \cdot \cos \vartheta [1 - \exp(-\frac{t}{\tau})] \tag{3}$$

i.e. for long pulses, the voltage approaches that given by Eq. (1), but pulses shorter than τ give progressively reduced membrane voltage. The derivation of the equations assumes that the membrane resistance is large enough to be ignored. This is generally the case for living cells (until breakdown occurs).

Two conclusions relevant to the practice of electro-fusion can be drawn from Eqs. (1) to (3):

(1) At a given field strength the induced membrane voltage is directly proportional to the radius of the cell. The breakdown voltage of the membrane has been shown to vary relatively little between cells. In

practice this means that smaller cells will require higher external field strengths than larger cells to undergo electrical breakdown.

(2) For any cell, the membrane voltage is a maximum where the direction of the overall electric field direction (vector) crosses the membrane perpendicularly ($\vartheta = 0°$ in Fig. 2), but is zero where the field direction is parallel to the membrane ($\vartheta = 90°$).

This means that as we increase the field strength, breakdown occurs first only in the two opposite areas which we term the "poles" of the cell (where $\vartheta = 0°$). At still higher field strengths breakdown is also possible over progressively wider areas described by the angle increasing. Even at very high field strengths (if these are attainable) the sites where the field direction is parallel to the membrane (where $\vartheta = 90°$) will not break down, because a uniform field induces no voltage there. The locus of these sites is termed the "equator" of the cell.

These conclusions hold only if the breakdown occurs at all sites at the same instant in time. If breakdown occurs sooner in time at some sites than at others the field distribution will change, because the breakdown areas will constitute low-resistance patches (Klee and Plonsey, 1974; Jeltsch and Zimmermann, 1979) and so these semi-quantitative conclusions no longer hold.

Similar conclusions can be reached for ellipsoidal cells, but only through the use of more complicated mathematics (Bernhardt and Pauly, 1973; Jeltsch and Zimmermann, 1979). The considerations of field distribution and breakdown do not change drastically in the case of two adhering cells (or a single straight pearl-chain) aligned in the field direction. When several cell chains are arranged parallel and close to each other (see below) the situation is much more complicated. Even in this case, the magnitude of the effects predicted is still a useful guide for setting up an experiment and for explaining the results.

It was explained above how breakdown can be understood on the basis of pore formation, which results in a large increase in membrane permeability. Field strengths which just give breakdown will only form these pores in the two small areas perpendicular to the field vector, but at higher voltages progressively larger areas will be made permeable and larger pores can be formed, so that larger molecules can pass through the membrane (Vienken et al., 1978; Zimmermann, 1982, 1983; Kinosita and Tsong, 1977a,b, 1978).

The resealing time of the field-induced pores depends on the degree of permeability initially induced, which is very dependent on the applied field strength and the pulse duration, and on the area of the membrane being considered. Resealing of biomembranes with the concomitant return of the original resistance and impermeability is observed when the applied field strength was not more than three to four times the critical value for pore generation in the pole region, and for pulse lengths not longer than 50 to

200 μs, assuming a cell radius of 10–40 μm. Larger cells will require proportionately longer pulse lengths due to their longer time constants (see Eq. (7)). In the case of the larger eggs, pulse lengths of some milliseconds may be required.

The pulse lengths applied to lipid bilayer membranes should not exceed 1 to 10 μs (Benz and Zimmermann, 1980a). The use of stronger fields or longer pulses results in the formation of so many pores of such a large area that irreversible breakdown and destruction of the membrane ensue. If the breakdown pulse is chosen with these points in mind, the pores formed in biomembranes will heal within seconds to minutes at 37°C, but require about 30 min at 4°C (Zimmermann et al., 1980c). The resealing time in artificial planar lipid bilayers is much shorter (e.g. for oxidized cholesterol bilayers 2 μs at 20°C and 20 μs at 4°C) (Benz and Zimmermann, 1981a).

The difference between the resealing times for artificial and biological membranes probably results from a difference in the mechanism of "the resealing process". In artificial lipid bilayer membranes resealing is due to free diffusion of lipid molecules. This can be said because the diffusion coefficient calculated from the resealing process (Benz and Zimmermann, 1981a) agrees well with that obtained from optical measurements (Webb, 1978). This mechanism is also applicable to resealing of biomembranes when the pores have been formed in lipid domains. The alternative, breakdown in lipoprotein regions (Pilwat et al., 1975, 1980a; Zimmermann et al., 1977a), will probably still result in pore resealing by lipid molecules because of the very low diffusion coefficient of protein molecules relative to those of lipid (Webb, 1978). However, the protein molecules in the neighbourhood of the pores may themselves have been disturbed by the high field values. This is a consequence of the large dipole moments of some proteins (Pethig, 1979) which will cause a force to be exerted on them by the field. This may result in conformational changes which have relatively long relaxation times due to hysteresis effects (see also Neumann and Katchalsky, 1972; Neumann, 1981).

The reversibility of membrane breakdown has been used in the encapsulation of drugs in erythrocytes and lymphocytes. The drug-loaded cells can either be used to target drugs to a specific site in the organism or to give extended release in the circulation (Zimmermann, 1977b, 1983; Zimmermann et al., 1976c, 1978, 1980c; Zimmermann and Pilwat, 1976; Kinosita and Tsong, 1979). This technique is therefore important in diagnostics and chemotherapy.

An important application of membrane breakdown may be in genetic engineering for the transfer of DNA sequences through the membrane. Schnettler et al. (1984) have recently shown this is possible in eukaryotes by transferring plasmids into genetically deficient yeast cells (see also Auer et al., 1976). It has even proved possible to move organelles through the plasma-

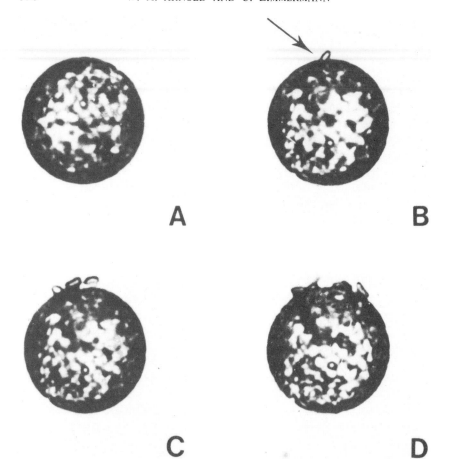

FIG. 3. Electric field pulse mediated release of chloroplasts (exocytosis) from a mesophyll cell protoplast of *Avena sativa*. The protoplast is shown before treatment in (a). (b), (c) and (d) show progressive release of chloroplasts in response to three successive field pulses at intervals of 5 s. (Field strength 4 kV/cm, pulses of length: 20 μs.) From Zimmermann *et al.* (1982b)

lemma by use of reversible breakdown (Fig. 3). Thus the release ("field-induced exocytosis") of chloroplasts from *Avena sativa* protoplasts was observed without apparent ill effect to the cell (Zimmermann *et al.*, 1982b). If movement of organelles in the opposite direction were possible, then such field induced uptake of chloroplasts or nuclei of other species would provide an alternative to fusion for the production of new hybrids. Another application of reversible electrical breakdown allows access to the cell interior in the study of the process of exocytosis (Baker and Knight, 1979).

B. ELECTRICAL BREAKDOWN BETWEEN CELLS IN CONTACT

So far we have considered the breakdown for a single cell by a field pulse and the induction or transfer of material between cell and medium. This has very important uses, but the main topic of this chapter requires consideration of at least two cells positioned such that breakdown can be made to occur where their membranes have contact. This results in a new effect—fusion.

In two cells which are arranged in line with the electric field pulse (see Fig. 4) and forced into close membrane contact, breakdown will not only lead to exchange of material with the medium at either free end, but also to intracellular exchange. This results in the formation of a cytoplasmic continuity between the two cells and it appears that the individual resealing of the membranes (as occurs in single cells) is prevented. The disorganized lipid molecules within this region take up the alternative configuration, namely the formation of a membrane bridge as shown in Figs 4 and 5. The membranes in this bridge are highly curved, so we would expect from thermodynamic considerations that the initial arrangement would eventually form a more spherical cell. In time, exactly spherical cells will form. In this way fusion of the cells has been achieved by an electric field pulse.

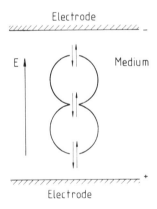

FIG. 4. Representation of two cells oriented parallel to the electric field (E) undergoing membrane breakdown. The opposed arrows indicate exchange of material through the temporarily permeable membranes. The size of the cells relative to the electrodes is much exaggerated. From Zimmermann *et al.* (1981b).

The electro-fusion technique has facilitated measurements that were difficult with earlier methods: namely those of the size of the particular cells that are united to form a particular fusion product. These measurements have shown that the volume of the fusion product is equal to the sum of the volumes of the cells fused (Vienken *et al.*, 1981; Büschl *et al.*, 1982). This implies that the total surface area has decreased. It can easily be calculated that, when two equal cells fuse at constant volume to form a similarly shaped product, the reduction in surface area is about 20%. As more cells are fused, the reduction in surface area becomes more severe. For eight cells the figure is 50%, for thousands of cells forming a giant cell it is more than 90%. The excess membrane material must go somewhere, and we can account for it if we postulate that:

(1) The membrane contact at fusion is both very close and extended over an appreciable area.

(2) The breakdown pulse induces a large number of apposed pores in this contact area (the pulse may need to be considerably higher than critical strength to do so).

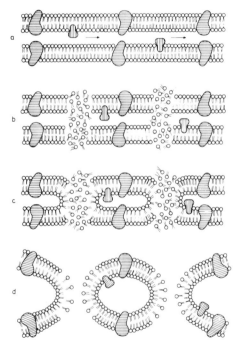

FIG. 5. Model for the formation of membrane bridges from the closely apposed membranes of two cells during the electro-fusion procedure. Before breakdown, (a) the alternating field causes protein aggregation and the formation of lipid domains. In (b) breakdown has just occurred resulting in pore formation and random orientation of the lipid molecules. The re-organization of the lipid molecules leads to membrane bridges (c) and inverted vesicles (d). Although many pores form, only two are shown for the sake of clarity. From Zimmermann and Vienken, 1982.

Under these conditions many membrane bridges will be formed at the edges of pores (Fig. 5). Pores will also be formed in the planes immediately above and below that shown in Fig. 5. If this three-dimensional picture is considered, it can be seen that vesicle formation is likely when the pore density is high enough for several pores to occur close enough to each other. It is interesting that these vesicles will be inverted (inside-out). The formation of the vesicles can account for some of the excess membrane material, and vesicles have indeed been observed under the conditions postulated here (Vienken *et al.*, 1983).

Before describing these findings in more detail, we would like to discuss how the required intimate contact could be achieved between cells that are to be fused.

With the exception of special cases, cells suspended in solution will not approach each other closer than approximately 20 nm (Knutton, 1979). A closer approach is prevented by Brownian movement and by the repulsion between the like surface charges of the outer membrane surfaces. If we assume that, for fusion to occur, the membranes must come within 1 to 2 nm of each other, additional energy will be required to remove some of the water bound by the surface charges of the membranes. This is equivalent to a force preventing the close approach of the membranes (Rand, 1981). In the electrofusion technique the force necessary to overcome the repulsion described above is provided by the attraction between electric dipoles induced in the two cells by means of an alternating field.

III. Pearl-chain Formation

When non-uniform electric fields are applied to cells in suspension, lines of cells are seen to form which usually start from an electrode (Figs 6–12). The fields may be direct, alternating or pulsed, $10-100$ V cm^{-1} being sufficient amplitude, depending on cell size. These are called pearl chains and were first directly observed by Muth (1927) by exposing emulsions of fat particles to high-frequency fields. The first report on cells was made by Liebesny (1939), who saw pearl chains of erythrocytes form in a high frequency field. An analysis of the phenomenon by Krasny-Ergen (1936, 1937) was unfortunately based on the false assumption that the cells can be considered conductive, which is not true as long as the membrane is intact. More careful and extensive considerations were made by Schwarz *et al.* (1965) and Saito *et al.* (1966) and the fundamental analyses necessary have proved most useful in looking at other interactions between cells and electric fields (see also Schwan and Sher, 1969). The formation of chains from a suspension means that the cells must somehow experience a force which brings them together and towards the electrode. The origins of this force are considered in Fig. 7a–e.

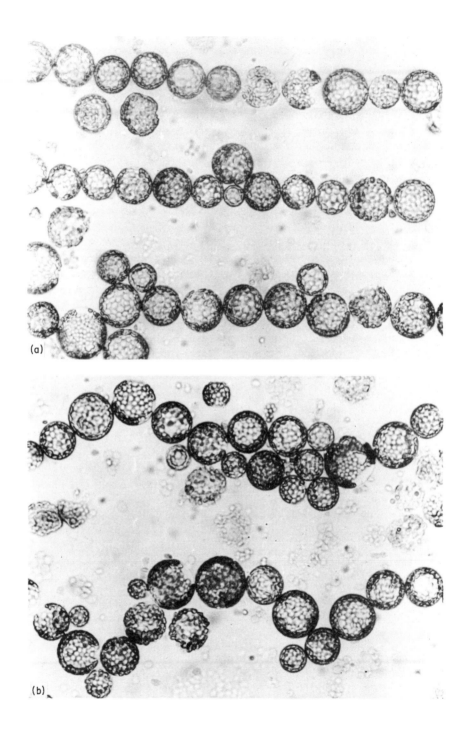

(a)

(b)

In (a), the cell is assumed to be charged and is moving in a homogeneous electric field (electrophoresis). This can give no net movement if the field is alternating, which means that this mechanism cannot be the general cause of pearl-chain formation.

In (b), the field polarizes the cell (assumed to be otherwise neutral), producing charge separation so that the opposite sides of the cell become charged. (This also occurs in Fig. 7a but this does not alter the argument so is not shown for the sake of clarity.) In a uniform field as shown, the forces on the charges are equal but opposite, and again no movement can occur. However, if the field is non-uniform as in (c), the charge in the stronger field experiences a stronger force than the charge at the opposite side of the cell, so the cell tends to move towards the region of higher field. The effect works equally well if the field is reversed (as in d), and so will also work in alternating fields. This effect was termed dielectrophoresis (Pohl, 1951, 1978) and can occur whether the cells are charged or not, whereas electrophoresis requires the cell to be charged.

The dielectrophoretic force F on a small sphere of radius a in a field of strength E is given by (Pohl, 1958; Mason and Townsley, 1971; Pethig, 1979):

$$F = 2\pi a^3 \cdot \varepsilon_0 \cdot \varepsilon_1 \left(\frac{\varepsilon_2 - \varepsilon_1}{\varepsilon_2 + 2\varepsilon_1} \right) \nabla |E|^2 \tag{4}$$

where ε_0 is the permittivity of free space, and ε_1, ε_2 are the relative dielectric constants of the medium and of the sphere, respectively. ∇ is the del vector operator, and its presence shows that the force is proportional to the divergence of the square of the field strength. That is, the force increases with the degree of non-uniformity of the field, and with the square of the applied voltage.

This equation shows that the dielectrophoretic force increases with the volume of the cell, and is sensitive to the dielectric constants of the medium and cell. It is usually assumed that $\varepsilon_2 > \varepsilon_1$ (positive polarizability), which condition results in movement of the cells into regions where the field is stronger. This is termed positive dielectrophoresis. When $\varepsilon_2 = \varepsilon_1$, the force is zero. It is however, possible that $\varepsilon_2 < \varepsilon_1$ (negative polarizability) in which case the dielectrophoretic force pushes cells away from the regions of maximum field. This is termed negative dielectrophoresis. The relationship

FIG. 6. Pearl-chain formation from mesophyll protoplasts of *Avena sativa*, in an alternating field of strength 66 V cm^{-1}. (a) Field frequency 2 MHz. The chains form parallel to the field, i.e. perpendicular to the two wire electrodes (not shown). (b) Field frequency 25 kHz, the optimum for rotation. The cells form chains oriented at approximately 45° to the field. From Holzapfel *et al.* (1982). In both cases, the electrodes (not shown) are to the sides of the pictures.

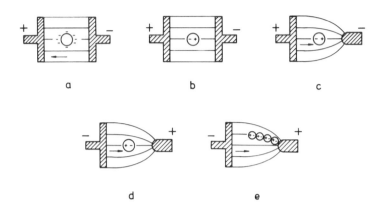

FIG. 7. Some possible interactions between a cell and the field it is freely suspended in. The arrows always indicate the direction of movement. (a) The cell has a permanent charge and therefore undergoes electrophoresis. (b) Although charge separation occurs in an uncharged cell, the net force is zero in a uniform field. (c) and (d) The action of a non-uniform field of either polarity gives movement towards greater non-uniformity. (e) A pearl chain showing the alignment of induced cellular dipoles that caused its formation. From Zimmermann *et al.* (1981b).

between ε_2 and ε_1 changes with frequency, which results in the direction of cell movement becoming frequency dependent.

In the above, both medium and cells are treated as pure dielectrics (insulators). In fact the conductivity of both components should also be considered (Schwarz, 1963) although only the terms in ε in Eq. (4) are affected (Sher, 1968). Indeed, as shown recently (Sauer, 1983) the presence of conductivity means that energy must be dissipated in the medium, and, in principle, the earlier calculations no longer hold. However, a qualitative description of the behaviour in media of low conductivity can still be given.

The dielectric constants that must now be used in Eq. (4) are termed complex, i.e. they include not only the "real" capacitive component, but also the "imaginary" conductive one. This is nothing more than a way of expressing the relative magnitude of the capacitive and conductive currents in one. The use of complex algebra merely reflects the phase difference between these currents. For example, if an ionic solution in water has conductivity σ, and the "capacitive" dielectric constant is taken to be 80, then

$$\varepsilon_1 = 80 - \frac{j \cdot \sigma}{2\pi f \cdot \varepsilon_0} \tag{5}$$

where f is the frequency and j is $\sqrt{(-1)}$. The imaginary component can become very large when f is small. Therefore cells may show positive or negative dielectrophoresis depending upon field frequency and medium conductivity.

The dielectric constant of the cell must take account of the different substances of which it is composed and of the structure of the cell and composition of the membrane. These factors result in the dielectric constant decreasing in stages as the frequency is increased. Each step is known as a dielectric dispersion. The most prominent three dispersions are known as the α-, β- and γ-dispersions (Schwan, 1957; Pethig, 1979; Bernhardt, 1979).

At very low frequencies (1 kHz or lower) counterions loosely bound to the charged surface of the cell (Schwarz, 1962) and the ions in the diffuse "atmosphere" of the cell (Dukhin, 1971), can move in response to the field, so that a dipole is generated. The inability of these movements to follow increasing field frequencies seems to be responsible for the α-dispersion (Schwarz, 1962).

Field-induced ion movements within the cell cause a build-up of charge at the membrane surfaces. The mechanism is that of the induced polarization already shown in Fig. 7b–d, except that counterions on the solution side of the membrane capacitor must be considered as well. The rate at which these charges can build up is limited by the conductance of the solutions as considered later in Section VI on rotating field-induced rotation. The change in dielectric constant (the β-dispersion) occurs between 100 kHz and 10 MHz for animal tissues (Schwan, 1957), but can be expected to occur at lower frequencies in low-conductivity suspensions.

Ion movements are not the only mechanism capable of affecting the dielectric constant. The field-induced alignment of dipoles inherent in the structure of the material is equivalent to the movement of charge. The time required for the dipoles to respond depends on the size and ability of the molecules or groups to rotate. Protein and other macromolecules rotate comparatively slowly. For example, the dispersion of myoglobin solutions which seems to result from molecular rotation is in the low MHz range (South and Grant, 1972). Smaller groups such as the side-chains of proteins cause dispersions in the 10–100 MHz range and water bound to proteins causes a dispersion near 500 MHz (Pennock and Schwan, 1969). The head groups of lipids are also dipolar and therefore show a dielectric dispersion. For bilayers made from various mixtures of cholesterol and dipalmitoyl phosphatidylcholine at 38°C, the dispersion lies in the 20–50 MHz range (Shepherd and Büldt, 1979). Carbohydrates and nucleic acids can also be expected to show dielectric dispersions. The extent to which these materials affect the properties of the whole cell will depend on the amounts in which they are present.

The γ-dispersion referred to above is due to free water and need not concern us further as it is not significant for frequencies below 1000 MHz. Studies of electro-fusion and rotation have not so far used frequencies approaching this figure.

A positive dielectrophoretic force acts to concentrate the cells by attracting them into a smaller volume. At higher concentrations the distance between cells is reduced, and the cells begin to exert a force on each other because of the local non-uniformity in the field that each cell produces. If, as shown in Figs 7c or d the cells are polarized by a non-uniform field, they will begin to move towards the greatest non-uniformity (the electrode). As they come together, the induced charges on each cell begin to influence each other and the cells attract each other. The cells therefore accelerate together and pearl chains form as in Fig. 7e. In principle, the later stages of this process can occur in uniform applied fields if the first step of dielectrophoretic concentration is made unnecessary by using a very dense suspension. In the practice of electro-fusion, the ability to see individual cells is important, so a low suspension density and a non-uniform field are used.

Saito and Schwan (1961) analysed the forces between particles forming pearl chains in terms of the induced dipoles. The speed of formation of the chain was shown to:

(1) Increase very gradually with the field at low field strengths;
(2) Increase practically with the square of the field at higher field strengths;
(3) Decrease as the cube of the radius of the cell increases. Therefore, large cells collect only very slowly at low field strengths; indeed, in these conditions the time constant of chain formation may be as long as minutes which makes the threshold very difficult to determine. The rate of pearl-chain formation also depends on the composition and structure of the cells, as these affect the polarizability. In practice, for cells of say 1 μm radius in moderate field strengths, the time constant of pearl-chain formation is approximately 1–10 s, while typical field strengths giving chain formation are in the range 10 to 100 V cm^{-1}.

The use of dielectrophoretic pearl-chain formation is very useful for arranging cells prior to electro-fusion because:

(1) the cells are arranged in line with the field, so that field-induced membrane breakdown will occur most readily where the cells touch (as described below);
(2) dielectrophoresis allows the use of alternating fields (or pulsed fields of alternating polarity), which do not cause problems with gas production at the electrodes (at field frequencies of a few kHz or higher).

The size dependence of the force given by Eq. (4) means that pearl-chain formation from cells of diameter less than 0.3 to 0.5 μm is not possible without using field strengths which may lead to membrane breakdown. Smaller cells are still affected by the field, but the minimization of energy that results from addition of one cell to a pearl chain is not large compared to the thermal "scrambling" energy, equal to kT.

The reduction in dielectrophoretic force with decreasing cell volume has been used to separate cells of different sizes (Pohl, 1978; Glaser *et al.*, 1979). It now seems quite feasible to use this technique to enrich a population of lymphocytes in those cells that are actively secreting antibodies, because "stimulated" cells have a larger volume than non-stimulated cells (Eshhar, Weizmann Institute, Israel, personal communication). This may be useful in the production of hybridoma cells where only stimulated lymphocyte cells will contribute to the yield of useful hybridoma cells (see below).

IV. Orientation of Cells

The aggregation of two or more cells in an electric field, as discussed previously, is a necessary prelude to electro-fusion. If the cells to be fused are non-spherical, their mutual orientation can also be important in fusion. It has been found that electric fields can determine the orientation in single or aggregated cells. This effect is frequency dependent (it requires a minimum field strength range), and varies with medium conductivity and between organisms. A minimum field strength is required for orientation because it is necessary to make the difference in energies between the orientations large compared with that of Brownian motion (kT), as in pearl-chain formation. Typical values from theory and experiment are approximately 10 to 100 V cm^{-1}, with smaller cells requiring more field strength. This value is usually slightly less than that required for pearl-chain formation (see above). Field strengths necessary to cause pearl-chain formation will therefore simultaneously give orientation, as is indeed utilized in the fusion of red blood cells.

In particles smaller than cells, higher field strengths are required. For polar macromolecules field strengths of several kV cm^{-1} can be calculated to be necessary (Schwan, 1977; Neumann, 1981). Purple membrane fragments (from *Halobacterium halobium*) were seen to orient at 15–20 kV cm^{-1} (D. Chapman and U. Zimmermann, unpublished data). Shinar *et al.* (1977) observed that pulsed fields applied to an aqueous suspension of purple membrane fragments produced transient linear dichroism effects. They interpretated these as being due to the alignment of the chromophore (and the protein bearing it) in response to the field.

As long as a century ago (Ludloff, 1895; Verworn, 1896; Mast, 1931), studies had begun on the orientation and movement of motile cells (e.g. *Paramecium* sp.). These have recently been started again (e.g. Martinac and Hildebrand, 1981). Teixeira-Pinto *et al.* (1960) described the effects of pulsed radio frequency (1–100 MHz) fields on various inert particles and motile unicells. *Euglena gracilis* was seen to move parallel to the electric field at lower frequencies, but pivoted through 90° at frequencies between 5 and 18 MHz, depending on the age of the preparation. The authors attribute the age dependence to a change in metabolite concentrations, which (from the modern viewpoint) is seen to act by changing the conductivities of the medium and cytoplasm.

It was also shown that in a given experiment, different species exhibited different frequencies at which they changed direction. It was thus possible to see simultaneously various organisms in a mixed culture travelling at right angles to each other. Living cells were the only dispersed phase seen to be able to orient perpendicular to the field, whereas inert particles always took up the parallel orientation. In perpendicularly oriented *Amoeba* cells, cytoplasmic inclusions were seen to orient parallel to the field. Griffin and Stowell (1966) used somewhat higher frequencies and were able to demonstrate a second frequency (100 MHz) at which *Euglena* cells changed orientation and direction, which resulted in a return to the parallel-field behaviour noticed below 5 MHz. The effect of conductivity was explicitly studied and it was found that high conductivities gave the parallel orientation at all frequencies for living or inanimate particles.

The results with motile organisms raised the question of whether the active processes giving locomotion are themselves triggered by the field to drive and steer the movement. The alternative view, that the field determines the orientation of any particle whether motile, non-motile or inanimate, and hence *directs* the movement of a motile cell has been put forward by Saito *et al.* (1966). If this is accepted, the reports of "movement along" a given axis can be read as "oriented along" for the usual case where a motile organism proceeds in the direction of its longest axis.

A. NON-SPHERICAL CELLS

Erythrocytes (human and chicken) were observed to change orientation with frequency in the MHz-range (Griffin, 1970). Once again the orientation was found to be affected by conductivity, low conductivity favouring the perpendicular orientation. Under conditions of very low conductivity (sucrose medium with EDTA as anti-coagulant), Griffin (1970) noticed that the perpendicular orientation persisted only for about 10 min. Studies on human erythrocytes in our laboratory (Pilwat and Zimmermann, unpublished results) have shown that the perpendicular orientation can be stabilized in the MHz range if albumin is present in the non-electrolyte solution, in which case EDTA is not necessary. These findings can be considered consistent with ion leakage through cell membranes which have been made leaky by the absence of calcium ions in the medium.

In alternating and pulsed fields orientation of *Fucus serratus* eggs was observed by Novak and Bentrup (1973), and of *E. coli* by Sher (1963) and other authors (Heller and Teixeira-Pinto, 1959; Jennings and Morris, 1974; Morris *et al.*, 1975). Recently, Vienken *et al.* (1984) have shown that sickle cells orient perpendicular to the field at 500 kHz whereas normal human erythrocytes orient parallel to the field. This effect may be useful in the diagnosis of sickle-cell anaemia.

The different orientations taken up by turkey erythrocytes (parallel or perpendicular to the field) are shown in Fig. 8. The parallel orientation is taken up in the kHz range, whereas frequencies in the high MHz range (about 60 MHz) give the perpendicular orientation (Zimmermann *et al.*, 1984a,b). We shall return to this point later when discussing intimate membrane contact as a prerequisite for fusion.

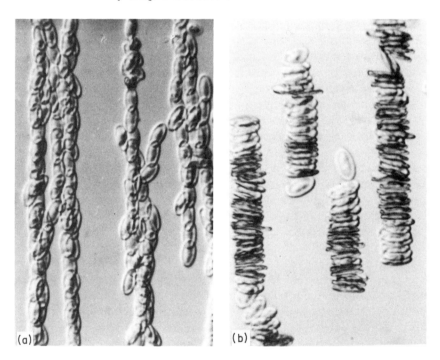

FIG. 8. Dielectrophoretic pearl chains formed from suspensions of turkey erythrocytes. (a) When the field frequency is in the kHz range, the cells line up with their major semi-axis parallel to the field, but overlapping slightly as do tiles on a roof. (b) At field frequencies of the order of 60 MHz, the cells form stacks with the major semi-axis perpendicular to the field. From Zimmermann *et al.* (1984a,b).

In cells, interpretations of the orientation effects are not easy to make on a quantitative basis. Füredi and Valentine (1962) suggested that the induced dipole effect was responsible, and that the cross-field orientation of biological particles was due to the outer membrane of these particles having a profound effect at certain frequencies. Schwarz et al. (1965), Saito et al. (1966) and Schwan and Sher (1969) analysed the energy of a generalized asymmetric particle in an alternating field. They showed that orientation was a sensitive function of frequency and conductivity using numerical solutions (for minimum energy) based on given particle and medium constants; this provided a theoretical basis for the observed phenomena. The change from one orientation to the perpendicular one was calculated to be of the all-or-nothing variety ("turnover") if certain conditions were met. The effect of a membrane was shown to provide one more turnover frequency than that of a theoretical membrane-less particle. Use of approximate electrical constants for cell and medium showed that turnovers were expected in the range of a few MHz to roughly 1 GHz. Turnovers in this region have been found by experiment, although the top of this range is yet to be thoroughly explored.

Calculating the properties of cells of particular species according to these methods should be possible if the large amount of data required is available. It would be necessary to know the conductivities and permittivities of all significant components which are not constants but functions of frequency. As shown when considering pearl-chain formation, the difference in dielectric constants between medium and cell interior may easily reverse as the frequency is raised.

All the above calculations are based on the assumption that the applied field is uniform, although there is no reason why orientation effects should not also occur in non-uniform fields, and such effects are indeed observed. Non-uniform fields can cause pearl-chain formation as described above, and the cells adopt the same orientation as do isolated cells. The orientation of the chain itself is usually directly in line with the field, although unusual cross-field chains were observed by Griffin and Ferris (1970) in the 10–190 MHz frequency range. A mixed suspension of Chlorella pyrenoida cells (regarded as a lossy dielectric) and polystyrene spheres (a good dielectric of low permittivity at these frequencies) was observed to form homogeneous pearl chains aligned parallel to the field. Short, mixed chains were also observed, and these formed transverse to the field. These formations did not show the frequency dependence of orientation found for asymmetric cells.

Pearl chains have also been observed to form at 45° to the field. This was true with human erythrocyte chains observed at 120 MHz by Füredi and Ohad (1964). Holzapfel et al. (1982) found a similar orientation at 25 kHz with pearl chains of Avena sativa protoplasts (Fig. 6b). It is significant that both these groups also report rotation of the cells under the same conditions. If rotation requires a negative polarizability as discussed below, it may be possible that the chains are repelled from the electrode and this leads to the 45° position. Other aspects of this phenomenon will be considered later.

Generally, the approach used by Schwan and his co-workers for cell orientation is also theoretically applicable to the analysis of pearl-chain orientation. Although it is certain that these arrangements are those giving an energy minimum, the complications caused by the presence of mutual interactions make the problem intractable.

Griffin (1970) has suggested that the orientations observed for asymmetrical particles may have been partly determined by the usual procedure of allowing the cells to settle on to a microscope slide. Minimum gravitational energy will cause the long axis to lie parallel to the slide. Hopefully, future work will avoid such complications by using the techniques developed in rotation work for the stable suspension of free cells (isopycnic suspension using "Percoll" or sucrose to adjust the medium density).

B. SPHERICAL CELLS

Orientation of cells has only been observed when the cells were obviously geometrically asymmetric. It is possible that even a spherical cell may have asymmetric dielectric characteristics if the asymmetric distribution of organelles is significant, or if current views on membrane structure are considered. Therefore, the orientation of spherical cells is also a possibility, even though more sophisticated techniques may be necessary to see this. Orientation of intracellular particles is frequently seen in the presence of an alternating field, as was first observed by Teixeira-Pinto et al. (1960). Richter et al. (1981) observed that the orange–red echinochrome particles of sea-urchin eggs moved in the alternating field applied in order to form pearl chains. Caps of pigment were observed to form at the points of contact between the cells in a chain. Most of the pigment capping disappeared after removal of the field.

Preferential orientation of spherical cells having asymmetric properties could be very useful in electro-fusion of plant protoplasts with subsequent regeneration to give a whole plant. Regeneration of plants from protoplasts has been reported in a few cases (see Reinert and Bajaj, 1977). On the other hand, regeneration of cereal plants from protoplasts has proved to be either very difficult or impossible. One reason seems to be that cell wall formation from cereal protoplasts is hindered. It seems quite conceivable that a removal of only a small area of the cell wall will still allow regeneration. Therefore, if two or more cells with only small areas free of cell wall (achieved by short enzymatic digestion) are brought together in such an orientation that they have intimate *membrane* contact, then fusion by an electric field pulse should be possible. The product will have a considerable area of cell wall, which may permit eventual growth of a new hybrid cereal plant.

C. ORIENTATION UNDER BREAKDOWN CONDITIONS

Experiments using pulsed alternating fields have shown that they are just as effective as continuous fields of the same average power in causing orientation. Theoretical considerations by Sher *et al.* (1970) have shown that indeed no difference is expected. It is, of course, assumed that pulsing the field does not cause membrane breakdown, as this may cause orientation of any shape of cell. Reversible breakdown causes the membrane to become locally permeable (see above). Klee and Plonsey (1974) have shown by theoretical mapping of the field lines that the presence of a low resistance patch in the membrane causes the potential of the interior of the cell close to the patch to approach that of the medium. Returning to the orientation of cells, we can say that the new potential profile may not correspond to the lowest energy state for the cell, which can therefore be expected to rotate towards the new minimum energy position. The re-orientation of cells could be observed following a breakdown pulse applied to cells aligned parallel to the field as in Figs 6–12 (Zimmermann, 1982).

The amount of re-orientation following a breakdown pulse depends on the moments of inertia of the cells, steric hindrance between adjacent cells, adhesion and friction between cells, and on the viscosity of the medium. Some of these factors vary from preparation to preparation and so it may be necessary to adjust the fusion conditions as necessary. The rotation can affect fusion because misalignment of the pores generated by breakdown will prevent formation of the membrane bridge (Fig. 4).

Re-orientation is expected to have the greatest effect when fusion of small cells or of vesicles is attempted, because of their small moments of inertia. Rotation may also cause a serious disturbance during giant cell formation, in which the three-dimensional fusion of many pearl chains is necessary. Experiments with breakdown pulses have shown that they can cause entire pearl chains to rotate by up to 20°. Pearl chains displaced in this way cannot be expected to fuse with others. It has been found that these difficulties can be avoided if the alternating field is switched off during the breakdown pulse and for a few milliseconds afterwards. This is not long enough to allow disruption of the pearl chains, but the absence of the field means that there is no longer a driving force for rotation. Sometimes the use of multiple shorter pulses over several seconds is also useful. When several pearl chains of erythrocytes are fused to form a giant cell, some re-orientation of cells within a given pearl chain seems to be required after the first pulse. It is observed that the use of a train of pulses is necessary to allow all cells to fuse to each other. It may be that pores are moved round to other locations between pulses. Lateral fusion of other non-spherical cells may proceed in a similar manner.

The influence of orientation and rotation effects on electro-fusion has been overlooked by many groups. This is unfortunate as neglect of these matters will only lead to frustration when electro-fusion does not occur as expected.

V. Fusion

When electrical breakdown is carried out on cells appropriately positioned by means of dielectrophoresis, fusion of cells results. These methods depend only on the general properties of cells and their external membranes, so the process can be applied to all living cells (including inter-kingdom fusion), to liposomes, to planar bilayers or to any combination of these (Figs 9–13).

The variety of methods used in earlier work to achieve the necessary intimate membrane contact required for fusion is interesting. Zimmermann and Pilwat (1978) were the first to report fusion of cells following electrical breakdown when they described the effects of discharging a high voltage capacitor through a dense suspension of erythrocytes by means of two flat platinum electrodes. It now seems that occasional membrane contact must have occurred in such suspensions, driven by the local field distortion created around each cell even in uniform fields (see above). Each cell attracts its neighbours so that at high suspension densities some aggregation can occur in response to the field. This effect may have also given cell–cell contact in suspensions between rocks subjected to naturally occurring field pulses, such as that due to a nearby lightning strike (Zimmermann and Küppers, 1983; Küppers and Zimmermann, 1983; Küppers et al., 1984).

Neumann et al. (1980) agglutinated cells of the eukaryotic micro-organism *Dictyostelium discoideum* by rolling suspensions of them in plastic tubes. Application of a field pulse to these agglutinated cells resulted in fusion. This procedure is not applicable to other organisms. Senda et al. (1979) used two electrodes to directly push two protoplasts together, and then injected a rather long current pulse (of 5 ms duration). Although fusion occurred, the fused cell was not viable, probably because this pulse length gives irreversible mechanical breakdown (Benz and Zimmermann, 1980a; Zimmermann and Benz, 1980). This procedure is not suitable for the fusion of large numbers of cells. Weber et al. (1981) have recently reported the fusion of yeast protoplasts using a combination of polyethylene glycol (PEG) and a field pulse. While PEG may establish close membrane contact, it is also itself an excellent chemical fusogen (Lucy, 1982), so there seems to be little advantage in using the combination. In addition, the outcome of this particular experiment is also not very clear because the haploid cells used (mating types a and α) can conjugate directly (quoted in Halfmann et al., 1982).

The electro-fusion method seems superior to all these methods and requires only very simple equipment (Zimmermann and Scheurich, 1981a). The fusion chamber consists of microscope slide which has two parallel platinum wires (of circular cross-section) glued to it. The separation of the wires depends on the size of the cells that are to be fused and the voltage capability of the pulse generator that will produce the breakdown pulse. Typical distances are between 100 and 500 μm. An electronic switch connects the pulse generator for the short time required, the rest of the time a sine-wave generator supplies the dielectrophoresis voltage (Pilwat *et al.*, 1981). This arrangement can fuse up to several thousand cells.

Several ideas for flow cells have been developed enabling continuous operation and production of fused cells on a larger scale (Zimmermann and Vienken, 1982; Zimmermann *et al.*, 1984a; Schnettler *et al.*, 1984). Vienken and Zimmermann (1982) have also described an electro-hydraulic procedure by which it is possible to obtain a high yield of AB heterokaryons from two given species A and B with the minimum of AA and BB contamination. The isolation of the hybrid is thereby simplified.

The successful use of electro-fusion may be hampered by the production of heat in the suspension between the electrodes. Whereas the field pulse is too short to cause significant heating, the alternating field used for dielectro-phoretic alignment leads to pronounced heat production in electrolyte solutions, due to current flow. The production of heat may cause general damage to the cell or prevent fusion by speeding up the resealing time of the membrane as mentioned above. Extreme cases of heating give turbulence in the medium and so destroy the alignment of cells in the pearl chains upon which fusion depends.

The current flow which causes this can be minimized by using a medium with a low conductivity, i.e. one containing predominantly non-electrolytes. A convenient conductivity is approximately 10^{-4} S cm^{-1} or less. The lack of ions seems to have no adverse effect on mammalian cells and yeast protoplasts for up to 1 h incubation The stability of plant protoplasts is also assured because they are isolated by enzymic digestion in almost ion-free solutions of mannitol or sorbitol.

With some cells precautions must be taken to prevent the pH of these unbuffered solutions from changing considerably during the dielectrophoresis and fusion precedure, for example by adding histidine to a final concentration of 10 mM. Histidine is used as buffer because it contributes little to the con-ductivity (Vienken and Zimmermann, 1982). This will maintain the pH at about 7, although pH variation is not usually very important for the kinetics of fusion, except in the case of mammalian cells. No significant difference was found between pH 6 or pH 8 when fusing plant protoplasts or sea-urchin eggs. However, it becomes progressively more difficult to establish intim-ate membrane contact when the pH falls below 7. In this range, human erythrocytes agglutinate and do not show normal pearl chains during dielectrophoresis. Optically homogeneous tubes form, particularly if the dielectrophoretic voltage exceeds the membrane breakdown voltage (Scheurich *et al.*, 1980b). These may seem to be fused, but removal of the field and addition of electrolyte leads to immediate re-appearance of cell boundaries within the tubes and soon the usual biconcave cells detach themselves again. Even the use of a breakdown pulse will not give true fusion of these agglutinated cells. The agglutination often disappears if the cells are simply allowed to stand 10 min in non-electrolyte solution. This is probably due to the potassium leakage which is considerably increased when erythrocytes are placed in low conductivity media (Donlon and Rothstein, 1969). By this means the cells supply enough ionic atmosphere to reverse the agglutination.

Permanent cell lines tend to be very active metabolically, so a low pH occurs readily when these cells are in unbuffered media. This may be the reason why fusion is often prevented with such cells: a prominent example occurs in the fusion of lymphocytes and myeloma cells when producing hybridoma cells (see below).

Although the presence of electrolytes in the medium is generally undesirable as discussed above, electro-fusion is still possible in conductive media as shown recently by Pilwat and Zimmermann (unpublished results). To reduce heating, the usual sine-wave dielectrophoresis voltage was replaced by a series of square pulses of alternating polarity so that the total "on" time was relatively small. The viability of the fused products has not yet been investigated. The conditions of the dielectrophoresis prior to the application of the breakdown pulse are important, and so the frequency and strength of the applied alternating field, as well as the duration of this treatment, must be considered carefully.

Work with plant and yeast protoplasts as well as with mammalian cells has shown that the frequency range between 10 kHz and 1 MHz is generally suitable. Liposomes made from pure lipids may require other frequencies as they can show negative dielectrophoresis (i.e. they are pushed away from the electrodes instead of forming pearl chains). Cholesterol containing vesicles show positive dielectrophoresis and therefore ready pearl-chain formation (Büschl *et al.*, 1982). Non-spherical cells can also be troublesome as the alternating field will orient them in a way which may not give intimate membrane contact. As mentioned earlier such orientations can be "turned-over" by changing the field frequency. It is interesting to note the application of this to erythrocyte fusion, which is as follows:

Both human and turkey red blood cells line up with their major semi-axis parallel to the field vector if the usual frequency range of the alternating field is used. The biconcave human cells contact end to end allowing the dielectrophoretic force to directly squeeze the membranes together: fusion results after the breakdown pulse. The turkey red blood cells are rigid and oval;

they form chains by means of slight overlap, resembling tiles on a roof (Fig. 8a). The main dielectrophoretic force acting along the "pearl chain" is not pushing the membranes directly together and no fusion is possible even when pulses of very high field strength are used (as described below). However, fusion of these cells can be achieved by turning the cells over so that they form an arrangement resembling a stack of coins (Fig. 8b) (Zimmermann *et al.*, 1984a,b). The compressive force in this arrangement can now give intimate membrane contact. This arrangement is produced by using an alternating field of 60–70 MHz frequency.

The field strength should not be more than is just sufficient for dielectrophoretic collection and this must not take longer than a few minutes for reasons described below. The strength of the alternating field used is very important and is constrained in two ways. The first condition is satisfied by using that field strength which only just gives point-to-point contact between cells (Figs. 6 and 9). The dielectrophoretic force between cells increases with the volume of the particles (Sher, 1968; Pethig, 1979; Pohl, 1978), so the field strength required will depend on which cells are to be fused. Point-to-point contact is not sufficient for fusion, for which the membranes must meet over a substantial area (see Figs 5 and 9) and this needs more field strength. Therefore, the alternating voltage is raised until the cells flatten in the contact area, just before the field pulse is applied (Figs 9a,b). This higher value of field strength must not cause membrane breakdown, which under these conditions leads to destruction or deterioration of the membrane. This will be immediately obvious from turbulence due to release of electrolyte, for example.

Breakdown by the dielectrophoretic field cannot be fully reversible because although a single half-cycle of the alternating field may be by itself short enough to give reversible breakdown, subsequent half-cycles follow too quickly to allow the pores time to reseal (Zimmermann *et al.*, 1982c). It is frequently observed that the value of field necessary for deformation is just below or above that expected to give breakdown. Spherical cells become elliptical and budding or vesicle formation can also be seen. Although removal of the field results in a return to the spherical shape, the cell membranes have been affected in some less obvious way because they can no longer be fused by appropriate gentle conditions. The whole process up to the flattening of the contact zone should only take a couple of minutes. Longer exposure to the alternating field seems to produce changes in the membrane structure, although these are partially reversible when the field is removed. There is some evidence that under the influence of the field component tangential to the membrane a redistribution of intramembranous particles (proteins) occurs so that aggregation and emergence of particle-free lipid domains occurs (see Fig. 5a) (Zimmermann and Vienken, 1982).

Such effects are well established in the presence of steady state (d.c.) fields (Brower and Giddings, 1980; Jaffe, 1977; McLaughlin and Poo, 1981; Orida and Poo, 1978; Poo, 1981; Poo and Robinson, 1977; Poo et al., 1979; Sowers and Hackenbrock, 1981; Zagyansky and Jard, 1979). It is possible that they can also be induced by an alternating field because the magnitude and direction of the field vector changes between the "equator" and the "pole" of the cell. The evidence that such effects occur is seen when subsequent breakdown is attempted.

Cells that have been exposed to the alternating field for a long time are more resistant to the field pulse than are untreated cells. They are able to withstand higher field strengths and breakdown pulses which are longer than those seen to destroy untreated cells (Pilwat et al., 1981; Zimmermann et al., 1982c). The explanation of this "field stability" seems to be that lipid domains emerge in the membrane contact zone due to the field-induced particle movement and aggregation already mentioned. The resealing time of a lipid bilayer is much faster than that of the lipoprotein matrix (Benz and Zimmermann, 1981a), so the damage to the cell due to interchange of cytoplasm and medium through the open pore is much reduced.

A similar increase in the field that can be withstood by cells occurs when a train of breakdown pulses of gradually increasing amplitude is applied (Pilwat et al., 1981). The first pulse is just above the breakdown threshold, and the interval between pulses is much longer than the membrane resealing time. Therefore, there is time for the membranes to be sealed by lipid diffusion. The second and subsequent pulses can then cause breakdown in the newly created lipid domains. An increase in the sustainable breakdown pulse field strength and duration is also noticed when cells are subjected to a pulse in the presence of pronase or dispase (Zimmermann et al., 1981a; Zimmermann and Vienken, 1982), or in the presence of very low calcium ion concentrations (Zimmermann, 1982). Either enzymatic action or lack of calcium ions seem to enhance the ability of the field to produce lipid domains. This interpretation is similar to the one used to explain the mechanism of fusion when viruses or chemicals are used as fusogens (Poste and Allison, 1973; Lucy, 1982). However, the emergence of lipid domains in the presence of fusogenic agents was recently questioned by Chandler and Heuser (1979, 1980; see also Shotton, 1980) who quote evidence obtained using a quick-freezing electron microscopy technique. It is also possible that in the presence of pronase (or other such conditions) and in an inhomogeneous field the breakdown voltage itself is increased. No such increase was found with pronase-treated 3T3 or Friend cells when the breakdown voltage was measured in the Coulter Counter (Zimmermann et al., 1982c) although this uses a uniform field. These effects have been discussed in more detail elsewhere (Zimmermann, 1982).

It is also conceivable that changes in membrane structure can occur by mechanisms other than the direct action of applied fields or the presence of enzymes. Gingell and Ginzberg (1978) have shown that the very close approach of two membranes could itself give rise to an aggregation of intra-membranous particles by direct electrostatic displacement, which may then be enhanced by the alternating field.

The field stability effect can be experimentally exploited when fusion of two cells of very different radii is to be attempted. We have already shown that the membrane voltage developed by a given field strength is proportional to the radius of the cell. Hence the field pulse that is just sufficient for fusion of the smaller cell is much higher than that required for the larger cell. It is likely that the larger cell will be destroyed by such a pulse unless it is first made field-stable, either by the treatment described above, or by the use of the enzymes mentioned.

Intermingling of membranes occurs within seconds in all cells so far investigated. The time taken to give the final spherical shape of the fusion product varies from species to species but also increases as more cells are included. The addition of medium which is 1 mM in calcium ions (but other-wise of low conductivity) is often required about 1 min after the breakdown pulse, or else the cells do not become spherical.

Typically this is achieved 3 min after breakdown for the fusion product of two protoplasts of *Avena sativa*, which increases to at least 15 min for the product from a pearl chain of 8 protoplasts (Zimmermann and Scheurich, 1981b; Vienken *et al.*, 1981; Salhani *et al.*, 1982). Similar data has been found for the fusion of cells from permanent cell lines and for other mam-malian cells (Zimmermann, 1982). On the other hand, vacuoles fuse within a few seconds (Vienken *et al.*, 1981), as do vesicles made from a mixture of phospholipid and cholesterol (Büschl *et al.*, 1982). In these cases it is hardly possible to follow the stages of fusion under the microscope. The alternating field must also be considered during and after breakdown. It must be switched off during the breakdown pulse to avoid reorientation and rotation of the cells (see above) which could otherwise displace the pores induced by the breakdown pulse in the closely apposed membranes. However, the alter-nating field must be present again soon after breakdown to create a tension in the fusing cell membranes which seems to be necessary for the fusion process. An automatic switch is therefore used to remove the field about 1 ms before the breakdown pulse and to re-apply it 1 ms afterwards (Pilwat *et al.*, 1981). The field strength is gradually decreased to zero over the course of the next 2 min, to enable the fusion products to become spherical. Using these conditions, 60–80% yields of fused cells are obtainable. The equipment now commercially available (GCA/Precision Scientific Group, Chicago, Illinois;

Krüss GmbH, Hamburg) includes the electronics to accomplish the above functions.

Fusion of cells of *Kalanchoë daigremontiana* has shed new light on the fusion mechanism discussed above (Vienken *et al.*, 1983, Zimmermann *et al.*, 1982d, see Fig. 5). The mesophyll cells of this plant are so large (80 μm or more in diameter) that detailed observations on the fusion of the protoplasts can be made. 20 s after the breakdown pulse has been applied, an ovoid vesicle is seen in the area of membrane contact (Fig. 9). The vesicle becomes progressively larger as the process of fusion continues (Fig. 9b). By focusing through these protoplasts, it can be seen that the vesicle is near to the centre of the contact zone. Occasionally two vesicles are seen and these are near the edge of this area (Fig. 9e). The vesicle is still visible after the disappearance of the boundary lines between the two protoplasts and after the product becomes spherical. As a rule the vesicle disappears a few minutes later. If the protoplast vacuoles were stained with neutral red before application of the field pulse the vesicles formed contained no observable dye. When a pearl chain of four or five protoplasts was fused, vesicle formation occurred in each contact area.

The formation of macroscopic vesicles can be explained in terms of the proposed fusion model (Fig. 5) if we assume that the sub-microscopic vesicles formed in the membrane contact zone fuse to form larger vesicles. This process could be mediated by the vacuoles, which seem (from observations on several types of plant protoplasts) to remain separate after fusion of the plasmalemma. The gradual transition of the fusion product from an ovoid to a spherical shape appears to indicate that the vacuoles are being squeezed together and thereby fuse. Submicroscopic vesicles trapped at the flat interface between them will perhaps fuse in sufficient numbers to give visible vesicles. Such a mechanism of fusion appears feasible in view of the similar fusion of phospholipid vesicles observed by Nir *et al.* (1982). The fusion of the vacuoles follows after perhaps 1 min, and a random distribution of chloroplasts and vesicles is left.

It is surprising that the sub-microscopic vesicles appear to fuse with one another but apparently not with the tonoplast membrane in this process. This conclusion seems to be supported by the result that the vesicles formed during fusion do not take up neutral red from the stained vacuoles. However, it can be argued that the vesicles contain a mannitol solution with a pH higher than that of the vacuoles. Furthermore, the electrical fusion of protoplasts with isolated vacuoles from the leaves of *K. daigremontiana* shows that tonoplast and plasma-membranes can fuse, i.e. that they are biocompatible (Vienken *et al.*, 1981). This finding is contrary to current opinion (Matile, 1975). It is therefore quite possible that the sub-microscopic vesicles fuse

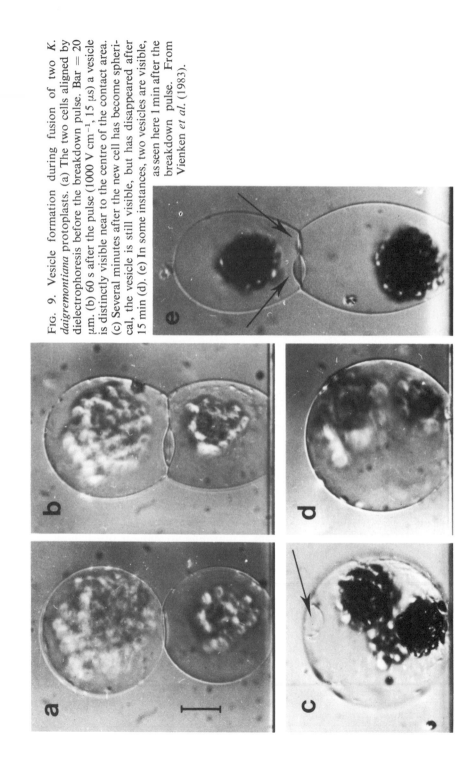

FIG. 9. Vesicle formation during fusion of two *K. daigremontiana* protoplasts. (a) The two cells aligned by dielectrophoresis before the breakdown pulse. Bar = 20 μm. (b) 60 s after the pulse (1000 V cm⁻¹, 15 μs) a vesicle is distinctly visible near to the centre of the contact area. (c) Several minutes after the new cell has become spherical, the vesicle is still visible, but has disappeared after 15 min (d). (e) In some instances, two vesicles are visible, as seen here 1 min after the breakdown pulse. From *Vienken et al.* (1983).

with the tonoplast membrane in the earlier stages of fusion. Such inter-membrane fusion should not contribute to the overall mass balance, because fusion of vacuoles should also lead to elimination of excess membrane material in the form of vesicles due to the reduction in the surface area of the fused vacuole. Thus, we cannot rule out the possibility that the vesicles visible by microscope contain constituents of the tonoplast membrane as well as of the plasma-membrane. Calculation of the surface area of a typical vesicle such as shown in Fig. 9 shows that about 20% of the excess membrane of the plasmalemma is represented by one vesicle. We have to assume that the rest of the residual membrane material is removed either by the formation

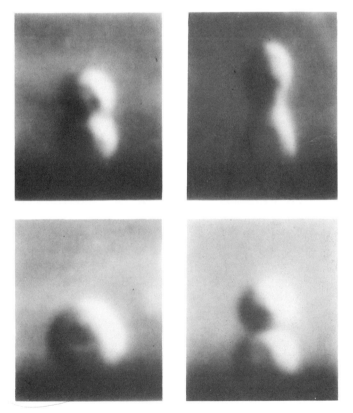

FIG. 10. Electro-fusion of yeast protoplasts. The sequence (running clockwise from top left) shows two dielectrophoretically aligned cells prior to fusion and the appearance of the fusion product at 3 stages after the application of a breakdown pulse. Field conditions: dielectrophoresis, 2 MHz and 630 V cm⁻¹; fusion pulse, 7.9 kV cm⁻¹ and 40 μs, repeated once 10 s later. The lack of definition of these micrographs is a consequence of the high magnification required (mean protoplast radius, 2 μm). From Halfmann *et al.* (1982).

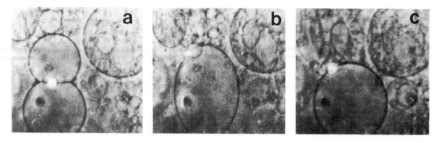

Fig. 11. Electro-fusion of large unilamellar vesicles prepared from equal amounts of cholesterol and a synthetic positively charged lipid. (a) Alignment of two vesicles (37 and 45 μm in diameter) using a 100 kHz field of 200 V cm^{-1}. (b) Fused liposome 1 s after the breakdown pulse (of 30 μs and 9 kV cm^{-1}), still subject to the alternating field. (c) After removal of the field, the vesicle forms a sphere. From Büschl *et al.* (1982).

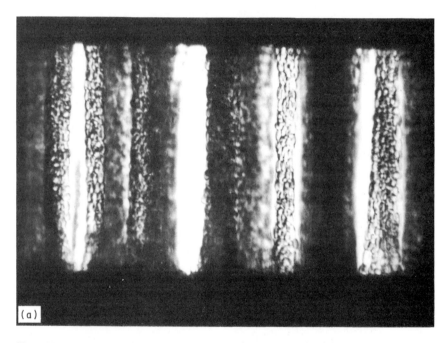

Fig. 12. Formation of a giant cell by fusion of human erythrocytes. (a) Many parallel pearl chains are formed close to each other by dielectrophoresis using 400 V cm^{-1} at 1 MHz field frequency. (b) After three pulses of electric field (each of 6 kV cm^{-1} and 5 μs) the giant cells form. (c) Giant cells which have been removed from the fusion chamber and transferred to electrolyte solution. The insertion of a micro-capillary into one gaint cell is demonstrated. (a) and (b) from Zimmermann *et al.* (1981b); (c) from Zimmermann and Vienken (1982).

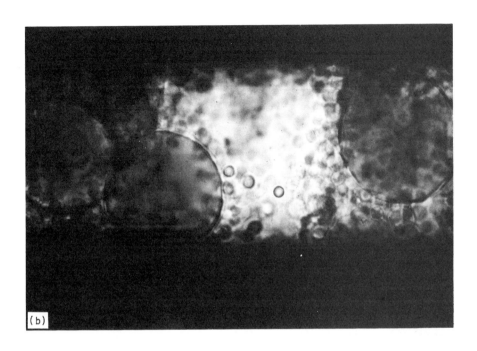

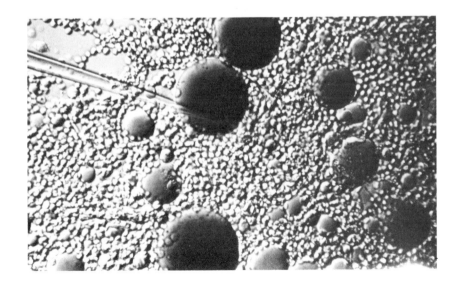

of more vesicles as shown in Fig. 9 or by the formation of more vesicles of smaller dimensions which cannot be seen in the light microscope. It is also quite conceivable that membrane material is incorporated into the interior of the visible vesicle, leading to inclusions or to the formation of multilamellar vesicles.

The evident disappearance of the vesicles some time after the cell has become rounded may be caused by enzymic action or by an instability of the vesicles themselves. It is known that artificial lipid vesicles are unstable, if the intra- or extravesicular electrolyte concentration exceeds 10 mM (Hub *et al.*, 1982a). The inverted vesicles formed during fusion will contain the same types of proteins as found in the plasmalemma (Fig. 5), although possibly disturbed by the action of the breakdown voltage as discussed above. They may thus be permeable, allowing exchange between the intravesicular mannitol solution and the electrolytes present in the cytoplasmic space. A conclusive answer regarding the fate of the vesicles after the rounding of the cells can only be given with the aid of the electron microscope.

Figures 9–13 show examples of fusion of yeast and plant protoplasts, cells from permanent cell lines and erythrocytes, and of liposomes.

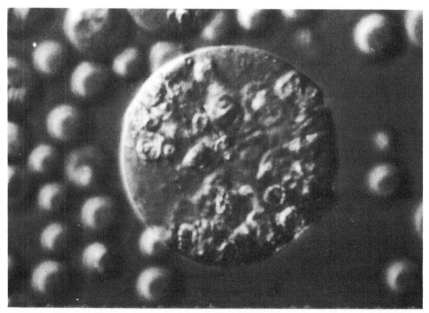

FIG. 13. Giant cell formed by fusion of a large number of parallel pearl chains of tumour cells (murine erythroleukaemic erythroblast) in a similar arrangement to that shown in Fig. 12a. The breakdown pulse of 2.7 kV cm^{-1} field strength and 20 µs length was followed about 1 min later by addition of calcium ions to induce the formation of a spherical cell. From Zimmermann and Vienken (1984).

For fusion of some cells enzyme pre-treatment may be necessary such as the enzymatic removal of the vitellin layer before fusion of sea-urchin eggs (Richter *et al.*, 1981) and removal of the glycocalix of erythrocytes by means of neuraminidase before application of the field (Zimmermann *et al.*, 1981b). Such treatment is also sometimes necessary before fusing other mammalian cells.

There is now considerable evidence that electro-fusion is a mild enough technique to give high yields of viable and reproductive hybrids. Fused sea-urchin eggs were capable of fertilization. The products underwent division, although only once (Richter *et al.*, 1981). Plant protoplast fusion products showed cell wall production and division (Zimmermann *et al.*, 1981b). Protoplasts of *Avena sativa* exhibited no change in their ATP/ADP ratios before and after fusion (Verhoek-Köhler *et al.*, 1983) although artificial lowering of the ATP levels (by inhibitors or uncouplers) caused progressive lengthening of the time required to complete fusion. One of the best examples to date is that of the fusion between protoplasts of *Saccharomyces cerevisiae* of different strains (Fig. 10) (Halfmann *et al.*, 1982). These were strain 2114 (a heterozygous diploid having a respiratory deficiency) and strain 3441 (which carries chromosomal markers). Compared with conventional methods, the yield of hybrids obtained by electro-fusion was high. (Isolation used standard selection media.) These hybrids were diploid, possessed respiratory competence and produced phototrophic spores. Therefore, these products contained only the chromosomal markers of strain 2114 and the cytoplasmic markers (of respiratory competence) from strain 3441. In this case electro-fusion gave mainly plasmogamy without karyogamy (i.e. fusion of the nuclei).

The absence of karyogamy may result from the use of heterozygous diploid yeast cells. It is known from work using PEG induced fusion of yeast cells that karyogamy only takes place if one of the strains is haploid (Maráz *et al.*, 1978; Russell and Stewart, 1979), and no fusion at all of heterozygous diploid strains has so far been reported using either the chemical or viral methods. It seems that the absence of karyogamy is an intrinsic property of the strains used, as under natural conditions, no conjugation occurs between hetero-zygous diploid strains. More recently, it has been shown (Halfmann *et al.*, 1983a) that fusion of at least some haploid yeasts can give karyogamy. When fusion of polyploid yeasts was carried out, the hybrids showed recombination, with later elimination of some chromosomes (Halfmann *et al.*, 1983b). This, and the production of so-called hybridoma cells in high yield has demonstrated that electro-fusion *can* result in karyogamy.

A. HYBRIDOMA CELLS

Hybridoma cells are capable of permanent production of antibody specifically directed against a given antigen and therefore have an enormous range of applications in medicine (serology, diagnostics and therapeutics) and technology (Goding, 1980; Galfrè and Milstein, 1981). Hybridoma production, first reported by Köhler and Milstein (1975), is based on the fusion of myeloma cells and spleen cells from appropriately immunized animals (usually mice or rats). This combination of cells was chosen in the hope that the myeloma cells would confer their ability to live indefinitely in tissue culture on the antibody-producing spleen cells (which soon die if culture is attempted).

PEG, or virus-induced fusion of individual clones of antibody-secreting cells with myeloma cells, yields immortal hybrids which constitutively secrete antibodies.

Selection of hybridoma cells depends on the use of mutant myeloma cells which lack either hypoxanthine guanine ribosyl-transferase (aza-guanine resistant) or thymidine kinase (bromodeoxyuridine resistant). Such mutants cannot grow in a medium containing aminopterin and supplemented with hypoxanthine and thymidine (HAT medium), because they cannot use the salvage pathway (Littlefield, 1964). Conversely, hybrids between such cells and spleen cells can be selected from their parents as they are the only cells that multiply in the HAT medium. Individual clones of these hybrids can be selected according to exactly which antibodies they secrete, and these antibodies are therefore of monoclonal origin. In this way, virtually unlimited quantities of homogeneous, exquisitely specific antigen can be produced, even when the immunizing antigen was not pure.

Monoclonal antibodies obtained from mouse (or rat) hybridoma cells are useful for a variety of diagnostic purposes and for the preparation of natural products useful in biotechnology (e.g. interferon). Unfortunately, the use of such antibodies in human immunotherapy is limited by possible adverse interspecies reactions. The use of immunization as a method of obtaining human lymphocytes of the required specificity is not practical since the number of lymphocytes with specificity against a particular determinant is small. The production of human derived hybridoma cells on a large scale (Croce et al., 1980; Olsson and Kaplan, 1980) therefore requires a fusion method giving a particularly high yield of hybrid cells.

The PEG method has only been able to give low yields of hybridoma cells, while the application of the electro-fusion technique has given an abundant yield. Moreover, the latter method has the considerable advantage that HAT sensitive mutants of myeloma cells are not required. This is a consequence of the facility with which the whole process may be observed under the microscope so that hybridoma cells can be individually removed from the fusion chamber by the use of micromanipulators.

The production of mouse hybridoma cells required the development of a special electro-hydraulic procedure (Vienken and Zimmermann, 1982).

Human hybridoma cells may be obtained using only the microscope slide fusion chamber described above (Bischoff *et al.*, 1982). When fusing a human myeloma with a lymphocyte cell, a field-pulse of particular parameters is used which can cause both myeloma/myeloma and myeloma/lymphocyte fusion, but not lymphocyte/lymphocyte fusion. Use of a 5 : 1 excess of lymphocytes over myeloma cells favours the fusion of myeloma/lymphocyte pairs only, and a very good yield of hybridoma cells can be achieved. The lack of lymphocyte/lymphocyte fusion may be due to the smaller radius of these cells or to less obvious differences such as in the membrane or in the cytoskeleton. With regard to the possible effects of the cytoskeleton, it must be said that no particular hindrance of fusion after the application of the field pulse has so far been seen. Even in the case of the relatively rigid turkey red blood cells, fusion followed breakdown if the cell orientation had first been made appropriate (Lucy, Vienken and Zimmermann, in preparation).

Following the initiation of fusion and after addition of mannitol warmed to 37°C, the human hybridoma cells became spherical within 10 min, and were then transferred to normal culture media after another 30 min. The first division occurred after 4 days, and three asynchronous divisions were seen in the first 3 weeks (Bischoff *et al.*, 1982). Ten hours after fusion, a portion of the cells was stained with acridine orange (Traganos *et al.*, 1977) to permit examination of their DNA content by taking cytofluorographs. Comparison was made with stained populations of myeloma cells and lymphocytes and a mixture (that had not been subjected to electro-fusion) of the two cell types. The results are shown in Fig. 14a–d. The DNA content is plotted as a function of the number of cells for B-lymphocytes in Fig. 14a, for myeloma cells in Fig. 14b, for the mixture in Fig. 14c and for the hybridoma cells in Fig. 14d.

The lymphocyte sample shows a peak with a narrow base which is characteristic of diploid cells in phases G0 or G1. As expected for lymphocytes, no significant numbers of S-phase cells can be distinguished in Fig. 14a. This should be compared with Fig. 14b where the G0/G1 peak is also found, but the large number of S-phase cells characteristic of fast-growing cultured cells (Andreeff, 1977) appears as an elevated plateau. The mixture (14c) shows exactly the pattern expected from linear addition of the two curves (14a) and (14b). This contrasts with the fusion products (Fig. 14d), which exhibit principally one peak and a slight shoulder (surviving excess lymphocytes).

The absence of the elevated plateau (which is prominent in Fig. 14b) means that myeloma cells are not present in significant numbers, and that this population is indeed predominantly hybridoma cells. The possibility that this peak may be the result of myeloma cells with a low DNA content fusing with one another was rejected by measurement of the DNA content as a function of the cell size (Bischoff *et al.*, 1982). This relationship is shown in Fig. 14e for the mixture and in Fig. 14f for the hybridoma cells. A comparison of the two figures reveals the presence of a population which is not detectable in the original cell populations. This population not only exhibits a totally different volume distribution from its parent cells, but is also characterized by a more homogeneous DNA distribution. It therefore seems reasonable to assume that this population can be attributed to hybridized cells. The volume of the hybrids indicates that fusion between two cells occurred, if the wide distribution in volume of the myeloma cells is considered.

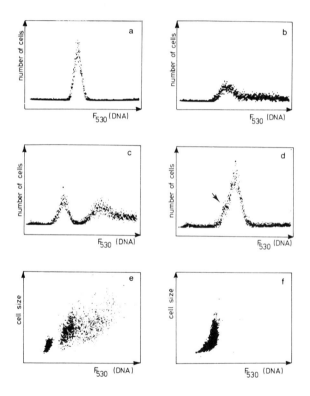

FIG. 14. Cytofluorographs of cell populations involved in the formation of hybridoma cells. The intensity of the fluorescence at 530 nm (F_{530}) of each individual cell is used to indicate the DNA content of that cell. Fluorographs (a–d) indicate the DNA distribution of the populations of (a) B-lymphocytes; (b) L 363 myeloma cells; (c) a mixture of B-lymphocytes and myeloma cells; (d) hybridoma cells assayed 10 hours after fusion. The shoulder due to surviving excess lymphocytes is arrowed. In (e) and (f) the distribution of cell size against DNA content is plotted, each dot represents one cell of (a) a mixture of B-lymphocytes and L 363 myeloma cells; (b) hybridoma cells assayed 10 hours after fusion. From Bischoff *et al.* (1982).

A final test was made to distinguish between hybridoma and fused myeloma cells using the indirect immunofluorescence technique. Hybridoma cells which have been cultivated for three weeks exhibited a positive fluorescence reaction with monoclonal anti-Ia sera. Fused L 363 myeloma cells, on the other hand, exhibit no fluorescence with anti-Ia sera.

We can therefore conclude that the cell fusion technique described here offers the possibility of hybridization of human lymphocytes producing antibodies of a given specificity with a non-immunoglobulin secreting myeloma cell line. The high yield of hybridized cells as well as the possibility of direct visualization of the cell fusion can be regarded as reasonable prerequisites for the production of human monoclonal antibodies of a given specificity.

B. FORMATION OF GIANT CELLS

It has been demonstrated that the electro-fusion technique can give high yields of hybrid cells. This is accomplished by the use of low suspension densities to obtain only nascent pearl chains having just two cells, by the use of special electrode arrangements as in the electro-hydraulic procedure, or else by taking advantage of differences in the geometric or electrical properties of the two cells (as was done when forming human hybridoma cells).

When it is required to fuse many similar cells, one possible method is to use the long pearl chains which form from high suspension densities. To accommodate longer pearl chains, a wider electrode gap becomes necessary Two factors limit this procedure to 10 or 15 cells: first, the electrode gap cannot be made wider than about 500 μm otherwise sufficient voltage for the breakdown pulse cannot be provided (pulse generators of the required speed and producing more than 150 V are not generally available*). Second, the fusion of a *pearl chain* containing many cells slows down as the number of cells included is increased. An alternative method that can fuse many, even thousands, of cells is based on the use of lateral fusion of cells between adjacent pearl chains as well as fusion within pearl chains (Scheurich and Zimmermann, 1981; Pilwat *et al.*, 1981; Zimmermann *et al.*, 1981b, 1982c; Zimmermann and Vienken, 1982; Zimmermann, 1982). At high suspension densities, pearl chains form close to each other in arrays parallel, so that intimate membrane contact now exists between cells in neighbouring chains. Lateral fusion is obtained by a rather powerful electric field pulse which is calculated on the basis of Eq. (1) to give breakdown in more than 80% of the area of a single cell. As in the case of single pearl chains, the equation is

*The apparatus available from GCA/Precision Scientific Group, Illinois, and from Krüss GmbH, Hamburg, includes a generator of pulses up to 260 volts.

used only as a guide. The breakdown and pore formation over such a large area, combined with the close contact between cells in horizontally and vertically adjacent chains, causes two- or three-dimensional fusion. It is possible in this way to produce giant cells of 100 to 1000 μm diameter. Examples of these formed from erythrocytes, plant protoplasts and tumour cells are shown in Figs 12 and 13. The giant cells become rounded within 3 to 5 min and are very stable, particularly if calcium ions are added about 1 min after the field pulse.

The cells can be mechanically transferred to electrolyte solution and subjected to microelectrode insertion. They may also be immobilized in a cross-linked polymeric matrix (Schnabl et al., 1980; Scheurich et al., 1980a) after which they can be made the subject of investigations using the pressure probe (Zimmermann, 1977a). The matrix serves to stabilize the membrane against internal excess pressure which is applied to simulate the turgor of a walled cell (see also Büchner and Zimmermann, 1982).

It has been found that the breakdown field should be applied as a series of short (less than 5 μs) pulses at intervals of about 1 s. This is also sometimes useful in the fusion of cells in pearl chains. The procedure seems to stop the rotation which may occur when the alternating field is reapplied (1 ms after the pulse).

The use of non-electrolyte during fusion is of particular importance when forming giant cells. As mentioned above, the fusion conditions may cause a large percentage of the membrane to be made permeable, so that the area available for exchange of material between cytoplasm and medium is much larger. The resulting loss of proteins from the cells could affect the viability of the cells. Non-electrolytes can prevent a substantial release of larger molecules because the secondary osmotic processes following breakdown are inhibited in the presence of non-electrolytes (Zimmermann et al., 1976b, 1980c). It is also possible that the potential for exchange of materials between cytoplasm and medium is not so great as might at first appear, because cells in the interior of a large cluster of pearl chains are surrounded by other cells rather than by medium.

Giant cells made according to the above considerations are stable and may be stored for several days without apparent change. Neither the products from fusion of 3T3 cells nor those from fusion of Friend cells (which are mouse erythroblasts transformed with Friend virus, Friend et al., 1971) are stained by Trypan blue, indicating that the membrane is at least a functional permeability barrier. A demonstration of the viability of the cells was recently provided by Pilwat et al. (1981). Friend cells will start haemoglobin synthesis 2 days after exposure to dimethylsulphoxide. Giant cells were found to react in the same way, showing that the cells certainly possess the control and metabolic mechanisms to initiate and carry out such a complex synthesis.

The controlled formation of giant cells is an ability unique to the electro-fusion method and opens novel and fascinating areas of research in cell biology and medicine. The formation of giant cells from sickled erythrocytes should be possible, so that the disorders of the membrane expected in these cells can be directly investigated using intracellular microelectrodes or the patch clamp technique (Neher *et al.*, 1978). Such a procedure could also be used to investigate the difference between the membranes of the giant cells formed from normal or tumour cells to elucidate the difference in membrane structure between these cells.

A similar procedure should allow investigation of the membranes of stomata guard cells, which are of great interest in plant physiology as they control water and CO_2 exchange. The giant cells would have to be made by fusion of protoplasts, the preparation of which has been described (Scheurich *et al.*, 1981).

VI. Rotation

As emphasized above, it is sometimes seen that rotation can occur during the electro-fusion procedure either during the arrangement of the pearl chains or immediately after the breakdown pulse. We are not going to discuss the latter effect further, but will confine ourselves to the rotation that occurs without membrane breakdown. This type of rotation is still relevant to the fusion procedure as any tendency of cells to rotate in the pearl chains will prevent the formation of stable and intimate membrane contact.

A. LINEAR FIELD

It is interesting to note from the historical standpoint that the first obser-vation of rotation was the result of a general investigation into the effect of pulses of radio frequency electric fields on cells. In these experiments, Teixeira-Pinto *et al.* (1960) observed that in a mixed suspension of *Euglena gracilis* and *Amoeba*, the *Euglena* cells were apparently drawn towards *Amoeba* cells and once they were "virtually in contact with the cell" started to rotate with an enormously rapid spin. It does not seem that the process was driven by the flagella of the *Euglena* as a similar spinning was seen when pseudopodic fragments of *Amoeba proteus* were in certain positions close to the parent cell. The spinning stopped if the field was removed, only to re-commence when the field was applied once more. Teixeira-Pinto *et al.* (1960) noted that the attraction between cells could be due to distortions of the field, but offered no specific explanation for the rotation. Füredi and Ohad (1964) found that erythrocytes subjected to a continuous 120 MHz electric field rotated. The rotation stopped if the sucrose medium was made 0.02% in NaCl.

Pohl and Crane (1971) observed spinning during dielectrophoresis experiments using yeast cells (*Saccharomyces cerevisiae*). A *small* percentage of the cells were observed to spin in response to field frequencies which varied from cell to cell, and spinning could occur in several changeable frequency bands in the range available (100 Hz to 600 kHz). Spinning was said to be possible for cells freely floating in the medium i.e. not close to an electrode or another cell, which contrasts with the earlier observation by Teixeira-Pinto *et al.* (1960) that spinning started only as a cell made a close approach to another.

Rotation of yeast cells was also reported by Mischel and Lamprecht (1980). The conditions are not explicitly stated, but it is possible to deduce that the cells were observed to rotate while dielectrophoretically adhering to an electrode, using continuous fields having frequencies up to 500 kHz. They attempted to correlate the rotation speed of the cells with the moments of inertia. This would appear to be a pointless exercise since the moment of inertia cannot affect steady rotation but only acceleration or deceleration. They also observed an apparent threshold voltage for rotation of these budded yeast cells, which probably originated in the difficulty such non-spherical cells had in starting to turn while stuck to the electrode or to other cells. Pohl (1978, 1979, 1980) put forward and developed several hypotheses to account for cell spinning. These hypotheses were those of ambipolar charging, cytoplasmic biochemical oscillators operating at radio frequencies, and "Bose-like" condensations of the charge waves resulting from these oscillators (Pohl and Braden, 1982). The proof of these hypotheses, and indeed the adoption of the cellular-spin technique by other groups, has been prevented by the difficulty of making uniform and reproducible observations.

Interest in cell rotation was stimulated when consistent and repeatable rotation of *all* cells in relatively narrow frequency ranges of an alternating field was reported (Zimmermann *et al.*, 1981c). The following experimental conditions were found to be necessary:

(1) high suspension densities;
(2) use of a medium equal in density to that of the cells;
(3) use of a cell population having only a small variation in size.

The high suspension densities gave many parallel pearl chains so that all cells had at least one other cell close and diagonally (with respect to the field) oriented to them (e.g. as in Fig. 6b). Use of isopycnic medium prevented gravity interfering with rotation. A narrow distribution of cell radius excluded the larger cells from the suspension. Larger cells undergo membrane breakdown at relatively low field strengths (see Eqs. (1) and (2)) and so stop rotation by releasing ions into the medium which raise the conductivity and thereby cause heating and turbulence.

Under these conditions it was found that *all* cells of a given type and species rotated provided that the field frequency was within relatively narrow limits. For example, in the kHz frequency range mesophyll protoplasts from leaves of *Avena sativa* and cells from permanent cell lines (e.g. Friend and 3T3 cells) gave maximum rotation in the range 20 to 40 kHz, while human erythrocytes required 80 to 100 kHz and yeast cells approximately 180 kHz. (In all cases the media used were almost non-conductive.)

Single, isolated cells were not observed to rotate. Rotation of human erythrocyte ghost cells was observed in the same frequency range as that of red blood cells. This observation and the fact that plant vacuoles and even uni-lamellar liposomes (Alonso, Chapman, Arnold and Zimmermann, unpublished results) were also observed to have the same rotation phenomenon was interpreted to mean that rotation was due to processes connected with the membrane alone. Interactions with organelles or cytoplasmic oscillators, such as postulated by Pohl (1978, 1979, 1980), are definitely not required. More detailed investigation of the multi-cell rotation (Holzapfel *et al.*, 1982) showed that:

(1) At reasonable field strengths (200 V cm⁻¹ for *Avena* protoplasts) the rate of cell rotation is a few Hz.

(2) The speed of rotation has a square dependence on the field strength, and not a linear one as claimed by Mischel and Lamprecht (1980) or as expected from the ambipolar charging theory of Pohl (1978).

(3) As far as could reasonably be seen, reduction of field strength merely resulted in the rotation speed becoming very slow and eventually indistinguishable from rotatory Brownian motion; that is, no threshold was observable.

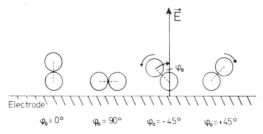

FIG. 15. The dependence of cell rotation on the orientation of two cells with respect to the field. The line joining the centres of the cells makes an angle defined as φ_0 with the electric field vector *E*. If φ_0 is exactly 0° or 90°, no rotation can result When $\varphi_0 = 45°$ both cells rotate with maximum speed in the same sense (clockwise if φ_0 is positive, anticlockwise if φ_0 is negative). From Holzapfel *et al.* (1982).

(4) Rotation was never observed for single free cells. A minimum of two is required, and they must be close together and not oriented either

exactly parallel or exactly perpendicular to the field (Fig. 15). Maximum rotation occurred when the pearl chains or doublets were oriented at 45° to the field direction (e.g. Fig. 6b). Indeed it was very often found that the chains took on this orientation when the alternating field was of the appropriate frequency for rotation (see above). Of the the two possible orientations, one always gives clockwise rotation of all the cells in a chain with that orientation, whereas chains orientated the other way contain cells which rotate anticlockwise (Fig. 15).

(5) A small cell rotating next to a large cell rotated much faster than did the large cell.

(6) The rotation rate of two cells was observed to decrease as the separation between them increased. Continuation of this process to the limit shows that a single free cell should not rotate. Single cells close to electrodes or polarizable debris cannot be regarded as "free" because they will be within the local field induced by those objects, and rotation can result solely from this induced field.

(7) The optimum frequency range observed for each species could be changed by chemical treatment of the cells (by glutaraldehyde or enzyme treatment) or by the addition of membrane-soluble ions (e.g. diphenyl-picrylamine). In the case of glutaraldehyde, rotation often stopped completely as the cells appeared to stick together.

The outcome of these findings was the conclusion that multi-cell rotation in the kHz range is the consequence of a polarization process at the cell membrane which is explicable in terms of established macroscopic physical processes. The proposal of the existence of intrinsic cellular R.F. oscillations (Pohl, 1982) and other hypothetical ideas are unnecessary.

Holzapfel et al. (1982) assumed that the dipoles induced in adjacent cells could interact with each other. Analysis of this interaction showed that within a certain range of frequencies, maximum average torque and hence rotation should result. The remarkable dependence of rotation on the orientation as shown in Fig. 15 and other results listed above were fully explained. The principles of the multi-cell analysis are too complicated to present here in detail, but the rather similar and much simpler theory of *single* cell rotation in a *rotating* field is presented in the following section.

Briefly, the theory of Holzapfel et al. (1982) shows that a torque acts on each of the two cells due to dipole-dipole interaction between adjacent cells when the period of the field (period is the inverse of frequency) is of the same order as the relaxation time (τ seconds) of the dipole induction. If the cells are free to turn, then the fastest spinning will occur at a field frequency (f Hz) given by:

$$2\pi f\tau = 1 \tag{6}$$

As discussed earlier, the predominant cellular relaxation mechanisms operating in low conductivity media between 10 kHz and 1 MHz are probably membrane charging processes. The relaxation time, τ, of this process for a single cell is given by (Schwan, 1957; Jeltsch and Zimmermann, 1979):

$$\tau = a\, C_m\, (\rho_i + \tfrac{1}{2}\rho_e) \tag{7}$$

where a is the radius of the cell in cm, C_m is the specific membrane capacitance in μF cm^{-2}, and ρ_i and ρ_e are the resistivities of the cell interior and the medium, respectively. The membrane resistance is assumed to be high enough to be neglected. Otherwise, see Zimmermann and Arnold (1983) and Arnold *et al.* (1984).

An equivalent circuit for the cell and medium is given in Fig. 16, and it can be seen that the cell is being treated as a simple resistance-capacitance charging circuit driven by the field.

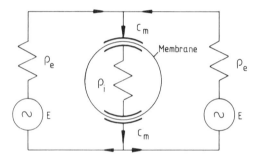

FIG. 16. Diagram of the equivalent circuit of a cell and surrounding medium to illustrate the factors that determine the time constant for induced charging of the membrane. The alternating field (E) set up in the medium drives current around the electrical circuit consisting of the external and internal solutions of resistivities ρ_e and ρ_i respectively which can be considered to be in series with two capacitances representing the membrane of the cell (having specific capacity C_m). This illustration is not intended as the basis for quantitative calculation.

For cells suspended in the low-conductivity media, as is usual in rotation experiments, then $\rho_i \ll \rho_e$ and so:

$$\tau = \tfrac{1}{2}a \cdot C_m \cdot \rho_e \tag{8}$$

If the membrane charging process is the only significant cause of dielectric dispersion at the frequency f, then the τ in Eq. (6) is the same quantity as in Eqs (7) and (8). Application of these equations derived for a single cell to multi-cell rotation is something of an approximation, but substitution of the appropriate values for *Avena* protoplasts into Eq. (8) gives agreement with the experimentally determined optimum frequency. When rotation of *Avena* protoplasts in media of relatively high conductivity was studied, the effects of internal and external conductivities expected from Eqs (7) and (8) were seen (Küppers and Zimmermann, unpublished results). During long experiments ions can leak out of the cells and change the internal or external conductivities, so that shifts in optimum frequency may be noticed.

In liposomes made from a special lipid polymerized *in situ*, the external and internal conductivities could be changed together (Hub *et al.*, 1982a). When the conductivities were changed, the optimum frequency for rotation was seen to shift in the expected direction.

According to Eqs (7) and (8) in combination with Eq. (6), larger cells should have lower optimum field frequencies than smaller cells. It is therefore important in multi-cell rotation to use cell populations of practically uniform size or the width of the optimum frequency ranges may become too great.

B. ROTATING FIELD

The theory of dipole-dipole interaction (Holzapfel *et al.*, 1982) can be interpreted to mean that the applied field is phase shifted by the dipoles in correctly oriented cells to produce a rotating field locally, and that this field produces the torque. This implies that application of a rotating field or a field having a rotating component to a single cell should result in rotation. This theory was recently verified by Arnold and Zimmermann (1982a,b) and by Pilwat and Zimmermann (1983). We would like to elaborate on this subject in more detail because the use of single cell rotation eliminated the complications caused by the interaction of two dissimilar cells which had made quantitative measurements using the multi-cell technique very difficult.

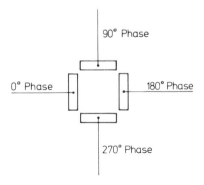

FIG. 17. Electrode configuration used to produce a rotating field by the superimposition of two sinusoidal fields differing in phase by 90°. Voltages of the correct phases are produced by the unit shown in Fig. 18. From Arnold and Zimmermann (1982b).

In the short time since rotating field apparatus has been available the method has not only made these measurements possible but has also made available additional information regarding the rotation direction of cells, which was not apparent from the earlier method. A rotating electric field is one which has a constant amplitude but rotates at audio or radio frequency (e.g. a 100 kHz rotating field makes 10^5 revolutions per second). It is somewhat similar to the "circular polarization" increasingly used in broadcasting at higher frequencies, which is electromagnetic rather than electric in character.

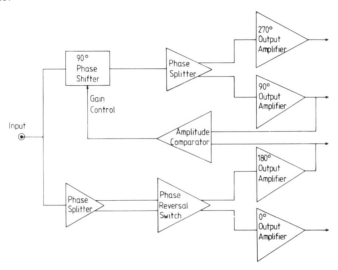

FIG. 18. Generator to provide four equal alternating voltages having progressive 90° phase shifts. Frequencies between 5 Hz and 5 MHz are covered by six decade ranges. From Arnold and Zimmermann (1982b).

A pure rotating field can be generated by several means, but the simplest is to apply two equal sinusoidal voltages of identical frequency (but having 90° phase difference) to two sets of electrodes set in a square configuration (Fig. 17). A block diagram of electronics to accomplish this over the 5 Hz to 5 MHz range is shown in Fig. 18. A full description is beyond the scope of this chapter. The chamber used is shaped like an open cube, the four vertical sides are electrodes which are bonded to a microscope slide. Cells are usually trapped half-way up the chamber by filling it with solutions of greater and lesser density (for *Avena* protoplasts, 0.5 M sucrose and mannitol are suitable). This method also prevents the cells having any contact with the microscope slide, and so prevents problems with friction or sticking. The cells are observed by microscope through the base of the slide; measurements are taken only on those cells equidistant from the electrodes. The reason for taking these precautions is that although rotation can be observed every-where in the chamber, the following theory concerning the rotating field only applies to the centre of the chamber.

A simplified mathematical treatment of the effects behind rotating-field induced rotation follows to allow semi-quantitative conclusions to be drawn. If a cell is suspended in medium in between the electrodes, it will be subjected to the sum of the two fields, each having frequency f and peak value E^0:

$$E_y = E^0 \cos 2\pi ft \tag{9}$$

and

$$E_x = E^0 \cos (2\pi ft - 90) = E^0 \sin 2\pi ft \tag{10}$$

The resultant amplitude can be found by squaring and summing these terms and is found to be equal to E^0 (independent of time). The angle (ϑ radians) of the resultant is found to vary as:

$$\vartheta = 2\pi ft \tag{11}$$

The resultant field therefore fulfils the description of the rotating field described. The amplitude of the rotating field can be easily measured by taking the peak value of either of the component fields (e.g. by use of an oscilloscope).

The effect of this field on a freely suspended cell depends on the frequency of the field and other factors. To see why this is true, we must consider the mechanism of dipole induction. Assuming, for the present, that the same membrane charging mechanism (Fig. 16) as described for the multi-cell rotation is operating, then calculation shows that maximum torque and hence rotation speed are once again developed when

$$2\pi f\tau = 1 \tag{12}$$

and Eqs (7) and (8) are directly applicable.

This can be qualitatively understood as follows. The resultant dipole moment vector M of the cell can be described as the sum of components (M_y and M_x) induced by the two component fields (Fig. 19). M lags behind the rotating field by an angle φ. This is a consequence of the membrane capacitance (Eq. (7)) taking time to charge through the resistance in the "circuit" of Fig. 16. The *torque* (N) on the cell is given by:

$$N = E \cdot M \cdot \sin \varphi \tag{13}$$

According to Ogawa (1961) the magnitude of the torque developed on a body by a rotating field will also depend on the absolute dielectric constant of the medium. Only the vacuum dielectric constant was considered by Holzapfel *et al.* (1982) and by Arnold and Zimmermann (1982b), therefore the relative dielectric constant of the medium should be used as a correction factor.

At low frequencies the induced voltage has a maximum value, but there is little phase lag (sin $\varphi = 0$). At high frequencies, M is very small (the membrane capacitance has a very low reactance) and the phase lag is 90°, so sin φ has its maximum value (unity). When Eq. (12) is satisfied, M is still large (0.707 of the low frequency value) and the phase lag is 45° (therefore sin $\varphi = 0.707$ as well). This gives the maximum torque.

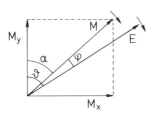

FIG. 19. Relationship between the rotating electric field E and the rotating induced dipole M having a phase lag φ. M is the resultant of components M_x and M_y which are induced by the two superimposed electric fields E_x and E_y (not shown, for the sake of clarity). The angles α and ϑ are both increasing with time at the rate $2\pi f$ radians per second, where f is the field frequency. From Arnold and Zimmermann 1982b). The theory of Ogawa (1961) implies that not only the magnitudes of M_x and M_y, but also the size and sign of the angle ϑ depend on the complex dielectric constants of the medium and cell.

Equation (8) can be re-arranged after substituting the conductivity of the medium (σ_e S cm^{-1}) for the reciprocal of the resistivity, giving:

$$f \cdot a = \frac{\sigma_e}{\pi \cdot C_m} \tag{14}$$

The theory predicts that the frequency giving fastest rotation (which we term the characteristic frequency of a particular cell) should increase linearly with conductivity, as long as the membrane charging mechanism is operating. In practice, the results are fully in accord with the above predictions (Arnold and Zimmermann, 1982a,b).

The following are observations made with this technique.

(1) Application of the rotating field to a chamber prepared as above results in the rotation of single cells. In the kHz range rotation is in the opposite direction to that of the field but reversing the direction of the field rotation immediately reverses the rotation of the cell. This is not possible with methods of cell rotation using two electrodes so far reported. The axis of rotation is perpendicular to the plane of the field.

(2) Field strengths of *only* 20 V cm^{-1} are adequate to produce rotation speeds of a few Hz. The increased sensitivity of this method over the multi-cell method is probably due to several factors such as the friction and hydrodynamic losses encountered when two cells are rotating in contact with one another. Indeed, in the case of cells that will not give multi-cell rotation because they are too "sticky", the use of single cell rotation avoids membrane contact and leads to free rotation.

It must also be remembered that the rotating field (considered to be present locally) in the multi-cell method is a consequence of the interaction between the two induced dipoles. The effective rotating field strength so produced cannot necessarily be compared with a directly imposed rotating field.

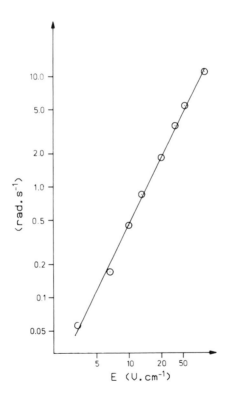

FIG. 20. Double logarithmic plot of the relationship between rotation speed and rotating field strength for a protoplast of *Avena sativa* suspended at the boundary between 0.5 M mannitol and 0.5 M sucrose solutions, both having conductivity 6.4 × 10^{-5} S cm^{-1}. The line is the least-squares fit to the logarithmic data and yields a gradient of 2.00 ± 0.04. From Arnold and Zimmermann (1982b).

(3) Once again a square law dependence of rotation speed on field strength is found (Fig. 20). This is to be expected from Eq. (13) because the induced dipole (which is itself proportional to the field) is multiplied with the field to give the torque (N). That is:

$$N = c \cdot E^2 \tag{15}$$

where c is a constant. Assuming that the frictional couple on such a sphere is proportional to the speed of rotation (Lamb, 1906), we can say that the rotation rate is also proportional to E^2.

(4) No threshold has been observed for single-cell rotation for free cells. If cells are allowed to settle on the microscope slide, they may after some time adhere to it. Under these conditions an apparent threshold is of course seen but the cell may not show the same "threshold" on reducing the voltage as on increasing it.

(5) The rotating field is symmetrical, so no orientation dependence is expected or seen. However, when the suspension density is high, cells often form aggregates of up to 20 or more cells. All cells continue to rotate as before at the same time as the whole aggregate rotates. Naturally, rotation of individual cells under these conditions is hampered, especially if they are not spherical. The aggregate is usually planar, i.e. it takes up the maximum area in the plane of the field. It is possible that this is due to the step in density used to stop the cells from sinking, although the observation that in the absence of field the distribution of cells extends over a considerable vertical range suggests that this is not true. Very highly aspherical objects show a form of orientation in which the long axis rotates in the plane of the field. However, long chains (10 or more cells) of *Bacillus megaterium* are occasionally seen to adopt a helical motion resembling that of an Archimedian screw.

(6) The characteristic frequency is found to be a linear function of conductivity over the range 30 kHz to almost 1 MHz (for *Avena* mesophyll protoplasts). Figure 21 shows the frequency plotted against medium conductivity. If the radius dependence is also considered (as indicated in Eq. (14)), the scatter of the points is much reduced. The axes are logarithmic to better display the wide range of data. The membrane capacitance may be determined from the gradient of a linear plot by use of Eq. (14). The value obtained for *Avena* mesophyll protoplasts from the data shown in Fig. 21 was $0.48 \pm 0.07 \ \mu F \ cm^{-2}$. It must be remembered in the case of mesophyll protoplasts that there are two membranes involved due to the presence of a large vacuole with at least 70% of the radius of the cell. If these two membranes are modelled as two equal capacitors in series the individual specific capacitances

are approximately 1 μF cm⁻² each. This agrees with the values found
for other biological membranes (Pauly, 1962, 1963; Cole, 1968;
Bernhardt and Pauly, 1974; Läuger *et al.*, 1981). Later measurements
on *Avena* protoplasts gave a slightly lower value of 0.39 ± 0.03 μF
cm⁻², whilst preliminary work on protoplasts from which the vacuole has
been removed indicate that the plasmalemma, taken by itself, exhibits
a capacitance of 0.6–0.7 μF cm⁻² (Mehrle, Wendt, Arnold, Hampp,
Zimmermann, in preparation). These results confirm the effective
series connection of the tonoplast and plasmalemma, at least in these
cells. In addition, measurements on vacuoles isolated from mesophyll
cell and guard cell protoplasts of *Vicia faba* (Arnold, Schnabl and
Zimmermann, in preparation) have shown that the tonoplast membrane capacitances are indeed in the range 0.9 to 1.1 μF cm⁻².

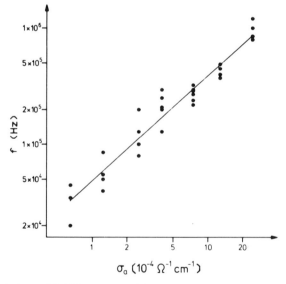

FIG. 21. The variation of field frequency giving maximum rate of cell rotation with
the conductivity of the medium for 37 mesophyll protoplasts of *Avena sativa*.
Double logarithmic plot, the line is the least squares fit to the logarithmic data.
The slope of the line of a linear plot leads to a mean membrane capacitance for the
protoplasts of 0.48 ± 0.07 μF cm⁻¹. From Arnold and Zimmermann (1982b).

Rotation of single cells in the rotating field has also been observed when
using Friend cells, large unilamellar liposomes prepared according to Hub
et al., 1982b, or the very large mesophyll protoplasts from *K. daigremontiana*.
In all these cases preliminary results indicate individual membrane capacitances of approximately 1 μF cm⁻². It has recently been shown that mobile

charges, which may be part of a transport system, occur in cell membranes and increase the capacitance (Benz and Zimmermann, 1983, Zimmermann *et al.*, 1982a). Therefore, in the future cell rotation may be a way of detecting these charges.*

It has been mentioned that the direction (clockwise or anti-clockwise) of the rotating field can be altered, and that the cells reverse direction in response. In all the observations in this section the cells were seen to rotate in the opposite direction to that of the field. If it can be assumed that the direction of rotation is determined by the polarizability of the cell, this suggests that the cells have negative polarizability as was also assumed in the theory of multi-cell rotation. It must be remembered that the polarizability of the cell is affected by the counterions in the medium. The interaction between the counterions and a rotating cellular dipole may be the cause of this negative polarization relative to the medium.

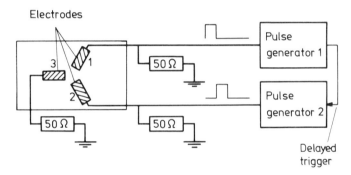

FIG. 22. Experimental apparatus for the generation of a discontinuous rotating electric field. The pulses from generator No. 2 follow at predetermined intervals after the pulses from generator No. 1 finish. Each generator is terminated by a 50 ohm resistor, as is the third electrode for reasons of symmetry. The electrodes are of platinum wire bonded to a "Perspex" microscope slide. From Pilwat and Zimmermann (1983).

The sine wave method described above is not the only way of generating a rotating field. It is also possible to sequentially apply pulses to several radially arranged electrodes. This discontinuous rotating field has also produced single-cell rotation (Pilwat and Zimmermann, 1983). Apparatus to produce this kind of field is shown in Fig. 22. The pulses are supplied to two of the three electrodes by two pulse generators, one of which triggers the other. The rise and fall times of the pulses are 10 ns.

*(Added in proof): The lipid-soluble ion formed from dipicrylamine has indeed been found to more than double the effective capacitance of swollen thylakoid vesicles (Arnold *et al.*, 1984).

In experiments with Friend cells allowed to settle on to the "Perspex" slide, the fastest rotation occurred when each pulse was 3–5 μs long and the pulse repetition rate (of each generator) was 5–10 kHz. Reversing the direction of field rotation by reversing the pulse sequence reversed the cell rotation. The field strength was in the range 120–480 V cm^{-1}, which is comparable to that used in multi-cell rotation, but having only a small heating effect as the "duty cycle" (percentage "field on" time) of the pulsed field is approximately 6%. Each pulse can be expected to polarize the cell membrane and induce a dipole as discussed above. The dipole will decay with time constant τ, which is the same as that described earlier (Eq. (7)).

The next pulse, which produces a field at an angle of 120° to the first, will result in the production of a torque the magnitude of which depends on the extent to which the dipole has decayed in the interval between the pulses. It is therefore expected that as the interval between the pulses is increased, the dipole will decay more between the "induction" and "drive" pulses, so the rotation rate will decrease.

In experiments where the interval between the two pulses was varied, this effect proved to be the case. The fastest rotation resulted when using pulses supplied to electrode 2 which immediately followed those applied to electrode 1. As the interval between pulses was increased, the rotation rate decreased until the pulses were equally spaced (50 μs spacing at 10 kHz rate per generator), when no rotation occurred. Increasing the separation still further was equal to reversing the direction of the field rotation, and indeed the cell rotation restarted but in the opposite direction. The cell rotation was always in the opposite direction to that of the field. This is consistent with the single cell rotation induced by a continuous rotating field in this frequency range.

The rate of rotation was found to depend on the field strength raised to approximately the power 2.5. This is a stronger dependence than was expected from the theories of multi- or single-cell rotation using continuous fields, where a square-law dependence was predicted and observed. It is thought that the difference in this case is that the motion of the cells is discontinuous, i.e. the cells "step" at a rate of 10 kHz. This appears to be a continuous rotation. The pulsed nature of the motion may increase the effects of friction at lower rates of rotation, leading to a stronger than expected dependence of rotation rate on applied voltage.

It is possible in principle to derive a value for the membrane charging time from the dependence of rotation rate on the time between the pulses. It must be remembered in this and other methods using square pulses, that the Fourier spectrum of frequencies equal to such pulses is very wide. It is therefore possible that other dielectric processes with shorter time constants than the membrane charging are also excited by the pulses, and these will produce their own dipoles and contribute to the rotation. It is even possible that such processes can be detected by the use of square pulses of variable separation, which would make the method a form of time domain frequency analysis. A discussion of some other possible dielectric processes follows.*

C. FURTHER APPLICATIONS OF CELL ROTATION

The vast majority of rotating field observations has been made at frequencies between about 1 kHz and 200 kHz. In this range, all single cells seen to rotate at a particular frequency have done so in a direction counter to that of the field. The weight of evidence presented above is that this rotation is due to a dipole induced by membrane charging, which is also the case in multi-cell rotation (Holzapfel *et al.*, 1982). The frequency giving maximum rotation is at the centre of the dispersion in dielectric constant due to this membrane charging effect (see Eqs (6) to (8)).

Rotation has also been observed at much lower (about 100 Hz) and much higher frequencies (up to 100 MHz) by several groups (Füredi and Ohad, 1964; Pohl and Crane, 1971; Hub *et al.*, 1982a; Zimmermann *et al.*, unpublished work). The membrane charging mechanism cannot be responsible for rotation in these frequency ranges, but as we have already described (see "Pearl-chain formation"), the dielectric constant of biological materials shows several dispersions at widely different frequencies. It may be that these other rotations each correspond to a different mechanism of polarization. If this is true, the equations derived above will still be true except that the appropriate function for τ must be used.

We have already discussed how the very low frequency α-dispersion seems to be due to the movement of counter ions close to the cell surface. In the case of bacterial cells, it seems that the movements of counter ions in the cell wall are important (Einolf and Carstensen, 1973). The same may be considered to apply to other walled cells, such as yeast.

We have detected rotation of *Saccharomyces cerevisiae* cells in both linear and rotating fields, and of *Bacillus megaterium* chains in the rotating field in the frequency range below 100 Hz. The rotation in these cases is in the same direction as that of the field ("co-field"). If it can be assumed that the direction of rotation is determined by the polarizability of the cell, this suggests the

---

*(Added in proof): After preparation of this manuscript, two other groups have reported studies using the rotating field technique. Glaser *et al.* (1983) report measurements on erythrocytes, protoplast and vacuoles from *Avena sativa*, and suspension-cultured *Beta vulgaris* cells. An effect of ionophores could be demonstrated. The results are largely in agreement with the theory and results presented here and in Zimmermann and Arnold (1983). Mischel *et al.* (1982) report that although dead yeast cells exhibited a single rotation peak between 1 and 100 kHz, live cells exhibit two peaks of rotation, apparently in the same direction as the field. Although the method of killing the yeast cells is not reported, it seems reasonable to suppose that the rotation exhibited only by living cells is dependent on the possession of an intact membrane.

cells have positive polarizability at these frequencies. Co-field rotation is also observed at frequencies between 500 kHz and at least 3 MHz for all cells so far examined, and also when using liposomes (uni- or multi-lamellar). In this frequency range, interference from the membrane-charging mechanism (giving counter-field rotation) is seen when the medium conductivity is relatively high (see Eq. (14)). The co-field mechanism may be of molecular dipole orientation, which can also account for rotation of erythrocytes or *Avena* cells seen at approximately 100 MHz. However, it must not be forgotten that the frequency-dependent, complex dielectric constant (see Eq. (5)) *of the medium* will also play a role in determining the sign and magnitude of the torque acting on the cell.

The mechanism of some of these rotation maxima may be made clearer if simpler model systems are examined. It has recently been observed (Küppers *et al.*, 1983) that "Chelex" particles (polystyrene beads modified with iminodiacetic acid, giving the equivalent of immobilized "EDTA") exhibit a high-frequency rotation which is specifically affected by the binding of divalent ions (e.g. calcium or cupric ions). This would seem to indicate that the rotation is due to the dielectric properties of the side chains, which presumably become relatively immobile when such ions are complexed. Studies of the effects of variables such as temperature, viscosity and pH on rotation can also be expected to distinguish between the various mechanisms underlying different rotation maxima. Once these mechanisms have been identified, studies of the dielectric properties of *single* cells will be possible for the first time. The use of the cell itself to indicate at which frequencies a dielectric effect is occurring removes the problem of electrode polarization, which is usually very troublesome in low frequency dielectric studies. The small amount of material necessary (one cell, if carefully handled) enables work to be done even when the preparation of material is very difficult. Therefore, it is expected that the study of rotation will yield new information on the membrane and other parts of cells.

VII. Conclusion

It is difficult to properly assess the potential of electro-fusion and cell rotation at this stage, but the applications already found in membrane research, medicine and technology are very promising. It is certain that an understanding of the effects of applied electric fields on the membrane structure at the molecular level will require further research, particularly in combination with such techniques as ESR or with optical methods (e.g. fluorescence depolarization or light scattering). Hopefully, such measurements will also give insight into the full role of the intrinsic electric field, as this is of the same order as that produced by the breakdown pulse leading to fusion.

We believe that many processes in the membrane are controlled by the intrinsic field, which is of unknown distribution across the membrane because only external measurements (i.e. those of the transmembrane potential) have been made so far. The superposition of external fields (the amplitude and frequency of which can be chosen selectively) may offer a solution to this problem.

The measurements and results reported here also include effects that involve weaker fields. Further research in this direction may help to explain some very important but unexplained effects of weak electric (and magnetic) fields such as bone healing, limb regeneration and the stimulation of plant growth (see e.g. Korenstein, 1984). The complications inherent in some of these studies are considerable, but we hope that the challenge will stimulate increasing numbers of laboratories to use electrical effects in the study of membrane biophysics.

VIII. Acknowledgements

We are very grateful to Dr G. Pilwat for helpful discussions and photographic assistance and to Mrs S. Alexowsky for typing the manuscript. This work was supported partly by grants from the BMFT (No. 03 7266) and from the Deutsche Forschungsgemeinschaft (Zi 99/8) to U. Z.

References

Abidor, I. G., Arakelyan, V. B., Chernomordik, L. V., Chizmadzhev, Yu. A., Pastushenko, V. F. and Tarasevich, M. R. (1979). *Bioelectrochem. Bioenerg.* **6**, 37–52.

Ahkong, Q. F., Tampion, W. and Lucy, J. A. (1975a). *Nature* **256**, 208–209.

Ahkong, Q. F., Fisher, D., Tampion, W. and Lucy, J. A. (1975b). *Nature* **253**, 194–195.

Ahkong, Q. F., Botham, G. M., Woodward, A. W. and Lucy, J. A. (1980). *Biochem. J.* **192**, 829–836.

Akeson, S. P. and Mel, H. C. (1981). *Biophys. J.* **33**, 76a.

Allan, R. S. and Mason, S. G. (1962). *Proc. Roy. Soc. A.* **267**, 45–61.

Allen, M. and Soike, K. (1966). *Science* **154**, 155–157.

Almers, W. (1978). *Rev. Physiol. Biochem. Pharmacol.* **82**, 96–190.

Anderson, A. K. and Finkelstein, R. (1919). *J. Dairy Sci.* **2**, 374–406.

Andreeff, M. (1977). "Zellkinetik des Tumorwachstums." Thieme Copythek, Thieme Verlag, Stuttgart.

Arnold, W. M. and Zimmermann, U. (1982a). *Naturwissenschaften* **69**, 297.

Arnold, W. M. and Zimmermann, U. (1982b). *Z. Naturforsch.* **37c**, 908–915.

Arnold, W. M., Wendt, B., Zimmermann, U. and Korenstein, R. (1984). *Biochim. Biophys. Acta* (submitted).

Auer, D., Brandner, G. and Bodemer, W. (1976). *Naturwissenschaften* **63**, 391.

Baker, P. F. and Knight, D. E. (1979). *TINS* **9**, 288–291.

Barraneo, S. D., Spadaro, J. A., Berger, T. J. and Becker, R. O. (1974). *Clin. Orthopaed. Rel. Res.* **100**, 250–255.
Beatlle, J. M. and Lewis, F. C. (1925). *J. Hyg.* **24**, 123–127.
Benz, R. and Conti, F. (1981). *Biochim. Biophys. Acta* **645**, 115–123.
Benz, R. and Zimmermann, U. (1980a). *Biochim. Biophys. Acta* **597**, 637–642.
Benz, R. and Zimmermann, U. (1980b). *Bioelectrochem. Bioenerg.* **7**, 723–739.
Benz, R. and Zimmermann, U. (1981a). *Biochim. Biophys. Acta* **640**, 169–178.
Benz, R. and Zimmermann, U. (1981b). *Planta* **152**, 314–318.
Benz, R. and Zimmermann, U. (1983). *Biophys. J.* **43**, 13–26.
Benz, R., Beckers, F. and Zimmermann, U. (1979). *J. Membrane Biol.* **48**, 181–204.
Berg, H., Forster, W., Jakob, H.-E., Jungstand, W. and Mühlig, P. (1978). *Studia Biophysica* **74**, 31–32.
Bernhardt, J. (1979). *Z. Naturforsch.* **34c**, 616–627.
Bernhardt, J. and Pauly, H. (1973). *Biophys.* **10**, 89–98.
Bernhardt, J. and Pauly, H. (1974). *Rad. and Environm. Biophys.* **11**, 91–109.
Bischoff, R., Eisert, R. M., Schedel, I., Vienken, J. and Zimmermann, U. (1982). *FEBS Lett.* **147**, 64–68.
Brower, D. L. and Giddings, T. H. (1980). *J. Cell Sci.* **42**, 279–290.
Büchner, K.-H. and Zimmermann, U. (1982). *Planta* **154**, 318–325.
Büschl, R., Ringsdorf, H. and Zimmermann, U. (1982). *FEBS Lett.* **150**, 38–42.
Chandler, D. E. and Heuser, J. E. (1979). *J. Cell. Biol.* **83**, 91–108.
Chandler, D. E. and Heuser, J. E. (1980). *J. Cell Biol.* **86**, 666–674.
Chernomordik, L. V. and Abidor, I. G. (1980). *Bioelectrochem. Bioenerg.* **7**, 617–623.
Chizmadzhev, Yu. A. and Abidor, I. G. (1980). *Bioelectrochem. Bioenerg.* **7**, 83-100.
Ciezynski, T. (1967). *Acta Morphol. Acad. Sci. Hung.* **15**, 309–312.
Cole, K. S. (1968). "Membranes, Ions and Impulses." University of California Press, Berkeley and Los Angeles.
Coster, H. G. L. (1965). *Biophys. J.* **5**, 669–686.
Coster, H. G. L. and Smith, J. R. (1980). In "Plant Membrane Transport: Current Conceptual Issues". (eds R. M. Spanswick, W. J. Lucas and J. Dainty), pp. 607–608, Elsevier/North-Holland Biomedical Press, Amsterdam.
Coster, H. G. L. and Zimmermann, U. (1975a). *Biochim. Biophys. Acta* **382**, 410–418.
Coster, H. G. L. and Zimmermann, U. (1975b). *J. Membrane Biol.* **22**, 73–90.
Coster, H. G. L. and Zimmermann, U. (1975c). *Z. Naturforsch.* **30c**, 77–79.
Coster, H. G. L. and Zimmermann, U. (1976). *Z. Naturforsch.* **31c**, 461–463.
Coster, H. G. L., Steudle, E. and Zimmermann, U. (1977). *Plant Physiol.* **58**, 636–643.
Croce, C. M., Linnenbach, A., Hall, W., Steplewski, Z. and Koprowski, H. (1980). *Nature* **288**, 488–489.
Crowley, J. M. (1973). *Biophys. J.* **13**, 711–724.
Doevenspeck, H. (1961). *Fleischwirtschaft* **13**, 986–989.
Doevenspeck, H. (1962). *Arch. Lebensmittelhyg.* **13**, 25.
Donlon, J. A. and Rothstein, A. (1969). *J. Membrane Biol.* **1**, 37–52.
Dukhin, S. S. (1971). *Surf. Coll. Sci.* **3**, 83–165.
Einolf, C. W., Carstensen, E. L. (1973). *Biophys. J.* **13**, 8–13.
Franz, W. (1956). In "Flügge's Handbuch, Band XVII, Dielektrica," pp. 155–263, Springer-Verlag, Berlin.
Friend, A. W. Jr., Finch, E. D. and Schwan, H. P. (1975). *Science* **187**, 357–359.

Friend, C., Sher, W., Holland, J. G. and Sato, T. (1971). *Proc. Natl Acad. Sci. USA* **68**, 378–382.

Füredi, A. A. and Ohad, J. (1964). *Biochim. Biophys. Acta* **79**, 1–8.

Füredi, A. A. and Valentine, R. C. (1962). *Biochim. Biophys. Acta* **56**, 33–42.

Galfrè, G. and Milstein, C. (1981). *Meth. Enzym.* **73**, 3–46.

Gauger, B. and Bentrup, F. W. (1979). *J. Membrane Biol.* **48**, 249–264.

Gingell, D. and Ginsberg, L. (1978). *In* "Membrane Fusion". (eds G. Poste and G. L. Nicolson), pp. 791–833, Elsevier/North-Holland, Amsterdam.

Glaser, R., Pescheck, Ch., Krause, G., Schmidt, K. P. and Teuscher, L. (1979). *Z. Allg. Microbiol.* **19**, 601–607.

Glaser, R., Fuhr, G. and Gimsa, J. (1983). *Studia Biophys.* **96**, 11–20.

Goding, J. W. (1980). *J. Immunol. Meth.* **39**, 285–308.

Griffin, J. L. (1970). *Exp. Cell Res.* **61**, 113–120.

Griffin, J. L. and Ferris, C. D. (1970). *Nature* **226**, 152–154.

Griffin, J. L. and Stowell, R. E. (1966). *Exp. Cell Res.* **44**, 684–688.

Gutknecht, J., Hartmann, F., Kirmaier, N., Reis, A. and Schöberl, M. (1981). *GIT Fachz. Lab.* **25**, 472–481.

Halfmann, H. J., Röcken, W., Emeis, C. C. and Zimmermann, U. (1982). *Curr. Genet.* **6**, 25–28.

Halfmann, H. J., Emeis, C. C. and Zimmermann, U. (1983a). *Arch. Microbiol.* **134**, 1–4.

Halfmann, H. J., Emeis, C. C. and Zimmermann, U. (1983b). *FEMS Microbiol. Lett.* **20**, 13–16.

Hamilton, W. A. and Sale, A. J. H. (1967). *Biochim. Biophys. Acta* **148**, 789–800.

Helfrich, W. (1973a). *Z. Naturforsch.* **28c**, 693–703.

Helfrich, W. (1973b). *Physics Letts* **43A**, 409–410.

Helfrich, W. (1974). *Z. Naturforsch.* **29c**, 182–183.

Heller, J. H. and Teixeira-Pinto, A. A. (1959). *Nature* **183**, 905–906.

Holzapfel, Chr., Vienken, J. and Zimmermann, U. (1982). *J. Membrane Biol.* **67**, 13–26.

Hub, H.-H., Ringsdorf, H. and Zimmermann, U. (1982a). *Angew. Chem.* **94**, 151–152; *Int. Ed. Engl.* **21**, 134–135.

Hub, H.-H., Zimmermann, U. and Ringsdorf, H. (1982b). *FEBS Lett.* **140**, 254–256.

Hui, S. W., Stewart, T. P., Boni, L. T. and Yeagle, P. L. (1981). *Science* **212**, 921–922.

Hülsheger, H., Potel, J. and Niemann, E.-G. (1981). *Radiat. Environ. Biophys.* **20**, 53–65.

Ingram, M. and Page, L. J. (1953). *Proc. Soc. Appl. Bact.* **16**, 69–87.

Jaffe, L. F. (1977). *Nature* **265**, 600–602.

Jakob, H. E., Forster, W. and Berg, H. (1981). *Z. Allg. Microbiol.* **21**, 225–233.

Jausel-Hüsken, S. and Deuticke, B. (1981). *J. Membrane Biol.* **63**, 61–70.

Jeltsch, E. and Zimmermann, U. (1979). *Bioelectrochem. Bioenerg.* **6**, 349–384.

Jennings, B. R. and Morris, V. J. (1974). *J. Colloid and Interface Sci.* **49**, 89–97.

Kao, K. N. and Michayluk, M. R. (1974). *Planta* **115**, 355–367.

Keller, W. A. and Melchers, G. (1973). *Z. Naturforsch.* **28c**, 737–741.

Kinosita, K. and Tsong, T. Y. (1977a). *Biochim. Biophys. Acta* **471**, 227–242.

Kinosita, K. Jr. and Tsong, T. Y. (1977b). *Nature* **268**, 438–441.

Kinosita, K. Jr. and Tsong, T. Y. (1978). *Nature* **272**, 258–260.

Kinosita, K. Jr. and Tsong, T. Y. (1979). *Biochim. Biophys. Acta* **554**, 479–497.

Klee, M. and Plonsey, R. (1974). *IEEE trans Biomed. Eng.*, **BME-21**, 452–460.

Knutton, S. (1977). *J. Cell Sci.* **28**, 189–210.

Knutton, S. (1979). *J. Cell Sci.* **36**, 61–72.

Knutton, S. and Pasternak, C. A. (1979). *TIBS* 220–223.

Köhler, G. and Milstein, C. (1975). *Nature* **256**, 495–497.

Korenstein, R. (1984). *In* "Electrical Effects in Membranes" (eds U. Zimmermann and R. Benz), Springer-Verlag, Berlin, Heidelberg and New York (in press).

Krasny-Ergen, W. (1936). *Hochfrequenztechnik u. Elektroakustik* **40**, 126–133.

Krasny-Ergen, W. (1937). *Hochfrequenztechnik u. Elektroakustik* **49**, 195–199.

Krauss, K. (1932). *Zentbl. Bakt. Parasit Kunde Abt. II* **124**, 64–77.

Küppers, G. and Zimmermann, U. (1983). *FEBS Lett.* **164**, 323–329.

Küppers, G., Diederich, K. J. and Zimmermann, U. (1984). *J. Memb. Biol.* (submitted).

Küppers, G., Wendt, B. and Zimmermann, U. (1983). *Z. Naturforsch.* **38c**, 505–507.

Lamb, H. (1906). "Hydrodynamics", 3rd ed., Cambridge University Press, Cambridge.

Läuger, P., Benz, R., Stark, G., Bamberg, E., Jordan, P. C., Fahr, A. and Brock, W. (1981). *Quart. Rev. Biophys.* **14**, 513–598.

Liebesny, P. (1939). *Arch. Phys. Therapy* **19**, 736–740.

Littlefield, J. W. (1964). *Science* **145**, 709–710.

Lucy, J. A. (1970). *Nature* **227**, 815–817.

Lucy, J. A. (1978). *In* "Membrane Fusion" (eds, G. Poste and G. Nicolson), pp. 267–304, Elsevier/North-Holland, Biomedical Press, Amsterdam.

Lucy, J. A. (1982). *In* "Biological Membranes". (ed. D. Chapman,) pp. 367–415, Academic Press, London and New York.

Ludloff, K. (1895). *Pflügers Arch.* **59**, 525–554.

Maráz, A., Kiss, M. and Ferenczy, L. (1978). *FEMS Microbiol. Letts* **3**, 319–322.

Martinac, B. and Hildebrand, E. (1981). *Biochim. Biophys. Acta* **649**, 244–252.

Mason, B. D. and Townsley, P. M. (1971). *Can. J. Microbiol.* **17**, 879–888.

Mast, S. O. (1931). *Z. wissenschaft. Biologie, Abt. C.: Z. vergleich. Physiol. (Berlin)* **15**, 309–328.

Matile, Ph. (1975). "The Lytic Compartment of Plant Cells." Springer-Verlag, Wien, New York.

McLaughlin, S. and Poo, M. (1981). *Biophys. J.* **34**, 85–93.

Mischel, M. and Lamprecht, I. (1980). *Z. Naturforsch* **35c**, 1111–1113.

Mischel, M., Voss, A. and Pohl, H. A. (1982). *J. Biol. Phys.* **10**, 223–226.

Mishra, K. P., Binh, L. D. and Sing, B. B. (1981). *Ind. J. Exp. Biol.* **19**, 520–523.

Morris, V. J., Rudd, P. J. and Jennings, B. R. (1975). *J. Colloid and Interface Sci.* **50**, 379–386.

Muth, E. (1927). *Kolloid Z.* **41**, 97–102.

Neher, E., Sackmann, B. and Steinbach, J.-H. (1978). *Pflügers Arch.* **375**, 219–228.

Neumann, E. (1981). *Topics Bioelectrochem. Bioenerg.* **4**, 113–160.

Neumann, E. and Katchalsky, A. (1972). *Proc. Natl Acad. Sci. USA* **69**, 993–997.

Neumann, E. and Rosenheck, K. (1972). *J. Membrane Biol.* **10**, 279–290.

Neumann, E. and Rosenheck, K. (1973). *J. Membrane Biol.* **14**, 192–196.

Neumann, E., Gerisch, G. and Opatz, K. (1980). *Naturwissenschaften* **67**, 414–415.

Nir, W., Wilschut, J. and Bentz, J. (1982). *Biochim. Biophys. Acta* **688**, 275–278.

Novák, B. and Bentrup, F. W. (1973). *Biophys.* **9**, 253–260.

Ogawa, T. (1961). *J. Appl. Phys.* **32**, 583–592.

Olsson, L. and Kaplan, H. S. (1980). *Proc. Natl Acad. Sci. USA* **77**, 5429–5431.

Orida, N. and Poo, M. (1978). *Nature* **275**, 31–35.

Pareilleux, A. and Sicard, N. (1970). *Appl. Microbiol.* **19**, 421–424.

Pauly, H. (1962). *IRE Trans Bio-Med. Electr.* **9**, 93–95.

Pauly, H. (1963). *Biophys.* **1**, 143–153.
Peberdy, J. F. (1980). *Enzyme Microb. Technol.* **2**, 23–29.
Pennock, B. E. and Schwan, H. P. (1969). *J. Phys. Chem.* **73**, 2600–2610.
Péqueux, A., Gilles, R., Pilwat, G. and Zimmermann, U. (1980). *Experientia* **36**, 565–566.
Peretz, H., Toister, Z., Laster, Y. and Loyter, A. (1974). *J. Cell. Biol.* **63**, 1–11.
Pethig, R. (1979). "Dielectric and Electronic Properties of Biological Materials." John Wiley and Sons, Chichester, New York, Brisbane, Toronto.
Petrov, A. G., Mitov, M. D. and Derzhanski, A. I. (1980). "Advances in Liquid Crystal Research and Applications." (ed. L. Bata), pp. 695–737, Pergamon Press, Oxford-Akadémiai Kiadó, Budapest.
Pilwat, G. and Zimmermann, U. (1983). *Bioelectrochem. Bioenerg.* **10**, 155–162.
Pilwat, G., Zimmermann, U. and Riemann, F. (1975). *Biochim. Biophys. Acta* **406**, 424–432.
Pilwat, G., Hampp, R. and Zimmermann, U. (1980a). *Planta* **147**, 396–404.
Pilwat, G., Zimmermann, U. and Schnabl, H. (1980b). *In* "Plant Membrane Transport: Current Conceptual Issues." (eds R. M. Spanswick, W. J. Lucas and J. Dainty), pp. 475–478, Elsevier/North-Holland, Amsterdam.
Pilwat, G., Richter, H.-P. and Zimmermann, U. (1981). *FEBS Lett.* **133**, 169–174.
Pohl, H. A. (1951). *J. Appl. Phys.* **22**, 869–871.
Pohl, H. A. (1958). *J. Appl. Phys.* **29**, 1182–1188.
Pohl, H. A. (1978). "Dielectrophoresis." Cambridge University Press, Cambridge.
Pohl, H. A. (1979). *Research Note* **98**, Oklahoma State University.
Pohl, H. A. (1980). *Research Note* **111**, Oklahoma State University.
Pohl, H. A. (1982). *Research Note* **130**, Oklahoma State University.
Pohl, H. A. and Braden, T. (1982). *J. Biol. Phys.* **10**, 17–30.
Pohl, H. A. and Crane, J. S. (1971). *Biophys. J.* **11**, 711–727.
Poo, M. (1981). *Ann. Rev. Biophys. Bioeng.* **10**, 245–276.
Poo, M. and Robinson, K. R. (1977). *Nature* **265**, 602–605.
Poo, M., Lam, J. W., Orida, N. and Chao, A. W. (1979). *Biophys. J.* **26**, 1–22.
Poste, G. (1972). *Int. Rev. Cytol.* **33**, 157–252.
Poste, G. and Allison, A. C. (1973). *Biochim. Biophys. Acta* **300**, 421–465.
Poste, G. and Pasternak, C. A. (1978). *In* "Membrane Fusion". (eds G. Poste and G. L. Nicolson), pp. 305–367, Elsevier/North-Holland, Amsterdam.
Power, J. B., Evans, P. K. and Cocking, E. C. (1978). *In* "Membrane Fusion." (eds G. Poste and G. L. Nicolson), pp. 369–385, Elsevier/North-Holland, Amsterdam.
Puchkova, T. V., Putvinsky, A. V., Vladimirov, Yu. A. and Parnev, O. M. (1981). *Biophysica* **26**, 265–270.
Rand, R. P. (1981). *Ann. Rev. Biophys. Bioeng.* **10**, 277–314.
Reinert, J. and Bajaj, Y. P. S. (1977). "Plant Cell, Tissue and Organ Culture," Springer-Verlag, Berlin, Heidelberg, New York.
Richter, H.-P., Scheurich, P. and Zimmermann, U. (1981). *Develop., Growth and Differ.* **23**, 479–486.
Riemann, F., Zimmermann, U. and Pilwat, G. (1975). *Biochim. Biophys. Acta* **394**, 449–462.
Ringertz, N. R. and Savage, R. E. (1976). "Cell Hybrids," Academic Press, New York and London.
Rowley, B. A. (1972). *Proc. Soc. Biol. Med.* **139**, 929–934.
Russell, I. and Stewart, G. G. (1979). *J. Inst. Brew.* **85**, 95–98.

Saito, M. and Schwan, H. P. (1961). In "Biological Effects of Microwave Radiation". (ed. M. F. Peyton), Vol. 1, pp. 85–97, Plenum Press, New York.

Saito, M., Schwan, H. P. and Schwarz, G. (1966). Biophys. J. 6, 313–327.

Sale, A. J. H. and Hamilton, W. A. (1967). Biochim. Biophys. Acta 148, 781–788.

Sale, A. J. H. and Hamilton, W. A. (1968). Biochim. Biophys. Acta 163, 37–43.

Saleemuddin, M., Zimmermann, U. and Schneeweiss, F. (1977). Z. Naturforsch. 32c, 627–631.

Salhani, N., Schnabl, H., Küppers, G. and Zimmermann, U. (1982). Planta 155, 140–145.

Sauer, F. A. (1983). In "Coherent Excitations in Biological Systems" (eds H. Fröhlich and F. Kremer) pp. 134–144.

Scheurich, P. and Zimmermann, U. (1981). Naturwissenschaften 68, 45–46.

Scheurich, P., Schnabl, H. and Zimmermann, U. (1980a). Biochim. Biophys. Acta 598, 645–651.

Scheurich, P., Zimmermann, U., Mischel, M. and Lamprecht, I. (1980b). Z. Naturforsch. 35c, 1081–1085.

Scheurich, P., Zimmermann, U. and Schnabl, H. (1981). Plant Physiol. 67, 849–853.

Schnabl, H., Scheurich, P. and Zimmermann, U. (1980). Planta 149, 280–282.

Schneeweiss, F., Zimmermann, U. and Saleemuddin, M. (1977). Biochim. Biophys. Acta 466, 373–378.

Schnettler, R., Zimmermann, U. and Emeis, C. C. (1984). FEMS Microbiol. Lett. (submitted).

Schulz, W., Kirmayer, N. and Reis, A. (1982). GIT Fachzeitung Lab. 26, 28–32.

Schwan, H. P. (1957). In "Advances in Biological and Medical Physics". (eds J. H. Laurence and C. A. Tobias), Vol. 5, pp. 147–209, Academic Press, New York and London.

Schwan, H. P. (1977). Ann. N.Y. Acad. Sci. 103, 198–213.

Schwan, H. P. and Forster, K. R. (1977). Biophys. J. 17, 193–197.

Schwan, H. P. and Sher, L. D. (1966). J. Electrochem. Soc. 116, 22C–26C.

Schwan, H. P. and Sher, L. D. (1969). J. Electrochem. Soc. 116, 170–174.

Schwarz, G. (1962). J. Phys. Chem. 66, 2636–2642.

Schwarz, G. (1963). J. Chem. Phys. 39, 2387–2388.

Schwarz, G., Saito, M. and Schwan, H. P. (1965). J. Chem. Phys. 43, 3562–3569.

Senda, M., Takeda, J., Abe, S. and Nakamura, T. (1979). Plant Cell Physiol. 20, 1441–1443.

Shepherd, J. C. W. and Büldt, G. (1979). Biochim. Biophys. Acta 558, 41–47.

Sher, L. D. (1963). Ph.D. Thesis, Univ. Pennsylvania, Philadelphia, Pa.

Sher, L. D. (1968). Nature 220, 695–696.

Sher, L. D., Kresch, E. and Schwan, H. P. (1970). Biophys. J. 10, 970–979.

Shinar, R., Druckmann, S., Ottolenghi, M. and Korenstein, R. (1977). Biophys. J. 19, 1–5.

Shotton, D. (1980). Nature 283, 12–14.

South, G. P. and Grant, E. H. (1972). Proc. Roy. Soc. Lond. A. 328, 371–387.

Sowers, A. E. and Hackenbrock, C. R. (1981). Proc. Natl Acad. Sci. USA 78, 6246–6250.

Steinbiss, H. H. (1978). Z. Pflanzenphysiol. 88, 95–102.

Sugár, I. P. (1979). Biochim. Biophys. Acta 556, 72–85.

Teissie, J. and Tsong, T. Y. (1981). Biochem. 20, 1548–1554.

Teixeira-Pinto, A. A., Nejelski, L. L., Cutler, J. L. and Heller, J. H. (1960). Exp. Cell Res. 20, 548–564.

Tell, R. A. and Harlen, F. (1979). *J. Microwave Power* **14**, 405–424.

Tien, H. Ti. (1974). "Bilayer Lipid Membranes", Marcel Dekker, New York.

Traganos, F., Darzynkiewicz, Z., Sharpless, T. and Melamed, R. (1977). *J. Histochem. Cytochem.* **25**, 46–56.

Tsong, T. Y. and Kingsley, E. (1975). *J. Biol. Chem.* **250**, 786–789.

Tsong, T. Y., Tsong, T. T., Kingsley, E. and Siliciano, R. (1976). *Biophys. J.* **16**, 1091–1104.

Turnbull, R. J. (1973). *J. Membrane Biol.* **14**, 193.

Verhoek-Köhler, B., Hampp, R., Ziegler, H. and Zimmermann, U. (1983). *Planta* **158**, 199–204.

Verworn, M. (1896). *Pflügers. Arch.* **62**, 415–450.

Vienken, J. and Zimmermann, U. (1982). *FEBS Lett.* **137**, 11–13.

Vienken, J., Jeltsch, E. and Zimmermann, U. (1978). *Cytobiol.* **17**, 182–196.

Vienken, J., Ganser, R., Hampp, R. and Zimmermann, U. (1981). *Physiol. Plant* **53**, 64–70.

Vienken, J., Zimmermann, U., Ganser, R. and Hampp, R. (1983). *Planta* **157**, 331–335.

Vienken, J., Zimmermann, U., Alonso, A. and Chapman, D. (1984). *Naturwissenschaften* **71**, 158–160,

Webb, D. (1978). *In* "Electrical Phenomena at the Biological Membrane Level". (ed. E. Roux), pp. 119–153, Elsevier/North-Holland, Amsterdam.

Weber, H., Förster, W., Berg, H. and Jakob, H.-E. (1981). *Current Genetics* **4**, 165–166.

Zagyansky, Y. A. and Jard, S. (1979). *Nature* **280**, 591–593.

Zimmermann, U. (1977a). *In* "Integration of Activity in the Higher Plant". (ed. D. Jennings), pp. 117–154, Cambridge University Press, Cambridge.

Zimmermann, U. (1977b). *Chemie f. Labor und Betrieb* **28**, 505–508.

Zimmermann, U. (1978). *Ann. Rev. Plant Physiol.* **29**, 121–148.

Zimmermann, U. (1980). *In* "Animals and Environmental Fitness." (ed. R. Gilles), pp. 441–459, Pergamon Press, Oxford and New York.

Zimmermann, U. (1982). *Biochim. Biophys. Acta* **694**, 227–277.

Zimmermann, U. (1983). *Polym. Biol. Med.* **2**, 153–200.

Zimmerman, U. and Arnold, W. M. (1983). *In* "Coherent Excitations in Biological Systems" (eds H. Fröhlich and F. Kremer), p. 211–221.

Zimmermann, U. and Benz, R. (1980). *J. Membrane Biol.* **53**, 33–43.

Zimmermann, U. and Küppers, G. (1983). *Naturwissenschaften* **70**, 568–569.

Zimmermann, U. and Pilwat, G. (1976). *Z. Naturforsch* **31c** 732–736.

Zimmermann, U. and Pilwat, G. (1978). Sixth International Biophysics Congress, Kyoto, **Abstr. IV-19(H)**, p. 140.

Zimmermann, U. and Scheurich, P. (1981a). *Planta* **151**, 26–32.

Zimmermann, U. and Scheurich, P. (1981b). *Biochim. Biophys. Acta* **641**, 160–165.

Zimmermann, U. and Vienken, J. (1982). *J. Membrane Biol.* **67**, 165–182.

Zimmermann, U. and Vienken, J. (1984) *In* "Cell Fusion: Gene Transfer and Transformation" (eds R. F. Beers and E. G. Bassett), pp. 171–187. Raven Press, New York.

Zimmermann, U., Schultz, J. and Pilwat, G. (1973). *Biophys. J.* **13**, 1005–1013.

Zimmermann, U., Pilwat, G. and Riemann, F. (1974a). *Biophys. J.* **14**, 881–899.

Zimmermann, U., Pilwat, G. and Riemann, F. (1974b). *In* "Membrane Transport in Plants". (eds U. Zimmermann and J. Dainty), pp. 146–153, Springer-Verlag, Berlin, Heidelberg, New York.

Zimmermann, U., Pilwat, G. and Riemann, F. (1975). *Biochim. Biophys. Acta* **375**, 209–219.

Zimmermann, U., Pilwat, G., Beckers, F. and Riemann, F. (1976a). *Bioelectrochem. Bioenerg.* **3**, 58–83.

Zimmermann, U., Pilwat, G., Holzapfel, Chr. and Rosenheck, K. (1976b). *J. Membrane Biol.* **30**, 135–152.

Zimmermann, U., Riemann, F. and Pilwat, G. (1976c). *Biochim. Biophys. Acta* **436**, 460–474.

Zimmermann, U., Beckers, F. and Coster, H. G. L. (1977a). *Biochim. Biophys. Acta* **464**, 399–416.

Zimmermann, U., Beckers, F. and Steudle, E. (1977b). *In* "Transmembrane Ionic Exchanges in Plants". (eds M. Thellier, A. Monnier, M. Demarty and J. Dainty), pp. 155–165, CNRS, Paris.

Zimmermann, U., Pilwat, G. and Esser, B. (1978). *J. Clin. Chem. Clin. Biochem.* **16**, 135–144.

Zimmermann, U., Groves, M., Schnabl, H. and Pilwat, G. (1980a). *J. Membrane Biol.* **52**, 37–50.

Zimmermann, U., Pilwat, G., Péqueux, A. and Gilles, R. (1980b). *J. Membrane Biol.* **54**, 103–113.

Zimmermann, U., Vienken, J. and Pilwat, G. (1980c). *Bioelectrochem. Bioenerg.* **7**, 553–574.

Zimmermann, U., Vienken, J. and Scheurich, P. (1980d). *In* "Biophysics of Structure and Mechanism" (ed. K. Gersonde), Vol. 6, p. 86, Springer International.

Zimmermann, U., Pilwat, G. and Richter. H.-P. (1981a). *Naturwissenschaften* **68**, 577–578.

Zimmermann, U., Scheurich, P., Pilwat, G. and Benz, R. (1981b). *Angew. Chem.* **93**, 332–351, Int. Ed. **20**, 325–344.

Zimmermann, U., Vienken, J. and Pilwat, G. (1981c). *Z. Naturforsch.* **36c**, 173–177.

Zimmermann, U., Büchner, K.-H. and Benz, R. (1982a). *J. Membrane Biol.* **67**, 183–197.

Zimmermann, U., Küppers, G. and Salhani, N. (1982b). *Naturwissenschaften* **69**, 451–452.

Zimmermann, U., Pilwat, G. and Pohl, H. A. (1982c). *J. Biol. Phys.* **10**, 43–50.

Zimmermann, U., Vienken, J. and Pilwat, G. (1982d). *Studia Biophysica* **90**, 177–184.

Zimmermann, U., Vienken, J. and Pilwat, G. (1984a). *In* "Investigative Microtechniques in Medicine and Biology" (eds J. Chayen and L. Bitensky), Vol. 1, Chap. 3, Marcel Dekker, New York.

Zimmermann, U., Vienken, J., Pilwat, G. and Arnold, W. M. (1984b). *In* "Cell Fusion", CIBA Symp. 103 (eds B. Pethica and J. Whelan), pp. 60–85. Pitman, London.

Author Index

(Numbers in italic indicate pages on which full references are given)

Subject Index